D1269491

Handbook of Reinforcements for Plastics

Handbook of Reinforcements for Plastics

Edited by

JOHN V. MILEWSKI
Consultant
Santa Fe, New Mexico

HARRY S. KATZ
Utility Research Co.
Montclair, New Jersey

VNR VAN NOSTRAND REINHOLD COMPANY
New York

Copyright © 1987 by Van Nostrand Reinhold Company Inc.
Library of Congress Catalog Card Number 86-28255
ISBN 0-442-26475-5

All rights reserved. Certain portions of this work © 1978 by Van Nostrand Reinhold Company Inc. No part of this work covered by the copyright hereon may be reproduced or used in any form or by any means—graphic, electronic, or mechanical, including photocopying, recording, taping, or information storage and retrieval systems—without written permission of the publisher.

Printed in the United States of America

Van Nostrand Reinhold Company Inc.
115 Fifth Avenue
New York, New York 10003

Van Nostrand Reinhold Company Limited
Molly Millars Lane
Wokingham, Berkshire RG11 2PY, England

Van Nostrand Reinhold
480 La Trobe Street
Melbourne, Victoria 3000, Australia

Macmillan of Canada
Division of Canada Publishing Corporation
164 Commander Boulevard
Agincourt, Ontario M1S 3C7, Canada

16 15 14 13 12 11 10 9 8 7 6 5 4 3 2 1

Library of Congress Cataloging-in-Publication Data

Handbook of reinforcements for plastics.

Includes bibliographical references and index.
1. Plastics—Additives—Handbooks, manuals, etc.
2. Reinforced plastics—Handbooks, manuals, etc.
I. Milewski, John V. II. Katz, Harry S.
TP1142.H377 1987 668.4′94 86-28255
ISBN 0-442-26475-5

Foreword

Plastics are playing an increasingly important role in our daily lives—in our homes, businesses, and environment. Until recently, their plentiful supply and low cost had been taken for granted. However, the oil shortage and subsequent materials shortages, plus escalating prices for monomers and polymers, have resulted in a rude awakening to the fact that new initiatives must be exerted in our field.

A more important role must now be accorded to the increased and more efficient use of fillers and reinforcements as a means for stretching the resin supply and lowering the cost of molding compounds.

In my past years of plastics engineering, I have strived to bring to the industry a greater awareness of the proper materials, design, fabrication methods, and economics. I consider this *Handbook* a giant step forward in that direction, and believe that it will accelerate the proper use of fillers and reinforcements and therefore be beneficial to the plastics industry.

J. HARRY DUBOIS

Morris Plains, New Jersey

Note: This was the Foreword of the first edition (1978), and we consider these comments by an outstanding person in the plastics field still pertinent to this second edition.

—The Editors

Preface

Prior to the publication of the first edition of *The Handbook of Fillers and Reinforcements for Plastics*, the plastics industry lived in a fool's paradise, where resins were inexpensive and plentiful; a condition that was expected to persist. Then came the rude awakening during the oil embargo of 1973, which caused a shortage of raw materials and resins, and initiated a series of escalations in the prices of polymers. This led to an increased interest in the use of fillers and reinforcements as a means of reducing the price of molding compounds and expanding the supply of resins. The editors were among those who considered it desirable, to make more extensive use of fillers and reinforcements, they realized that this goal would be aided by a unified compilation of information and data that would enable the rapid choice of the correct filler or reinforcement. As a consequence, they initiated *The Handbook of Fillers and Reinforcements for Plastics*. Prior to publication of the handbook, the choice of a filler and reinforcement usually involved many contacts with materials suppliers, compounders, design engineers, and molders in order to select candidate materials and formulations.

The great increase in technical articles related to new reinforcements, and other factors indicated a growing need for a continued updating of the information on reinforcements in this type of handbook. We believe that it is timely and useful to update the handbook that has served as a standard reference of information for everyone involved in the plastics industry. As the editors gathered information for the new edition, it soon became apparent that both the fillers and reinforcements industry have grown considerably and that it would be cumbersome to compile all of the information into one handbook. Therefore, it was decided that the revision would be two handbooks, one on fillers and the other on reinforcements. *The Handbook of Reinforcements for Plastics*, is the first handbook on reinforcements specifically devoted to the plastics industry. There has been a considerable expansion of the information found in the first edition and many new topics are included that have not been treated before in any major text.

This handbook is directed toward all individuals involved in the production, design, or specification of a molded end product. This includes design engineers, materials scientists, polymer chemists, compounders, and molders. The editors will welcome comments from every profession so that future editions of the handbook will provide the information required for more efficient utilization of reinforcements.

Since the publication of the first edition, much progress and growth have been made in the understanding and use of reinforcements. We have received many favorable comments on the original edition and believe that our effort and the efforts of all the chapter contributors have played an important role in accelerating the growth of the polymer composites industry. A more intensive insight into the science and technology of reinforcements is being used in their selection and application. Greater sophistication in the selection process has created a need for more complete information. In this edition, we have added new information on each reinforcement, and we have also added chapters on recently developed materials.

Compounders and end users are now more demanding in their request for specific properties of new materials. Complex design requirements often demand a combination of properties, rather than a few requirements such as cost, strength, and modulus. Thermal expansion, thermal conductivity, impact resistance, and many other properties may be essential for a specific application. Electromagnetic shielding and electronic applications require stringent control of electrical conducting or insulating properties, and these can be tailored to the desired end use by the judicious choice of reinforcements.

Multi-fiber and filler-fiber combinations are receiving more attention. During recent years, fiber suppliers have accepted the importance of packing concepts, and are marketing short fibers and fillers in multi-sized mixtures and combinations designed for good packing, efficient molding and reduced resin demand. The surface treatment of these materials is now receiving more attention and sales of surface treated products are growing rapidly.

Anyone attending a recent Exhibit and Conference of the Society of the Plastics Industry's Composite Division will realize the amazing recent advances in the art and science of the reinforcements field. We and the other authors of this handbook are proud that we have made a contribution that has played a role in stimulating the growth of this industry. We hope that these revised editions will continue to guide and encourage the increased and more effective use of reinforcements.

John V. Milewski
Harry S. Katz

Contents

Handbook of Reinforcements for Plastics

Section I

1

Introduction

John V. Milewski Ph.D.
Consultant
Santa Fe, New Mexico

Harry S. Katz
Utility Research Co.
Montclair, New Jersey

Expanding horizons in industrial activities create a continual demand for improved materials that will satisfy increasingly more stringent requirements, such as higher strength, modulus, thermal and/or electrical conductivity, heat distortion temperature, and lower thermal expansion coefficient and cost. These requirements, which often involve a combination of many difficult to attain properties, may require the use of a composite material, whose constituents act synergistically to solve the needs of the application. As we crossed the threshold into the "composite materials age," it became increasingly important to understand the properties, performance, cost, and potential of the available composite materials.

Reinforcements have always played an important role in the plastics industry. The early growth of the phenolic plastics industry would not have been possible without the enhancement of properties by the use of fillers and reinforcements such as cotton fibers. The commodity resins, such as polyvinyl chloride, polystyrene, polyethylene, and polypropylene, have properties that meet the requirements of high volume end uses; thus, they have been sold and used as essentially pure resins. There was no incentive to use reinforcements in the resins. This situation has changed. Price escalations combined with the sporadic and possible future shortages of resins and petroleum feed stocks have established the urgent need for widespread utilization of fillers and reinforcements. This will apply especially to engineering resins, but also to the commodity resins. Composite systems afford a means of extending the available volume of resins while improving many of their properties. These improvements are often associated with economic advantages such as lower raw material cost, faster molding cycles as a result of increased thermal conductivity, and fewer rejects due to warpage.

At present, most molded products do not contain any fillers or reinforcements in spite of the fact that the judicious choice of a filler and reinforcement can result in a lower cost product with equivalent or improved properties. Recent technological advances, such as improved dispersion and new packing concepts, have not been widely applied and, as a consequence, much current information on filled and reinforced plastics indicate poorer qualities than the true potential of these composites. Also, there have been recent advances in equipment and procedures for compounding and molding highly filled polymers. Currently, there are few valid technical justifications for the use of an unfilled resin in most molded products, from considerations of physical properties, moldability, and cost. Moreover, as this technology advances, there will be increased benefits from the use of fillers and reinforcements.

The purpose of *The Handbook of Reinforcements for Plastics* is to present in one volume, all the basic and much in-depth information on reinforcements being used in industry today. It is the most complete and up-to-date volume published in this area of technology, and is a very convenient desktop reference book. The first edition of the *Handbook of Fillers and Reinforcements for Plastics*, has been called the "Bible" of the industry. This same high standard has been maintained in this second edition.

The very significant growth in the areas of both filler and reinforcements was the reason for this two-volume format. These new volumes are more than fifty percent larger than the first edition, with major revisions in every chapter.

There are new chapters on new materials. Six chapters have been added on materials that have not been well presented in previous books. These chapters cover short metal fibers and flakes, phosphate fibers, chopped and milled fibers, technology of cutting fibers, short organic fibers, and new polyethylene filaments.

The handbook is composed of five major sections comprising 21 chapters. In the Section I, Chapter 1 is an Introduction. Chapter 2 is Concise Fundamentals of Fiber Reinforced Composites; this chapter contains a brief, but significant coverage of the basic theory of fiber reinforcement with many references to more detailed theoretical considerations. Chapter 3 covers the classic theories of fiber-filler packing concepts. This volume and the first edition are the only two sources of this extremely vital information for making more efficient and economical, short-fiber-reinforced composites.

Section II is devoted to flake and ribbon reinforcements and presents an in-depth discussion of these materials.

Section III is about short fibers—the fastest growing area of new reinforcements. There are 8 chapters in this section, four are new chapters on materials which have not been discussed in detail in any previous book. These chapters cover short metal fibers and metal flakes; phosphate fibers; chopped and milled fibers; short organic fibers and the technology of cutting fibers. The other chapters are on Wollastonite, asbestos, inorganic micro-fibers and whiskers. All of these chapters have been substantially revised for this new edition.

Section IV is on fiber glass and basalt glass fiber; both have been brought up to date with the latest in-depth information.

Section V is about high modulus filaments and contains a new chapter on polyethylene, plus major revisions on aramids, boron and silicon carbide filaments, carbon/graphite, ceramic filaments, and continuous metallic filaments.

One of the most significant subjects that is covered by this book is filler-fiber combinations. This information has revolutionized the short fiber industry by improving performance, efficiency, and economics.

The short metallic fiber information is especially important to the rapidly growing static charge elimination and electro-magnetic shielding applications, and for improved thermal conductivity applications.

The information on the new phosphate fiber will be of special interest to all those looking for asbestos replacements. Chapter 8, Wollastonite, and Chapter 10, Inorganic and Micro Fibers, provide information on other potential asbestos replacements. Those interested in the technology of chopping and cutting fibers will find the first really complete, well-illustrated work on this subject in Chapter 11.

Those interested in short organic fibers as a reinforcement will find the first detailed coverage of this subject in Chapter 12. The expanding growth in whiskers is well-documented in the wealth of new information and sources contained in Chapter 13.

A material for water cables, advanced composites, and armor is the new polyethylene filament. Many detailed examples of applications of this material are given.

This Handbook satisfies the need of the user. It is a handy complete reference on the subject of reinforcements presenting detailed data from hundreds of companies as well as thousands of products in an extremely well-organized manner. There is no other book that is specifically devoted to this subject.

The book will be of interest to the compounder guiding him in the selection of the most efficient and economical source of raw materials. Although primarily intended for those in the reinforced plastics market, this Handbook will be useful for everyone involved in the rapidly expanding fields of metal and ceramic matrix composites.

Recently, continuous filaments and fibers like boron, silicon carbide, and aluminum oxide and short micro fibers and whiskers are being used for new applications in both metallic and ceramic matrix composites.

As noted above, Sections II through V present all of the significant materials used as reinforcements. The editors presented the contributors with a suggested outline that might be used as a standard format for the chapters. However, it is apparent that they required some latitude for variation from the standard format. In spite of some of these differences, the reader will note that most chapters follow the same basic format in presenting their data, which will facilitate locating specific information on a particular material.

The field of reinforcements is growing rapidly and encompasses a wide variety of materials. This handbook has been designed to present a complete and orderly discussion of pertinent information in this field. However, there have been some arbitrary choices and, undoubtedly, some omissions. Nevertheless, the editors are confident that the reader will find this a useful and rewarding book. In addition, it is hoped that this presentation will act as a catalyst to inspire the increased use of reinforcements.

The editors anticipate that there will be periodic revisions of this handbook. Therefore, any comments and suggestions on a more useful format and additional data on current and new materials will be appreciated.

2

Concise Fundamentals of Fiber-Reinforced Composites

Harry S. Katz
Utility Research Co.
Montclair, New Jersey

Harold E. Brandmaier
Consultant
Harrington Park, New Jersey

CONTENTS

1. INTRODUCTION

Fibers have extremely high tensile strengths and moduli. A fiber can be defined as a particle longer than 100 μm with a length to diameter, or transverse dimension, ratio greater than 10. The tensile strength of a fiber is many magnitudes greater than the strength of the same material in bulk form. For example, bulk graphite is a brittle material with a tensile strength below 10,000 psi, whereas commercial graphite fibers have tensile strengths to 800,000 psi, and laboratory research on graphite whiskers yielded a 3,000,000 psi tensile strength. As another example, nylon plastic has a tensile strength of about 12,000 psi, whereas nylon fibers have tensile strengths in the order of 120,000 psi. Similarly, the tensile strengths of E-glass at 500,000 psi, boron fibers at 500,000 psi, and alumina fibers at about 150,000 psi are far above the strength levels of the bulk materials. Thus, it is apparent that when these high-strength fibers are used effectively in a matrix, composites with strength levels above those of bulk materials can be produced.

Since fibers are the highest-strength materials, they are the best reinforcements for many applications, ranging from improvement of the strength of low-cost materials or composites to acceptable levels, to the production of superior structural aerospace components. Therefore, everyone involved in the production, design, specification, or use of plastics should be familiar with the fundamental characteristics of fibers as reinforcements.

There are many good reasons for adding a high volume fraction of low-cost fillers to plastics, especially from the standpoint of stretching the resin supply. However, high particulate filler loadings will usually result in a substantial reduction of physical properties. For the many end uses where the reduced physical properties would be unacceptable, the addition of fibers to the compound formulation can be a means for increasing the properties to a satisfactorily high level. Many subtle considerations are involved in the science and art of formulating and fabricating a fiber-reinforced composite. There have been many programs in this field during the past two decades, and continued advances may be anticipated. An important aspect that will form the basis for future progress in improved molding compounds is the study of the packing of fibers with spheres and particulate fillers, as presented in Chapter 3.

The use of fibers as a reinforcement can lead to superior materials, in some cases, at a premium cost; or they can be a means for achieving improved properties while maintaining low costs and stretching the resin supply. The choice of the fiber reinforcement will depend on the requirements of the different end products. At the present time, the highest volume fiber reinforcement is glass fiber, but the near future will undoubtedly lead to the increased use of competitive materials, such as carbon-graphite filaments, whiskers, and microfibers. As industrial competition becomes more sophisticated, and end uses more demanding in performance and cost, it will be essential to continue the rapidly expanding use of fiber-reinforced composites.

This chapter is intended to give the reader insight into the principal characteristics of fiber-reinforced composites. A number of excellent books give comprehensive treatments of important topics, including theoretical analyses of the strength and modulus of various composite structures, crack propagation theories, and the stress analysis of advanced composite materials. These books are listed in the references at the end of this chapter.

2. COMPOSITE ELEMENTS

There are three basic elements in a fiber-reinforced composite: fiber, matrix, and the fiber–matrix interface. Each of these elements must have appropriate characteristics and function both individually and collectively in order for the composite to attain the desired superior properties.

The fiber contributes the high strength and modulus to the composite. It is the element that provides resistance to breaking and bending under the applied load.

The main roles of the matrix are to transmit and distribute stresses among the individual fibers, and to maintain the fibers separated and in the desired orientation. The matrix also provides protection against both fiber abrasion and fiber exposure to moisture or other environmental conditions, and causes the fibers to act as a team in resisting failure or deformation under load. The maximum service temperature of the composite is limited by the matrix. Other desirable features of the matrix are resistance to liquid penetration and freedom from voids.

The fiber–matrix interface is the critical factor that determines to what extent the potential properties of the composite will be achieved and maintained during use. Localized stresses are usually highest at or near the interface, which may be the point of premature failure of the composite. The interface must have appropriate chemical and physical features to provide the necessary load transfer from the matrix to the reinforcement. The use of a coupling agent can provide improved interfacial conditions. The interfacial bond must resist stresses due to differential thermal expansion of fiber and matrix, and shrinkage of the resin during cure. The interface also assists the matrix in protecting the fibers.

As indicated above, the presence of a coupling agent at the fiber–matrix interface can be the means for obtaining the optimum physical properties of the composite, and retaining these properties after environmental exposure or aging. This fact became apparent in early studies of fiberglass-reinforced polyester laminates, when the simple test of putting a panel into boiling water for 2 hours showed an amazing difference in strength retention between the composite utilizing a fiberglass treated with a silane coupling agent and a control containing untreated fiberglass.

Silanes have been the most frequently used type of coupling agent. The silanes provide dual functionality in one molecule, such that one part of the molecule forms a bond to the glass filament surface while another part forms a covalent bridge with the matrix molecule. There are many different silanes, and the choice must be based upon the matrix to be used. Silanes and other coupling agents such as the titanates are discussed in detail in the *Handbook of Fillers for Plastics*.

3. COMPOSITE PROPERTIES

As composite technology has advanced, it has become imperative that the properties of a composite be predictable from a knowledge of the constituent matrix, fiber, fiber volume, and fiber orientation. This is especially true in the case of advanced composites, where fiber and fabrication costs may be high, and the end product must be extremely reliable in performance, so that a trial-and-error approach cannot be tolerated. There has been good progress in the theoretical analysis and design of composite materials, so that even critical aerospace structures can be designed with assurance that they will safely meet the material property requirements.

In the analysis of the theoretical properties of different types of composites, various analytical models and failure theories have been used. There are different considerations involved in the case of a randomly oriented discontinuous fiber composite than for a continuous-filament unidirectional or angle-ply laminate.

3.1 Rule of Mixtures

A first approximation of the tensile modulus and strength of a fiber-reinforced composite can be obtained from the rule of mixtures. The modulus of a continuous filament composite with filaments oriented longitudinally can be estimated, with good precision, for the direction parallel to the fibers, by:

$$E_L = V_f E_f + V_m E_m \qquad (2\text{-}1)$$

where:

E_L is the longitudinal modulus of the composite.
V_f is the volume fraction of the filaments.
E_f is the tensile modulus of the filaments.
V_m is the volume fraction of the matrix.
E_m is the tensile modulus of the matrix.

Equation 2-1 assumes that all filaments are perfectly bonded to the matrix.

Equations analogous to Eq. 2-1 may be used to express either the longitudinal Poisson's ratio or the tensile strength of the composite. This prediction of tensile strength will not be as accurate as in the case of the modulus. Actual strengths are usually lower than the predicted strengths, and this ratio has been used as an indication of the fiber efficiency in the composite. In the case of some metal matrix composites, however, the indicated fiber efficiency has been greater than 100%, because of factors such as favorable residual stresses in the matrix.

In a discontinuous fiber composite, the stress along the fiber is not uniform. There are portions along the fiber's ends where the tensile stresses are less than that of a fiber that is continuous in length. This region is often called the fiber ineffective length. The tensile stress along the fiber length increases to a maximum along the middle portion of the fiber. If the fiber is sufficiently long so that the ratio of length to diameter, or aspect ratio, equals or exceeds the critical aspect ratio, the middle portion stress will be equal to that of a continuous filament. The shear stress at the fiber–matrix interface is a maximum at the ends of the fiber, as shown in Fig. 2-1.

The critical aspect ratio $(l/d)_c$ which would result in fiber fracture at its midpoint, can be expressed as:

$$\left(\frac{l}{d}\right)_c = \frac{S_f}{2Y} \qquad (2\text{-}2)$$

where:

l and d are the length and diameter of the fiber.
$(l/d)_c$ is the critical aspect ratio.
S_f is the tensile stress of the fiber.
Y is either the yield strength of the matrix in shear, or the fiber–matrix interfacial shear strength; whichever value is lower will determine the critical aspect ratio.

If the fiber is shorter than the critical length, the stressed fiber will de-bond from the matrix

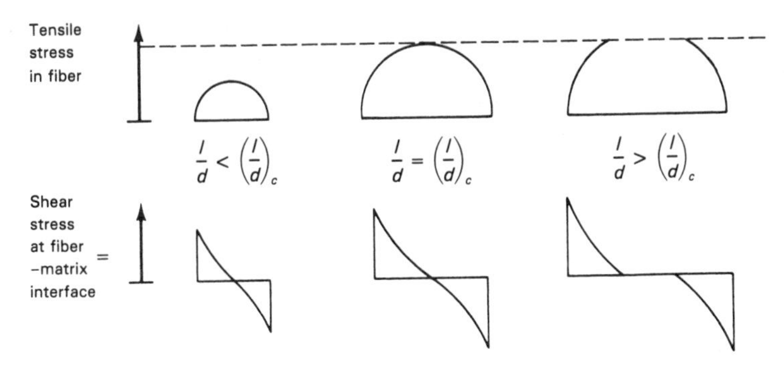

Fig. 2-1. Shear stresses at fiber–matrix interface.

and the composite will fail at a low strength. When the length is greater than the critical length, the stressed composite will lead to breaking of the fibers and a high composite strength.

The rule of mixtures for discontinuous fiber composites may be expressed as:

$$S_c = V_f S_f\left(1 - \frac{l_c}{2l}\right) + V_m S_m \qquad (2\text{-}3)$$

where:

S_c is the tensile strength of the composite.
V_f and V_m are the volume fractions of the fiber and matrix.
S_f and S_m are the tensile strengths of the fiber and matrix.
l_c is the critical length of the fiber.
l is the length of the fiber.

The rule of mixtures does not apply at low volume percentages of fibers. In order for the composite to have a higher strength than the matrix, a minimum V_f must be exceeded. This value may be 0.1 or greater for plastic matrix composites, but can be much lower for metal or ceramic matrix composites. Also, at high V_f, on the order of 0.7, the composite properties may decrease sharply, due to practical problems such as the difficulty in avoiding fiber–fiber contact, which results in stress concentrations that initiate failure. If the load is compressive, the theoretical problem is more difficult because of the possibility of fiber buckling. A number of theories have been proposed to predict the strength of discontinuous fiber composites. These theories indicate that, primarily because of high stress concentrations at discontinuities that occur at the fiber ends, the tensile strength of a discontinuous fiber composite will be from 55% to 86% of the tensile strength; and the modulus can approach 90% to 95% of the tensile modulus of the corresponding continuous filament composite.

3.2 Micro/Macromechanics

Analytical methods for composite materials can be divided into two main types, micro- and macromechanics.

Micromechanics analysis is applied to the basic filament–matrix microstructure in order to determine the strength and elastic properties from the properties and arrangement of the constituent filament and matrix materials. Micromechanics attempts to predict fracture behavior of the composite from the state of stress and strain at the micro level. Much progress has been made during the past 10 to 15 years in developing mathematical procedures and corresponding computer programs for estimating the elastic moduli and strength of composites.

Macromechanics encompasses the analytical methods that are used to predict the strength and elastic properties of an advanced composite laminate from the known properties of the individual unidirectional monolayers or plies. Macromechanic studies have been essential for design guidance in the rapid progress of advanced composite development in the aerospace industry.

The elastic properties of a plate consisting of

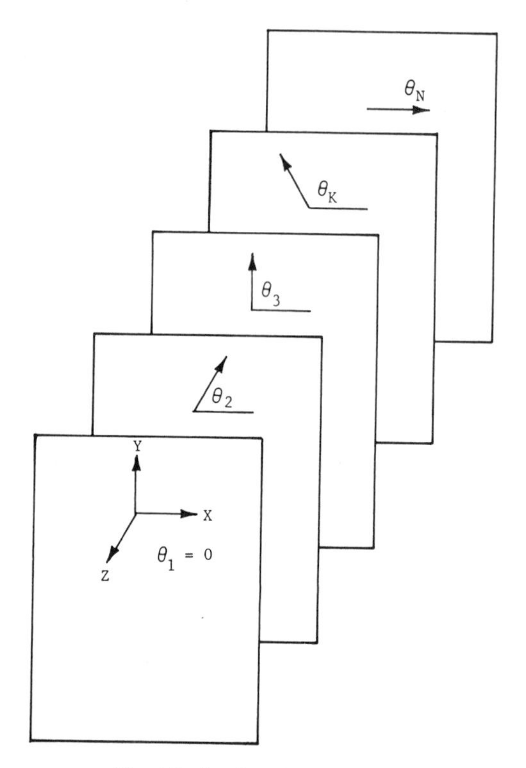

Fig. 2-2. Laminated plate, *N* plies.

N plies, in which the filament orientation may vary from ply to ply as shown in Fig. 2-2, may be obtained by summing the contribution of each ply. As the elastic properties of each ply vary with direction, angle θ in Fig. 2-2, these properties must first be transformed to the same coordinate system. Details of this procedure are contained in the references at the end of this chapter.

There is a large body of literature on composite failure criteria. The most used criterion at present is based on strain energy. Its advantage, compared to the maximum stress and maximum strain criteria, is that it considers the entire stress state at a point in a composite.

Only a few years ago, the complex calculations required to estimate a composite's elastic and strength properties required a large-scale computer. The recent development of programmable calculators and microcomputers has resulted in the availability of a number of inexpensive programs for laminate analysis. Some of these programs and their sources are:

1. "Advanced Composite Materials Laminate Plate Programs," for Hewlett-Packard 41 series programmable calculators, from:
   Intellicomp, Inc.
   292 Lambourna Ave.
   Worthington, OH 43085
   (614) 846-0216
2. "LAMINATE," for IBM PC/XT, AT, or their clones, from:
   Engineering Software Company
   Three Northpark East, Suite 901
   8800 North Central Expressway
   Dallas, TX 75231
   (214) 361-2431

In addition, members of the "Think Composites Software Users Club" (Box 581, Dayton, OH 45419) receive a set of Apple MacIntosh or IBM PC software. Finally, Texas Instrument Model 59 programmable calculator programs are available from the authors of *Introduction to Composite Materials* (see bibliography at the end of this chapter).

One advantage of these programs to the designer and materials engineer is that they enable a material selection based on the properties of realistic ply orientation patterns rather than the unidirectional properties that are provided by materials suppliers. In addition, they reduce some of the time-consuming material testing that otherwise would be required.

3.3 Design

The design of a part that will be fabricated from an isotropic material, such as a structural metal, has become relatively simple and straight-forward, as the result of a long history of use and complete design data that have been generated. In contrast, composite materials and their fabrication processes are generally unfamiliar to the designer, and the available information may not be complete. In spite of these drawbacks, which are rapidly being overcome, the use of composites has been expanding in all markets, and excellent progress has been made in the design of more intricate and larger parts. Notable accomplishments have been attained in the automotive, aerospace, and sporting goods industries. The use of sheet molding compounds has led to the design and mass production of large components with complex contours, for automobiles and other forms of transportation. The aerospace industry has led the way with high-performance composites that surpass the specific strength and stiffness of the usual structural metals. These advanced composites offer the designer the opportunity to produce aircraft whose outer appearance may not even resemble current aircraft design. Tennis racquets, golf clubs, and fishing rods fabricated from graphite/epoxy composites are becoming commonplace.

In order for the potential of fiber-reinforced composites to be fully realized, an interdisciplinary effort is required. There must be inputs and liaison between experts in materials, chemistry, adhesive bonding, crystal structure, metallography, stress analysis, engineering mechanics, fracture mechanics, vibration analysis, structural design, composite fabrication, quality control, and testing.

The design of a composite part must take into consideration the complete operational environment and the ability to meet environmental requirements for the anticipated life of the structure. Important factors include physical properties, fatigue and creep resistance, materials and manufacturing costs, maintenance and repairability, attachment method, and means for quality assurance.

4. TESTING

Methodical testing procedures are an essential supporting activity for the utilization and improvement of fiber-reinforced composites. Testing has guided the research and development of each composite material and structure, and must be used for quality control in production programs. The use of small test samples as a means of predicting the performance of large structures is often necessary. In this situation, the test should be conducted by simulating, as closely as possible, the end-use conditions. However, it is seldom feasible to apply to a test part the complex load conditions and extended time to which a component may be subjected. Therefore, good judgment must be exercized in the choice of tests that will provide either valid comparison data that will permit ranking of candidate materials, or meaningful data for designers and materials engineers. Standard tests, such as those proposed by the ASTM, will usually be preferable to specially designed tests that cannot be directly compared with similar testing by other investigators. In reporting the results of these tests, it is important that the exact test conditions be specified, since minor changes in test conditions can result in great differences in the data produced.

In each composites project, careful planning must include the choice of the testing procedures that will be used. A brief discussion of pertinent test methods is presented in the following paragraphs.

Flexural testing has been used frequently in composite development programs. This test is convenient and easy to perform. The data obtained can be used for grading specimens in materials evaluation, process development and improvement, and quality control. However, a true correlation between flexural and other mechanical properties has not been established in general. There are so many subtle variables affecting bending test strengths that the problem of correlating flexural data with the more basic properties can outweigh all advantages of the test when meaningful design data are required. In flexural testing, either four-point loading or the more frequently used three-point loading fixtures can be employed. During the application of load, one face of the sample is under compression and the other face is in tension, and the failure of fiber-reinforced composites usually occurs on the tension side. Flexural strength values for fiber-reinforced composites are usually much higher than the tensile strength values for the same material. The percentage difference in the flexural and tensile moduli is not generally as large as the difference in the strength values.

Many different tensile test procedures have been proposed for individual fibers and fiber bundles or strands. Tensile tests of fiber-reinforced composites have been performed by use of various-shaped specimens, including the dogbone configuration, NOL ring, filament-wound pressure vessels, and straightsided panels with bonded end tabs. The choice of method is important, especially if data are to be compared with similar information from other sources in order to choose the best candidate material for a specific application. A common mistake in the reported tensile strength data is neglecting to record the exact test conditions, such as sample geometry, gauge length, rate of jaw separation, and method of gripping the specimen.

A great deterrent to the use of advanced composites for aerospace structural applications has been the concern of designers that the use of a defective part would cause failure of a vehicle and mission. As a result, the use of Non Destructive Testing (NDT) methods is essential for the safe utilization of the new composite structural components. These methods provide a means for detecting flaws such as voids, inclusions, and delaminations. Great progress has been made in NDT methods during recent years, and there is now a high confidence level in the capability of detecting defects.

Visual inspection is used to detect surface defects and irregularities. Methods applicable to internal defects are x-ray (including three-dimensional), sonic transmission, pulse echo, and infra-red void detection. By the use of through-transmission ultrasonics, the ultrasonic longitudinal velocity can be correlated with the tensile and compressive moduli, as well as the ultimate tensile, compressive, and interlaminar shear strengths of boron–epoxy, carbon–carbon, and other composites, thus providing a means of physical property measurement as well as flaw detection. Other NDT methods include the use of liquid fluorescent or dye penetrants; magnetic-particle, liquid rubber penetrants; ultrasonic resonance; optical, acoustic, and laser holography; and thermoluminescent coating, liquid crystals, microwave, eddy current, eddy sonic, beta back scatter, neutron radiography, and acoustic transmission methods. Acoustic emission shows great promise as an NDT technique, particularly in real-time investigations of impact damage.

The applications of composite materials have progressed to the point where the repair of composite material structures is important. Readers involved or interested in this aspect of composites should refer to *Composite Repairs*, listed in the biliography, for information.

REFERENCES

General

Broutman, L. J. and Krock, R. H. (eds.), *Modern Composite Materials*, Addison-Wesley, Reading, Mass., 1967.

Brown, H. (ed.), *Composite Repairs*, Monograph No. 1, SAMPE, Covina, Calif., 1985.

Grayson, M. (ed.) *Encyclopedia of Composite Materials and Components*, Wiley, New York, 1983.

Hancock, N. L. (ed.), *Fibre Composite Hybrid Materials*, Macmillan Publishing Co., New York, 1981.

Harrigan, W. C., Jr., Strife, J., and Dhingia, A. K., (eds.), *Fifth International Conference on Composite Materials*, The Metallurgical Society, Warrendale, Pa., 1985.

Holister, G. S. and Thomas, C., *Fibre Reinforced Materials*, Elsevier, New York, 1966.

Holliday, L., *Composite Materials*, Elsevier, New York, 1966.

Hull, D., *An Introduction to Composite Materials*, Cambridge University Press, New York, 1981.

Lubin, G. (ed.), *Handbook of Fiberglass and Advanced Plastics Composites*, Van Nostrand Reinhold, New York, 1969.

Lubin, G. (ed.), *Handbook of Composites*, Van Nostrand Reinhold, New York, 1982.

Manson, J. A., and Sperling, L. H., *Polymer Blends and Composites*, Plenum, New York, 1976.

Nielsen, L. E., *Mechanical Properties of Polymers and Composites*, Marcel Dekker, New York, 1974.

Oleesky, S. S. and Mohr, J. G., *SPI Handbook of Technology and Engineering of Reinforced Plastics/Composites*, Van Nostrand Reinhold, New York, 1973.

Parratt, N. J., *Fibre-Reinforced Materials Technology*, Van Nostrand Reinhold, London, 1972.

Schwartz, M. M., *Composite Materials Handbook*, McGraw-Hill, New York, 1984.

Schwartz, R. T. and Schwartz, H. S., *Fundamental Aspects of Fiber Reinforced Plastic Composites*, Interscience, New York, 1968.

Seymour, R. B. (ed.), *Additives for Plastics*, Vol. 2, Academic Press, New York, 1978.

Springer, G. S., *Environmental Effects on Composite Materials*, Vol. 2, Technomic, Lancaster, PA, 1984.

Theory

Ashton, J. E., Halpin, J. C., and Petit, P. H., *Primer on Composite Materials: Analysis*, Technomic, Stamford, Conn., 1969.

Calcote, L. R., *The Analysis of Laminated Composite Structures*, Van Nostrand Reinhold, New York, 1969.

Chamis, C. C. Prediction of Fiber Composite Mechanical Behavior Made Simple. Rising to the Challenge of the 80's. Society of the Plastics Industry, Inc. New York, 1980, pp 12-A-1 to 12-A-10.

Donaldson, S. L., "Revised Instruction for TI-59 Combined Card/Module Calculations for In-Plane and Flexural Properties of Symmetric Laminates," AFWAL-TR-82-4081, June 1982.

Halpin, J. E., *Primer on Composite Materials: Analysis*, 2nd ed., ASM, Metals Park, Ohio, 1984.

Jones, B., *Design, Fabrication and Mechanics of Composite Structures*, Technomic, Lancaster, PA.

Levy, S. and Dubois, J. H., *Plastics Product Design Engineering Handbook*, Van Nostrand Reinhold, New York, 1977.

Tsai, S. W. *Composites Design—1985*, Think Composites, Dayton, Ohio, 1985.

Tsai, S. W. and Hahn, T. H., *Introduction to Composite Materials*, Technomic, Stamford, Conn., 1980.

Tsai, S. W., Halpin, J. C., and Pagano, N. J., *Composite Materials Workshop*, Technomic, Stamford, Conn., 1968.

Whitney, J. M., Daniel, I. M. and Pipes, R. B., *Experimental Mechanics of Fiber Reinforced Composite Materials*, Prentice-Hall, Englewood Cliffs, N.J., 1982.

3

Packing Concepts in the Use of Filler and Reinforcement Combinations

John V. Milewski

Consultant
Santa Fe, New Mexico

CONTENTS

1. INTRODUCTION

Chapter 2 explained some of the basic theory of the effect of using fillers and reinforcements in plastic composites. For years, it has been found that combinations of filler and reinforcement are complementary and, in many cases, synergistic,[1-4] but until recently there was no theory to explain these effects or to predict and find the optimum combination of fillers and fibers. It is now proposed that the major part of the improvement is due to better packing of filler and reinforcements.[5,6] This chapter will review the most recent information available on the subject of bimodal packing.

First, a review will be given on the bimodal packing of spheres. Then the packing of fiber at various length-to-diameter ratios (L/D's) will be given, followed by the bimodal packing of fibers and spheres. Next, experimental data and examples will be given to explain how to utilize this information for the design of a material containing nearly sphere-shaped particles and fibrous reinforcements.

2. BACKGROUND (PACKING)

The packing of particulate solid materials enters into many aspects of applied science, and has therefore been studied more or less continually for a number of years. This great interest in packing particles has attracted many investigators, and as Graton and Fraser[7] pointed out in 1935, it might be expected that the subject of particle packing would have been thoroughly explored by this time. Such is not the case, however; the empirical approach adopted by many investigators has produced data for specialized applications, but has produced little precise, quantitative information, directly applicable to all packing circumstances. In 1961, McCrae and Gray[8] reported that researchers have tended to be concerned either with theoretical, unrealistic situations, or else those that are entirely empirical. The former are mainly concerned with the geometry of simplified packing situations (although restricted supporting experimentation is not uncommon) and are principally the study of systems of perfect spheres.

The author has been involved in studies of the packing of fiber and combinations of spheres and fibers. These studies have shown that there are specific conditions where the packing obtained would not have been anticipated from prior theories that evolved from the study of packing of spheres alone.

In 1973, Milewski[5] introduced a system for the study of the packing of fibers and spheres. This original paper covered minimum packing conditions for fiber L/D's from $4/1$ to $15/1$, and gave only general direction for obtaining maximum packing conditions. His 1974 paper[9] was much broader in scope, and identified the minimum packing conditions for fiber L/D from $4/1$ to $37/1$ plus the theoretical and experimental maximum packing conditions at various values for parameters such as percent fiber loading, fiber L/D, and R ratios (the ratio of the diameter of the sphere to the diameter of the fiber). Also, a new parameter called critical L/D was identified, as well as the relationship of this critical L/D to fiber loading in fiber–sphere systems.

By utilizing the above information, the density of packing of various fiber–sphere compositions can be calculated, and the void volume or resin demand can be determined. This also permits the economics of any fiber–sphere system (in any chosen combination) to be calculated for wide ranges of fiber loading L/D, R, and resin and fiber cost. Recent high resin costs and resin shortages make these packing concepts appear worthy of serious consideration because they point the way to more efficient utilization of systems with higher filler loadings.

3. REVIEW OF PACKING

3.1 Fiber Packing

The packing combinations that were studied in a thesis[6] at Rutgers University are illustrated in Fig. 3-1 as (a) fiber packing at various L/D's, (b) fibers packed into fibers, (c) fibers packed into spheres, and (d) spheres packed into fibers. The experiments were run on two scales. The larger scale experiment used wooden rods; the smaller scale experiment used fiberglass. Figures 3-2 and 3-3 visually illustrates that bulk density varies with L/D. The data obtained in measuring the relative bulk volume for known L/D's of both wooden rods and fiberglass are plotted on Fig. 3-4 and show very good volume packing agreement between the exact L/D's of wooden rods and the numerical average L/D's for the fiberglass. Relative bulk volume is defined as:

$$RBV = 100/\%\ \text{Volume accupied by solids}$$

The significance of this relationship is that it has been found that three-dimensional random packing values of fiber vary with the L/D and can be predicted from the given curve. Thus, with this relationship, the average fiber L/D's can be determined from the bulk volume data

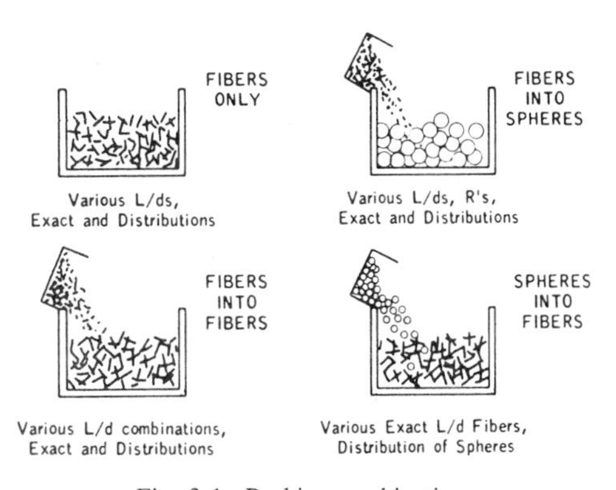

Fig. 3-1. Packing combinations.

Fig. 3-2. Wooden rods. Experimental packing of wooden rods.

Fig. 3-3. Milled fiberglass. Experimental packing of milled fiberglass.

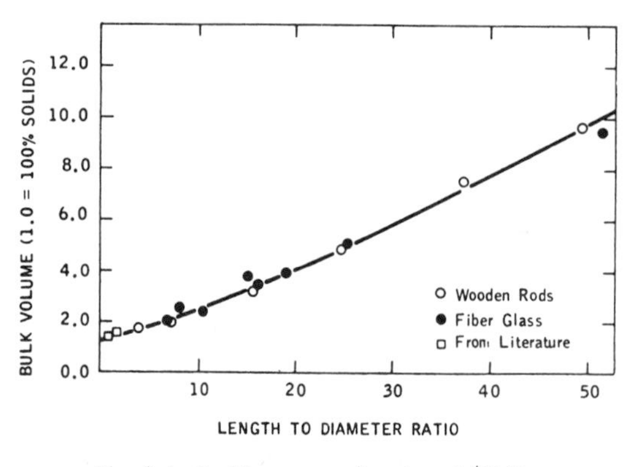

Fig. 3-4. Packing curve of various L/D fibers.

of the fiber, without the time-consuming expense of photographing, counting, and averaging hundreds of fibers to obtain a statistically significant sample of the fiber L/D.

3.2 Review of Bimodal Packing Theory

Before we study the packing of fibers and spheres, the bimodal packing of spheres will be reviewed so that the detailed concept of bimodal packing of fibers and spheres can be more easily understood.

3.3 Theoretical Maximum Density Boundary Conditions for Sphere–Sphere Packing

Figure 3-5 illustrates the densification that occurs when small spheres are added to large spheres. Maximum density is obtained when the small spheres are packed to their maximum density within the voids of the larger spheres. In Figs. 3-5A and 3-5B, each illustration represents the same volume of solid material; thus, the relative bulk volume is shown decreasing as densification occurs. In Fig. 3-5A, this occurs because in each step toward greater density, a large sphere is removed, and the same amount of solid material is replaced as small spheres within the voids of the remaining large spheres. Figure 3-5B illustrates how densifica-

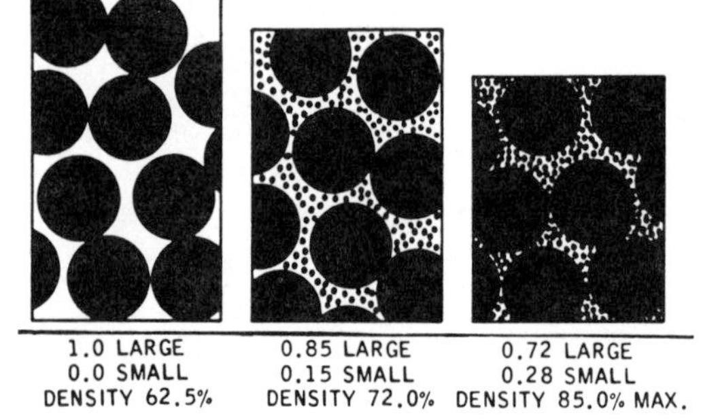

Fig. 3-5A. Addition of small spheres to large.

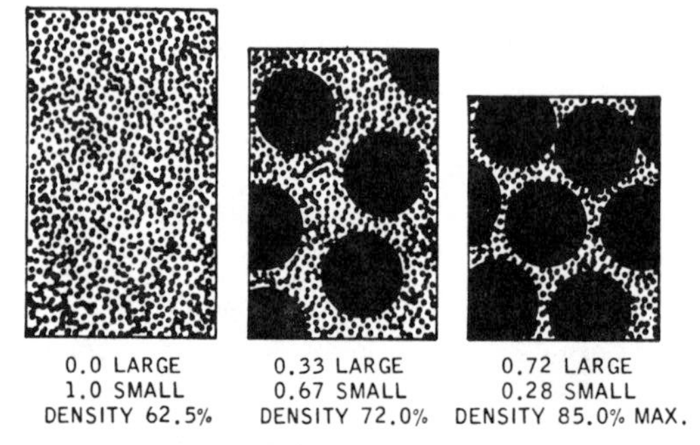

Fig. 3-5B. Addition of large spheres to small.

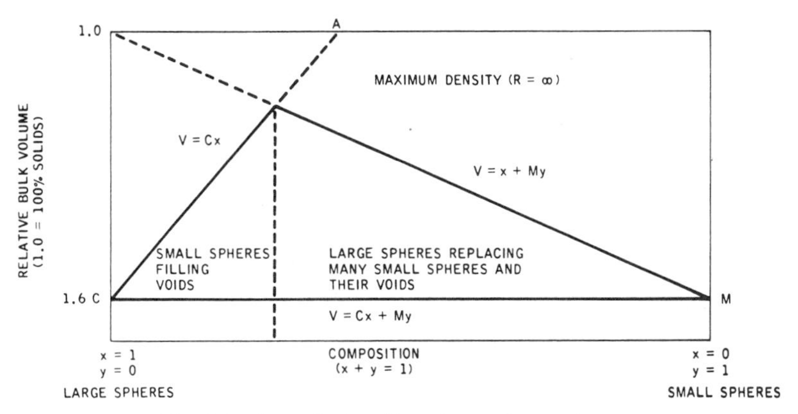

Fig. 3-6. Theoretical packing of two sphere systems.

tion occurs by the opposite process, in which a number of small spheres and their associated voids are removed, and the same amount of material is replaced as one large solid sphere. The solid line in Fig. 3-6 is a theoretical packing curve for an infinite size ratio $R = \infty$. For this example, the ratio R is the diameter of the large sphere divided by the diameter of the small spheres. The maximum density point represents the condition illustrated in Figs. 3-5A and 3-5B at the extreme right. In Fig. 3-6, the composition of the mixture is shown by the horizontal scale X, the volume fraction of the large spheres, and Y, the volume fraction of the small spheres, for a total volume of unity. The left-hand ordinate is the relative bulk volume and is defined such that 1.0 is equal to 100% solid material (100 divided by the percent theoretical density). A material with a relative bulk volume of 1.6 would be 62.5% theoretically dense. Thus, C gives the experimentally determined packed volume of the large spheres, and M gives that of the small spheres. By using relative bulk volume, rather than percent theoretical density, the packing curves for the infinite size ratio become two straight lines.

3.4 Maximum Density Boundary Conditions for Sphere–Fiber Packing

The theory of bimodal packing must be modified for packing two differently shaped materials because it is then possible to have R values less than 1 (R now being defined as the diameter of the sphere divided by the diameter of the fiber). This leads to the development of two theoretical maximums, both having the same relative bulk volume but at different fiber loadings. This concept is illustrated in Fig. 3–7. An example is given as follows:

S = Spheres packing at 50% solids + 50% voids.
Relative bulk volume = 2.0.

F = Fibers packing at 25% solids + 75% voids.
Relative bulk volume = 4.0.

x = Volume fraction of S.

y = Volume fraction of F.

$1 = x + y$.

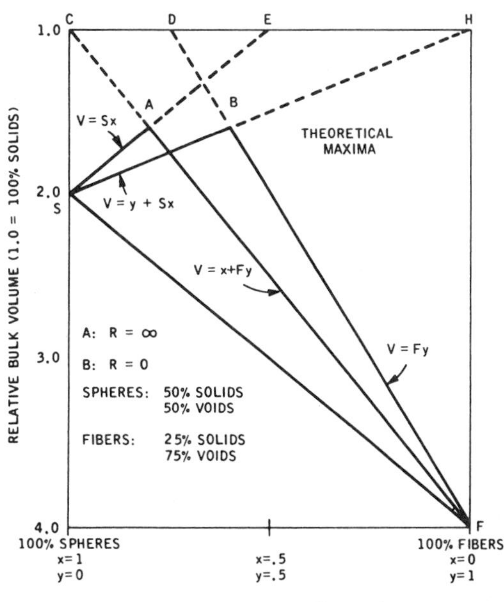

Fig. 3-7. Theoretical packing of fiber–sphere systems.

The curves are constructed in this manner: Line *SE* represents the filling of voids within the spheres, similar to Fig. 3-6, except that the small component is a fiber instead of other spheres. This line *SE* is represented by equation $V = Sx$. Line *FC* represents the fiber analog of Fig. 3-5B, with large-diameter spheres replacing voids and small-diameter fibers; this line is represented by equation $V = x + Fy$. The intersection of these two lines at point *A* represents the theoretical maximum for packing conditions for large *R* values where fibers fit within the voids of spheres.

For cases when the sphere diameter is smaller then the fiber diameter, a different construction is required. Line *FD* represents the void-filling case with small spheres filling the voids between the fibers, and *SH* is again the analog of Fig. 3-5B, but with large fibers replacing small spheres and their associated voids. The equation for these two lines are $V = Fy$ for *FD* and $V = y + Sx$ for *SH*. These lines intersect at point *B*. This point is the theoretical maximum packing density for $R = 0$.

3.5 Less than Maximum Packing Conditions

In reference to Fig. 3-6, it was shown in a paper by Westman and Hugill[11] that for a ratio equal to unity ($R = 1$), the packed volume, v, was given by the line *CM*, which has the equation:

$$v = C_x + M_y \tag{3-1}$$

and for an infinite size ratio ($R = \infty$), the values of v are given by the line joining *M* with unity at the left side of the diagram, and the line *CA* is obtained by joining *C* with the zero point at the right side of the diagram, the equations of these lines being:

$$v = C_x \tag{3-2}$$

and:

$$v = X + M_y \tag{3-3}$$

When the size ratio *R* is less than infinity, maximum density is not attained because the small spheres begin to dilate the packing of the larger spheres before all the voids are filled. This concept is graphically illustrated in Fig.

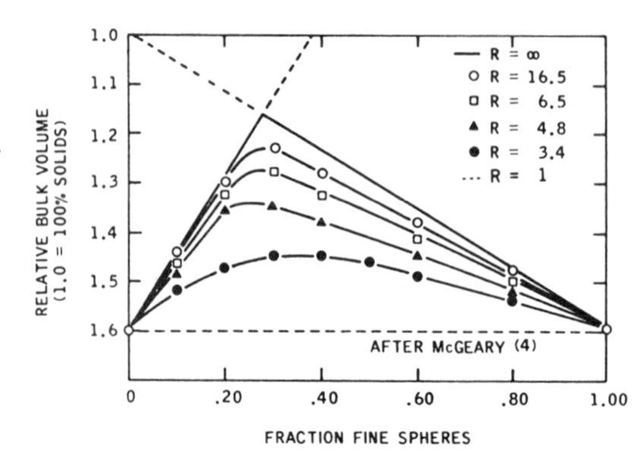

Fig. 3-8. Sphere–sphere packing curves.

3-8, which shows curves for several *R* values. At an *R* value of 1, the two spheres are identical, and obviously no packing advantage is seen. In this case, the packing of the right component with the left follows the law of mixtures, that is, a straight line connecting the two components. This is of no consequence when the two components are spheres, but when one component is of a different shape, such as a fiber, the straight line defines the worst possible packing conditions, also defining 0 packing efficiency.

In sphere–sphere packing for diameter ratios between the two limits of $R = 1$ and $R = \infty$, packing curves were developed by McGeary,[10] as illustrated in Fig. 3-8 and in the sphere–fiber packing system by Milewski[5] in the 1973 SPI paper. These curves were developed by packing fibers and spheres over a wide range of *R* values (*R* now defined as sphere diameter divided by the fiber diameter), while varying the L/D of the fiber component. The results were viewed in two basic ways, illustrated in Figs. 3-9 and 3-10.

In Fig. 3-9, the L/D of the fiber component (milled fiberglass) is held constant while *R* is varied. This is done with graded sizes of glass beads. The general effect is the same as that previously discussed for sphere packing, although the two components no longer have the same initial bulk volume.

Figure 3-10 shows the effect of a changing L/D at a constant *R* value. In the example, for $R = 10$, it can be seen that the short fibers fit easily in between the larger spheres, producing an increase in density, or decrease in bulk vol-

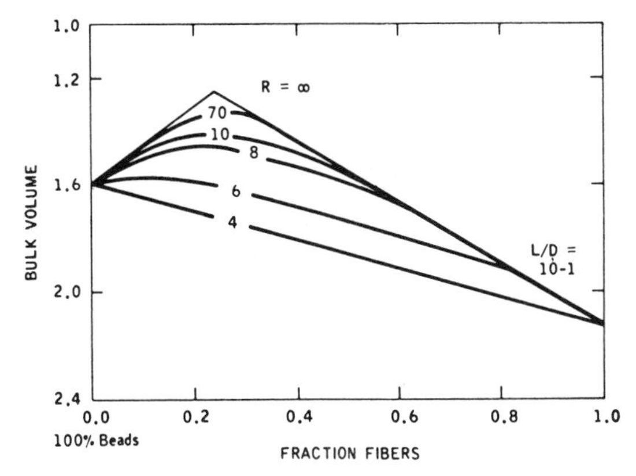

Fig. 3-9. Packing of fiberglass and beads at constant L/D.

ume, while the longer fibers nearly follow the linear law of mixtures.

The concept of packing efficiency is defined as the maximum deviation of the bulk volume from the mixture line H_{max} divided by the theoretical maximum deviation H^{∞} (see Fig. 3-11). The packing efficiency approaches zero for the long fibers and approaches 100% for the short fibers in the example of Fig. 3-10. Plots similar to this for wider ranges of fibers and spheres for various R's are shown in Fig. 3-12. Note the increased curvature, or higher packing efficiency, at the largest and smallest R's.

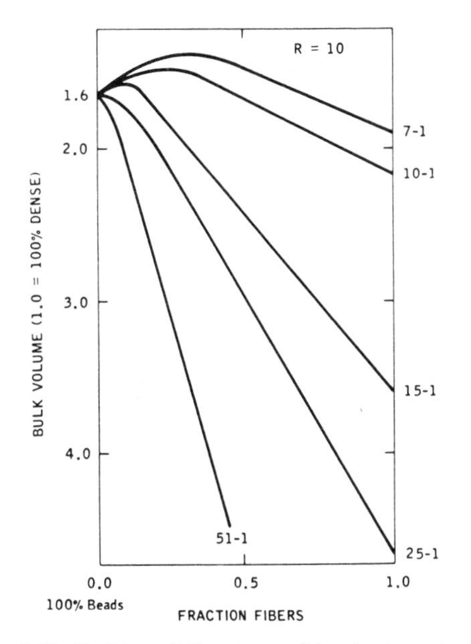

Fig. 3-10. Packing of fiberglass and beads at constant R and varying L/D's.

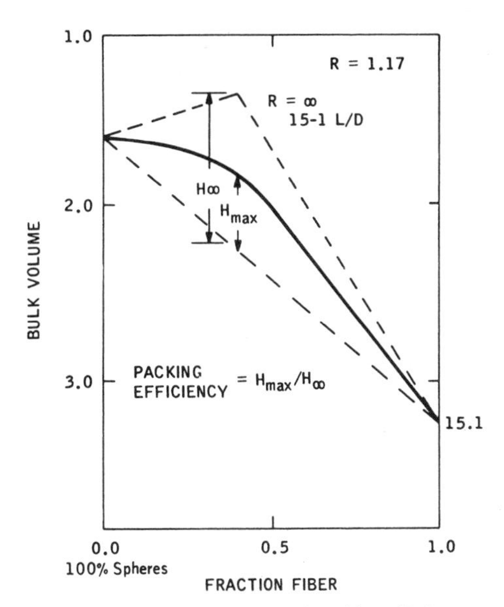

Fig. 3-11. The calculation of packing efficiency.

3.6 Experimental Minimum Packing Efficiency for Fiber Sphere Systems

Using the concept for packing efficiency, the bulk of the experimental packing work with fibers and spheres (from Fig. 3-12) can be viewed in a single graph (Fig. 3-13).

Each fiber of different L/D exhibits a minimum packing efficiency. The longer fibers show the minimum at higher R, as can be seen by the minimum in the 37/1 L/D curve at $R = 13.5$, compared to the minimization between $R = 1$ and 2 for the 4/1 L/D fiber. The packing efficiency increases on either side of the minimum, and eventually reaches 100% as R approaches ∞, or 0, as defined by the system (Fig. 3-11). Although the maximum packing density or minimum bulk volume for either very large or very small R could have been predicted, the exact location of the minimum conditions for the different L/D fibers must be determined by experiment.

From Fig. 3-13, the R values at the minimum packing conditions were determined for each fiber L/D; this relationship is illustrated in Fig. 3-14. The curve is of importance to future investigators, in that it points out what R values are to be avoided if good packing is desired, or at what R values lowest packing density can be obtained within the range of fiber L/D studied.

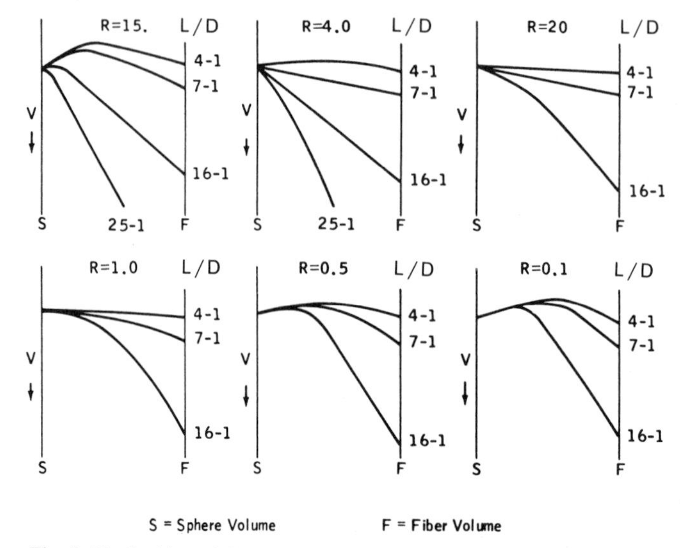

Fig. 3-12. Packing of fibers and spheres with varying L/D's and R-ratios.

3.7 An Example of Good Packing Combinations

By utilizing the above information, the density of packing of various fibersphere compositions can be calculated, and the void volume or resin demand can be determined. This also permits the economics of any chosen combination of fiber–sphere system to be calculated for wide ranges of fiber loading, L/D, R, and resin and fiber cost.

A clear example of the effectiveness of good packing combinations is illustrated in Fig. 3-15. This photograph depicts four cylinders, all containing the same solid volume of fiber and spheres at two different R values. The two cylinders on the left are in the unmixed condition with the fiber and spheres completely filling them. The two cylinders on the right are in the mixed condition and illustrate no packing advantage at an $R = 4$ value, where the void volume or resin demand is 50%. No packing advantage occurs here because the spheres are too large to fit within the average voids within the fibers, and the fibers are too large to fit within the average size void within the spheres. In the other cylinder, where $R = 0.5$, the small spheres easily fit and pack within the voids of

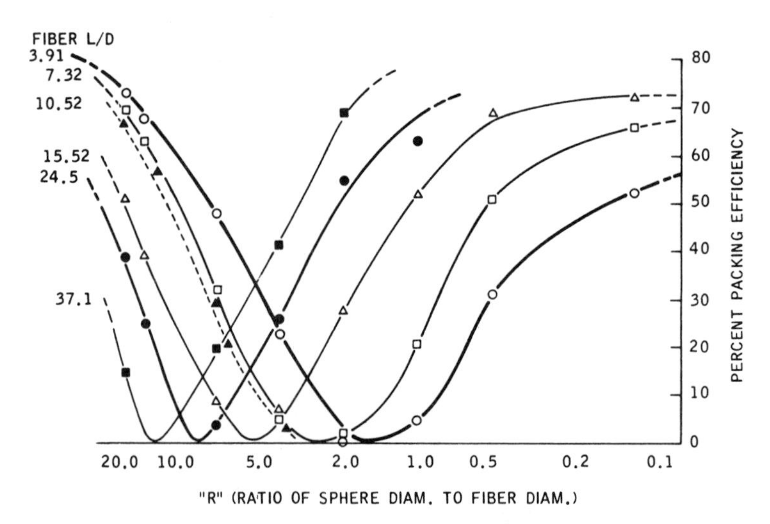

Fig. 3-13. Location of minima in packing curves.

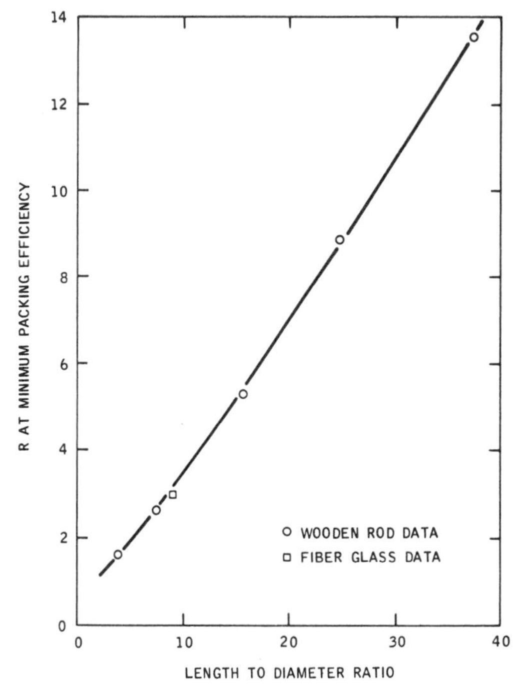

Fig. 3-14. Relationship between L/D and R at minimum packing efficiency.

the fiber, and a significant packing advantage is realized with a reduction of the void volumes from 50% to 35%. This also permits a uniform distribution and mixing of both fibers and spheres.

3.8 Theoretical Maximum Packing Densities

The theoretical maximum packing percentages were determined graphically by a method given above in bimodal packing theory for Fig. 3-7, except that the relative bulk volume data for the sphere and fiber L/D studied were used in construction of the graphs. These data are given in Table 3-1. Then experimental maximum densities were read from the data points of the packing curve for various R's and L/D's at 25, 50, and 75% fiber. These data are given in Tables 3-2 and 3-3. The 25% fiber data are plotted on Fig. 3-16. An examination of this figure tells what percent solids are the theoretical limits for packing spheres with 25% fiber of various

Fig. 3-15. The effect of proper fiber to sphere diameter ratio on packing efficiency. All cylinders contain the same volume of solids, 3 parts spheres and 1 part fibers at $15/1$ L/D.

Table 3-1. Theoretical percent solid content for fiber–sphere packing.

% Fiber loading	R	L/D 1.00	2.00	3.91	7.31	15.51	24.49	37.10
10	$R = \infty$	68.5	68.5	68.5	68.5	68.5	68.5	61.0
	$R = 0$	64.1	64.1	64.1	64.1	64.1	64.1	64.1
20	$R = \infty$	77.0	77.0	77.0	77.0	69.0	56.2	44.3
	$R = 0$	66.7	66.7	66.7	66.7	66.7	66.7	66.7
30	$R = \infty$	87.7	87.0	83.3	77.5	60.0	46.2	34.5
	$R = 0$	69.5	69.5	69.5	69.5	69.5	68.5	45.5
40	$R = \infty$	85.5	83.3	78.7	72.0	52.9	39.1	28.6
	$R = 0$	72.5	72.5	72.5	72.5	72.5	51.3	34.2
50	$R = \infty$	82.7	80.6	74.6	67.2	47.2	34.1	24.2
	$R = 0$	76.4	76.4	76.4	76.4	62.5	41.2	27.5
60	$R = \infty$	80.0	77.5	71.0	62.9	42.7	30.2	21.0
	$R = 0$	80.0	80.0	80.0	80.0	51.8	34.2	23.0
70	$R = \infty$	77.5	74.6	67.6	59.2	39.1	27.0	23.0
	$R = 0$	84.0	84.0	84.0	72.5	44.5	29.4	19.7
80	$R = \infty$	74.6	72.0	65.0	56.2	36.0	24.5	16.7
	$R = 0$	88.5	84.0	74.6	63.3	39.0	25.8	17.5
90	$R = \infty$	72.5	69.5	62.2	53.2	33.4	22.4	15.1
	$R = 0$	78.7	74.6	66.2	56.2	34.5	23.0	15.4
100% fibers		70.5	67.1	59.5	50.3	31.1	20.7	13.8
100% spheres		61.5	61.5	61.5	61.5	61.5	61.5	61.5
Theoretical max.		88.5	87.4	84.4	80.6	73.5	69.5	66.7

Table 3-2. Fiber loading at maximum solid content of various fiber–sphere systems.

R		L/D 3.91	7.31	15.51	24.49	37.10
0.11	% Solids	74.6	73.0	67.6	61.5	61.5
	% Fiber loading	60.0	40.0	30.0	0.0	0.0
0.45	% Solids	67.1	68.0	63.7	61.5	61.5
	% Fiber loading	50.0	45.0	30.0	0.0	0.0
0.94	% Solids	61.5	61.5	61.5	61.5	61.5
	% Fiber loading	0.0	0.0	0.0	0.0	0.0
1.95	% Solids	61.5	61.5	61.5	61.5	61.5
	% Fiber loading	0.0	0.0	0.0	0.0	0.0
3.71	% Solids	64.5	61.7	61.5	61.5	61.5
	% Fiber loading	30.0	8.0	0.0	0.0	0.0
6.96	% Solids	71.0	65.0	61.5	61.5	61.5
	% Fiber loading	30.0	22.0	0.0	0.0	0.0
14.30	% Solids	75.8	73.5	63.7	61.7	61.5
	% Fiber loading	31.0	25.0	8.5	3.5	0.0
17.40	% Solids	78.1	75.1	64.9	62.5	61.5
	% Fiber loading	33.0	26.0	12.0	4.0	0.0
*	% Solids	84.7	80.6	73.5	69.5	68.0
	% Fiber loading	27.5	24.0	16.5	11.3	9.6

*Theoretical Values $R = 0$ or $R = \infty$.

Table 3-3. Experimental solid contents at 25%, 50%, and 75% fiber loading for fiber–sphere packing.

Fiber L/D	Percent Fibers	*R* Value									
		0	0.11	0.45	0.94	1.95	3.71	6.96	14.30	17.40	∞
	25	68.5	68.5	65.4	61.7	61.0	64.5	70.0	74.6	76.4	82.0
3.91	50	76.4	74.6	67.2	61.7	60.2	64.1	67.5	72.5	74.5	75.7
	75	78.2	69.5	64.5	61.0	59.5	62.5	64.2	66.7	67.2	67.1
	25	68.5	68.5	64.5	61.0	58.5	59.9	64.5	73.5	74.6	80.6
7.31	50	76.4	71.4	67.5	58.8	55.5	56.6	58.8	65.4	67.1	67.1
	75	66.3	61.7	60.0	55.0	52.8	53.5	54.6	57.2	58.2	57.4
	25	68.5	66.7	63.7	59.9	54.6	50.3	50.5	54.1	57.5	65.0
15.52	50	61.7	55.6	51.8	50.7	45.5	42.0	42.4	44.3	44.3	48.1
	75	41.0	40.4	37.9	38.2	37.3	35.7	35.5	36.0	36.8	38.2
	25	68.5	66.5*	61.5*	55.5*	47.5	45.5	40.2	42.7	44.7	50.5
24.50	50	40.0	39.0*	38.0*	36.0*	34.0*	32.7	30.3	31.8	31.8	33.5
	75	26.4	26.3*	26.2*	25.8*	25.5*	25.2	24.3	25.0	25.6	26.2
	25	50.0	48.0*	45.0*	42.0*	39.4	37.7	33.8	33.1	39.2	41.3
37.10	50	25.7	—	—	—	—	—	22.6	22.6	22.6	25.6
	75	—	—	—	—	—	—	—	—	—	—

*Estimated values (extrapolated data)

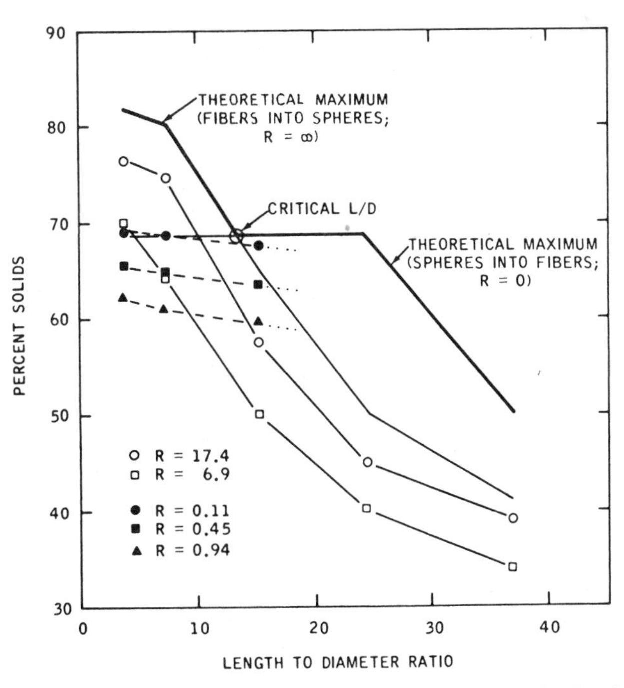

Fig. 3-16. Theoretical maximums and experimental percent solids at 25% fiber loading for various *R* values.

L/D's at different R values. For example, with a fiber L/D of $13/1$, it will be impossible to pack these fibers with any size spheres to a density greater than 68.5%, since this is more than the theoretical value. However, 74% solid packing is not only possible, but practical, and can be obtained when packing fiber with an L/D of $7/1$ or less with spheres with R values of 17.4 or greater. These data also indicate that R values much less than 1 give higher actual packing densities at intermediate L/D's (greater than the critical L/D) than R values much greater than 1. For example, R of 0.11 will pack to 68% solid with a $15/1$ L/D fiber, while an R of 17.4 will pack to 57.5% solid for the same $15/1$ L/D fiber. This effect can also be clearly seen by the crossing of theoretical maximum curves for $R = \infty$ and $R = 0$. The points at which these two curves cross is defined as the critical L/D. Theoretically, the highest percent solids are obtained with very-large R's when the L/D's are less than critical, and highest percent solids are obtained when using very small R's with fiber of L/D's greater than critical.

The location of the critical L/D was studied and determined by plotting its theoretical maxima for $R = \infty$ and $R = 0$ versus the L/D's, at various percentages fiber loading, as given in the data from Table 3-1. Figure 3-17 locates the critical L/D from the crossover points of theoretical curves for $R = 0$ and $R = \infty$ at various volume loading of fibers.

A relationship between the critical L/D's and volume percent fiber loading is shown in Fig. 3-18. When fibers and spheres are packed, Fig. 3-18 will determine whether large R's or very small R's should be used if maximum theoretical solid packing is desired, and the percent fiber loading and fiber L/D are given. For example, in using a 25% fiber loading with a $30/1$ L/D, small R's are most advantageous, while, on the other hand, if the L/D is 10, a large R is necessary. A limit of 60% fiber loading is indicated for large R's from Fig. 3-18.

4. GENERAL APPLICATION OF PACKING CONCEPT

To approach the maximum packing condition of the fiber–sphere mixture, in general, fibers should be chosen with L/D's ranging from $1/1$ to $100/1$, and preferably about $5/1$ to $50/1$, fiber loadings should be between 5% and 40%, and R should be chosen to be either large or small.

To maximize packing efficiency in choosing small R values, R should be less than 2, preferably less than 0.01, but practically about 0.1. If large R values are chosen, then R should be

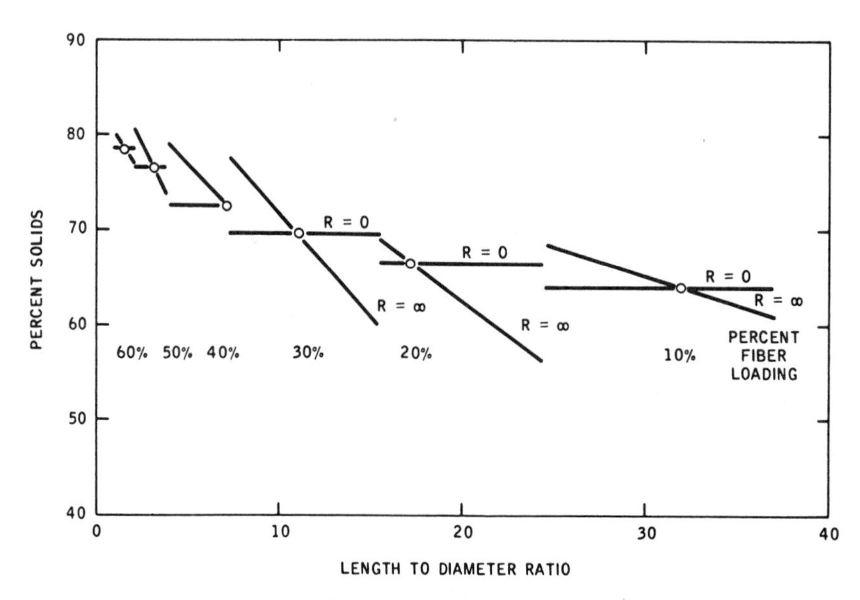

Fig. 3-17. Location of critical L/D.

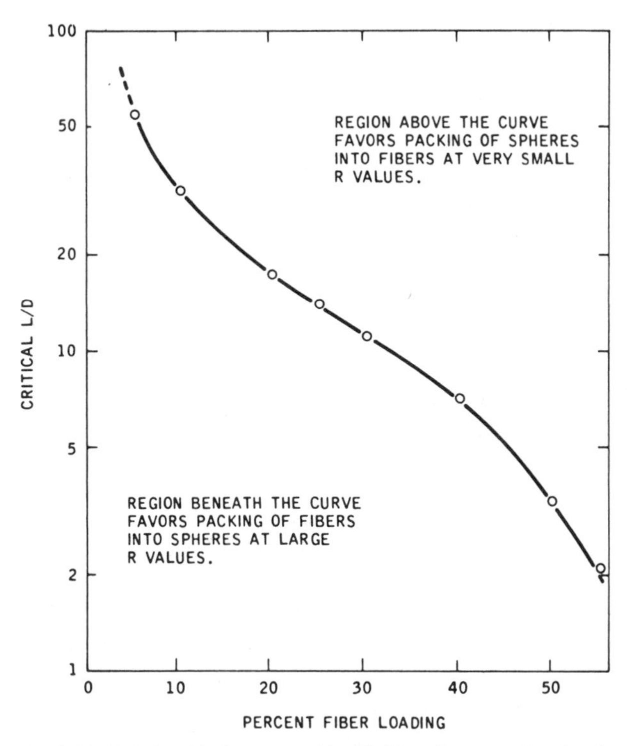

Fig. 3-18. Relationship between critical L/D and percent fiber loading.

greater than 5, preferably greater than 100, and more practically, about 20.

In a typical fiber–sphere formulation in which the fiber L/D is 40 or greater with a fiber loading range of 10% to 80%, R should be chosen to be about 0.1.

In a typical fiber–sphere formulation utilizing large R values, for L/D's ranging from 2/1 to 19/1 and fiber loadings of 1% to 25%, R at about 20 is very desirable.

In fiber loadings greater than 60%, for all L/D values, R should be small or about 0.1.

Where a moldable article is desired, the L/D's preferably range from 15/1 to 100/1, the fiber loadings from 5% to 75%, and R should be considerably less than 5 and as small as 0.5, if practical.

5. SPECIFIC EXAMPLES OF HOW TO APPLY PACKING CONCEPTS IN FIBERGLASS–SPHERE COMBINATIONS

Nearly spherical particle fillers can be used instead of perfect spheres, leaving the major packing concepts essentially valid.

Select the fiber to be used and determine its average L/D from its relative bulk volume (Fig. 3-4). For example, a commercial 1/16-in. milled fiberglass has an average L/D of about 25/1, and an average fiber diameter of about 14.0 μm. If one wishes to replace half the fiberglass with glass beads of −325 mesh (average diameter 30 μm), this would give an R value of about 2. From the data in Table 3-2, 50% fibers of 24.5/1 L/D and R value 1.94 will pack to about 33% solid volume or 67% void volume. To make a fully dense plastic composite, this 67% voids must be completely filled with resin; any lesser amount of resin would result in a partially dry mix with molded in voids.

A more practical combination would be three parts beads (filler) to one part fiber, or 25% fibers by volume; in this case, 1/32-in. milled fiberglass, which has an average L/D of about 10/1, will be considered. Table 3-2 does not contain data for 10/1 L/D, but it does contain 7.3 and 15.5/1 L/D data. Extrapolation for the 10/1 L/D data at 25% fiber loadings and R values of 1.95, yields 57% solids. This leaves the 43% voids to be filled with resin. If the resin is the most expensive component, this combination would be more economical than the previous mixture of fibers and beads. Since there are a wide variety of fiber and bead sizes, as well as resin costs, one cannot determine the optimum formulation without specifying particular needs. But much can be done to reduce resin demand and percent fibers, if proper sizes of beads are chosen to match specific packing parameters.

6. SAMPLE CALCULATION OF ECONOMICS IN PACKING-FORMULATIONS

Case 1. Assume that a batch of fiberglass-reinforced nylon (20/80 by volume) will be prepared, and the following components will be used:

A. Fiberglass, 1/16 milled, average $L/D = 25/1$, specific gravity 2.5, estimated price 50¢/lb. The theoretical percent solid content for random packing of this material, as derived from Table 3-1, is 20%. Thus, the resin demand will be 80% by volume.

B. Nylon resin, specific gravity 1.1, price 80¢/lb.

C. Composite:

20 volume percent milled fiberglass at specific gravity 2.5; 20 × 2.5 = 50 parts by weight.
80 volume percent nylon, 80 × 1.1 = 88 parts by weight.
Total batch weight = 138 lb.
Total batch cost = 50 × $.50 = $25 for fiberglass and 88 × $.80 = $70.40 for nylon; total $95.40.
Cost/lb = $95.40/138 = $0.69/lb materials cost.

Case 2. Assume that a 50/50 mixture of glass fibers and glass beads will be used as a filler for nylon resin.

A. Fiberglass, 1/16 milled, average $L/D = 25/1$, specific gravity 2.5, fiber diameter 14.0 μm, estimated price $0.50/lb.

B. Glass beads, −325 mesh (average diameter 28 μm), $R = 2.0$, specific gravity 2.5, price $0.15/lb.

C. The theoretical solid content of a packed combination of 50–50 fiberglass (25/1 L/D) plus glass beads with R value 1.95, as shown in Table 3-3, is 34%. Thus, the resin demand will be 66%.

D. Composite:

34% mixed filler volume:
17% fiberglass × 2.5 = 42.5 parts by weight;
17% glass beads × 2.5 = 42.5 parts by weight.
66% nylon × 1.1 = 72.6 parts by weight.
Total batch weight = 157.6 lb.
Total batch cost = 42.5 × $0.50 + 42.5 × $0.15 + 72.6 × $0.80 = $85.71.
Materials cost = $85.71/157.6 lb. = 54¢/lb.
Summary of Materials cost:
Nylon = 80¢/lb
20 wt % fiberglass = 69¢/lb
34 wt % fiberglass bead mixture = 54¢/lb

Additional costs would be associated with the blending of the fillers with the nylon resin.

It is apparent that the glass bead and fiberglass mixture can offer considerable economic advantages in materials costs. Additional advantages would be reduced shrinkage and part distortion, and a high modulus due to the higher loading. The addition of glass beads to a molding compound has frequently improved flow, and this may be a further advantage.

7. PACKING GUIDE

From an understanding of basic principles of packing, the following information can be realized:

1. For fiber packing alone, the average L/D can be determined from its relative bulk volume (Fig. 3-4).

2. From the relationship between the critical L/D and percent fiber loading (Fig. 3-18), a positive guide is available for selecting bead (particle filler) sizes at various combinations of fiber loading and fiber L/D.

3. In making more exact formulations, the experimental solid contents are given in Table 3-2; this gives precisely the void volume or resin demand for a specific fiber–bead combination.

4. For designing maximum solid filled systems, percent solid values are given over a wide range of R's and L/D's in Table 3–3.

5. By increasing solid content (more filler and fiber), void volume or resin demand can be lessened. In cases where the resin is the most expensive component, there is a definite economic advantage to using a better packed system of fillers and reinforcements.

6. It is believed that these packing concepts of replacing some of the fibers with spheres and reducing resin demand can be applied profitably to many aspects of the reinforced plastics business, in such areas as thermoplastic molding compounds, BMC, SMC, casting resins, tool resins, floor coating, furniture casting, reinforced foams, and so on, or in almost any composition where fillers are used and more strength is needed, or where reinforcements are used and less cost and improved processing efficiencies are desired.

8. THEORETICAL AND EXPERIMENTAL INFORMATION ON BIMODAL AND MULTIMODAL PACKING

8.1 The Relationship of Fine Particle Packing to Fiber Packing

Figure 3-4 shows the relationship between relative bulk volume (*RBV*) of a fiber and its aspect ratio (L/D). This is also given in Fig. 3-19, with an extended range. A similar relationship exists with fine particles when size is substituted for aspect ratio, as seen in Fig. 3-20. This curve shows that as the particles get smaller, especially less than 20 μm in diameter, their *RBV* increases rapidly. This occurs because the smaller particles stick to one another and string out in a random fashion to form a lower-density structure.

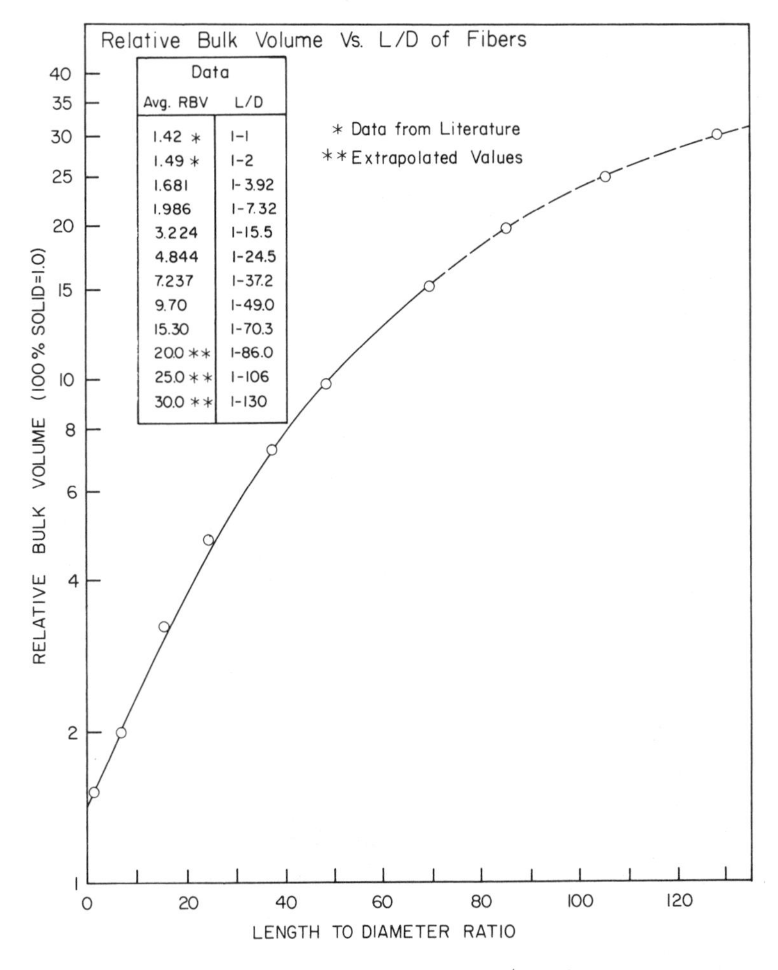

Fig. 3-19. Relative bulk volume vs. L/D of fibers.

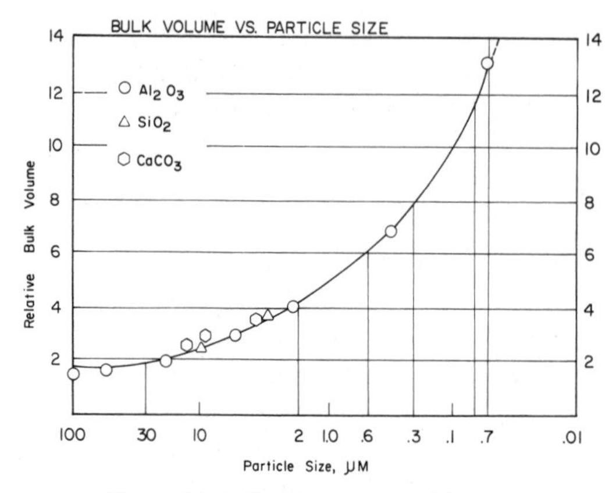

Fig. 3-20. Bulk volume vs. particle size.

Now the bulk volume data from both curves can be combined to show a new relation, as given in Fig. 3-21. From this curve, one can see that a particle with an average diameter of 0.1 μm has about the same *RBV* as a fiber with a 50:1 aspect ratio. It appears that there is a relationship between fine particles and fibers through their *RBV*.

It has been proposed that this relationship is based on the average number of particles that are strung together and act like a fiber, producing a nested structure of agglomerates of particles and raising their bulk as shown in Fig. 3-22. This is very similar to an increase in the fiber aspect ratio. It would be interesting to study this relationship in detail with a number of SEM photographs of fine particles of different sizes, counting each agglomerate and determining the average number of particles per agglomerate, to see if there is a direct relationship between the number of particles per string and their equivalent aspect ratio or L/D as determined by the curve in Fig. 3-21. If this re-

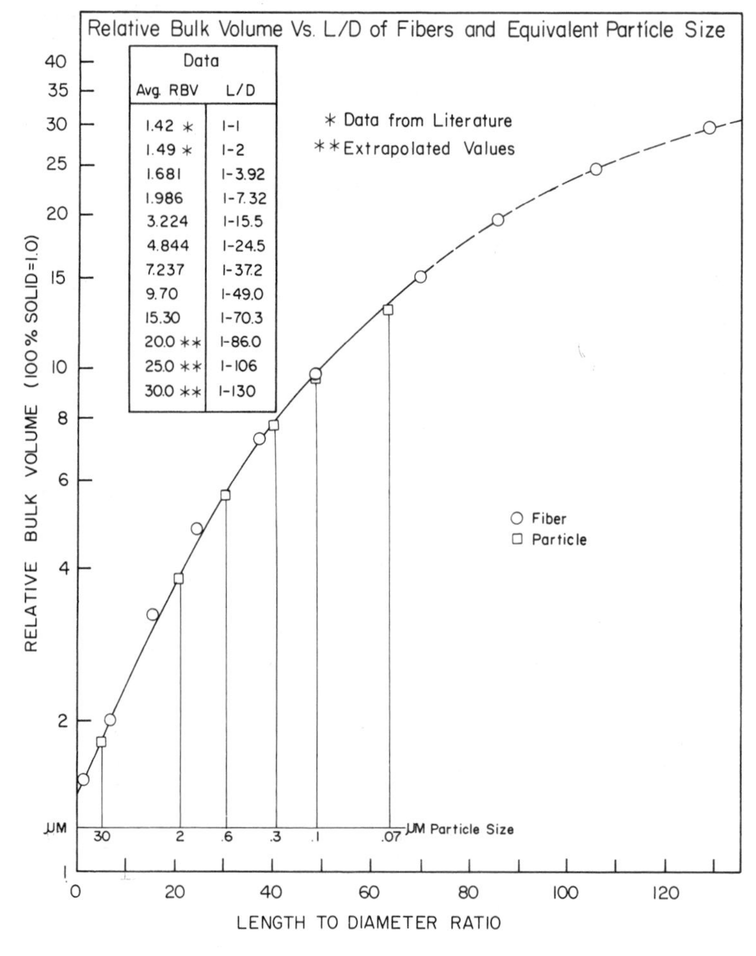

Fig. 3-21. Relative bulk volume vs. L/D of fibers and equivalent particle size.

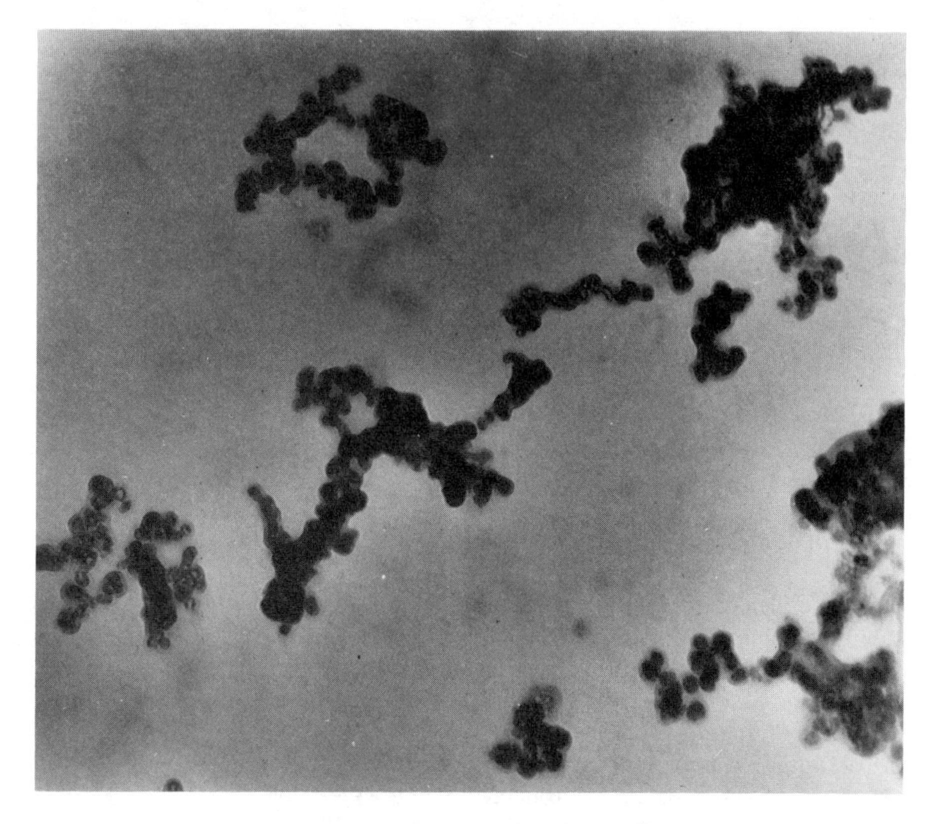

Fig. 3-22. Fine particles stringing together like a fiber structure.

lationship were true, the curve in Fig. 3-21 would predict that a particle 0.1 μm in diameter giving an *RBV* of 10 would have an aspect ratio of 50 : 1, with an average of 50 particles stuck together per agglomerate. This work apparently has not been done yet, and would make a good research project.

8.2 Stepwise Packing of Particles

The curves in Fig. 3-23 illustrate the *RBV* of sphere-to-sphere packing for both experimental and theoretical data. The figure is very similar to Fig. 3-8 except for the addition of the parts ratio line that runs horizontally across the graph at the 1.6 *RBV* value. This line connects the *RBV* of large spheres to small spheres at an *RS* (size ratio) of one, that is, when both spheres are the same size.

From these curves, one can easily see that the maximum packing occurs at an *RV* (volume ratio) of large spheres to small spheres of 3 : 1. Also from these curves it can be seen that the theory predicts that the *RBV* for bimodal packing of two spheres with an *RS* of infinity is 1.16, or 86.2% solids. If one chooses an *RS* of about 10, the *RBV* is 1.25, or about 80% solids. Using the theoretical and experimental information on these curves, in order to develop 92% packing efficiency for a bimodal mixture one should use an *RS* of about 10 and a volume ratio, *RV*, of about 3. These data were developed with a model system that used relatively large spheres (about the size of marbles); but one should keep in mind that actual experimental data for bimodal mixtures that contain fine spheres or particles in the micrometer size range get distorted by surface area effects that cause agglomeration and stringing together of fine particles, which in turn interferes with good packing.

These data and the suggested corrections for

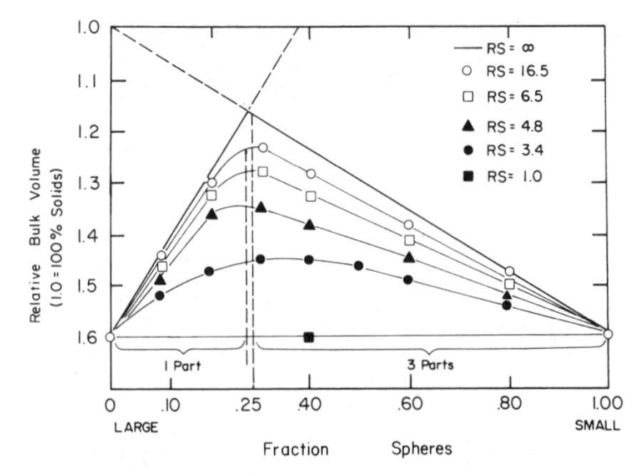

Fig. 3-23. Theoretical and experimental data for sphere-sphere packing. Experimental data indicate 25% small spheres for maximum packing (i.e., 3 parts large, 1 part small).

these effects will be discussed in these next few paragraphs.

Another way of determining what size small spheres will pack within the voids created by larger spheres is by geometry, as illustrated in Fig. 3-24. This figure shows a projection of the largest dimension of five equally sized spheres, with four packed in a cubic array and three in a hexagonal array. From geometric consideration, one can determine that the maximum-size small sphere that will fit within a cubic array is one-third of the diameter of the larger spheres ($RS = 3$). Similarly, when one considers a hexagonally packed array, the smallest dimension will only permit a small sphere size, one-eighth of the diameter of the larger spheres.

Randomly packed spheres contain about a 50/50 mix of both hexagonally and cubically packed spheres. From these geometric considerations, one can conclude that in order to have good packing efficiencies, the smaller sphere should be equal to or less than one-eighth the diameter of the larger sphere.

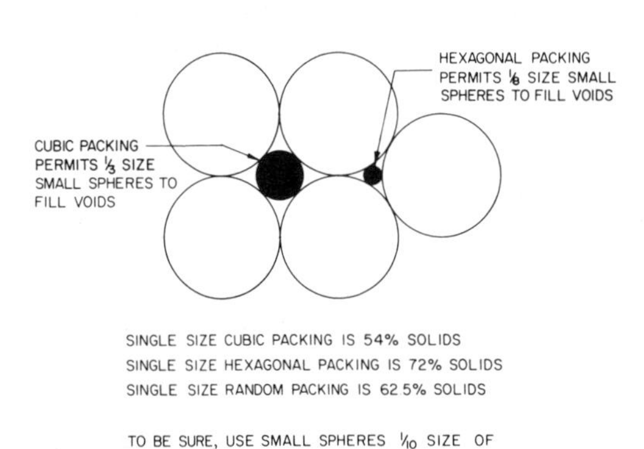

Fig. 3-24. Sphere sizes for ideal packing geometries.

8.3 Experimental Testing of *RS* Ratio

An opportunity for testing the *RS* ratio became available to satisfy the requirements for a very highly filled plastic (greater than 90% by volume). To accomplish this, it was decided that at least a trimodal packing combination should be used. Unfortunately, the exact sphere size combinations required for optimal trimodal packing were not available, so a quadmodal packing experiment was conducted. It was thought that addition of the fourth component would compensate for this lack, and it did.

The experimental data are given in Fig. 3-25. In this case, *RV* is the volume ratio of adjacent sizes. As an example, at $RV = 2$, the volume of 35 μm particles is twice that of 7 μm particles, the volume of 600 μm particles is twice that of 35 μm particles, and the volume of 4700 μm particles is twice that of 600 μm particles; thus for $RV = 2$, there are 1 part of 7 μm, 2 parts of 35 μm, 4 parts of 600 μm, and 8 parts of 4700 μm particles. The experiments were started by molding specimens with an *RV* of 3, which was previously predicted to be optimal in the analysis of the data of Fig. 3-23. However, the experimental data did not confirm this, with only a 79.4% solid loading obtained. So smaller and smaller *RV*'s were tried, with increasing success. The optimal *RV* is in the neighborhood of 2.0 (at a 93.8% solid filled system) instead of 3.0, as previously predicted.

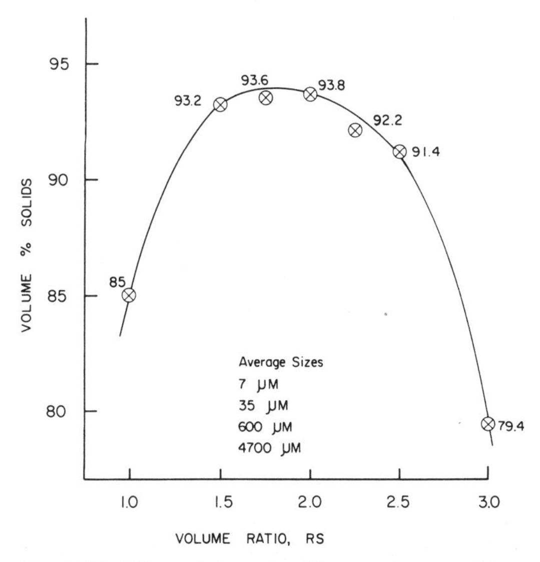

Fig. 3-25. Effects of size ratio, *RS*, on maximum packing.

Consideration of these results and the previous information about the behavior of smaller particles made it obvious why the theoretical ratio *RV* does not apply. It was seen from the data in Figs. 3-20 and 3-22 that small particles stick to one another and also to the sides of larger particles. This phenomenon is associated with their higher surface area and generally greater surface activities. When this is taken into account, Fig. 3-24 can be redrawn to look like Fig. 3-26. In Fig. 3-26, one can easily see

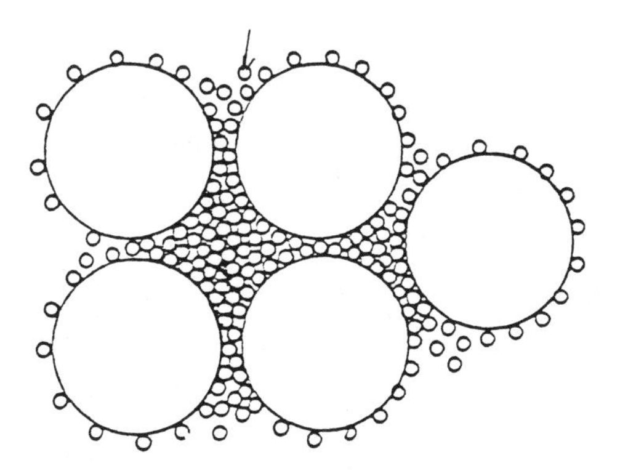

Fig. 3-26. A problem with actual packing of spheres. When the small and larger spheres are mixed together, the small spheres generally stick to the sides of the larger spheres, dilating them—creating a larger volume to be filled with small spheres, thereby changing the packing ratio from an ideal ratio of 3/1 to an actual ratio of 2/1. This is an especially acute problem when the small particles are less than 10 μm in diameter.

that if the small particles stick to the sides of the larger ones, the larger particles cannot pack as closely together as is indicated by ideal cubic or hexgonal packing. Therefore, the voids between these larger spheres are dilated, and the resulting volume that is made available for inclusion of the finer spheres is increased. This means that a higher percentage of finer-size spheres is required to fill these voids, and the *RV* decreases.

8.4 Additional Comments on Molding Highly Filled Systems

A suggestion that was found helpful in molding highly filled resin systems is this: When systems are filled to over 90% loading, the amount of resin is small, and uniform distribution is difficult. This results in dry spots on some of the fillers that increase friction and restrict flow, thus preventing full densification. To overcome this, about 10% excess resin is added to permit good flow and to lubricate all particles, which in turn permits redistribution of the particles during molding in the direction of optimal packing.

The molding die was reworked with deep undercuts close to the molding surface to receive the flow of excess resin during squeeze-out, thus permitting full densification. This is shown in Fig. 3-27.

8.5 Continuous Distribution for Optimized Packing

Apparently, no experimental work has been done to produce curves depicting the optimized continuous distribution of fillers in plastics. Specifically, this study should be done for castable plastic with resins of various viscosity such as those found in the usual castable polyester and epoxy formulations, with, say, a viscosity of 200 to 500 centipoise.

However, extensive work has been done in the determination of optimized particle size distribution in the ceramics industry using water slips; and much can be learned from a review of what has been done in this industry, as a guide to what should be done in the higher-viscosity resin systems found in cast plastic formulations.

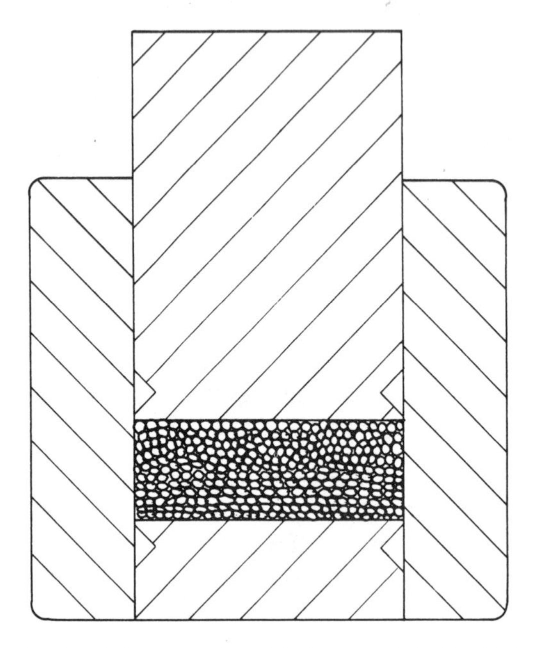

Fig. 3-27. How to mold to full density with highly filled systems. Add 5% to 10% excess resin to allow good flow and lubrication for redistributing particles to a good packing mode. The die undercut receives the excess resin squeeze-out, permitting full densification.

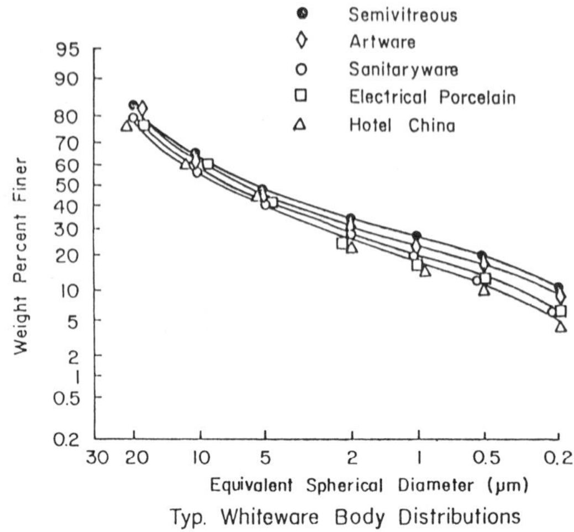

Fig. 3-28. Particle-size distribution of typical clay-based whiteware body compositions.

Phelps et al.[12] discuss the curves given in Fig. 3-28. These curves depict the optimized particle distribution that has been developed for various whiteware body compositions found in semivitreous ware, art ware, sanitary ware, electrical porcelain, and hotel china. The significant point is that these industries generally operate independently of each other, yet they all produced similar optimized distribution curves. The average data from these curves have been replotted using a linear scale for the *y* coordinate, as shown in Fig. 3-29. A significant feature of this curve, which does not make it immediately applicable to plastics, is the high percentage of submicron particles that are required to complete the distribution; for example, 7% less than 0.2 μm and 13% less than 0.5 μm. It is believed that the main purpose served by these very small particles is to increase the viscosity and to form a thixotropic gel-like network to help hold the larger particles in suspension. In such high-viscosity resin systems, this viscosity increase is not necessary and is in fact undesirable. In resin systems, there is less need for a thixotropic filler, and the situation is quite different from that of the water system used in ceramics. Major adjustments must be made in the distribution curve for the amount and size of the submicron component. In general, however, an educated guess as to how to change this distribution curve would be that the whole curve should be shifted to the left as the viscosity of the resin system increases with a minimum change in slope. The development of optimized curves for continuous distribution packing in resin systems is another case of research in this area of packing that has not yet been done.

9. AUTHOR'S FOOTNOTE

The author realizes that the final selection of the formulation in making a plastic composite is often directly related to the resulting physical properties and processability of the composite. This chapter suggests many maximum loading combinations of fibers and beads that should produce highly desirable properties. In general, the preferred properties are obtained with higher percents of reinforcements, as long as a void-free composite can be made. Very few of these maximum packing combinations have been tried, because of the great variety of plastics, as well as fibers and beads.

The author welcomes questions and comments on new combinations as they are evaluated.

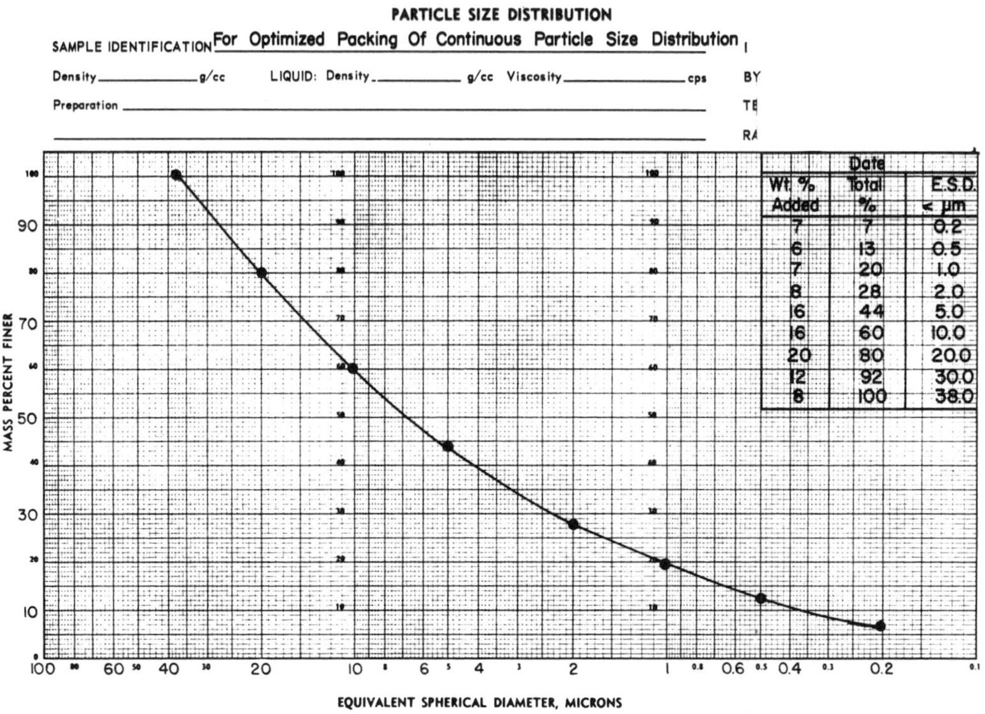

PARTICLE SIZE DISTRIBUTION

SAMPLE IDENTIFICATION For Optimized Packing Of Continuous Particle Size Distribution

Wt % Added	Data Total %	E.S.D. < μm
7	7	0.2
6	13	0.5
7	20	1.0
8	28	2.0
16	44	5.0
16	60	10.0
20	80	20.0
12	92	30.0
8	100	38.0

Fig. 3-29. Particle size distribution for optimized packing of continuous particle size distribution.

REFERENCES

1. Ritter, J., "Sphere-Filled Plastic Composites Theory," Testing Applications 25th Annual Tech. Conf. SPI, Composites Div., 1970.
2. Potter's Industries Inc., Carlstadt, N.J., Technical Bulletin #T24-1.
3. Wells, H., "Preliminary Investigation into the Use of Glass Beads in Fiberglass Reinforced Plastics," 22nd Annual Tech. Conf. SPI, Reinforced Plastics Div., 1967.
4. Potter's Industries Inc., Carlstadt, N.J., Technical Bulletin #PB24-1, June 1969.
5. Milewski, J. V., "A Study of the Packing of Milled Fiberglass and Glass Beads," 28th Annual Tech. Conf. SPI, Reinforced Plastics Div., Feb. 1973.
6. Milewski, J. V., Ph.D. thesis, "A Study of the Packing of Fibers and Spheres," Ceramics Dept., Rutgers Univ., New Brunswick, N.J., 1973.
7. Graton, L. C. and Fraser, H. J. "Systematic Packing of Spheres—with Particular Relation to Porosity and Permeability," *J. Geology* XLIII (8): 785–909, 1935.
8. Macrae, J. C. and Gray, W. A., "Significance of the Properties of Materials in the Packing of Real Spherical Particles," *British J. Appl. Phys.* 12: 164–172, 1961.
9. Milewski, J. V., "Identification of Maximum Packing Conditions in the Bimodal Packing of Fibers and Spheres," 29th Annual Tech Conf. SPI, Reinforced Plastics Div., Feb. 1974.
10. McGeary, R. K., "Mechanical Packing of Spherical Particles," *J. Am. Ceram. Soc.* 44 (10): 513–521, 1961.
11. Westman, A. E. R. and Hugill, H. R., "The Packing of Particles," *J. Am. Ceram. Soc.* 13: 767–779, 1930.
12. Phelps, G. W., Silwanowicz, A., and Romig, W., "Role of Particle-Size Distribution in Nonclay Slip Rheology," 72nd Annual ACS Meeting, May 1970.

Section II
Flake and Ribbon Reinforcements

4

FLAKES

George C. Hawley

George C. Hawley & Associates Ltd.
Saranac, New York

CONTENTS

1. INTRODUCTION

Flakes have many of the reinforcing characteristics of fibers, including the ability to enhance the strength, stiffness, and heat resistance of polymers, when used in composites. Unlike fibers, however, flakes do not align along flow lines during processing, but align themselves parallel and perpendicular to the flow and cross-flow directions. This reduces any flow-induced differential thermal expansion and thus inhibits warping in those polymers that are prone to it, especially PBT and PET. Any strong, stiff flake has these characteristics, but in practice commercial flakes are only available in mica and in C glass. C glass flakes have a low aspect ratio and are very brittle, so that they pulverize during processing. They have little application in plastics, apart from a long-time preeminence in epoxy linings and a growing interest in their use in RRIM polyurethanes.

Aspect ratio, the ratio of diameter to thickness, is an important parameter for flakes because it relates to the efficiency of stress transfer from the weak polymer matrix to the strong reinforcement. Until recently, mica had not been considered as a reinforcement for plastics, because of its low aspect ratio.

Dr. Woodhams and his students at the University of Toronto showed that high aspect ratio mica could be produced by ultrasonic delamination, and that the product was a valuable reinforcement. This work was carried further by Drs. Maine, Shepherd, Osborne, and Moskal at Fiberglas Canada Ltd. The author and his team at Marietta Resources International (MRI) developed methods to make high aspect ratio mica in commercial tonnages.

Stress transfer also relies on good coupling between mica and polymer. Hydrophilic mica is not wetted by olefins. Coupling agents to overcome this problem have been developed by MRI, in conjunction with coupling agent manufacturers. A milestone was the discovery of low-cost coupling agents by F. Meyer and S. Newman at Ford Motor Co.

Considerable work remains to be done in the development of new, lower-cost coupling agents in the optimization of composite properties by increasing the flake aspect ratio, reducing flake damage in processing, and improving flake orientation.

Impact resistance is the Achilles' heel of mica because, unlike fibers, flakes cannot entangle, and exercise only short-range effects. Improvements have been attained by the use of coupling agents, polymer impact modifiers, and combinations of flakes with other fillers and reinforcements such as glass and organic fibers and calcium carbonate. Mica-reinforced polypropylene and polyethylene are now accepted materials for such automobile end uses as heater valve housings, seat backs, load floors, glove compartment boxes, inner fender liners, and so on. These compounds provide better performance than talc-filled polyolefins, approaching the performance of glass-fiber, reinforced polymers.

Mica overcomes the warping tendency of PBT and PET and has allowed these compounds to be used in glue-gun housings, computer keyboards, rear quarter panels and cowl vents for automobiles.

Use of mica-reinforced polymer composites is currently about 25 million pounds annually and growing at a rapid rate. It will likely increase with the development of mica-reinforced RIM, polyamide, and PVC.

2. CHEMICAL AND PHYSICAL PROPERTIES OF MICA

Mica is a generic term for a family of phyllosilicates—leaf-like aluminosilicates, mainly of potassium, in which some of the aluminum ions can be replaced by magnesium and iron and part of the chemically bonded water may be replaced by fluorine.

The most common form of mica in commerce is muscovite. Production of phlogopite mica flakes began nearly two decades ago, and they have proved to be especially useful as a reinforcement for plastics. Biotite, a common mineral, is an adulterant in the other micas, but has not yet found any commercial use. Fluorphlogopite is a synthetic mica used in electronic applications, where no combined water can be tolerated. The chemical compositions of the forms are:

Muscovite	$K_2Al_4(Al_2Si_6O_{20})(OH)_4$
Phlogopite	$K_2(Mg,Fe)_6(Al_2Si_6O_{20})(OH,F)_4$
Biotite	$K_2(MgFe^{2+})_4(Al_2Si_6O_{20})(OH)_4$
Fluorphlogopite	$K_2Mg_6(Al_2Si_6O_{20})F_4$

These compositions are subject to an infinite number of variations by isomorphous substitution. Not only are such variations found from mine to mine, but small variations occur from one part of a mine to the next and from the surface to the depths. This occurs because micas are created by the alteration of a host rock by infiltration of mineral-rich ground waters. A common variation in muscovite and phlogopite is substitution with ferrous and ferric iron. Iron, chromium, and other chromophores produce the various colors commonly seen in micas—ruby, green, brown, black. Pure white mica is seen only in synthetic fluorphlogopites made from chromophore-free chemicals. Micas such as phlogopite and biotite can also become hydrated to form vermiculite. In vermiculite, water content as high as 20% is possible, whereas the parent micas typically contain only 3–5% combined water. Vermiculite flakes expand to 12 times their thickness when heated rapidly to 800–2000°F. Certain muscovites and phlogopites may expand up to 300% when heated to 600°C. Others, like Suzorite phlogopite, do not expand on heating, nor do they lose the combined water until a temperature of 1200°C is reached. Muscovites in general lose the combined water at lower temperatures, even as low as 200°C.

The micas are distinguished from other silicates by the ease with which they can be split into thin flexible sheets. It can be seen from Fig. 4-1, that mica consists of a sandwich with outer layers of silicon–oxygen tetrahedral sheets in which one-fourth of the silicon atoms have been replaced by aluminum and a filling of aluminum, magnesium, iron, hydroxyl, and fluorine ions, arranged in octahedral fashion. In phlogopite, the octahedral layer is primarily brucite, $Mg(OH)_2$, and it is gibbsite, $Al(OH)_3$, in muscovite. These three-layer units, which are about 10 Å thick, are held loosely to one another by potassium ions that are linked, in 12-fold coordination to the six oxygen atoms at the surface of one unit, and to the equivalent six oxygen atoms of the next three layer unit. The strength of the interlayer van de Waals forces is quite weak, so that the sheets of mica can be readily cleaved. On a macroscopic level, it has been found that there are certain weak planes in a mica crystal that are readily split. During manufacture of mica flakes, splitting occurs at these weak planes. The remaining

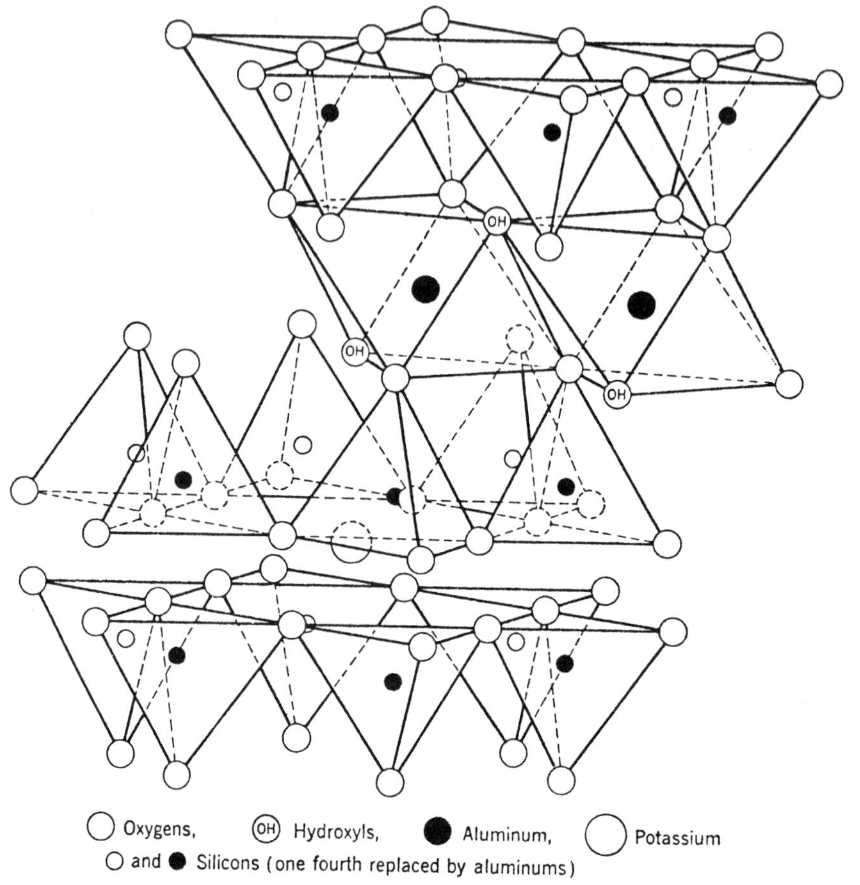

Fig. 4-1. A diagrammatic sketch of the crystal structure of muscovite shows a portion of the planar layers held together with potassium ions, after Jackson and West (1930).

planes are held together more strongly than this by forces superior to the cohesive strength of most polymers and to the adhesive strength of the bond between the polymer and mica. If this were not so, mica could not be used as a reinforcement for plastics.

The strength of the weak planes also appears to vary from mica to mica. Muscovite and biotite are much more difficult to delaminate into high aspect ratio flakes than phlogopite.

A typical chemical analysis of phlogopite, muscovite, and biotite is shown in Table 4-1. It should be remembered that variations occur from source to source, and that processing can change the valency of the iron and the water content.

Table 4-2 summarizes the physical and chemical properties of phlogopite and muscovite micas. Both phlogopite and biotite are decomposed by strong mineral acids such as sulfuric and hydrochloric, whereas muscovite is unaffected. All micas are dissolved by hydrofluoric acid.

The table does not show that the permeability of mica to liquids and gases is very low.

2.1 Properties Related to the Utility of Mica in Plastics

Table 4-2 shows that mica has high strength, and stiffness; a low coefficient of thermal expansion; high thermal conductivity and temperature resistance; excellent dielectric properties; good chemical resistance and low solubility in water; low coefficient of friction; and low hardness.

By the rule of mixtures, these properties are conferred on any composite in which mica is used, in proportion to the volume fraction that is mica.

Table 4-1. Chemical analysis of commercial phlogopite and muscovite micas.

	Phlogopite[a] [wt %]	Muscovite[b] [wt %]	Muscovite[c] [wt %]
SiO_2	40.74	47.90	48.30
Al_2O_3	15.76	33.13	31.55
MgO	20.56	0.69	1.99
FeO	7.83	–	–
Fe_2O_3	1.21	2.04	2.20
K_2O	9.96	9.80	7.86
Na_2O	0.46	0.80	1.20
BaO	0.45	–	–
CaO	<0.01	0.50	1.25
TiO_2	0.42	0.65	1.04
Cr_2O_3	0.14	–	–
MnO	0.07	0.01	0.01
F	2.16	–	–
P	–	.03	–
S	–	.01	–
H_2O^+ [combined]	3.01	4.30	4.20
H_2O^- [free]	0.01	0.10	0.10

[a]Suzorite variety, Marietta Resources International Ltd.
[b]The English Mica Co.
[c]Whittaker, Clark and Daniels, Inc.

The effects on composites are summarized in Table 4-3. The data are generalized to include all polymeric composites. Mica is more effective in some polymer types (polyolefins, thermoplastic polyesters, polyamides, polar styrenics) than in others (polycarbonate, polystyrene, ABS, PPS). It is especially effective in polar crystalline polymers. In nonpolar polymers, coupling agents must be used.

Mica is the most effective mineral filler for reducing warpage and increasing stiffness and heat deflection temperature. These effects are all related to the high modulus of the flake itself and to the fact that the flakes orient themselves along flow lines, in a plane, to confer isotropic reinforcement. In contrast, fibers orient themselves in the flow direction, creating differences in the coefficient of thermal expansion between this direction and the cross-flow or transverse direction that lead to distortion of the part on demolding and when placed in high temperature service. The high modulus of the flake composites persists at high temperatures, as measured by the heat deflection temperature, leading to higher service temperatures.

Low warpage, high temperature modulus, low weight, and low cost are the main needs of the automotive industry for parts other than body panels. Meyer and Newman[1] demonstrated that a composite containing 40% high aspect ratio mica, surface-treated with chlorinated hydrocarbon, in polypropylene could replace a steel part of equal stiffness; they used a plastic part 2.88 times as thick as the steel.[41] The plastic part weighed half as much as the steel, and cost the same. Meyer noted that this was the first time the result had been achieved with a polymeric system. Table 4-4 compares steel with mica/polypropylene and other candidates.

The high modulus of mica also increases the velocity of sound in composites. This effect allows the production of speakers with a wider cone angle thus reducing their depth and yielding greater sound fidelity, as well as the production of sound and vibration damping composites.[2]

Mica has long been used in phenolics and epoxies, because of its high dielectric properties,[3] which are related to the iron content. Muscovite usually has a lower iron content than other micas, and is preferred unless other factors are also important.

Low permeability to gases has led to investigations of the use of mica to reduce the high permeability of low-cost packaging films.[4]

Mica is a soft mineral, slightly harder than talc, but much harder than most polymers. It does not cause abrasion in polymer processing equipment as happens with harder reinforcements such as glass fibers. However, many muscovite micas are produced as fine-particle-size by-products that contain hard minerals such as quartz and feldspar, which cannot readily be removed. The Suzorite phlogopite occurs at 85–95% purity and is purified while still coarse enough for the removal of abrasive by-products.[5] The finer grades are then produced by grinding the purified mica flakes.

2.2 Health Aspects

The mica industry has a long history and is relatively well documented. Lusis reviewed this literature, which indicates that mica is a relatively harmless mineral dust, but that health

Table 4-2. Properties of muscovite and phlogopite micas.[100]

Property	Natural Muscovite	Natural Phlogopite
Density, g/cc	2.76–3.2	2.74–2.95
Hardness, Mohs	3–4	2.5–3.0
Transparency	Ruby [varies]	Amber-Brown [varies]
Shape	Roughly hexagonal or irregular flakes	Roughly hexagonal or irregular flakes
Cleavage	Basal, eminent	Basal, eminent
Crystal structure	Monoclinic	Monoclinic
Optical axial angle, 2V, degrees	38–47	30–10
Refractive index	1.552–1.611	1.54–1.69
pH	6.5–8.5	7.0–8.0
Melting point, °C	Decomposes	Decomposes-[>1300][1]
Water of constitution, %	4.5	3.2-[1.01][1]
Maximum temperature with little or no decomposition, °C	500–530	850–1000
Tensile modulus, GN/m² [psi]	172 [25 × 10⁶]	172 [25 × 10⁶]
Tensile strength, MN/m² [psi]	255–296 [37 × 10³–43 × 10³][a]	255–296 [37 × 10³–43 × 10³][a]
	3100 [450 × 10³][b]	–
	690–900 [100 × 10³–130 × 10³][c]	690–900 [100 × 10³–130 × 10³][c]
Thermal conductivity [⊥ to cleavage], cal/cm²/sec/cm/°C	16 × 10⁻⁴	16 × 10⁻⁴
Linear coefficient of thermal expansion per °C [⊥ to clevage]		
20°C to 100°C	15 × 10⁻⁶–25 × 10⁻⁶	1 × 10⁻⁶–1 × 10⁻³
100° to 300°C	16 × 10⁻⁶–25 × 10⁻⁶	2 × 10⁻⁵–2 × 10⁻³
Linear coefficient of thermal expansion per °C [‖ to cleavage]		
0° to 200°C	8 × 10⁻⁶–9 × 10⁻⁶	13 × 10⁻⁶–14.5 × 10⁻⁶
Specific heat [25°C]	0.206–0.209	0.206–0.209
Radiation resistance: Max. use temp. with 5 × 10¹⁴ thermal neutrons/cm², °C	500	550
Specific resistivity ohm cm [25°C]	10¹²–10¹⁶	10¹⁰–10¹³
Dielectric strength, V/mil [1.3 mils thick]	6,000–3,000	4,200–2,100
Dielectric constant, ε [1 MC]	6.5–9.0	5.0–6.0
Dissipation factor, tan δ [25°C]		
60 cycles	0.0008–0.0009	–
1 MC	0.0002	0.004–0.070
Loss factor, ε tan δ [1 MC]	0.0013	0.020–0.042
Chemical resistance	Very good[d]	Good[e]
Water solubles	Trace	Trace
Coefficient of static friction	0.2–0.4	0.2–0.4

[1] Suzorite variety [Marietta Resources International Ltd.].
[a] Measured on sheets [edges stressed[100]].
[b] Measured on sheets [edges unstressed[102]].
[c] Calculated value for the effective strength of high aspect ratio flakes in plastics.
[d] Attacked by HF and molten alkalis.
[e] Attacked by HF, H_2SO_4, H_3PO_4 and molten alkalis.

problems may be caused by mineral impurities.[6] A common mineral found with mica is crystalline quartz, which may not be removed by the mica processor. If such a micaceous product is used, the threshold limit value should be calculated from the quartz content, as shown in Table 4-5 which gives values for pure mica and compares them with talc and glass fibers and flakes.

Finely divided mica is a nuisance dust. Some

Table 4-3. Mechanical properties of compression-molded mica-reinforced thermoplastics.

Polymer	Volume Fraction	FLEXURAL STRENGTH		FLEXURAL MODULUS		System Used (Reference)
		MN/m^2	(10^3 psi)	GN/m^2	(10^6 psi)	
Poly(ethylene)	0.5	124	(18.0)	31.1	(4.51)	powder
Poly(propylene)	0.5	173	(25.1)	38.0	(5.15)	powder
Poly(styrene)	0.5	166	(24.1)	44.9	(6.51)	powder
Poly(styrene)	0.5	172	(24.9)	48.3	(7.00)	encapsulation[56]
Poly(styrene-co-acrylic acid)	0.5	165	(23.9)	44.0	(6.38)	latex[30]
Poly(styrene-co-acrylonitrile)	0.5	207	(30.0)	53.1	(7.70)	powder
Nylon-6,6	0.5	186	(27.0)	44.9	(6.51)	powder
Nylon-11	0.5	173	(25.1)	37.3	(5.41)	powder
Poly(butylene-terephthalate)	0.5	186	(27.0)	47.6	(6.90)	powder

temporary eye irritation has been reported from time to time. In dusty areas, a dust respirator and eye goggles should be worn.

3. PARAMETERS INFLUENCING PROPERTIES

The physical properties of mica-reinforced composites depend on the following:

- Polymer matrix
- Volume fraction of mica
- Aspect ratio of the mica *in the final part*
- Flake size distribution
- Coupling efficiency between the mica flakes and the polymer
- Orientation effects
- Voids

These factors are described in the following sections.

Table 4-4. MICA/PP as a replacement for steel in automotive applications.[41]

Reinforcement/ polymer	Cost per in.^{3}	Thickness relative to steel for equal thickness	Cost for equal stiffness	Specific gravity relative to steel	Weight for equal stiffness
Steel, for comparison	\$0.044	1.00	\$0.044	1.00	1.00
40% mica/ 60% PP	0.015	2.88	0.043	0.157	0.45
40% glass fibers/ 60% PP	0.035	3.10	0.109	0.157	0.49
40% talc/ 60% PP	0.015	3.88	0.058	0.157	0.61
40% talc/ 60% PP	0.015	4.21	0.063	0.157	0.66
40% $CaCO_3$/ 60% PP	0.026	3.16	0.082	0.157	0.50
40% mineral/ PA	0.047	3.68	0.173	0.191	0.70
45% glass fibers/ polyester	0.064	2.46	0.157	0.215	0.53

Table 4-5. Health aspects of mica, talc, glass fibers and flakes, and quartz.

	Mica dust (<5% free silica)	Talc	Glass fibers and flakes
Toxicity:			
Acute local	inhalation 2 irritant 2	inhalation 1	inhalation 1 irritant 2
Acute systemic	0	U	0
Chronic local	inhalation 2	inhalation 2	irritation 2
Chronic systemic	0	inhalation 3	U
Threshold limit value	20 million* particles/ft^3 6 mg/m^3 Total dust** 3 mg/m^3 Respirable dust	20 million particles/ft^3	5 mg/m^3

Ratings:

No toxicity	0
Slight toxicity	1
Moderate toxicity	2
Severe toxicity	3
Unknown	U

*ACGIH, 1982.
**OSHA, 29 CFR 1910.1000, Table 2-3.

3.1 Polymer

The polymer matrix contributes its own characteristics according to its volume fraction. At the same mica content, a PBT or nylon composite will be stronger than a PP composite.

Mica reinforces crystalline polymers such as PP, HDPE, TPX, PBT, PET, and PA better than it reinforces amorphous polymers. Garton[7] has shown that PP crystallizes preferentially at the edges and discontinuities of mica flakes, but when the mica is surface-treated, this nucleation does not occur. Since the treated mica gives better mechanical properties, nucleation probably is not beneficial.

Mica degrades PP so that stabilizers have to be used for long-term stability. It also lowers the exotherm in epoxies and in thermosetting polyesters, so that the promoter content must be increased.

Any absorbed water or amine flotation reagent remaining on the mica may adversely affect RRIM polyurethane.

3.2 Volume Fraction of Mica

Increasing the mica volume fraction increases the mechanical properties. Volume fractions up to 50% have been injection-molded, and those up to 80% have been made with thermosetting polymers using lamination and impregnation and other special techniques. The volume fractions that can be used are limited by processing considerations. The addition of mica increases viscosity to a much greater extent than the addition of calcium carbonate. In thermosets, this effect may be offset by using low-viscosity reactive diluents, but usually at the expense of physical properties. Defoaming agents and surface treatments such as lecithin[8] or titanates also reduce viscosity, but add to cost.

Thermoplastic composites may be processed at higher temperatures to reduce viscosity, but at the risk of degradation of the polymer, or a high-melt-flow polymer may be used with possible reduction in strength. Flow is improved by silicone fluids, stearic acid, surface active agents, and titanates. The silane coupling agents also reduce viscosity.

At high loading, above 60% by weight, flake degradation becomes severe because of crowding, with a consequent loss of mechanical properties.[9] This is particularly so with high shear equipment such as sigma blade and screw mixers.

Also at high loading levels, air easily becomes entrapped. If it is not removed by defoamers or a vacuum, the resulting voids reduce mechanical properties.

3.3 Aspect Ratio

Woodhams and his students at the University of Toronto have worked with polyester composites of 0.35 and 0.40 volume fractions of mica. They found that flexural strength and modulus increase rapidly as the aspect ratio of the mica is increased up to a value of about 100, after which a plateau forms (Figs. 4-2 and 4-3).[10] The actual values fall short of various modified rule of mixtures relationships.

The basic rule of mixtures equation is:

$$E_y = KE_1V_1 + E_2V_2$$

where:

E_1, E_2, E_y = modulus of flake, matrix and composite respectively
V_1, V_2 = volume fractions of flake and matrix
K = constant = 1 for long ribbons with large overlapping areas

Padawar and Beecher[11] predicted for short ribbons in uniform array:

$$K = 1 - \frac{\tanh(u)}{u}; \; u = \alpha\left[\frac{G_2V_1}{E_1V_2}\right]^{1/2}$$

Riley[12] developed an expansion for the random case, including the effect of secondary flake interactions:

$$K = 1 - \frac{\ln(u + 1)}{u}$$

where:

G = polymer matrix shear modulus
α = flake aspect ratio

Although composites have been made using mica of aspect ratios as high as 300, there appears to be no benefit in doing so because the above data show minor improvement once the aspect ratio reaches 100. As higher aspect ratio flakes have larger surface areas, they demand more polymer to wet them and more coupling agent, if any is required. Processing becomes difficult because of higher viscosities. Also, these thinner flakes are fragile and easily broken during processing, thus reducing the aspect ratio in the final molded part.[9]

Muscovite micas, produced as by-products, have aspect ratios as low as 10. This explains why mica, long known as a filler, has only recently been used as a reinforcement. Using processing techniques, to be described later, phlogopite mica flakes are now being made commercially with aspect ratios in the range of 30 to 120.

3.4 Flake Size Distribution

The rule of mixture equation does not predict any change in reinforcing effects due to the dis-

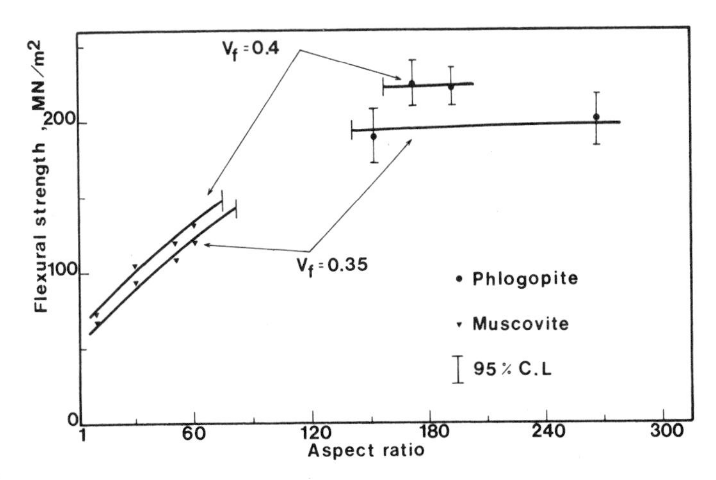

Fig. 4-2. The influence of flake aspect ratio on the flexural strength of a thermoplastic (▼) and a thermoset (●) resin containing two volume fractions of mica (0.35 and 0.40) are summarized. The thermoplastic resin was polystyrene copolymer containing a minor proportion of acrylic acid. The thermoset was a commercial polyester resin (Derakane 411-45).

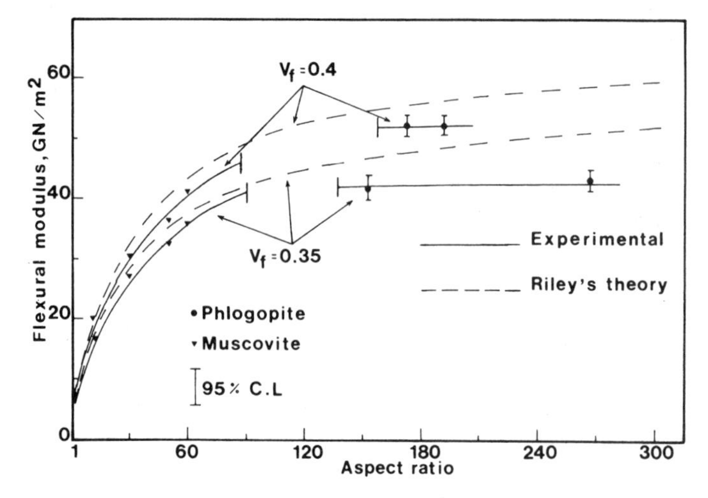

Fig. 4-3. The experimental data are compared with the theoretical predictions of Riley (12). The dependence of flexural modulus on aspect ratio is plotted for two mica volume fractions (0.35 and 0.40), and two different resin systems, using both muscovite and phlogopite micas.

tribution of flake size as found in commercial products.

Large flakes have been found to create a rough surface texture that may provide sites for crack initiation. They also give the surface a rough texture and tend to entrap air, thus creating voids.

Very fine flakes tend to agglomerate. They feed poorly into the compounding equipment and are difficult to wet and disperse. The aspect ratio of these flakes is usually lower than that of the larger flakes, so the mechanical properties are usually poorer. However, fine flakes, about 2 μm in diameter, improve impact resistance, with some loss in flexural strength and modulus.[13] The finer the flake is, the lighter the color and the smoother the surface. In the United States, for maximum mechanical properties, high aspect ratio phlogopite flakes of mesh size −40 + 100 (150–425 μm) are preferred. For improved surface appearance, smaller flakes, −60 + 140 (106–250 μm) and −100 + 325 (44–150 μm), with slightly lower aspect ratios are used. Their use results in slightly lower strength and modulus. In Japan, the finer grades are preferred.

3.5 Orientation Effects

Ideally, the flakes should be arranged in parallel planes, edge to edge, overlapped to the maximum extent. When they are oriented in this way, using special techniques, very high mechanical properties are achieved.

In everyday processing, optimal flake orientation is not achieved. During injection molding, the flakes orient according to the velocity profile of the melt. Near the mold surface, the flakes are aligned parallel to each other; at the center of the molded part, the flakes are randomly aligned to form a core with little strength (see Fig. 4-4). The size of this random core is a function of thickness. Thinner parts have less core and thus are stronger. In an extruded sheet, the core can be reduced, if not destroyed, by stretching. Woodhams and co-workers[14] have demonstrated that orientation in PP and HDPE/mica composites is improved by hot forging and calendering techniques, with consequent large increases in tensile strength, flexural modulus, and drop weight impact resistance.

When mica/polyester liquids are sprayed, the impact on the mold surface tends to orient the flakes parallel to the mold. Unfortunately, air is also entrapped. Improved results are obtained rolling the composite.

3.6 Voids

Voids represent a reinforcement with zero strength and stiffness. They must be prevented unless low density is sought. Because of the high viscosity and fish-scale effect of mica

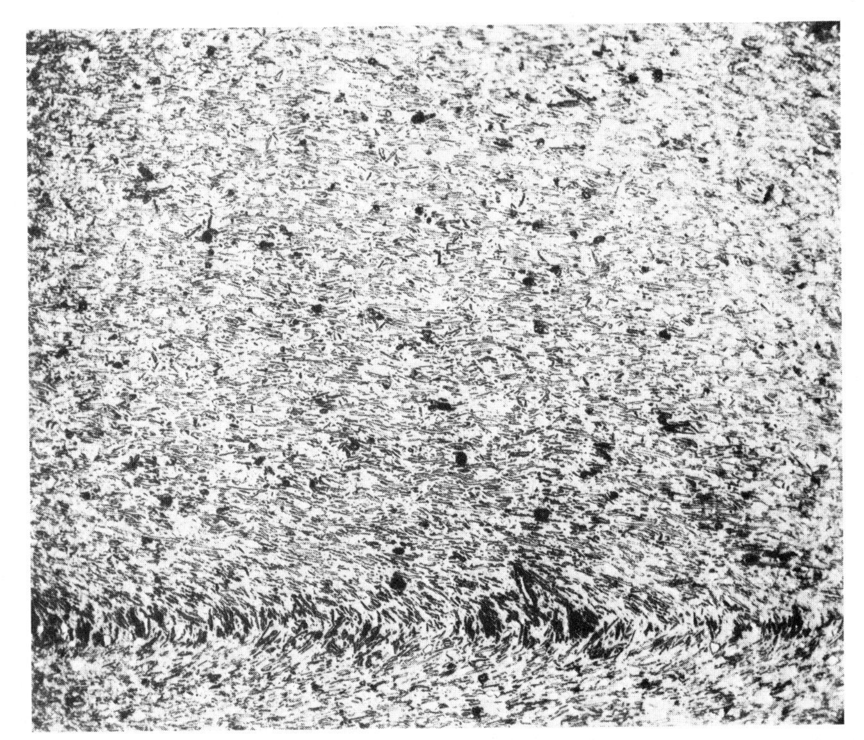

Fig. 4-4. The photograph shows a polished section cut parallel to the direction of injection of a tensile specimen comprising 40% by weight silane-treated mica in poly(butylene-terephthalate). The flake orientation is good except for a thin region near the center of the specimen. The dark spots are largely voids.

flakes, air entrapment is a problem. Techniques to overcome it include vacuum venting of extruder barrels and the addition of defoamers to liquid resins.

Voids are not always visible to the naked eye, but may be seen microscopically. A quick way to measure void content is to measure the specific gravity of the composite, by pycnometer, and compare it with the theoretical specific gravity, calculated from a knowledge of the ingredients and their volume fractions.

4. METHODS FOR PRODUCING HIGH ASPECT RATIO MICA

4.1 Introduction

For practical use in plastics, only small-diameter flakes of mica (20 to 500 μm) can be used. The aspect ratio should be in the range of 50 to 200. Such flakes are processed from larger (6–50 mm diameter) flakes. The processing methods are designed to maximize delamination rather than breakage perpendicular to the plane, to obtain high aspect ratios.

It is easier to cleave phlogopite than muscovite mica, probably because of the difference in van der Waals forces in the two types of mica. The energy required to separate mica sheets has been calculated and appears to confirm this view.[15]

A larger proportion of high aspect ratio mica is obtained if a large-flake feed is used. This causes difficulty in producing high aspect ratio mica from by-product sources (such as feldspar production) because all the ore is first ground to a smaller diameter than 500 μm to facilitate extraction of the primary mineral.

The methods for producing high aspect ratio mica may be divided into wet and dry processes.

4.2 Wet Processes for the Production of High Aspect Ratio Mica

4.2.1 Chaser Mills (also called edge runner, end run, muller mills). This, the most ancient process, is still the main one used commercially for production of "water ground" mica.

The mill is in the form of a cylinder, typically 10 ft in diameter by 3½ ft deep. A revolving vertical central pillar carries horizontal cross-bars, at each end of which are heavy rollers. A ton of mica scrap, from other sheet mining and processing operations, is charged to the mill with about ½ ton of water. The rollers revolve at 30 to 40 rpm on the surface of the mica, with a smearing action, for 8 hours. The mill contents are then flushed out. The heavier thick flakes settle out readily and are returned with the next charge. The well-delaminated mica overflows into large settling tanks. The settled pulp is dewatered, dried, and screened to size.

This method produces mica with different physical characteristics from those of mica processed by dry grinding. Wet ground mica has a polished surface of extremely uniform thickness, and the edges of the flakes are rounded. In contrast, dry ground mica flakes show step-like variations in thickness, and the edges show a degree of feathering (partial delamination).

Because of these characteristics, wet ground mica has better lubricity and sheen than the dry ground equivalent and so is preferred for use in lubricants and in the print industry.

Unfortunately, the wet ground process is costly, has very low capacity, and produces slimes that can cause water pollution problems. The annual production of wet ground mica, 3000 to 4000 ton, is less than the current demand, because of the closing of some production facilities and the acquisition of producers by end-users for captive consumption.

In England, commercial wet grinding is done in a log mill. This is a long, 6 ft diameter, steel cylinder that revolves on its axis. Inside is a steel log that rolls and slips in a bed of mica as the cylinder turns.

A variant of this process is vibro-energy milling. A steel cylinder, which may be horizontal or vertical, contains alumina grinding media, usually cylindrical. A charge of mica slurry is forced through the mill while the media are vibrated at high frequency by an unbalanced rotor. This process has prepared high aspect ratio mica on a laboratory scale, but when it is scaled up to production size, the product is heavily contaminated with pulverized alumina.

4.2.2 Other Wet Processes. The only other commercial wet process involves passing 25 to 50 mm diameter mica scrap between opposing high-pressure water jets.[16] Delaminated mica is removed by elutriation, and the rest is recycled. Aspect ratios in the range of 200 to 600 are achieved. The product is fed into a modified paper-making machine to produce mica paper that is destined for use as electrical insulation. A low-capacity process, it produces 1000 to 2000 tons annually.

Other methods have been developed, but have not become commercial. They include various high shear processes, such as the Ruzicka[17] and Stavely[18] methods. Aspect ratios as high as 10,000 have been claimed.

Various pre-treatments of the mica are used to improve the efficiency of delamination. Some muscovite and phlogopite flakes expand in thickness by 150–300% when heated as high as 800°C (similar to the exfoliation of vermiculite, a highly hydrated form of mica). This process aids delamination; however, not all micas show this property. The only commercial flake phlogopite, Suzorite™ mica, does not expand in this manner.

Other pre-treatments include alternate treatment with acid and carbonates, with strong acids, wetting agents, and soluble salts.

Ultrasonic delamination of mica has received much attention.[19-21] Mica flakes in suspension are passed through a chamber containing an ultrasonic transducer. Delamination of mica is accelerated by adding harder particles, such as silica or calcium carbonate, which must later be removed. Aspect ratios in excess of 100 are easily produced, and ratios in the range of 300 to 600 are achievable. The process has not yet become commercial.

4.2.3 Separation of High Aspect Ratio Flakes

4.2.3.1 Sedimentation. Commercial wet grinding processes rely on sedimentation. After grinding, the mica that is not well delaminated settles out readily in troughs or in a thickener, which is a large cylindrical tank in which revolving scraper arms turn very slowly. The well-delaminated mica overflows to other thickeners where settling is very slow. Some very fine and well-delaminated mica from the secondary overflow is lost as a slime and may cause a water pollution problem. This overflow is passed into settling ponds, sometimes many acres in size, before discharge of the effluent. This lost mica may be recovered and sold.

4.2.3.2 Water Elutriation. Elutriation with water is a proven laboratory technique,[22,23] but it is not used commercially to produce mica.

A current of water passes up a column, and mica flakes are introduced into it. The upward velocity of the water is adjusted to exceed the sedimentation velocity of some of the mica flakes, and the overflow is collected. The collected fraction contains a mixture of thick and thin mica flakes of specific diameters and thicknesses. By varying the upward velocity of the water and by sieving, mica flakes with a narrow range of diameter and aspect ratio can be collected; this is shown in Fig. 4-5.

Xanthos[24] used a cylindrical glass column, of 15 cm inside diameter and 2 m length, to effect separation, and then wet-sieved the product into a narrow range of mica products.

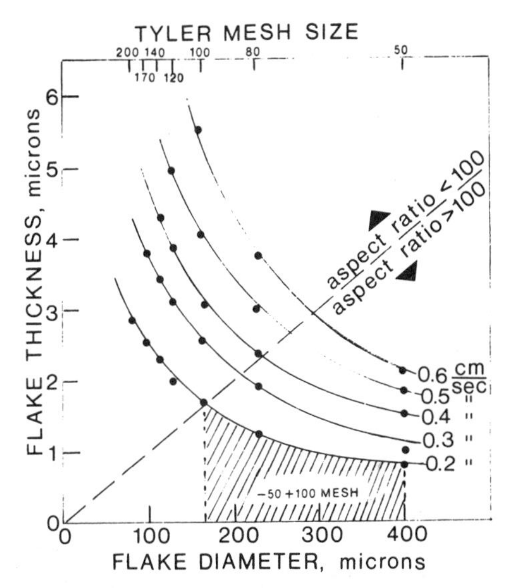

Fig. 4-5. The cross-hatched region represents the dimensions of all mica flakes (thicknesses and diameters) that would be isolated from a mixture using a water elutriation rate of 0.2 cm/sec and a combination of a 50-mesh and a 100-mesh Tyler screen. This mica fraction would have an average aspect ratio above 100.

4.2.3.3 Other Methods. Methods that have been used to recover large flakes of mica, in the mining and processing of mica ores, include the use of trommels (revolving horizontal wet sieves) and Humphreys spirals. The spiral consists of a vertical spiral trough, down which a thin film of water flows. The thinner mica flakes are washed to the outer run of the spiral, where they overflow through collector holes. The method once was very common for separating mica from granular ores, but has been largely superseded by froth flotation. The author has tried this to separate high from low aspect ratio mica, but found it efficient only with large flakes.

4.3 Dry Processes for the Production of High Aspect Ratio Mica

4.3.1 Impact Milling. Unlike the wet processes that mainly rely on shear, dry processing of mica involves impact processes. The resultant flakes tend to have varying thicknesses and feathered edges.

The impact processes used involve various modifications of hammer mills, cage mills, pin mills, and rotor mills. Impact of the flakes with the hammers and walls of the mill is an important factor for breakdown of larger flakes. For smaller flakes, impact of flake on flake in the fluid medium is the major event. For this reason, production of small flakes requires very high-speed milling.

Aspect ratios in the range of 50 to 120 can be produced by these processes with phlogopite mica. In general, the smaller the flake diameter of the feed, the lower the aspect ratio is achievable. Also, the smaller flakes of the products tend to have lower aspect ratios than the larger flakes.

When processed in the same equipment, those muscovites that have been tested do not delaminate as well as the phlogopite of both Canadian and South African sources. Aspect ratios of impact-milled muscovite are typically half those obtainable with phlogopite.

In practice, Suzorite™ phlogopite ore is passed through a series of impact mills. After primary milling, the ore is purified to remove mineral impurities such as feldspar and pyroxene, which exist as granules between the "books" of mica. Further milling takes place in closed circuit with air classifiers that remove well-delaminated flakes and return thicker flakes to the mills for reprocessing. The delaminated product is sized by screens into various commercial products.[25]

4.3.2 Fluid Energy Milling. This process is very commonly used in the production of both commercial dry ground muscovite and phlogopite micas. Mica flakes are entrained in high-velocity gas streams and caused to impact on each other. The gas used is usually heated compressed air, or high pressure superheated steam.

In the Micronizer type equipment, the gas is caused to flow in a doughnut-shaped raceway. In the Majac mill and others of its kind, two jets of fluid blast in opposition in a chamber into which mica is falling.

Generally, these machines are used to produce small-diameter flakes of mica, at low aspect ratios (10 or less), but conditions can be adjusted to produce larger flakes with high aspect ratios (diameters of 150 μm with an aspect ratio of 65).

The same provisos apply as for impact milling; that is, a coarse feed gives a higher aspect ratio product, and smaller-flake fractions have lower aspect ratios.

4.3.3 Separation of High Aspect Ratio Flakes

4.3.3.1 Air Classification. Regardless of the method of production, the mill product contains a range of flake diameters and thicknesses. The diameters are usually screened into ranges according to end use.

Because of its flake like nature mica tends to block screens. The screens employed are vibratory, circular, or gyratory in action. Screening becomes more difficult for flakes smaller than 100 μm, which are usually recovered by air classification and cyclones.

Air classifiers of various types, vertical or horizontal, are highly efficient in the separation of mica. Essentially, mica is entrained in air and is centrifuged. The outflung mica stream is hit by a side wind that removes the well-delaminated mica, while the thicker mica, with greater inertia, passes on and is recycled to the mills.

When fed the mill product, the air separator collects large, thin flakes and small, thick flakes, which must then be separated by screening. Alternatively, the mill product is screened first, and then each mesh size is fed separately to an air classifier.

4.3.3.2 Elutriation, Winnowing, and Other Processes. Other methods of separation include air elutriation and winnowing processes. Haultain[26] designed the Infrasizer, which consists of a series of vertical cones of increasing cone angle and decreasing length, through which a stream of air passes. Lusis[27] used this technique to separate mica into four fractions, with aspect ratios ranging from 15 to 60. Again, the feed must be screened into a narrow range of diameters before processing.

Kaye and Jackson[28] developed a method called "felvation" that combines screening and elutriation. Fluidized mica passes up a conical column through a series of sieves. This method was devised as an analytical tool.

Work with the Infrasizer, at the University of Toronto, led to development of the Tervel separator, a low-speed wind tunnel. Mica is dropped through the roof, is winnowed downwind, and is collected in a series of trays or bins.

The method has been proved to separate phlogopite mica into a range of aspect ratios when prescreened.[29]

The air elutriation and winnowing techniques hold great promise in that no moving parts are involved, and the equipment can be cheaply made. However, these processes are not yet known to be commercial.

A similar technique is the use of a Zig-Zag classifier. Mica is entrained in air, which is fed at the bottom of a rectangular section steel column arranged in a series of zig-zags. The device functions somewhat like a fractionating column and can be set to collect better delaminated mica. The process is used commercially to separate mica from granular minerals using a coarsely crushed feed.[30] The product is still somewhat impure, but is suited to its end use in drilling fluids. The method has been tried with small-flake micas, but is not efficient.

Air jigs are used to separate coarse, well-delaminated flakes from "blocky," undelaminated mica and gangue minerals.

5. PREPARATION OF MICA FOR USE IN PLASTICS

The surface energy of mica flakes is very high. Mica is readily wetted by polar compounds such as water, and such polar polymers as acrylics, epoxies, polyesters, and polyamides, but is not wetted by hydrophobic polymers such as polypropylene, polyethylene, and styrenics.

5.1 Drying

As a result of its high surface energy, mica can be used as a reinforcement for polar polymers without pre-treatment other than drying. However, drying may not be necessary because the manufacturing processes dry the mica, and pure mica is not hygroscopic and will not absorb large amounts of water from the atmosphere during storage.

In the case of water-sensitive polymers such as nylon and polycarbonate predrying may be a wise precaution.

Also, micas that are of by-product origin, as most fine-flake commercial muscovite micas are, have surfaces that may be contaminated with flotation reagents such as sulfuric acid, caustic soda, diesel oil, and fatty amines. Some of these chemicals are hygroscopic.

5.2 Surface Cleaning

Flotation reagents are known to have deleterious effects on polymers. The heat aging resistance of polypropylene is impaired by acids, amines and some coupling agents. Flotation reagents may be removed by heat treatment or washing with suitable solvents.

5.3 Surface Treatments for Polar Polymers

Treating a pure mica surface with silanes will increase its dry mechanical properties, such as tensile and flexural strength, in polar composites, but only by 10–15%. Where the properties of such composites are reduced by the action of a liquid or water vapor, the effect of silanes is more marked. Retention of mechanical properties of nylon 66/mica composites can be increased from 50% to 85% by the use of a suitable silane.[31]

Those agents found most suitable for this end use include amino silanes[32] and amino styryl silanes.[33] Both types are used in commercial coupled mica products, and they are recommended for use in nylon, PBT, unsaturated polyesters, and RIM polyurethane.[34]

Amino-silane-treated mica is considered acceptable in food contact composite products.

5.4 Surface Treatments for Nonpolar Polymers

With its high surface energy, mica is not wetted by such nonpolar polymers as polyolefins. Without a coupling agent, its tensile and flexural strength enhancement is of the same order of magnitude as with talc or calcium carbonate. However, with proper coupling, tensile and flexural strength is markedly improved and tends toward the levels achievable with glass fiber reinforcement. Other properties such as heat deflection temperature are also improved. Small additional improvements are seen in flexural modulus and in Izod impact strength.

The agents found to be most useful in coupling mica to polyolefins include:

- Silanes
    - azido functional[35]
    - amino styryl functional[36]
- Organofunctional silicone compounds[37]
- Chlorinated hydrocarbons[1]
    - plus silane[36,38]
- Modified polyolefins
    - acrylic modified[39]
    - maleic anhydride modified[40]
- Monomers
    - microwave-induced plasmas[42,43]
    - in situ polymerization[44]
- Proprietary treatments
    - "Q" treatment[45]

All the above are commercial products and processes except for monomer treatment. Monomers such as ammonia, ethylene, and PDMS have been grafted onto the mica surface in the form of gaseous products, activated as plasmas, by microwaves. The surface modification results in improved coupling and thus enhanced mechanical properties.[46]

5.4.1 Mechanisms of Coupling

5.4.1.1 Azido Silane. The azido silane bonds to the mica by the normal method of hydrolysis of the three alkoxy groups, R–$(SiOH)_3$, then reacts with hydroxyl groups present on the mica surface to give Si–O–Si bonds. The organic moiety loses nitrogen gas on heating to polymer processing temperatures (above 400°F) to give a short chain terminated by a nitrene group that can react with the backbone of the polypropylene polymer. Thus the mica/silane is grafted onto the polypropylene macromolecule.

This simplistic view is probably not wholly true. Garton[7] has shown that polypropylene crystallizes preferentially on active sites, for example, edges, of the mica flakes when the mica is untreated, but crystallization is not induced when the mica has been pre-treated with azido silane. The alteration of regular crystallization patterns by untreated mica may lead to microdefects that cause early mechanical failure.

Other interactions occur. Mica, per se, like many silicates, reduces the resistance of polypropylene to heat aging. Surface treatment of mica with azido silane reduces this resistance still further.[47]

In practical terms, the resistance to heat aging can be raised to tolerable levels, in both cases, by the addition of suitable stabilizers.[48] Azido silane, other silanes, and other coupling agents have been observed to act as internal lubricants. Part of the effectiveness of these coupling agents is due to their improving flow characteristics so that the flakes orient themselves more evenly in the plane of extrusion or injection molding.

5.4.1.2 Chlorinated paraffin. Meyer and Newman[1] indicated a reaction mechanism that involved dehydrochlorination of C_{20}–C_{26} paraffin with a chlorine content of 70% by weight. This produces a chlorinated residue with conjugated double bonds that, presumably, grafts onto the polypropylene. Meyer and Newman reported that the induction time was reduced when mica was pre-treated compared with the direct addition method; when finer mica flakes were used; and when stabilizers were omitted. He postulated that stabilizers compete with the chlorinated paraffin for the high-energy mica surface. The process must be related to the diffusion rate of the unsaturated residue and to the reaction rate with active sites on the mica.

This does not explain the short induction time required for polyethylene versus polypropylene. Garton[7] has shown that mica accelerates thermal degradation of chlorinated hydrocarbon, and that the degradation products at the mica/PP interface react with PP.

5.5 Methods for Surface Treatment of Mica

For maximum effectiveness, the coupling agent must be present at the mica/polymer interface. It must either be placed there (pre-treatment) or migrate there (direct addition).

5.5.1 Pre-treatment

5.5.1.1 Polymer Modification. The polymer is totally modified so that it is compatible with the mica, as in the case of polar modified propylenes (e.g., Profax PCO 72). However, the method is inefficient because the entire polymer is treated in an expensive manner to ensure that the small amount in contact with mica is polar.

5.5.1.2 Mica Pre-treatment. The coupling agent is applied directly to the surface of the flake and remains there until it is able to react with the polymer during compounding or injection molding. This implies that the coupling agent must be stable.

If *liquid*, the coupling agent may be applied as received or may be diluted with a solvent to reduce its viscosity and to ensure a more uniform surface treatment. This solvent, and the low-molecular-weight alcohols produced by mica–silane reaction, are usually removed before processing with polymers.

Solid coupling agents may be added to the mica in liquid form as solutions or emulsions and treated as above, or they may be added in the solid state. Powdered coupling agent is blended with mica, and the temperature is then raised until the coupling agent melts onto the mica surface.

5.5.2 Direct Addition. In these processes, the mica, coupling agent, and any other additives are blended with the polymer just prior to compounding or injection molding. It is assumed that the coupling agent will migrate to the mica/polymer interface. Such migration has been found to occur, but the coupling efficiency has been observed not to be as good as with the pre-treatment method in liquid polyester with amino- and methacryl silanes,[49] with polypropylene; and with chlorinated paraffin.[1]

Disadvantages of the direct addition method are segregation; dusting and agglomeration; and dispersion.

5.5.2.1 Segregation. Because of large differences in specific gravity, bulk density, and shape, mica flakes tend to settle to the bottom of the hopper when blended with pelletized polymer. Then, the first material entering the barrel is high in mica content, which gets progressively lower. To overcome this problem, machines are used that meter and mix polymer with mica just before the mix enters the hopper.[50] Segregation is minimized when mica is blended with powdered or flake polymer. However, such a mix is not stable for long-term storage, particularly in the presence of vibration. Powder blends should be made just before use or reblended prior to addition to the hopper.

5.5.2.2 Dusting. Their peculiar shape causes mica flakes to settle slowly when suspended in air. They may travel long distances and contaminate other products and molds. Therefore mica/polymer dry blends are best handled in closed conveying and metering systems.

5.5.2.3 Agglomeration. Larger mica flakes feed easily, but flakes smaller than 100 μm tend to agglomerate and hang up in the feed hopper. Vibration or stirring mechanisms are used to overcome this problem.

5.5.2.4 Dispersion. The screw and barrel of an injection-molding machine are not designed as a mixer, but as a heat transfer device. Therefore, the mica may not be well compounded with the polymer and other additives. Generally, a large aspect ratio favors good mixing.

Woodhams has shown that good dispersion is aided by a restricted orifice nozzle. However, there is danger of damage to the flakes, and high injection pressures may be necessary.

5.6 Special Precautions and Other Factors Related to Coupling Agents for Mica

5.6.1 Silanes

5.6.1.1 Health Considerations. Silanes are not generally regarded as highly toxic, but they are known to cause eye irritation. Once coupled to the mica, the silane is partially deactivated, but the user still should take precautions.

Chlorinated and aromatic solvents are known to have long-range effects on health and may also create an explosion hazard.

Before using any chemical products, one should become acquainted with the safe-handling documentation available from the manufacturers of the chemicals and the coupled mica.

5.6.1.2 By-products.

Warning: The by-products from reaction of silanes with any mineral, including mica, consist of up to 25%, by weight of the silane, of highly volatile, inflammable alcohols that form explosive mixtures with air. All equipment used to react silane with minerals should be freely vented to the atmosphere.

The reaction is not instantaneous; the coupled mica will still give off some of these vapors for a short time after removal from the reactor.

Care should be taken to ensure that reaction is total, and that the explosive fumes have dissipated before the coupled mica is shipped in a closed container, or is placed in a non-explosion-proof environment.

5.6.1.3 Factors Influencing Coupling Efficiency.

Water: A small amount of water is required to hydrolyze any silane so that it can react with a mineral. Normally, any silicate has enough absorbed water on its surface to complete this reaction. Some suppliers recommend that a small amount of water be added to the silane prior to use, to ensure complete reaction.

Catalysts: Free-radical agents have been found to be effective in catalyzing the reaction of silanes with polypropylene.[36] Cumyl peroxide is effective at concentrations of 1% of the silane concentration.

Solvents: When silanes are applied from silane solution, the solvent used may have an effect on the coupling efficiency. The choice of solvent is related to the silane type.[33]

pH: The pH of the solution has an effect on the hydrolysis of the silane, a necessary step before reaction with the mica surface can occur. It is not possible to generalize on the best pH for all micas, because of differences in the types, but it is probable that the reaction should take place at an acid pH.[36]

5.6.1.4 Stability. On storage, certain silanes tend to polymerize, to form oligomers; they may precipitate and form a sludge at the bottom of the shipping container. Oligomerization results in less silane being available to react with the mica surface. To avoid such problems, silanes should be shipped and stored at low temperatures and should not be kept long in storage.

Azido silanes are reported to decompose rapidly when mixed with strong mineral acids, with amines, and with metals such as potassium, powdered aluminum, zinc, and magnesium.[51]

It has been suggested that azido silanes may also be deactivated by reaction with chemicals used as stabilizers in polyolefins, such as thioesters, hindered phenols, epoxidized oils, residual monomers, and atactic polymers.

5.6.2 Chlorinated Paraffins

5.6.2.1 Health Considerations. The *chlorinated paraffins* are not regarded as harmful per se. However, upon excessive heating with mica, the chemicals degrade, producing hydrochloric acid in a highly toxic gaseous form, which is extremely irritating to eyes and lungs. Equipment used to heat a chlorinated hydrocarbon should be equipped to deal safely with the acid by-product.

Solvents used may also create health and explosion hazards (see section 5.6.1.2).

5.6.2.2 Corrosion. The acid gas produced by degradation of the chlorinated paraffin will attack base metals. Construction materials for the coupling equipment and for plastics processing equipment (such as compounding and injection-molding equipment, including molds) should be acid-resistant. The quantity of acid produced can be greatly reduced by processing at as low a temperature as possible and by the use of acid getters (see section 5.6.2.4).

5.6.2.3 Mica Type. Meyer and Newman[1] showed that phlogopite treated with chlorinated paraffin gave much higher flexural yield strength than muscovite treated in the same manner.

5.6.2.4 Acid Getters. Acid getters such as magnesium oxide are recommended to reduce the evolution of hydrochloric acid gas.[1,36]

5.6.2.5 Catalysts. Cumyl peroxide is recommended to improve the efficiency of coupling with polypropylene.[36]

5.6.2.6 Induction Time. Meyer and Newman[1] showed that polypropylene reinforced with mica, coupled with chlorinated paraffin, required a certain residence time (up to 10 minutes) at processing temperatures to achieve improved mechanical properties. These investigators also reported that the induction time in polyethylene is very short.

5.6.2.7 Stability. Meyer and Newman[1] have reported that chlorinated hydrocarbons may be deactivated by BHT used in some PP stabilizer packages.

Mica accelerates thermal degradation of chlorinated hydrocarbons.[7] Traces of metals such as iron and zinc catalyze degradation.

5.7 Measurement of Effectiveness of Surface Treatment

For the coupling agent to be effective, it must be present in the correct amount, must be concentrated at the mica/polymer interface, and must have reacted with both mica and polymer in the proper manner.

5.7.1 Quantity of Coupling Agents. Since the quantity of coupling agent is typically in the range of 0.25% to 3% by weight of mica, measurement of the amount of agent present is difficult.

Methods have been used that are adequate for certain silanes; for example, DSC/TGA, gas evolution, elemental analysis, and isotope labeling of silanes. Garton used XRF analysis to measure the quantity of chlorinated hydrocarbon on mica surfaces.[7] In 1983, Berger[52] described the method of diffuse reflectance infrared fourier transform spectroscopy (DRIFT), which can be used to measure coupling agent concentrations in the range 0.4% to 1.2% by weight, with a precision of ±0.1%, directly on the filler surface.

5.7.2 Reaction at the Mica Interface. Reactions at the mica interface are characterized by the evolution of alcohols, which can be measured in the gaseous phase. Presumably the DRIFT technique could be used to detect residual (unreacted) alkoxy groups. Baum has reported SIMS (Secondary Ion Scattering Spectrometry) as a tool for characterizing mica surfaces.[104] This technique could be used to measure changes induced by coupling agents.

Althouse and her colleagues at Batelle Columbus Laboratories have used the torque rheometer to evaluate the effectiveness of surface treatments on fillers such as alumina, iron, and steel powders, as well as aluminum fibers.[105] For fibers, they suggest that the shape factor would overshadow the rheological effects of coupling agents.

Another possible technique is ESCA (electron emission spectroscopy for chemical analysis). The sample is bombarded with monochromatic X rays, and the spectrum of emitted photo-electrons is examined. Concentrations of elements as low as 10^{-3} atomic fraction can be detected. The depth of penetration is 0.5 to 100 Å; so surface treatment can readily be measured. Total carbon peaks can be expanded to resolve into carbonyl, alcohol, and amine peaks.

5.7.3 Reaction with the Polymer. Garton[7] used DSC to measure crystallization exotherms and optical microscopy, with crossed polarizers, to observe crystallization of polypropylene in contact with coupled and uncoupled mica. He also measured orientation and crystallization parameters by means of internal reflection infrared spectroscopy. He found that both silane and chlorinated hydrocarbon coupling agents change the morphology at the mica/polypropylene interface. The presence of chlorinated hydrocarbons at the surface of the polymer was measured by infrared spectroscopy and XRF techniques.

The practical method is to compound and mold test bars of the coupled mica/polymer composite and test for mechanical properties. Tensile and flexural strengths and heat deflection temperature are those properties most sensitive to the effects of coupling.

6. PROCESSING OF MICA COMPOSITES

6.1 Thermosetting Polymers

6.1.1 Polyester

6.1.1.1 Composition. The polyester should be of low viscosity, in the range of 100–200 centipoise. Because mica reduces the exotherm, the level of cobalt or amine promoter should be doubled. MEKP catalyst is used at a level of 0.75–1.00%.

To remove entrapped air and reduce viscosity, a defoamer such as 1% lecithin may be added.[8] Adding glass beads at 10% of the mica weight also reduces viscosity. It is best to leave a mica/polyester mix overnight to de-aerate. If used for spray-up, the resin should contain little or no silica aerogel thickener. If sagging on vertical surfaces becomes a problem, a small amount of ethylene glycol added to the catalyst is effective. The mica used may be −20 +40 mesh (425–850 μm) for casting; −40 +325 mesh (150–425 μm) and −100 +325 mesh (44–150 μm) grades are used for spray-up because they will pass through the spray nozzle. Polyester wets mica very well. Coupling agents will improve wetting, and thus wet strength retention, and may increase dry strength by 10% to 20%.

6.1.1.2 Hand Lay-up. A paste of 30% to 50% mica in polyester is prepared in a slow-speed paddle mixer, and then troweled in place and rolled out. Conventional disc or grooved rollers, used in FRP construction, block and tear up the surface; so a short mohair roller is used instead.

6.1.1.3 Hand Casting and Centrifugal Casting. In liquid resins, large mica flakes tend to settle out and give a mica-rich bottom layer, degrading composite properties. Fine mica flakes stay in suspension; thixotropic agents may also be used. The settling effects are especially marked with centrifugal casting.

6.1.1.4 Spray-up. Early studies were made with a Venus Driadder,[53] a machine that sucks mica from a hopper and blows it into a fanlike stream of catalyzed polyester resin. Glass fibers may be added simultaneously by a conventional chopper gun. It was found that the machine had a capacity to feed no more than 15% by weight of mica, the operation was dusty, and mica wetout was poor; so researchers switched to premixing the mica with promoted resin, and spraying the slurry by means of an airless pump through a modified gun of the external-mix type. Glass fibers were chopped into the mix as required. Table 4-6 compares a mica/glass laminate with a commercial all-glass-fiber laminate.

A maximum of 30% by weight mica may be used to maintain a sprayable viscosity. The spray equipment may be any internal-mix type. This is modified by adding large-diameter material hoses and large nozzles to handle the high viscosity and large flakes. Binks and Glas-Craft have both developed suitable modifications. The airless pump used must have a low ratio, not greater than 10. If higher ratios are used, the styrene in the polyester tends to flash off when the pressure is released during spraying. Because mica flakes prevent release of the styrene vapor, a foam is produced.

6.1.1.5 SMC and BMC. Mica is little used in SMC and BMC because it increases viscosity far more than calcium carbonate and alumina hydrate do. Thickening agents such as magnesium oxide or diisocyanates are used, as in glass-reinforced SMC and BMC.

Xanthos invented a novel type of SMC using powdered resins,[54] but it has not been commercialized. By this and impregnation techniques, extremely high strength and stiffness are achievable, as shown in Table 4-7.

6.1.2 Epoxy. Low-viscosity epoxy resins are used to reduce viscosity. The catalyst level must be increased to overcome the loss of exotherm caused by mica.

Table 4-6. The effect of mica on high-performance boat laminates

	Mica/glass laminates	All-glass laminates commercial
Reinforcement composition, total	50	43
% wt: Mica 60S	20	—
Glass rovings	19.2	NA
Glass mat	10.8	NA
Flexural strength, psi	50,500	44,200
Flexural modulus, psi	2,250,000	1,660,000
Izod impact, unnotched, ft lb/in.^{2}	60.2	67.1
Specific gravity (SG)	1.54	1.45
Specific modulus (modulus ÷ SG), 10^6 psi	1.46	1.14
Specific strength (strength ÷ SG), 10^3 psi	32.8	30.5
Materials cost (1978), cents per lb.	44	56

Table 4-7. Flexural properties of HAR mica reinforced thermostats (50 vol % mica).[54,101]

	FLEXURAL STRENGTH		FLEXURAL MODULUS	
	MN/m^2	(10^3 psi)	GN/m^2	(10^6 psi)
Polyester resin	159	(23)	47	(6.8)
Epoxy resin	166	(24)	44	(6.4)
Phenolic resin	145	(21)	52	(7.5)
Commercial phenolic resin (low aspect ratio mica)	62	(9)	21	(3.0)
Commercial glass fiber reinforced bulk moulding compound (BMC)	100	(15)	13	(1.9)

Other considerations, as with polyesters, include the effect of coupling agents. Liquid epoxies should be "B" staged to ensure adequate back pressure in compression molding.

Very high strengths have been achieved by impregnation of mica papers with epoxies.

Powdered epoxy systems may be dry-blended with mica and compression-molded.

6.1.3 Polyurethanes. RRIM equipment, as used with milled glass fibers, glass flakes, and other fillers, may also be used for mica. Considerable work has been done with Krauss-Maffei and Cincinnati-Milacron machines.[55-57] For best stability, the mica should be added to the polyol. If it is added to the isocyanate, some thickening may occur due to water absorbed onto the mica flakes. In work with Suzorite phlogopite, which has low absorbed water and clean surfaces, the viscosity buildup was very slow, and the polyisocyanate was usable for up to 48 hours. Some muscovite may be contaminated with amine flotation reagents that could catalyze the system.

The mica should be as coarse as can be processed to maximize properties, especially CTE and deflection under load. Suitable grades are −40 +100 mesh (150–425 μm), −60 +140 mesh (100–250 μm) and −100 +325 mesh (44–150 μm). The finer grades will tend to build viscosity more than others. Only pure mica grades should be used; hard mineral impurities will cause wear of equipment and of mold release coating. Coupling agents should also improve mechanical properties.

New mica grades with low aspect ratios have been produced for use in RRIM. These micas give viscous products that are processible; both their viscosity and their mechanical properties are superior to those of glass flake RRIM PU.[58]

Used alone, mica may reduce the composite impact properties too much. Combinations of mica with glass fibers or organic fibers are recommended.

6.1.4 Other RRIM Polymers. The above considerations apply to polyester and nylon RRIM. For nylon RRIM, it may be necessary to dry the mica flakes. Xanthos has found, in an unpublished preliminary study, that mica is a reinforcement for NYRIM.™

6.2 Thermoplastics

The general considerations are similar for all thermoplastic polymers, but special treatment may be necessary for specific polymers. As even pure mica tends to degrade polypropylene, stabilizer packages must be used. The same stabilizers and stabilizer levels that are used for talc have been found suitable for mica. Mica flakes are predried only when used in moisture-sensitive systems such as nylon and PC.

6.2.1 Powder Blending. Best results are obtained when mica is blended with a powder or flake resin.[60] Such resins are available in PP and HDPE. They may be mixed and added to the hopper of the extruder or injection-molding machine to be fed normally. No segregation of flakes is seen under normal conditions, but if such blends are vibrated for long periods, as during shipping, segregation may occur. In that case, the units should be retumbled prior to use. Mixing should be by low-speed, low-shear devices such as drum tumblers or paddle mixers. High-intensity mixers such as blenders and Henschel mixers will break down the mica flakes.

A major advantage of the powder-blending techniques is that the mix can be fed directly into an injection-molding machine, avoiding the costly compounding step. However, it is important that adequate mixing occur when the polymer is molten. The injection-molding machine screw should have a high aspect ratio and the residence time should be relatively long, especially for surface-treated systems.

Fine flakes (44 μm) tend to aggregate and are difficult to mix. Larger flakes such as −40 +100 mesh, −60 +140 mesh, and −100 +325 mesh present no such problem.

6.2.2 Mica/Polymer Pellet Blends. Mica/pellet dry blends may also be fed directly into an injection-molding machine. However, the mica flakes sift to the bottom of the hopper so that the first shots are highly rich in mica. To overcome this difficulty, mica flakes and pellets are fed separately by a metering/mixing/dispersing device such as that produced by Colortronics Inc.[50] Such a device is also useful to feed a compounding extruder.

6.2.3 Compounding. Mica composites generally are compounded to ensure proper mixing and uniform properties and color. Pigments, stabilizers, impact modifiers, lubricants, and coupling agents may be added. Compounding companies have built up a body of expertise in the preparation of mica composites. These proprietary techniques tent to offset the loss of flake aspect ratio incurred by the extra processing step.

Compounding equipment should be chosen to mix efficiently but gently. In order of best performance, such machines are the twin screw, Transfermix, and single-screw extruders and Banbury mixers. Henschel and other high-speed blending machines are not recommended. Flake breakdown is most severe for large, high aspect ratio flakes. Fine, low aspect ratio flakes are not affected even by Henschel mixing. Advantage has been taken of this effect by processing large mica flakes in a K mixer. This machine simultaneously breaks down and delaminates the mica flakes, wets the active fresh surfaces, and brings about homogeneous mixing.[61]

Mica, like talc, is more abrasive than the polymer, but is much less abrasive than glass fibers or wollastonite. Screws and barrels and other contact parts should be hardened. If the mica contains considerable amounts of high hardness impurities, screw wear could be even higher than that found in the processing of glass fibers. As mica flakes tend to float in air, compounding eliminates this dust problem in the molding shop.

6.2.4 Injection Molds. The best orientation and least flake damage have been found using hot runner valve-gated molds. However, cold runners also give satisfactory results.

Sprues and gates should be large. The melt flow should be streamlined. The mold should be designed to avoid knit lines, which can be 50% weaker than the body of the composite.[62] If unavoidable, the knit line should be arranged so that material flows past it. Addition of a cold well and/or a cartridge heater near the knit line is also beneficial.

6.2.4.1 Injection Molding Conditions. Mica increases the melt viscosity;[63] so all temperatures must be increased by an amount dependent on the mica content, mica type, and polymer type. High-melt-flow resins may be necessary to achieve flow in large parts.

Rapid injection speed is also preferred. Too slow or too cold a shot may result in a frosty surface appearance (called "mica splay"). This may also be caused by overcooling the mold.

Mica composite parts set up quickly because of their high modulus, high HDT, and high specific heat; so cycle time often can be reduced.

Considerable experimentation often is required to obtain the best combination of temperatures and cycling conditions.

6.2.5 Extrusion. Extrusion of mica-reinforced composites has not grown as fast as injection molding. The surface of extruded parts tends to be rough (''shark skin''), apparently because of rotation of the mica flakes as they pass out of the die. The phenomenon is not well understood, but is under considerable study. The surface can be made smooth by calendering or hot stamping. Die-swell of mica-reinforced composites is very low.

Researchers at NRC/IMRI[64] have found that starve-feeding the extruder improves the surface properties of extruded PP/mica.

6.2.6 Blow-Molding. Blow-molding has been a recent development with mica-reinforced HDPE. These compounds contain 15–30% by weight of phlogopite mica. Parisons as long as 60 in. are dropped with as little as 1½ in. variation is parison diameter. The final parts have wall thicknesses of 2.4 mm and weigh 10 lb.

6.2.7 Rotomolding. Tanks of 1000-gallon capacity have been rotomolded with dry blends of −40 +100 mesh mica and powdered HDPE. Stiffness is increased substantially, but surface texture is poor due to poor wetout of the mica in this no-pressure process. This problem could be overcome by using mica/HDPE as the core of a sandwich structure, but would require that the molds be charged from the outside during processing.

7 PROPERTIES AND APPLICATIONS OF MICA-REINFORCED PLASTICS

Table 4-8 summarizes the properties conferred on plastics by high aspect ratio mica.

Table 4-8. Properties conferred on plastics by high aspect ratio mica.

Improvements
- Low cost.
- High rigidity at normal and high temperatures.
- Broadened service temperature—high and low.
- Excellent warp resistance on demolding and in high temperature service.
- Low shrinkage on demolding.
- Excellent dimensional stability to heat and humid cycling.
- Increased surface hardness.
- Low coefficient of thermal expansion.
- Higher thermal conductivity.
- Faster molding cycles.
- Little effect on tensile strength.
- Increased flexural strength.
- Excellent wet strength retention.
- Excellent water and chemical resistance.
- Greatly reduced permeability to gas.
- Increased resistance to UV (phlogopite).
- Good dielectric properties.
- Reduced friction and wear.
- Higher notched Izod impact resistance (polyolefins) at room and low temperature.
- Higher Gardner impact resistance (polyolefins) at room and low temperatures.
- Increased adhesion of plated metals and adhesives.
- Excellent nucleation in structural foams (reduces density and increases impact resistance).
- Transparent to microwaves.

Drawbacks
- Greatly reduced tensile elongation.
- Low unnotched Izod impact resistance.
- Increased density (except structural foams).
- Less smooth surface appearance.
- Increased viscosity (thermosets).
- Increased processing temperature (thermoplastics).
- Reduction of knit-line strength.

7.1 Thermosetting Polymers

7.1.1 Polyesters. Because of its low impact strength, mica has been used mainly in low-impact applications, such as spray-up and hand lay-up processes.

Spray-up of mica is used in preparing high-stiffness backing for gel-coated FRP and acrylic thermoforms. Its applications include bathtubs, showers, spas, delivery chutes, truck bodies, boat decks, swimming pool steps, laboratory benches, and revolving-door housings.

A new area, requiring high dimensional stability, is the production of dish antennas for satellite ground TV.

Mica is seldom used alone; 25 to 45 phr is sprayed up with 10 to 20 phr of chopped glass fibers.

Mica applied by *hand lay-up* is used mainly as a stiffening agent for the bottom of dinghies and for the half-decks of small sailing craft. A mica/polyester paste is used as a high-strength stiff adhesive to join the halves of the hull of a large FRP motor vessel. A major effect is to stiffen the transom. A similar mix is used to replace asbestos-filled polyester as the reinforcing filler used to fill the false keel of large FRP sailboats.

Boat hull laminates can be tailored so that 30% of the glass mat and rovings is replaced by a mica core with a consequent increase in specific strength and stiffness, no loss of impact strength, and a 25% reduction in materials cost, as shown in Table 4-6.

7.1.2 Epoxies. Mica is mainly used in epoxy casting and potting compounds for electrical end-uses.[65,66] Considerable work in this area has been done in Japan.[67,68] Epoxy composite properties are shown in Table 4-9.

7.1.3 Polyurethanes. As early as 1977, Ford and General Motors were working on the reinforcement of RIM polyurethane with milled glass fibers and mica. The objectives were to increase stiffness and the coefficient of thermal expansion so that PU could be used in stiff exterior parts such as fenders and doors. The coefficient of thermal expansion (CTE) was of special concern, since a door of polyurethane (CTE unfilled 60–90 $\times$ 10^{-6} in./in./°F), must fit in a steel frame (CTE 3–6 $\times$ 10^{-6} in./in./°F). At 25% by weight of mica the CTE was reduced to 26.7, a value similar to 1/64 in. milled glass fibers at the same loading.

At that time, because of equipment limitations, low-viscosity blends were necessary. The only mica that could be used was a low aspect ratio type of small size (44–150 μm). It also contained hard impurity minerals that caused wear of nozzles and external mold release agent.

Work at Marietta Resources had shown superior mechanical properties with larger, high aspect ratio flakes (150–250 μm), which, however, could not be processed on production equipment. Table 4-10 shows that mica substantially increases flexural strength and modulus but reduces impact resistance. Blends of mica with milled glass fibers resulted in increased impact resistance and stiffness, as shown in Table 4-11.

Milled glass became the principal reinforcement in RRIM PU, but when large parts such as the Oldsmobile Omega fender were molded, it was found that the glass fibers orient along flow lines causing warping. Both mica and glass flakes reduce this warping because of their isotropy.

Ferranini and Cohen[70] have confirmed this, and showed that mica and glass flake are both equally effective in producing the lowest CTE and heat sag when compared to milled and chopped glass fibers. Unfortunately, both produce low notched Izod impact resistance.

Similar results have been obtained by Union Carbide,[56] Krauss Maffei, and Bayer.[57] No work has been reported using surface-treated micas that would be expected to improve stiffness, CTE, and heat sag. The impact problem probably could be solved by modification of the polymer and by using blends of mica with fibers such as glass, nylon, or polyester. It would be interesting to measure impact resistance of blends of mica with precipitated calcium carbonate.

A joint study by NRC/IMRI, Marietta Resources, and GCHA showed that low aspect mica flakes were processible on an Accuratio machine and gave better mechanical properties than glass flakes.[58]

Wong and Williams[71] showed that polyure-

Table 4-9. Effect of mica particle size on mechanical properties of an epoxy composite (compression molded).

Mica grade	Particle size (μm)	Processing method	Flexural strength ($\times 10^{-3}$ psi)	Flexural modulus ($\times 10^{-6}$ psi)
20H	420–840	Standard	23.5–23.7	6.54–6.42
20A	420–840	Impact mill	23.6–28.7	6.18–6.49
20U	420–840	Ultrasonic (α = 190)	26.2–31.2	5.77–6.93
60S	150–420	Standard	19.1–19.4	4.77–5.28
60HK	150–420	Impact mill	25.0–26.1	6.12–6.27
200S	150	Standard	17.8–21.3	4.70–4.72
Unfilled epoxy resin (Vibro-Flo E 8100-E-2)			13.2	0.46

Table 4-10. Properties of composites reinforced with mica-RIM flakes compared with glass flakes at loadings of 6.2–20.2% by weight.[69]

	% Filler		Flexural strength, psi		Flexural modulus, psi × 10^{-5}		Notched Izod Impact		Shrinkage %		
Filler type	In polyol	In cured composite	Flow	Cross flow	Flow	Cross flow	Flow	Cross flow	Flow	Cross flow	Density, g/cm^3
None	—	—	3557	3550	0.73	0.73	3.45	3.89	1.16	1.18	1.08
Glass flake	10	6.2	4170	4000	1.12	1.02	3.46	3.00	0.90	0.95	—
Glass flake	20	12.9	5190	5020	1.60	1.43	2.35	2.26	0.65	0.76	—
Glass flake	30	20.2	5170	5330	1.83	1.69	1.79	1.76	0.52	0.62	—
Mica-RIM	10	6.2	4262	4085	1.01	0.96	3.05	2.74	0.86	0.90	1.15
Mica-RIM	20	12.9	5044	4725	1.43	1.25	2.50	2.10	0.58	0.65	1.21
Mica-RIM	30	20.2	5679	5396	1.88	1.71	1.96	1.76	0.54	0.62	1.24

thanes have good vibration damping properties and that reinforcement with mica increased modulus, shifted the peaks of damping to higher temperatures, and raised the loss factor. A mica/polyurethane/epoxy composite was particularly effective.

7.1.4 Other Thermosets. Unpublished work by Xanthos has shown that mica flakes reinforce Nyrim. In addition, the presence of mica would reduce sensitivity to water. Also, mica has long been used in phenolics, where excellent electrical properties are desired.

7.2 Thermoplastics

7.2.1 Polyolefins. The greatest use of mica in plastics is in polyolefins. Commercial polymers that are reinforced with mica include polypropylene, high density polyethylene, copolymers, and polymethylpentene (TPX). Table 4-12 shows that mica reinforces polypropylene with greater efficiency than talc, calcium carbonate, and wollastonite and, at a 40% loading level, is equivalent to 30% glass fiber reinforcement.

Only in impact resistance is mica surpassed by calcium carbonate and glass fibers. Heinold[72] and Naik et al.[73] have found that optimum properties can be obtained by blending mica with glass fibers, calcium carbonate, or talc.

Meyer[41] has shown that injection-molded surface-treated mica at a 40% level in PP is the only composite that can replace stamped steel parts with equivalent properties, at equal cost, but at 45% of the weight of steel (see Table

Table 4-11. Property comparison of composites reinforced with mica-RIM alone and in combination with milled glass fibers (total reinforcement—20.2% by weight).[69]

	% Filler		Flexural strength, psi		Flexural modulus, psi × 10^{-5}		Notched Izod Impact ft. lb./in^2		Shrinkage %		
Filler type	In polyol	In cured composite	Flow	Cross flow	Flow	Cross flow	Flow	Cross flow	Flow	Cross flow	Density, g/cm^3
Mica-RIM	30	20.2	5679	5396	1.88	1.71	1.96	1.76	0.54	0.62	1.24
Milled glass fiber	30	20.2	7619	5510	2.25	1.31	2.54	2.21	0.31	0.88	1.20
Mica-RIM + milled glass fiber	20 10	12.9 6.2	6527	5849	2.14	1.69	2.24	1.87	0.50	0.61	1.23
Mica-RIM + milled glass fiber	10 20	6.2 12.9	7217	6026	2.15	1.62	2.15	1.95	0.49	0.83	1.23

Table 4-12. Comparison of PP, 40% reinforced with mica vs. other fillers.

	Virgin PP (Profax 6523)	Suzorite mica 200 HK (untreated)	40% filled Suzorex mica (surface-treated) 200 NP	Talc	Calcium carbonate	Wollastonite	30% Glass fibers
Tensile strength, 10^3 psi	4.7	4.1	6.2	4.3	2.8	4.5	6.3
Flexural strength, 10^3 psi	4.5	6.5	9.5	6.4	4.7	6.7	10.1
Flexural modulus, 10^6 psi	0.18	0.93	1.10	0.68	0.42	0.60	0.93
Izod impact strength, ft lb/in.							
Notched	0.45	0.60	0.65	0.45	0.75	0.90	1.4
Unnotched	no break	3.8	4.4	4.50	23.0	—	9.4
Distortion temperature @264 psi, °C	56	89	108	78	84	81	125
Mold shrinkage, in./in.	0.02	0.008	0.008	0.012	0.014	—	0.003

6-4). He points to the importance of warp resistance, and suggests that mica/PP will replace steel brackets, plates and retainers, and 20% glass reinforced styrene–maleic anhydride in crash pad armatures.

The first commercial use of mica/PP has been in the replacement of the die-cast zinc air-conditioner/heater housing of General Motors A and B bodied intermediate passenger cars. The weight saving is 1 lb per part. Other mica/PP parts that are commercial or close to commercialization include instrument panels, crash pad retainers, glove boxes, inner fender liners, battery boxes, battery trays, and fan shrouds.

Under-the-hood parts take advantage of the higher heat distortion temperature (HDT) that is obtainable with mica compared to talc or calcium carbonate. Table 4-13 shows the properties of commercial mica/PP compounds.

Mica/PP is largely used in the automotive industry, but is making some inroads into home appliances. Uses include dishwasher pump housings, attic fan blades, air-conditioner squirrel cage fans, portable TV housings, and speaker cones.

The most recent success has been with the blow-molded mica/HDPE seat backs/load floors used by the Fisher Body Division of General Motors for the Firebird and Camaro.[74,75] Ford uses the same compound in the EXP and LN7. A combination of steel and aluminum parts weighing 47 lb has been replaced by mica/HDPE parts weighing only 10 lb. This solution is being adopted in many hatch-backs and sports bodies. Table 4-14 shows the properties of three commercial grades.

Ten percent mica-reinforced PP/PE copolymer is foamed by the Union Carbide process to make large-sized speaker cabinets for professional musicians. Mica reduces shrinkage; and it gives better part-to-part uniformity and higher impact resistance at low temperatures.[76]

Finally, mica increases the heat deflection temperature of TPX, so that it can be used in such microwave cookware as chicken broilers and in hot dog cookers.

7.2.2 Polyesters

7.2.2.1 PBT. One of the first polymers to benefit from mica reinforcement was PBT. Mica was found to overcome the warping of PBT that results when parts cool on demolding and that may occur during high-temperature service or annealing. It is believed that the high isotropic stiffness of the mica flakes, which orient in the flow plane, resists the uneven shrinkage of the polymer. The warping of PBT is magnified when glass fibers are used to reinforce it because the fibers orient along flow

Table 4-13. Properties of commercial mica-reinforced polypropylene composites.

Manufacturer:	Washington Penn Plastics			Wilson-Fiberfil	
Code:	PP-G2MF-4	PP-G2MF-5	PP-G2MF-6	PP-60/MI/40	PP-60/MIB/40
Mica content, %	40	50	60	40	40
Tensile strength, psi	5250	6500	5400	5500	6000
Flexural strength, psi	8644	11,000	8900	8500	9500
Flexural modulus, 10^6 psi	1.2	1.3	1.9	0.8	0.9
Izod impact strength, ft lb/in.					
Notched	0.55	0.52	0.61	0.5	0.5
Unnotched	2.28	1.9	1.38	—	—
Gardner impact, in. lb.					
RT	2.4	2.3	2.2	—	—
−20°F	2.0	1.9	1.7	—	—
Deflection temperature					
66 psi	313	307	306	295	295
264 psi	245	268	267	230	230
Mold shrinkage in./in.	0.0032	0.0030	0.0018	0.007	0.006
Density, g/cm^3	1.23	1.37	1.53	1.25	1.25
Shore hardness	A93/D77	A95/D78	A97/D80	—	—

Table 4-14. Mica-reinforced HMW HDPE for blow-molding.

Compounder	A. Schulman	Phillips	American Thermoplastics
Grade	LP 514-01*	Marlex ER9-001	ATC 030 K40
Melt index, condition F	3–7	—	11
Flexural modulus, psi	210,000	410,000	1,000,000
Tensile strength,			
RT, psi	4050	5900	5200
82°C, psi	1550	—	—
HDT,			
66 psi °C	105	212	136
264 psi °C	—	175	73
SG	1.16–1.20	1.19	1.22
Notched Izod impact, ft lb/in.	0.9 min	1.5	0.4
Unnotched Izod impact, ft lb/in.	—	10.4	2.1
Gardner impact, in. lb	—	10	8
Mold shrinkage, $\frac{1}{8}''$, in./in.	—	—	0.002 − 0.006
Water absorption, 24 hr, %	—	—	<0.001
Shore hardness	—	—	82

*Approved by Fisher Body, under specifications FBMS 2-14.
Type IV Grade A for blow-molding.

lines. With mica, warpage can be brought to the same level as with mineral-filled nylon, or even lower.

In addition to warpage control, mica increases tensile and flexural strength, flexural modulus, heat deflection temperature, and dielectric strength, as shown by Table 4-15.

GAF[77] has shown that mica confers more warp resistance than glass spheres, kaolin, novaculite, wollastonite, PMF, or diatomaceous earth.

GAF LW (low warp) mica-reinforced PBT has been approved under automotive specifications FBMS 2-19 Type II Grade MI (Fisher Body Division of General Motors), ESB-M4D 422-A1 (Ford), and 5166 (American Motors). UL V-0 flame-retardant grades are available from GE, GAF, and Celanese.

Impact resistance is lowered by reinforcement with mica. This problem is overcome by the use of polymeric impact modifiers and combination with glass fibers. The glass fiber content is kept low to minimize warpage. A combination of 20–30 parts mica with 10–15% glass fibers gives a good compromise of properties.

Silanes have been found relatively ineffective with PBT, which wets mica well.[34] Improvements in tensile and flexural strength and modulus, using silanes, are of the order of 10–15% only. The additional cost is not justified except when improved stability to humid conditions is desired.

Table 4-15. Properties of mica-reinforced thermoplastic polyesters.

	Polybutylene terephthalate (PBT) (Valox 310)	
	Unfilled	40% Suzorite mica 325 HK
Tensile strength, psi	7,280	9.900
Flexural strength, psi	9,110	16,000
Flexural modulus, psi	347,000	1,680,000
Izod impact resistance, ft lb/in.		
Unnotched 23°C	no break	3.1
Heat deflection temperature, °C at 264 psi	53	169
Mold shrinkage, %		
Flow direction	2	0.6
Transverse direction	2	1
Dielectric strength, volt/mil	360	533 (558 with Suzorex mica 60 PO)

General Electric supplies Fisher Body with glass/mica PBT for the exterior cowl/vent hood—a paintable part. Ford and Chrysler use a similar compound for high-energy distributor parts. Non-OEM end uses include the Bostik glue gun housings, where strength, stiffness, and warp resistance at high temperatures are required. GAF supplies a compound for computer keyboard facings where dimensional stability is the primary concern.[78] Gafite LW is used in the solder side circuit board cover of Allen Bradley's Mini-PLC-2 microprocessor programmable control for industrial applications.[79]

7.2.2.2 PET. Table 4-16 restates some DuPont data showing the effects of mica replacement of part of the glass fiber reinforcement in PET. DuPont has reported reduction of warpage of glass-fiber-reinforced PET from 110–125 to 5 by substitution of mica for a good portion of the glass fibers.[80] Mica does not increase strength as much as glass fibers do, but the flexural modulus and Izod impact resistance are in the same range for both. The E coil on the Ford Lynx and Escort is made from mica- and glass-fiber-filled PET.[81] Mica improves the electrical, mechanical, and dimensional stability properties and confers good adhesion when the coil is potted with epoxy resin. These compounds are now commercially available as Rynite 935 and 940.[82]

7.2.3 Polyamide. LNP compared various fillers at 40 wt % loading level in Nylon 66 and showed that mica confers the greatest tensile strength, flexural strength, flexural modulus, and heat distortion temperature.[83] The results are shown in Table 4-17. Surprisingly, the Izod impact resistance was almost as good as that using calcium carbonate and wollastonite, and was superior to that with glass beads and alumina trihydrate.

Fiberfil has reported favorably on a combination of 40% short glass fibers and 20% mica flakes in Nylon 66. The results are shown in

Table 4-16. PET reinforced with glass fibers and with mica/glass fibers.[80]

Rynite #:	530 30% glass fiber	545 45% glass fiber	RE 5060 mica/glass fiber	Nylon 10B
Tensile strength psi × 10^{-3}	23	28	15	14.2
Flexural modulus, psi × 10^{-6}	1.30	2.0	1.54	1.05
Izod impact, notched ft lb/in.	1.9	2.4	1.4	0.6
Snake flow, in.	32*	42*	30	—
Relative warpage (4 × 0.125 in. annealed disc)	110 (PBT 160)	125	5	5

*Extrapolated from DuPont data to 18,000 psi to 0.1 × 0.5 in. Channel, 200°F, 565°F melt temperature.

Table 4-18. Work at MRI on silane coupling agents showed improvements in tensile and flexural strength, and modulus, and HDT, as shown in Table 4-19. Wet properties were particularly enhanced, as one would predict.

7.2.4 Rigid PVC. Little has been achieved with rigid PVC because of the surface roughness of extrudates. Substitution of mica for up to 40% of the calcium carbonate filler improves the fluxing rate by a factor of 3.[84] PVC compound of this rigidity has a CTE and specific stiffness close to that of aluminum.

7.2.5 Other Thermoplastics. Mica has been found to enhance the properties of styrenics such as HIPS, ABS, SAN, and SMA, but the surface quality and impct resistance are reduced. Other polymers where mica has been found to increase strength and stiffness include polyether sulfone, polyphenylene sulfide, polyvinylidene fluoride, and other fluoropoly-

Table 4-17. Properties of mineral-filled nylon 66, LNP data.[93]

Property	ASTM method	Units	Mica	$CaCO_3$	Wollastonite	Glass Beads	ATH	Clay
% by weight of filler	—	%	40	40	40	40	40	40
Specific gravity	D792		1.50	1.48	1.51	1.46	1.45	1.47
Mold shrinkage—$\frac{1}{8}$ in.	D955	in./in.	0.003	0.012	0.009	0.011	0.008	0.004
Water absorption—24 hr	D570	%	0.6	0.5	0.5	0.6	0.7	0.4
Ultimate tensile strength,	D638	psi	15,260	10,480	4800	9780	9200	10,850
Ultimate tensile elongation,	D638	%	2.7	2.9	3.0	3.2	2.8	2.5
Flexural strength	D790	10^3 psi	26.0	16.5	7.7	15.8	14.7	23.7
Flexural modulus	D790	10^3 psi	1540	660	790	615	645	1010
Impact—notched $\frac{1}{4}$ in.	D256	ft. lb./in.	0.6	0.5	0.6	0.4	0.5	0.3
Impact—unnotched $\frac{1}{4}$ in.	D256	ft. lb./in.	8.1	9.6	9.4	5.5	6.4	12.3
Heat distortion @264 psi	D64	°F	460	388	430	407	396	391
Thermal expansion	D696	10^{-5}/°F	2.2	2.8	2.3	2.8	3.0	2.6
Thermal conductivity	C177	$\frac{\text{Btu} \cdot \text{ft}^2}{\text{hr} \cdot \text{in.} \cdot °\text{F}}$	3.9	3.9	4.9	3.4	4.9	3.9
Oxygen index		%	31	27	27	27	36	28
"K" factor			230	180	715	460	560	1250

Table 4-18. Mica/glass-fiber-reinforced nylon 66 compared to talc-filled nylon 66 (Nylafil data).

	40% short glass fibers, + 20% mica	30% talc
Tensile strength, psi	23,200	11,000
Elongation, %	2.9	3.2
Tensile modulus, psi × 10^{-5}	11.5	9.0
Flexural strength, psi	33,000	19,000
Flexural modulus, psi × 10^{-5}	23.2	8.0
Izod impact, ft lb/in.		
Notched	1.6	0.7
Unnotched		9
Water absorption in 24 hr, %	0.6	
Linear mold shrinkage, in./in.		
$\frac{1}{8}''$ av. section	0.002	0.007
$\frac{1}{4}''$ av. section	0.002	0.010
$\frac{1}{2}''$ av. section	0.004	0.013
Heat deflection temperature, °F		
@ 264 psi	484	360
Specific gravity	1.69	1.39
Cost per lb (Aug. 14, 1981) TL, $	1.37	1.47
Cost per in^3, ¢	8.37	7.37

mers. Fenton and Hawley summarized these results.[99]

8. AVAILABILITY OF MICA

Mica is a mineral that is plentiful in every country. Large, perfect mica sheets are difficult to find and process, but flake mica is found as a by-product of many metallic and nonmetallic ores. Mica is usually recovered from such ores only when the concentration is high and when the primary product is of little value (feldspar and clay). Since the process is tailored to obtain the best processing conditions for the pri-

Table 4-19. Effects of silane treatment on the properties of nylon 66 Zytel MP 101, reinforced with Suzorite™ mica.

% Mica	0	40										
Silane	N/A	None	A1100	A1100	A1102	A1102	A1160	Z6020	Z6032	Z6076	A187	S3076
% by wt. of mica		(control)	0.5	1.0	0.5	1.0	1.0	1.0	1.0	1.0	1.0	1.0
Tensile strength, psi × 10^{-3}												
Dry	11.30	9.90	11.2	11.1	11.5	14.2	12.3	11.3	11.3	12.1	11.9	10.8
Wet	8.46	6.71	8.22	9.07	9.96	11.0	9.56	9.71	9.18	8.80	—	8.58
Flexural modulus, psi × 10^{-6}												
Dry	0.443	1.20	1.35	1.48	1.49	1.42	1.38	1.46	1.51	1.50	1.36	1.46
Wet	0.126	0.553	0.634	0.713	0.845	0.872	0.890	0.891	0.845	0.759	0.747	0.86
Flexural strength, psi × 10^{-3}												
Dry	12.3	16.5	18.8	20.1	19.4	19.7	19.8	19.5	19.7	19.9	19.1	19.3
Wet	Bends	8.89	11.5	12.8	13.4	13.7	14.2	13.9	12.8	12.3	11.7	12.3
Heat deflection temperature @264 psi, °F	165	361	392	403	414	419	394	—	—	—	—	—
Izod impact strength, unnotched, ft lb/in.												
Dry	Bends	4.0	4.0	4.4	4.2	4.2	3.8	4.1	3.8	4.7	—	3.9
Wet	Bends	3.5	4.7	4.5	4.6	5.0	4.4	5.1	4.8	4.6	—	4.8

mary product, the mica that is recovered is finer than is desired for efficient purification and delamination. Such mica grades are ground to fine, but low aspect ratio, products of low mica content for low-value, low-performance end uses such as joint cement and drilling mud.

Some mica ores are mined for drilling mud use. Simple crushing, beneficiating, and screening operations yield products with a mica content as low as 40%. Producers of these types of mica are attempting to purify and delaminate their products to meet the specifications of the plastics industry.

Some small tonnage of muscovite mica with an aspect ratio of 50 is made by wet-grinding of sheet mica scrap.

The world's largest operating source of phlogopite mica is the Bedard phlogopite mine in Northern Quebec, Canada. Reserves of 88–96% purity ore exceed 8 million tons. This ore is purified and delaminated to produce grades with aspect ratios as high as 120 at the Marietta Resources International plant near Montreal. An apatite mine, in Finland, has a high content of phlogopite that mica producers intend to extract.

Fine-grained muscovite is available in the tailings of many large clay and metal mines. Boliden Mineral discards 1 million tons of muscovite and 2 million tons of biotite annually. Backed by the Swedish government, the company has worked to produce a pure muscovite grade that will be further processed into high aspect ratio mica. A pilot plant is in operation.

Repco/Koizumi Ltd. of Japan mines mica schist in Kaladar, Ontario, and ships it to a beneficiation and grinding plant in Japan. The company also controls phlogopite reserves near the Bedard mine.

Thus, a great deal of mica is generally available, but there are few suppliers of grades suitable for plastics. Prices for these grades of mica in 1983 were:[98]

Suzorite mica (uncoated)	10.75–15¢/lb
Suzorex treated mica	26.5 –28.5¢/lb
Wet ground muscovite (uncoated)	16–19¢/lb

Table 4-20 lists the major suppliers.

9 AVAILABILITY OF MICA-REINFORCED PLASTICS

Mica-reinforced plastics are not difficult to compound and are available from any compounder on a custom basis. Those suppliers that have done substantial research and development, alone or in concert with RM suppliers and automobile companies, or have produced proprietary compounds, are listed in Table 4-21.

10 OTHER FLAKE MATERIALS

10.1 Glass Flakes

Flake glass is a commercial product supplied by Potter's Industries Inc. and by Owens/Corning Fiberglas Ltd. These products are used in decoration and to create moisture barriers in corrosion-resistant coatings. OCF offers type C (chemically resistant) and type E (electrical) glass flakes in nominal sizes of 1/8, 1/32, and 1/64 in. Measurement of thickness on the 1/64 in. flakes, by the film method, gave an aspect ratio less than 20. The price of these flakes in 1986 was $1.26–1.74/lb.

Aclin et al.,[85] in 1962, reported on a development program at Olin Matheson Chemical Corporation on manufacturing methods for glass-flake-reinforced plastics. This was the culmination of a series of developmental programs by Olin Matheson, Narmco, and others, funded by the U.S. Air Force, to develop high-strength, high-modulus composites for such applications as rocket cases, fins, and exhaust nozzles, aircraft and missile radomes and moisture barriers, and corrosion-resistant coatings. Good composites were found to be achievable with epoxy/polyamides and with thermosetting polyesters; phenolics and silicones were unsuitable.

Processing was critical to obtain wetout with as little flake degradation as possible. Flake degradation was severe despite all precautions and resulted in composites with lower mechanical properties than anticipated. The Abbe blender was found suitable for wet premixes and a Patterson-Kelley V Shell for dry blends.

Calendering and centrifugal casting were recommended for flake orientation and did not

Table 4-20. Major commercial North American and European producers/suppliers of mica flakes and powders for use in plastics.

	Mine location	Grades available
Phlogopite:		
Suzorite Mica Products Ltd.	McCarthy, Quebec	(20) untreated grades plus 10 surface-treated for plastics, $\frac{1}{8}''$ to 20 μm surface-treated for use in plastics
Repco Ltd. (Canada Mica Co. Ltd., Div. of Koizumi) Tokyo	Parent, Quebec	(6) 840–20 μm (not offered in North America)
Kemira Oy	Helsinki, Finland	(7), in development
Muscovite/Sericite:		
North Carolina Producers		
The English Mica Co. Ltd./ U.S. Mica Co. Ltd.	Kings Mountain	(4) 2 for plastics
U.S. Gypsum Co.	Kings Mountain/ Spruce Pine	(4) 3 for plastics
Deneen Mica Co.	Burnsville	(2) drilling mud only
Harris Mining Co./Nyco Ltd. (exclusive marketer of plastic grades)	Spruce Pine	(5) 5–140 μm, 3 for plastics
Franklin Mineral Products (Div. of Mearl Corp.)	Franklin	(3) 2–140 μm wet ground
Asheville Mica Co.	Asheville	
United Mica & Feldspar Co.	Spruce Pine	
Other North American Locations		
Repco Ltd.	Kaladar, Ontario	(9) 840–20 μm not offered in North America
Eagle Quality Products/Pacer Corp.	Custer, ND	(9 grades) 2 for plastics
M.I.C.A	Santa Fe, NM	joint cement, plastic grades in development
Thompson-Hayward Chemical Co.	Kansas City, KS	(4)
Whittaker, Clark & Daniels Inc.	South Plainfield NJ	(5)

Table 4-20. (*Continued*)

	Mine location	Grades available
Western Europe		
Microfine Minerals & Chemicals Ltd.	Plant, Derby U.K.	(27 grades, 9 plastic,) custom surface treatment with 6 chemicals
Comptoir des Mineraux et Matieres Premieres.	Ploemeur, France	
Norwegian Talc	Plant, Bergen	
Boliden Mineral AB	Gallivare, Sweden (by-product of copper)	pilot plant

cause additional flake breakdown, as compression and transfer molding did. Two-roll milling gave poor results.

Glass flakes have been produced by rolling glass beads, or tapes, blowing extruded ribbons, drawing tape from a lip, flame-spraying of glass frit, and enameling glass on various surfaces. Owens/Corning blows bubbles in molten tubing and crushes them into irregular shards, often triangular in shape.

The low tensile strength and impact resistance of glass-flake composites discouraged further work.[86]

However, there has been recent work on these flakes in reinforced RIM polyurethane.[58] When parts are reinforced with milled glass fi-

Table 4-21. Suppliers of mica-reinforced compounds.

Name	Head Office
A. Schulman Inc.	Akron, OH
Aclo Compounders	Cambridge, Ont.
American Thermoplastics Corp.	Houston, TX
Argo Plastics, div. of Grandview Industries	Brampton, Ont.
Canadian Plastics Concentrates Ltd.	Sarnia, Ont.
Celanese Corporation	Trenton, NJ
Comalloy	Brentwood, TN
Complas Industries Inc., div. of Ferro Corp.	Evansville, IN
DuPont Engineering Plastics	Wilmington, DE
GAF Chemical Group	New York, NY
General Electric, Valox Div.	Pittsfield, MA
Gulf Oil Chemicals	Houston, TX
LNP	Malvern, PA
Phillips Chemical Co.	Bartlesville, OK
Polyfil Inc.	Woonsocket, RI
Rhe-Tech Inc	Whitmore Lake, MI
RTP Co.	Winona, MN
Thermofil Inc.	Brighton, MI
Wash Penn Plastics Ltd.	Washington, PA
Wellman Inc.	Boston, MA
Wilson-Fiberfil, div. of Dart Industries	Evansville, IN

bers, flow orientation causes warping. This can be overcome with mica or glass flakes, which reinforce isotropically within the plane of the molding. In addition, these flakes increase the flexural modulus and reduce the coefficient of thermal expansion. Glass flakes are used commercially in the RRIM polyurethane vertical body panels of the Pontiac Fiero.[87]

10.2 Metal Flakes

10.2.1 Ball Milled Pigments. Metallic pigments are made in flake form by wet ball milling of atomized aluminum metal powder or foil trim, using steel balls. Stearic acid is used as a lubricant to prevent oxidation and to enable the pigment to float on the paint surface so that it imparts a metallic sheen.

Other metallic pigments are produced from copper/zinc alloys, sometimes with aluminum, to give a range of colors from copper to green gold.

The flakes produced are in the range of 12 to 50 μm in diameter. The thickness varies from 0.1 to 0.3 μm, corresponding to aspect ratios in the range 120 to 170.

Aluminum and SS flake pigments are used mainly in anticorrosive paints to form a vapor barrier and/or to give a decorative appearance. Their use in plastics is mainly to give a metallic appearance.

Fine silver and graphite flakes are used as fillers to render plastics conductive. Nickel flakes are used in EMI shielding applications.

10.2.2 Melt-Extracted Metal Flakes. Batelle Columbus Laboratories developed a process whereby molten aluminum is picked up by a grooved wheel and is rapidly cooled and thrown off in the form of square flakes 1 mm × 1.27 mm × 30 μm thick.[88] The aspect ratio is thus on the order of 17 to 33.

These flakes are used as fillers at volume fractions of 18% to 22% to produce electrically conductive composites. The flakes are far more effective in EMI shielding than conductive carbon black or graphite fibers.[88]

10.2.3 Cut Aluminum Foil-Glitter. Aluminum foil, coated with pigmented, light-fast, baked epoxy is cut into flakes of various sizes from 0.004 × 0.004 m to ⅛ in. square. The thickness is 0.00045 in. or 0.0008 in. This glitter is cast into the gel coat of fiberglass articles. Suppliers include Glitterex Corp., Meadowbrook Inventions Inc., and Atlantic Powdered Metals.

10.3 Hybrid Flakes

10.3.1 Metallized Flakes. Metallized glass flakes have been produced by Batelle Scientific Advances (aluminum) and by Potter's Industries (silver).

Silvered glass flake is normally used for decorative purposes. Heinze and Riiter[89] found that silvered glass flakes, at 75 phr, in a cast polyester, produced a resistivity of 0.6 ohm-cm. At the same loading, silvered spheres gave a resistivity of 633,000 ohm-cm, showing the superior effectiveness of the flake shape.

Mica sheet and film has been silvered commercially for many years for use in capacitors and other electrical applications. Fifty to eighty peicent silver coated mica flakes are now available from Potter's Industries. The average flake diameter is 30 μm. Aluminized mica flakes would be stronger and stiffer than aluminum flakes. Nickel-surfaced flakes may perform best in EMI applications. Such flakes are now available semicommercially from Kuraray Corp., Japan and from Marietta Resources International Ltd.

Conventional electroless techniques work with either mica or glass flakes. Deonath[90] plated mica with electroless copper to improve wetting in mica/aluminum composites.

10.3.2 Oxide-Coated Flakes (Nacreous Pigments). Mica flakes are coated with titanium dioxide or iron oxide to make a sandwich flake that produces pearlescence by means of optical interference.[91] The mica flakes must have a high aspect ratio, in excess of 50, and must be flat without steps. The only acceptable process at present is wet grinding.

The flakes are screened into relatively narrow size ranges (e.g., 3–20 μm, 15–40 μm, 15–100 μm, coated by precipitation from titanyl sulfate, and then calcined at 900°C. The mica is pre-treated with tin compounds to ensure that the titanium dioxide produced is in the rutile form, for better light stability, reflectivity, and chemical inertness.[92]

As the thickness of the rutile layer is increased from 20 μm to 200 μm, the interference colors change, from silver, through gold, red, violet, blue, and green to second-order gold and violet.

Carbon may be deposited on the treated flakes to intensify the color.[93] Transparent dyes and pigments are also used to produce specific aesthetic effects.

Dark colors are produced by similar processes, with iron oxide deposition on the mica flakes.

The pearlescent pigments are incorporated into transparent or translucent polymers, such as acrylic, polycarbonate, polystyrene, and LDPE at about 25% by weight, by conventional techniques.

Price ranges from $4.30/lb to $20.00/lb.

10.4 Single Crystal Flakes

Sparked by work on single crystal fibers (whiskers) in the 1960s, considerable work was done on various single crystals that could take flake form. These included aluminum diboride[94] alumina,[95] and silicon carbide.[96]

Aluminum diboride was recrystallized from a melt of aluminum boride/aluminum, and the thin flakes were recovered by filtration. By elutriation, flake fractions between 37 and 430 aspect ratio were separated, which were evaluated in epoxy and phenolic resin systems. At 60 vol % the aluminum boride flakes conferred flexural strength and modulus equal to that of mica flakes, at aspect ratios below 50 to 80. At aspect ratios over 100, however, AlB_2 flakes far surpassed mica and glass flakes in effectiveness (see Figs. 4-6 and 4-7).

As is usual with high aspect ratio flakes, it is difficult to make composites without voids, which reduce mechanical properties. Alpha alumina flakes have been made by the vapor transport process that occurs when a solution of alumina in lead fluoride is evaporated freely between 1260°C and 1380°C. Hexagonal flakes of 0.5–10 μm thickness were formed from the vapor phase.[95]

Triangular and hexagonal crystal flakes of beta silicon carbide have been grown from a

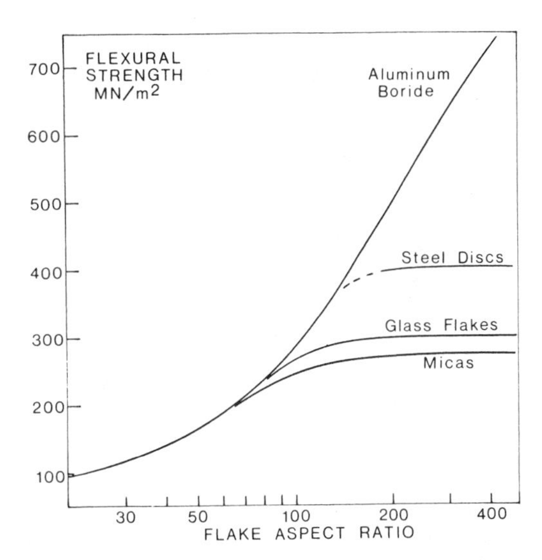

Fig. 4-6. The available experimental data have been used to predict the strength dependence of four flake composites as a function of the flake aspect ratio in an epoxy resin matrix. Each set of data has been normalized to a flake content of 60% by volume. At low aspect ratios, the reinforcement is governed by the relationships for spherical fillers. At high aspect ratios, the composite strength is limited by the strength of the flake.

solution of silicon saturated with carbon at 1530–1670°C.[96] Zapf et al.[97] (Fig. 4-8) did not find the expected high level of reinforcement, possibly because of the low aspect ratio and

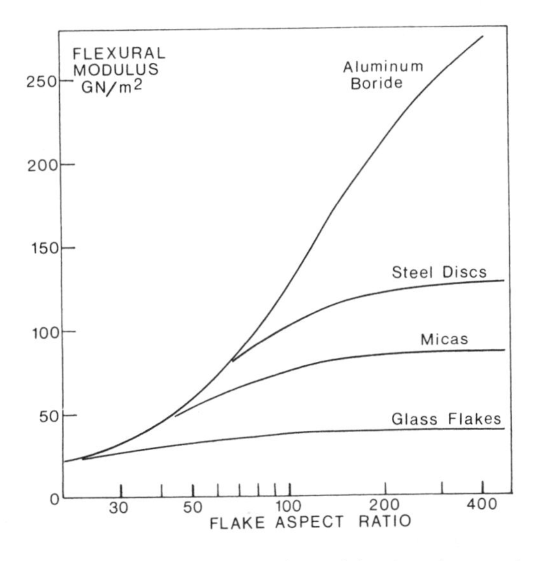

Fig. 4-7. The available experimental data have been used to estimate the aspect ratio dependence for four flake composites with respect to flexural modulus. The data have been normalized to a flake content of 60 vol %. The attractive properties of aluminum diboride composites are particularly evident.

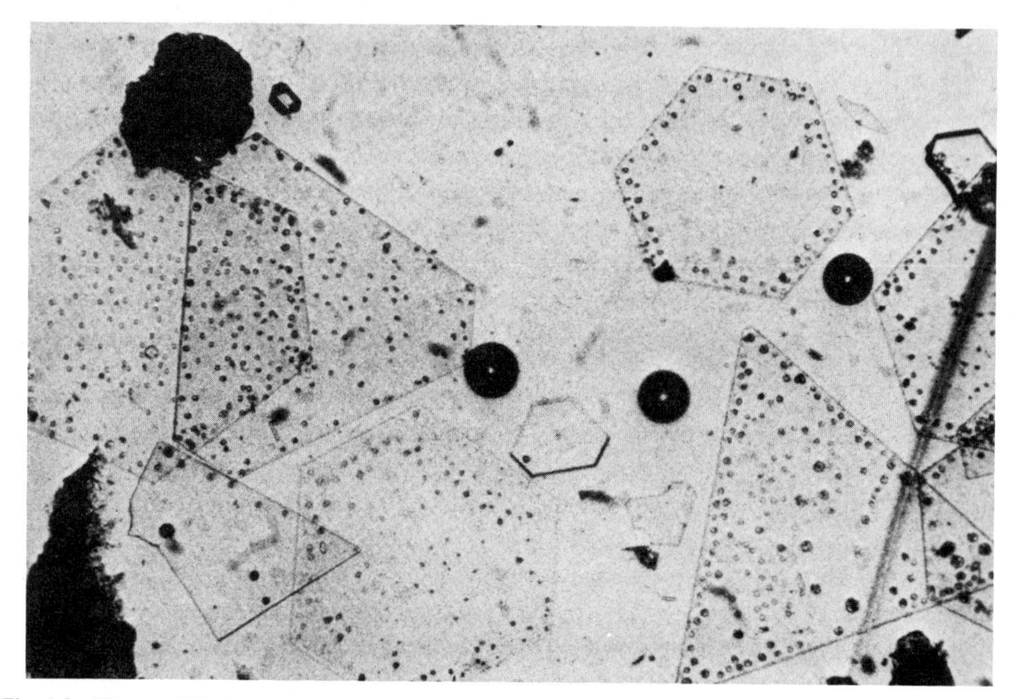

Fig. 4-8. Silicon carbide single crystals in platelet form reproduced from the studies of Zapf, Shaffer, and Corbett (97).

wide distribution of flake sizes and aspect ratios.

REFERENCES

1. Meyer, F. J. and Newman, S., "Improved Strength Mica-polypropylene Composites," 34th Annual Conference, Society of the Plastics Industry, Reinforced Plastics/Composites Institute, Section 14G, 1979.
2. Hawley, G. C. "The Effects of Suzorite Mica Reinforcement on the Acoustical and Vibrational Properties of Composites," Technical Bulletin T21, Marietta Resources International Ltd., 1978.
3. Kadotani, K., Aki, F., and Watanabe, T., "Mechanical and Electrical Properties of Mica Flake Integrated Sheet/Epoxy Laminates," *Composites*, pp. 260–264, Oct. 1981.
4. Zboril, V. G. and Harbourne, D. A., "Paper-like Filled Polyethylene Films," "Composites—81," National Research Council of Canada, IMRI Mini-Symposium, Dec. 1981.
5. Bedard, L., Hawley, G. C., Lavallée, J., "Suzorite Mica, New Low Cost Mineral Reinforcement for Structural Components," 79th Annual General Meeting, Canadian Institute of Mining and Metallurgy, Ottawa, 1977.
6. Lusis, J., "Review of the Health Effects of Mica," National Workshop on Substitutes for Asbestos, US EPA/CPSC/IRIG Conference, July 14–16, 1980.
7. Garton, A., "Characterization of the Interface: Some Initial Results for Mica–Polypropylene Composites," "Composites—81," National Research Council of Canada/IMRI Mini-Symposium, Dec. 1981.
8. Hawley, G. C., Xanthos, M., and Antonacci, J., "Suzorite Mica Filler Reinforcement in Spray-Up Applications," 32nd Annual Conference, Society of the Plastics Industry, Reinforced Plastics/Composites Institute, Section 6D, 1977.
9. Xanthos, M., Hawley, G. C., and Antonacci, J., "Hot Runner Injection molded Mica Reinforced Polypropylene," *Proc.* 34th ANTEC, SPC 22: 582, 1976.
10. Lusis, J., Woodhams, R. T., and Xanthos, M., "The Effect of Flake Aspect Ratio on the Flexural Properties of Mica Reinforced Plastics," *Polymer Eng. Sci.* 13(2), Mar. 1973.
11. Padawer, G. E. and Beecher, N., *Polymer Eng. Sci.* 10(3): 185, 1970.
12. Riley, V. R., "Interaction Effects in Fibre Composites," Polymer Conference Series, University of Utah, June 1970.
13. Nercessian, D., "Gardner Impact Resistance of Various Mica/PP Compounds, compared with Commercial Filled PP," Technical Bulletin T30, Marietta Resources International Ltd., Jan. 1979.
14. Lee, I., Turner, S., and Woodhams, R. T., "Forging and Calendering of Mica-Filled Plastics," Center for the Study of Materials, University of Toronto, Sept. 1981.
15. Donald, H., Jenkins, B., and Hartman, P., "Application of a New Approach to the Calculation of Electrostatic Energies of Expanded Di- and Trioctahedral Micas," *Phys. Chem., Minerals* 6: 313–325, 1980.
16. Heyman, M. D., U.S. Patent 2,405,576, 1946.

17. Ruzicka, J., U.S. Patent 3,719,329, 1973.
18. Stavely, D. C., U.S. Patent 3,570,755, 1971.
19. Kunz, E. C. and Ensminger, D., U.S. Patent 2,798,673, 1957.
20. Dye, Charles H., U.S. Patent 3,240, 203, 1966.
21. Kauffman, S. H., Leidner, J., Woodhams, R. T., and Xanthos, M., "The Preparation and Classification of High Aspect Ratio Mica Flakes for Use in Polymer Reinforcement," *Powder Technology* 9: 125–133, 1974.
22. Heywood, H. *J.I.C. Chem. Eng. Soc.* 4: 17, 1948.
23. Haller, W. K., U.S. Patent 3,087,482, 1958.
24. Xanthos, M., "High Aspect Ratio Mica Reinforced Thermosets," Dept. of Chem. Eng. and Applied Chemistry, University of Toronto, Ph.D thesis 1974.
25. Hawley G. C. and Lavallée, J., "Expanding Markets and Technologies for Mica from Suzor Township, Quebec," Annual Meeting, Society of Mining Engineers of AIME, Chicago, Feb. 1981.
26. Haultain, H. E. T., *The Infrasizer*, 2nd ed., University of Toronto Press, Toronto, Ont., 1961.
27. Lusis, J., "The Importance of Flake Aspect Ratio in Mica Reinforced Plastics," Dept. of Chem. Eng. and Applied Chemistry, University of Toronto, M.A. Sc. thesis, 1970.
28. Kaye, B. H. and Jackson, M. R., "Felvation Speeds Powder Fractionation,"*Chem. Eng. News*, p. 50, Mar. 6, 1967.
29. Etkin, B., Haasz, A. A., Raimondo, S., and D'Eleuterio, G. M. T., "Air Classification of Thin Flakes," *I. Chem. E. Symposium Series No. 59* 5(3):1, (1980).
30. Jordan, C. E., Sullivan, G. V., and Davis, B. E., "Pneumatic Concentration of Mica," U.S. Bureau of Mines, RI 8457, 1980.
31. Hawley, G. C., "The Role of High Aspect Ratio Mica in Reinforced Engineering Thermoplastics," Society of the Plastics Engineers, ANTEC, New Orleans 1984.
32. Organofunctional Silane A-1100," Union Carbide Product Information F-41929C, 1980.
33. "Silane Coupling Agents," Dow Corning 23-012A, 1981.
34. Collins, Ward T. and Kludt, Jari L., "New Silane Coupling Agents for Sphere and Mineral Engineering Thermoplastics," 30th Annual Conference, Society of the Plastics Industry, Reinforced Plastics/Composites Institute, Section 7D, 1975.
35. "Az-Cup MC Azidosilane Coupling Agent," Technical Information, Hercules Inc., HER 27477 Rev. 5-84.
36. Plueddemann, E. P. and Stark, G. L., "Adhesion to Silane-Treated Surfaces, through the Interpenetrating Polymer Networks," 35th Annual Conference, Society of the Plastics Industry, Reinforced Plastics/Composites Institute, Section 20B, 1980.
37. Godlewski, R. E., "Organosilicon Additives for Highly Filled Polyolefin Composites," 37th Annual Conference, Society of the Plastics Industry, Reinforced Plastics/Composites Institute, Section 6F, 1982.
38. Hartlein, Robert C., U.S. Patent 3,630,827, 1971.
39. Ludwig, Peter J., "Adhesion and Property Enhancement via Acrylic-Acid Grafted Polyolefins," *Modern Plastics*, p. 78, Jan. 1983.
40. Gaylord, Norman G., U.S. Patent 4,317,765, 1981.
41. Meyer, F. J., "Metal Replacement with Mica-Filled Polypropylene," *Body Engineering Journal*, pp. 57–62, Fall 1979.
42. Schreiber, H. P., Wertheimer, M. R., and Lambla, M., *J. Appl. Polymer Sci.* 27: 2269, 1982.
43. Schreiber, H. P., Tewari, Y., and Wertheimer, M. R., *J. Appl. Polymer Sci.* 20: 2663, 1976.
44. Iglesias, C., "Coupling Effects in Microencapsulated Mica Flakes," Dept. of Chem. Eng. and Applied Chemistry, University of Toronto, M.A. Sc. thesis, 1971.
45. "Suzorex Mica 60 Q, 200 Q Bulletin," Marietta Resources International, 1982.
46. Richard, C., Hing, K., and Schreiber, H. P., "Interaction Balances and Properties of Filled Polymers," "Polyblends—84," National Research Council of Canada, IMRI Mini-Symposium, Nov. 1984.
47. "Heat Aging and Creep Characteristics of Polyolefins Reinforced with Suzorite Mica (phlogopite)," Technical Bulletin T4, Marietta Resources International Ltd., 1979.
48. Albee, N., "A Review of Anti-oxidants," *Plastics Compounding*, pp. 70–82, May/June 1984.
49. "Union Carbide Silane Adhesion Promoters in Filled Thermosetting Resin Systems," Union Carbide Product Information Bulletin F-42028, July 1968.
50. Colortronic Systems, Inc., Dayton, Ohio.
51. "Material Safety Data Sheet—Az-Cup MC Additive," Hercules Inc., MSDS OR777, 1984.
52. Berger, S. C. and Desmond, C. T., "Silane Content of Particulate Mineral Fillers—Determination by Fourier Transform Infrared Spectroscopy," 38th Annual Conference, Society of the Plastics Industry, Reinforced Plastics/Composites Institute, Section 8D, 1983.
53. Xanthos, M., Hawley, G. C., and Antonacci, J., "Suzorite Mica/Glass Polyester Laminates for Marine and General Purpose Applications," 31st Annual Conference, Society of the Plastics Industry, Reinforced Plastics/Composites Institute, Section 22B, 1976.
54. Woodhams, R. T. and Xanthos, M., "Polyester Resins Reinforced with HAR Mica," 29th Annual Technical Conference, Society of the Plastics Industry, Reinforced Plastics/Composites Institute, Section 4E, 1974.
55. MacGregor, C. J. and Parker, R. A., "Controlling RIM Properties with Reinforcements," *Plastics Compounding*, p. 53, Sept./Oct. 1979.
56. "Fillers for Reinforced Reaction Injection Molded Urethane Composites (RRIM)," Union Carbide Processing Information F-47252A, 1980.
57. Seel, K. and Klier, L., "Reinforced Polyurethane for Car Body Parts,"*Reinforced Plastics*, p. 270, Sept. 1982.

58. Rémillard, B., Vu-Khanh, T., Fisa, B., and Naik, S., "Reaction Injection Molding of Mica Reinforced Polyurethane," "Composites—85," NRCC/IMRI, Mini-Symposium, Section 17, Nov. 1985.
59. McBrayer, R. L., "Variables in Reinforced RIM," *Elastomerics*, p. 33, July 1980.
60. Xanthos, M., Hawley, G. C., and Antonacci, J., "Parameters Affecting the Engineering Properties of Mica Reinforced Thermoplastics," 35th ANTEC, Society of the Plastics Engineers, p. 352, 1977.
61. Chu, K. C., Wright, A. N., and Woodhams, R. T., "Comparison of Gelimat Processing with Other Compounding Techniques for Mica Reinforced Polypropylene composites," NATEC, Society of Plastics Engineers, Apr. 29–May 3, 1985.
62. Burditt, N., King, A., Leonard, B., and Scheibelhoffer, A., "The Knit-Line Strength of Mica-Filled Polypropylene," *Plastics Compounding*, p. 64, Mar./Apr. 1985.
63. Boaira, M. S. and Chaffey, C. E., "Effects of Coupling Agents on the Mechanical and Rheological Properties of Mica Reinforced Polypropylene," *Polymer Eng. Sci*. 17(10): 715, 1977.
64. Utracki, L. A., Favis, B. D., and Fisa, B., *Polymer Composites* 5: 227, 1984.
65. Lee, H. and Neville, K., Handbook of Epoxy Resins, McGraw-Hill, New York, 1967, Chapter 21, p. 29.
66. Harper, Charles, A., *Electronic Packaging with Resins*, McGraw-Hill, New York, 1961, Chapter 6, Tables 6-1, 6-2, Fig. 6-9.
67. Tanaka, T., Hayashi, S., and Shibayama, K. J., *Appl. Physics*, 48(8): 3478, Aug. 1977.
68. Iisaka, K. and Shibayama, K., *J. Appl. Polymer Sci*. 22: 1845–1852, 1978.
69. Hawley, G. C., "New Mica-RIM Flake Reinforcement for RRIM PU," Preliminary Technical Bulletin, Marietta Resources International Ltd., 1985.
70. Ferranini, J. and Cohen, S., "Reinforcing Fillers in Polyurethane RIM," 37th Annual Conference, Society of the Plastics Industry, Reinforced Plastics/Composites Institute, Section 5D, 1982.
71. Wong, D. T. H. and Williams, H. L., *J. Appl. Polymer Sci*. 28: 2187–2207, 1983.
72. Heinold, R. H., "Recent Developments in Broadening the Capabilities of Polypropylene by Reinforcements and Fillers," Reinforced Thermoplastics II, Plastics and Rubber Institute, Manchester, 1977.
73. Naik, S., Fenton, M., and Carmel, M., "Blending of Mineral Fillers for Balanced Reinforcement of Thermoplastics," RETEC SPE, Polymer Modifiers and Additives Div., p. 119, Nov. 6–8, 1985.
74. Galli, E., "Blow Molding Case Book," *Plastics Design Forum*, pp. 87–88, Mar./Apr. 1982.
75. Miller, B., "Blow Molding," *Plastics World*, p. 87, July 1983.
76. "Bose designs Compact Portability into New Speaker Systems," *Plastics Design Forum*, pp. 36–37, Sept./Oct. 1982.
77. Charles, J. J. and Gasman, R. C., "Mica–Fiberglass Reinforced Thermoplastics Polyester," 37th Annual Conference, SPI, RP/C Institute, Section 25D, 1982.
78. "Non-fibrous Reinforcements," *Modern Plastics*, p. 45, July 1983.
79. *Plastics Design Forum*, p. 43, Jan./Feb. 1984.
80. Fleming, R. A., "New Engineering Plastics Resins Based on Polyethylene Terephthalate," NATEC, SPE, p. 27, Nov. 1979.
81. *Modern Plastics*, p. 59, May 1983.
82. "Rynite Thermoplastic Polyester Resin: General Guide to Products and Properties," DuPont E36602; "Rynite 940," DuPont E47649.
83. Theberge, John E., "Recent Product Advances in Thermoplastics Composites," NATEC, SPE, p. 177, Nov. 1979.
84. Bataille, P. and Bui, T. V., "Use of Mica in PVC," *Polymer Composites* 2(1): 8–12, Jan. 1981.
85. Aclin, J. J., Klahs, L. J., Manemeit, F. E., and Snyder, A. D., "Development of Manufacturing Methods for Glass Flake Reinforced Plastics," Aeronautical Systems Div., U.S. Air Force, Project 7-788 AD 299 351, June 1962.
86. Adams, E. R., "Bibliography on Glass-Reinforced Plastic Laminates and Reinforcing Glass Fibers and Flakes," Technical Information and Library Services, Ministry of Aviation (U.K.) TIL/BIB/97, 1965.
87. Galli, E., "RIM Extends Its Versatility," *Plastics Design Forum*, p. 17, Sept./Oct. 1983.
88. Simon, R. M., "EMI Shielding Can Be Made of Conductive Plastics,"*Industrial Research & Development*, June 1982.
89. Heinze, R. E. and Riiter, J. R., "Effects of Particle Type on The Conductivity of Silvered Glass in Reinforced Composites," 32nd Annual Technical Conference, SPI, RP/C Institute, Section 8D, 1977.
90. Deonath, Rohatgi, P. K., "Cast Aluminum Alloy Composites Containing Copper-Coated Ground Mica Particles," *Materials Science* 16: 1599–1606, 1981.
91. Rieger, C. J., "Use of Non-Metallic Pearlescent Pigments to Achieve Metallic Appearance," 37th ANTEC, SPE, New Orleans, May 9, 1979.
92. Deluca, C. V., Miller, H. A., and Waitkins, G. R., U.S. Patent 4,038, 090, 1977.
93. Klenke, E. F. and Stratton, A. J., U.S. Patent 3,087,827, 1963.
94. Economy, J., Worher, L. C., and Matkovich, V. I., "AlB_2 Flake Reinforced Composites," *SAMPE J.* 18: 340–368, 1973.
95. White, E. A. D. and Wood, J. D. C., "The Growth of Highly Perfect Alumina Platelets and Other Oxides by Solvent Vapor Transport," *J. Materials Sci.* 9: 1999–2006, 1974.
96. Nelson, W. E., Rosengreen, A., Bartlett, R. W., Holden, F. A., and Mueller, R. A., Stanford Research Institute, AFCRL-66-579, 1966.
97. Zapf, C. F., Shaffer, P. T. B., Corbett, W. J. et al., "Silicon Carbide Platelet Reinforced Epoxy Composite Investigation," AFML-TR-65-282, 1965.
98. Fenton, M., Naik, S., and Hawley, G. C., "New

Surface Treatments, Technical Developments of Mica and Other Micaceous Minerals in Reinforced Plastics,'' 39th Annual Conference, SPI, RP/C Institute, Section 12C, 1984.

99. Fenton, M. and Hawley, G. C., ''Mechanical Properties of Mica Reinforced Plastics—Related to Processing Conditions and Other Parameters,'' ''Composites—81,'' Mini-Symposium, NRCC/IMRI, Dec. 1981.
100. Shell, H. R. and Ivey, K. H., ''Fluorine Micas,'' Bulletin 647, Bureau of Mines, U.S. Dept. of the Interior, pp. 187–188, 1969.
101. Maine, F. W. and Shepherd, P. D., ''Mica Reinforced Plastics: A Review,'' *Composites*, p. 193, 1974.
102. Orowan, V. E., *Zeitschrift für Physik* 82: 235, 1933.
103. Golemba, F. and Woodhams, R. T., U.S. Patent 3,799,799, 1974.
104. Baun, W. L. and Solomon, J. S. ''ISS-SIMS and AES-SIMS Characterization of Mica Surfaces,'' Technical Report AFML-TR-79-4203, Air Force Materials Laboratory, July 1979.
105. Althouse L. M., Bigg, D. M., and Wong, W. M., ''Evaluating the Effectiveness of Filler Surface Treatment,'' *Plastics Compounding*, 6(2), 71–81, 1983.

5
Ribbons

Suresh T. Gulati
Corning Glass Works
Corning, New York

CONTENTS

1. ABSTRACT

The unique properties afforded by glass ribbon as a reinforcing filament are described. They include mechanical properties (strength and stiffness), thermal properties (expansion coefficient), chemical properties (durability and corrosion resistance), and transport properties (permeation coefficient). Experimental results obtained from glass-ribbon plastic composites and their respective advantages and disadvantages are pointed out.* As an example of a specific application, the performance of a glass-ribbon composite pipe is compared with that of a conventional steel pipe. A brief review of ribbon and film composites, including materials other than glass, is given.

*The high planar isotropic strength and modulus of ribbon composites result in advantages in some applications over continuous fiber reinforced composites.

2. INTRODUCTION AND MANUFACTURING METHODS

The versatility of glass as a structural material is largely a result of the many shapes into which it can be formed. Among the many forming processes of interest is the drawing of continuous microsheet. Corning Glass Works has, for many years, formed microtape or glass ribbon with widths from 0.12 to 1.5 in. and thicknesses from 0.001 to 0.003 in. from Code 8871 (alkali lead silicate) glass for capacitor application. Glass ribbon is highly flexible and can be rolled up on spools much like textile ribbon.

Most of the commercially produced ribbon can be wound safely to a diameter between 0.25 and 0.50 in.

The potential of glass ribbon as a reinforcement for plastics was first reported by Humphrey,[1] and shortly thereafter by Halpin and Thomas,[2] who established a firm theoretical foundation for ribbon reinforcement. The most appealing facet of ribbon as a reinforcing material, due solely to its geometry, is the isotropic stiffening and strengthening effect in the plane of the ribbons. This effective planar isotropic reinforcement is similar to the properties of randomly oriented fiber mat composites. Many applications requiring two-dimensional strengths and stiffness are found in the aerospace, marine, automotive, material handling, and building materials industries. Specifically, ribbon composites may have a performance advantage in aircraft structures such as wings, ship hulls, tank cars, pipelines, automobile bodies, belted tires, building cladding, and other structural members in the construction industry.

The above advantage of ribbons in providing isotropic mechanical properties is realized at significant savings in fabrication cost and reinforcement volume. Ribbons do not need to be oriented like fibers to provide isotropic properties; thus fabrication cost is reduced. Furthermore, owing to their rectangular cross-section, they can be packed to a much higher volume fraction than the fibers. For a given volume fraction, a quasi-isotropic fiber composite[3] is only 37% as stiff and 50% as strong as the ribbon composite. For similar mechanical properties, therefore, ribbon reinforcement would require 50–60% lower volume loading than fiber composites. Alternatively, ribbons only half as strong as fibers can result in composite strength comparable to that of a quasi-isotropic fiber composite.

Usually, in order to make a quasi-isotropic continuous fiber composite, a relatively large number of unidirectional plies must be stacked together in a multidirectional pattern, such as 0°, ±45°, 90°. However, since some applications, especially in the aerospace industry, require a thinner isotropic structure than could be attainable in this manner, ribbon reinforcement becomes extremely attractive. Each unidirectional filament ply ranges from 0.002 to 0.006 in. in thickness, and three to four orientations are needed to obtain quasi-isotropic properties. In contrast, planar isotropic structures can be attained in very thin ribbon composites because each layer of ribbon can be as thin as 0.0005 in.

The isotropic reinforcement provided by the ribbons is primarily due to their large aspect ratio (width/thickness), similar to the unidirectional properties of fiber composites. Transverse and shear properties as high as 90% of the longitudinal properties[1] were measured for aspect ratios > 100. Transverse strength of up to 60% of the rule of mixture value was measured by Lewis[4] with stainless steel ribbon of aspect ratio = 187 in SURLYN A matrix. Gray[5] worked with steel ribbon with an aspect ratio of 30 in silver matrix and obtained a transverse strength that was 50% of the longitudinal strength. Similar values were reported by Anderson et al.[6] and Anderson,[7] who used steel ribbon with an aspect ratio of 50 in aluminum matrix. More recently, Pollock and Arthur[8] have measured transverse strength of 50% of the longitudinal strength for 1% carbon steel ribbon having an aspect ratio of 40 in Sn-70% Pb matrix. Theoretical estimates of mechanical properties of ribbon composites have been reported by Padawar and Beecher,[9] Chen and Nielson,[10] and Chen and Lewis.[11]

One of the principal disadvantages of glass-ribbon composites is their notch sensitivity. In the presence of holes and slots, the strength degrades substantially. Transverse strength of 2 in. diameter tubes with two diametrically opposite holes of 1/8 in. diameter and of bars with freshly cut long edges was reduced by a factor of 2 to 5 due to stress concentration and the presence of secondary microcracking in the vicinity of holes and cut edges. However, etching in HF/HNO_3 solution reduces stress concentration by rounding off the flaw tip radius, and restores the strength to a value corresponding to unnotched specimens. Another method of minimizing strength loss is to use smaller aspect ratio ribbon (~90). Notch sensitivity is not a major problem in the case of boron film composites. In fact, they have been used effectively to strengthen[12] bolt holes in filament-reinforced structures.

Aside from mechanical properties, ribbon reinforcement offers superior expansion and transport properties in comparison with fiber reinforcement. As a direct consequence of isotropic elastic moduli and the low thermal expansion coefficient of glass ribbon, the composite expansion is found to be isotropic. Moreover, in view of the large stiffness of glass in comparison with that of the plastic matrix, a small volume fraction of ribbon (~ 0.3) is sufficient to reduce the expansion coefficient of the matrix by one order of magnitude.[13] This is a highly desirable feature in those applications where the dimensional stability of composites is of importance.

The permeation coefficient of glass-ribbon reinforced composites is also considerably lower[14] than that of fiber composites, due to the tortuous permeation path afforded by large aspect ratio ribbon filaments. Brydges et al.[15] recently developed a permeation model that indicates that glass ribbons of readily attainable aspect ratios and volume fraction offer 100–1000-fold improvement over the fibers. The permeation model was verified through mass spectrometer measurements of helium permeation through ribbon composites of various constructions. High permeation resistance is a key factor in applications such as storage containers, pressure vessels, and pipelines that store and transport precious fluids.

Yet another nonmechanical advantage of glass-ribbon composites is their chemical durability and corrosion resistance. In experiments conducted at Corning Glass Works, ribbon composites exposed to a 30% sulfuric acid environment for several days gained weight at a lower rate than either the fiber composites or plain matrix (epoxy) specimens.

In view of the unique features of glass-ribbon composites discussed above, an immediate and practical application of such composites is a glass-ribbon reinforced plastic pipe which can be fabricated economically by winding matrix impregnated ribbon over a mandrel in a coil-winding machine. The properties of such a pipe are compared with conventional steel pipe in section 7. It is shown that, except for stiffness, the glass-ribbon composite pipe surpasses the steel pipe in properties, although in terms of specific stiffness, the two are equivalent.

The chapter concludes with a brief literature review of work done on boron and graphite film composites.

2.1 Manufacturing Methods for Glass Ribbon

Glass ribbon is manufactured by two different methods, the choice of each being determined by the desired precision of cross-sectional geometry. In the first method, known as hot-draw, ribbon is drawn from the hot glass through an orifice of appropriate geometry made of platinum or other highly refractory material.

The hot-draw apparatus is shown in Fig. 5-1 and consists of a furnace, orifice assembly, air jet carriage, ribbon guide block, and pulling rollers. Pulling speeds are governed by the glass temperature and desired ribbon geometry. Higher glass temperature requires faster pulling speed and higher air jet pressure to maintain uniformity in section geometry. Figure 5-2 shows section geometry of ribbons of three different aspect ratios.

The second method of making ribbon is the redraw process, where an oversize blank with desired section geometry is heated to its softening point and then pulled. This process generally gives precise section geometry. Ribbons of several different aspect ratios are easily manufactured by this process. However, it involves

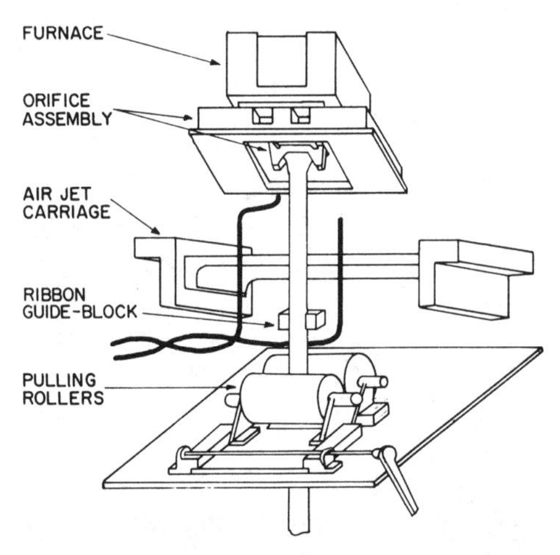

Fig. 5-1. Hot-draw apparatus for glass ribbon. (Courtesy Corning Glass Works.)

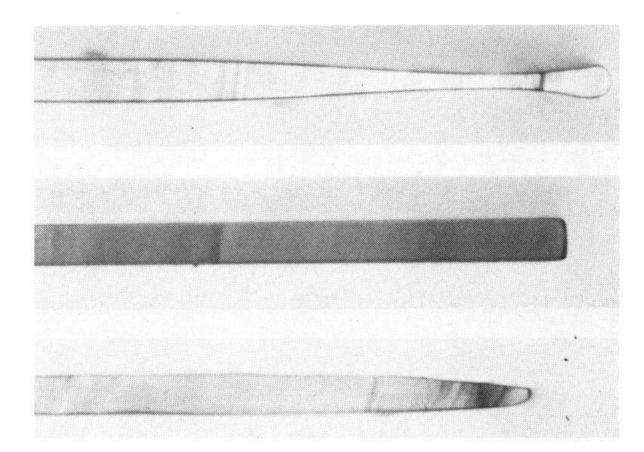

Fig. 5-2. Photomicrograph (100×) of cross-sectional geometry of glass ribbons of different aspect ratio (top: 230, middle: 120; bottom: 90).

two steps, one for preparation of a blank or preform and the second for the redraw operation.

2.2 Fabrication of Glass Ribbon Composites

Flat specimens in the form of bars and plates can be fabricated by the conventional hand layup technique. Ribbons are laid side by side with minimal gap between the edges and are temporarily held together by Scotch tape at the two ends of the ribbon sheet. Several sheets are prepared in this manner. A convenient matrix for construction of flat plates is SCOTCHWELD® or SURLYN®, available from 3M Company or E. I. DuPont Co. The former is a modified thermoset type matrix, while the latter is a thermoplastic. Both are available as films with varying widths and thicknesses. The ribbon sheet mentioned above is placed in a mold with Scotch tape ends sticking out. Then a film of matrix is placed, and this process is repeated with alternate layers of ribbon and matrix until the desired thickness of the composite is achieved. The mold is closed and placed in a press with heated platens for curing.

One can also use prepregged ribbons that have been dipped in the liquid matrix, such as epoxy. The matrix, in such a case, is semicured to B stage to permit handling during fabrication. Ribbon sheets are made as discussed above and placed in the mold. The positioning of successive sheets of ribbon is critical in that it determines the stacking sequence and optimum "gap pattern" between ribbon edges. This is referred to as the stacking parameter, γ, discussed later in the chapter. It has a pronounced effect on the transverse strength of ribbon composites. Figure 5-3 shows a section through a ribbon composite plate fabricated in the above manner using Epon 828 resin. The random sequence of gaps should be noted.

Fabrication of glass ribbon–epoxy tubes can be carried out in a coil winder, shown in Fig. 5-4. The mandrel may be made of steel, or aluminum with a Teflon coating, to facilitate removal of the tube from the mandrel. Prior to

Fig. 5-3. Photomicrograph (18×) of polished section through ribbon composite plate (Epon 828 resin).

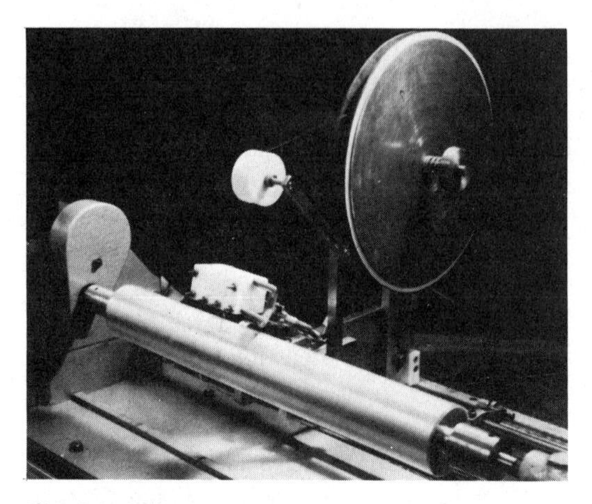

Fig. 5-4. Coil winder assembly for fabrication of ribbon composite tubes.

winding, the mandrel is preheated to 50°C which speeds up subsequent curing. Also, during curing the mandrel expands less than the tube in the oven since it is preheated to 50°C. After the tube is cured, it is removed from the mandrel by quenching the assembly in cold water, which shrinks the mandrel. The tubes made in this manner exhibit good transparency, as shown in Fig. 5-5, and allow large volume fractions of glass. The ribbon used for tube fabrication was Corning Code 8871 with an aspect ratio of 120. Approximately 500 ft of this ribbon (0.219 in. wide × 0.0018 in. thick) are needed for a 10 in. long, 20-layer tube having a diameter of 2.5 in.

The wet-winding technique described above

Fig. 5-5. Twenty-layer glass ribbon/epoxy wet-wound tube.

is not limited to tubular geometry. Winding can also be carried out on a square mandrel to obtain flat plates.

3. PHYSICAL PROPERTIES OF GLASS RIBBON

3.1 Glass Composition

The properties of glass ribbon are related to the glass composition. To date, ribbons have been made from three different glasses: soda lime (Corning Code 0080), alkali lead silicate (Corning Code 8871), and potash-soda-zinc borosilicate (Corning Code 0211). For comparison purposes we will also include E glass, which is a lime aluminoborosilicate glass used by Owens-Corning for manufacturing fibers. The compositions of these four glasses are given in Table 5-1. It should be pointed out that ribbon glass compositions shown in Table 5-1 do not necessarily define the optimum glass for ribbon application as a reinforcement.

The discussion of properties in sections 3.2 through 3.7 follows closely that given by Hutchins and Harrington,[16] where additional references may be found.

3.2 Mechanical Properties

The elastic modulus, Poisson's ratio, and specific gravity of the glasses mentioned in section 3.1 are given in Table 5-2. The strength is not included for two reasons. Since it is an extrinsic property, strength is a strong function of surface condition of the ribbon, which depends on the manufacturing process and the type of coating used to preserve the virgin surface. Secondly, any flaws introduced during manufacturing tend to orient themselves parallel to the draw or redraw direction, thereby making strength property anisotropic (i.e., the strength along the ribbon exceeds that in the width direction). The degree of anisotropy depends on the severity of oriented flaws. Of course, the ultimate goal is a flaw-free surface, which is coated immediately after the draw to minimize further degradation. If flaws do occur, they can be etched prior to coating. This step makes the process uneconomical but increases the strength

Table 5-1. Approximate glass compositions.

Composition (Wt %)	Corning Code 0080	Corning Code 8871	Corning Code 0211	Owens-Corning E Glass
SiO_2	73	42	65	54
Al_2O_3	1	–	2	14
B_2O_3	–	–	9	10
Li_2O	–	1	–	–
Na_2O	17	2	6.5	–
K_2O	–	6	7	–
MgO	4	–	–	4.5
CaO	5	–	–	17.5
PbO	–	49	–	–
ZnO	–	–	7	–
TiO_2	–	–	3	–

Courtesy: Corning Glass Works.

significantly and makes it isotropic. Strengths of up to 300,000 psi (tensile) have been measured in properly protected ribbons of Code 0211 glass made by Corning.

The commercially available ribbon, made from Code 8871 glass, is primarily produced for capacitor application in the electronics industry, where strength is not a critical requirement. This ribbon, which is not coated on the draw, has an average strength of 35,000 psi. Similarly, ribbon made from soda-lime glass by the redraw process is expected to have low strength unless its surface is protected by a suitable coating. Fortunately, ribbon reinforcement being three times as efficient as fiber, strength requirements for ribbon are not as stringent as for fibers. Because the surface area of a typical ribbon is several thousand times that of the fiber, ribbon strengths are lower than those of fiber. Table 5-1 also shows that in view of high lead content, Code 8871 glass has high density and low stiffness, both of which make this glass unsuitable for reinforcement.

Table 5-2. Mechanical properties.

Glass Code	Elastic Modulus (10^6 psi)	Poisson's Ratio	Specific Gravity
0080	10.2	0.22	2.47
0211	11.2	0.21	2.53
8871	8.4	0.26	3.84
E Glass	10.5	0.22	2.58

Courtesy: Corning Glass Works.

3.3 Rheological Properties

Viscosity plays an important role in all of the glass manufacturing processes. Ribbon or sheet drawing, in particular, is carried out at viscosities between 20,000 and 100,000 poises. The four important reference viscosity temperatures are listed in Table 5-3. Until recently, the temperatures corresponding to the four viscosities (from left to right) were defined as the strain point, annealing point, softening point, and working point, respectively. Of the four glasses listed, Code 8871 glass is the softest and E glass is the hardest.

3.4 Thermal Properties

The coefficient of thermal expansion of most glasses is significantly smaller than that of plastics. Also, glass is an order of magnitude stiffer than plastics. Thus, glass-ribbon reinforcement is very effective in lowering the expansion coefficient of plastics. The average expansion coefficients from 0 to 300°C and from room temperature to setting point (strain point + 5°C) are given in Table 5-4.

The room temperature specific heat for most glasses is between 0.075 and 0.230, a typical value being 0.19 cal/g°C. It is higher at elevated temperatures and approaches zero at 0°K. The room temperature thermal conductivity of common glasses ranges from 0.0012 to 0.0030 cal/sec cm °C. Room temperature values for

Table 5-3. Rheological properties of glasses.

Glass Code	REFERENCE VISCOSITY TEMP, °C, FOR VARIOUS VISCOSITIES IN POISE			
	Strain point $10^{14.5}$	Annealing point 10^{13}	Softening point $10^{7.65}$	Working point 10^4
0080	473	514	696	1005
0211	510	550	720	1010
8871	350	385	525	785
E Glass	625	665	845	1070

Courtesy: Corning Glass Works.

specific heat and thermal conductivity are given in Table 5-4.

3.5 Electrical Properties

The volume resistivity of glass depends primarily on the amount of alkali present and, to a lesser degree, on the glass network structure and previous thermal history of the glass. The volume resistivity is inversely proportioned to the alkali content. Similarly, the more slowly glass is cooled, the higher its volume resistivity. Volume resistivity at three different temperatures is given in Table 5-5. Values greater than 10^{17} can not be measured accurately at room temperature. Also shown in Table 5-5 are the dielectric properties at 10^6 Hz and 25°C.

3.6 Corrosion Resistance

Glass is generally considered to have excellent corrosion resistance in comparison with plastics. In many applications, this property is of primary importance. Table 5-6 compares the cor-

Table 5-4. Thermal properties.

Glass Code	THERMAL EXPANSION[a] (10^{-7} in./in./°C)		Specific Heat (cal/gm °C) at 25°C	Thermal Conductivity (cal/cm sec °C) at 25°C
	0-300°C	25°-Setting pt		
0080	93.5	104	0.19	0.0025
0211	74	84	0.18	0.0024
8871	102	110	0.11	0.0018
E Glass	52	70	0.18	0.0031

[a]See Table 5-11 for other thermal expansion data.
Courtesy: Corning Glass Works.

Table 5-5. Electrical properties of glasses.

Glass Code	Log_{10} VOLUME RESISTIVITY (Ω cm)			Power Factor	Dielectric Constant	Loss Factor
	25°C	250°C	350°C	(at 10^6 Hertz and 25°C)		
0080	12.4	6.4	5.1	0.009	7.20	0.065
0211	>17	7.95	6.45	0.0043	6.23	0.0268
8871	>17	11.10	8.80	0.0005	8.40	0.0042
E Glass	>17	13.6	11.8	0.0012	6.4	0.0075

Courtesy: Corning Glass Works.

Table 5-6. Corrosion resistance.

Glass Code	Weathering	Water	Acid
0080	3	2	2
0211	2	1	2
8871	2	1	4
E Glass	1	1	2

Courtesy: Corning Glass Works.

rosion resistance of four ribbon glass candidates. Glasses with high weathering resistance are rated 1; they show little, if any, weathering effects. Those rated 2 may occasionally be troublesome, and a rating of 3 indicates that careful consideration of the glass's use is necessary.

3.7 Optical Properties

The refractive indexes of the four glasses are given in Table 5-7. For most silicate glasses, the value is about 1.5. Code 8871 glass, however, with its high lead content, and therefore high density, has a higher index of refraction. The reflection loss per surface is related to the refractive index and is typically 4% for most silicate glasses. Consequently, a perfectly clear glass transmits about 92% of the incident light. The absorption losses, in view of the thinness of ribbons, are also negligible. The values of surface reflectance and overall transmittance for the four glasses are given in Table 5-7.

4. MANUFACTURERS AND SUPPLIERS

Corning Glass Works is the primary manufacturer of glass ribbon. The composition that is readily available in ribbon form is Code 8871, alkali lead silicate, known for its superior dielectric properties. This ribbon is commercially produced by the hot-draw process for making capacitors. It is stacked alternately with aluminum foil, heated, and passed through rollers to fuse the assembly into a monolithic, hermetically sealed capacitor. Table 5-8 shows the cross-sectional dimensions of various commercially drawn ribbons of Code 8871 glass. Aspect ratios range from 80 to 800. Ribbon length available in a spool is also listed in Table 5-8. Additional information may be obtained by contacting Electronic Materials Department, Corning Glass Works, Corning, N.Y. 14830.

Corning also manufactures microsheet from Code 0211 glass for microscope cover slides. It is available in thicknesses from 0.002 to 0.020 in. and in widths up to 16 in. It has good durability and is easily drawn into sheets. Corning has produced, in limited quantity, microsheet as thin as 0.0003 in. × 20 in. wide for use in moisture barrier in plastic and paper containers. Glass ribbon, in limited quantity, has also been redrawn from Code 0211 blanks for experimental studies. It was 0.0014 in. thick and 0.125 in. wide. OEM Sales, Science Products, Corning Glass Works, Corning, N.Y. 14830, is a supplier of microsheet glass products.

Eakins and Humphrey[17] obtained a patent in 1969 that was assigned to DeBell and Richardson, Inc., Hazardville, Conn. This patent dealt with manufacture of glass ribbon by the redraw process and utilizing these ribbons for fabricating glass–resin structures by the filament winding technique. Use of ribbon of an aspect ratio of 30–50 creates a tortuous path for permeation of gases and fluids contained in the filament-wound structure. They used preforms of Code 0080 glass, approximately 3 in. wide and 0.050 in. thick, and produced ribbon with an aspect

Table 5-7. Optical properties.

Glass Code	Refractive Index (5893A°)	Surface Reflectance	Transmittance
0080	1.512	0.041	0.919
0211	1.526	0.043	0.915
8871	1.656	0.061	0.882
E Glass	1.548	0.046	0.909

Courtesy: Corning Glass Works.

Table 5-8. Code 8871 glass ribbon manufactured by Corning.

Article Code	Width (in.)	Thickness* (in.)	Length per Spool (ft)
679173R	0.121	0.0015	2000
679174R	0.121	0.0018	1600
679153R	0.121	0.0028	1100
679175	0.219	0.0014	2000
679176R	0.219	0.0018	1600
679187R	0.219	0.0028	1200
679232R	0.330	0.0013	1600
679188	0.462	0.0019	1600
679189	0.462	0.0024	1500
679013	1.000	0.0013	–
679075	1.500	0.0027	–

*These are nominal values. The thickness varies slightly from midwidth to the edge. For details contact Electronic Materials Department Corning Glass Works, Corning, N.Y. 14830.
Courtesy: Corning Glass Works.

ratio of 37. They also produced glass ribbons with corrugated section, ribbons with grooved or ridged surfaces, and hollow ribbons, using a preform of appropriate geometry. The hollow ribbons have the advantage of providing a minimum-density, glass-reinforced structure.

5. FUTURE COST AND AVAILABILITY

Capacitor ribbon, made from Code 8871 glass, is commercially available in most sizes shown in Table 5-8. However, its present cost is considerably higher than that of E-glass fiber. It should be recalled that these ribbons are produced in limited quantity for a highly specialized application. Code 8871 glass is not intended for reinforcement of plastics. There is no reason why E glass ribbon cannot be manufactured at high speeds and lower prices. Assuming large-scale production, glass ribbon can be produced at prices comparable to fiberglass. It should be noted that the typical ribbon with a width of 0.25 in. and a thickness of 0.002 in. is equivalent to 4000 glass fibers in terms of cross-sectional area. Also, ribbon reinforcement offers some composite properties that are two to three times better than those afforded by fiber reinforcement. In view of these considerations, draw speeds for ribbon can be considerably lower than those for the fibers, and yet the price of ribbon composites can be cost-effective.

6. PROPERTIES OF GLASS RIBBON COMPOSITES

The inherent advantage of ribbon reinforcement is most easily understood by comparing their theoretical properties with those of the random fiber mat—which has isotropic properties in its plane. Table 5-9 shows the expected planar properties—elastic modulus (E), shear modulus (G), and strength (σ)—of random fiber mat and the ribbon composite.[3, 18–20] It is clear that for a given volume fraction (V_f) of reinforcement, the moduli of random fiber mat are only one-third of those for the ribbon composite and the strength is only one-half of that of the ribbon composite. Thus, in those structural applications where planar isotropy is desirable, the use of ribbon reinforcement can

Table 5-9. Ribbon vs fiber composites.[a]

Property	Random fiber mat	Ribbon Composite
E (Elastic modulus)	$\frac{3}{8} V_f E_f$	$V_f E_f$
G (Shear modulus)	$\frac{1}{3} V_f G_f$	$V_f G_f$
σ (Tensile strength)	$\frac{1}{2} V_f \sigma_f$	$V_f \sigma_f$

[a] In these expressions, the contribution of the matrix has been neglected since it is small.

result in substantial savings of volume loading. The additional advantage of ribbon reinforcement lies in the cost saving resulting from elimination of orientation of fibers in fabricating quasi-isotropic composites.

6.1 Elastic Properties

Glass ribbon composites are transversely isotropic and are characterized by five independent elastic constants (Fig. 5-6):

E_L: Elastic modulus for all directions in the plane of isotropy (x, y)

E_{TT}: Elastic modulus in the transverse direction (z).

G_{LT}: Shear modulus in the longitudinal-transverse plane (x, y)

G_{TT}: Shear modulus in the transverse plane (z)

ν': Poisson's ratio relating elongation in the plane of isotropy to contraction in the thickness direction or vice versa

The above elastic constants are related to constituent properties as follows:[2,10,21]

$$E_L = V_f E_f + V_m E_m \tag{5-1}$$

$$E_T = E_m \left[\frac{1 + J_e \eta_e V_f}{1 - \eta_e V_f} \right], \text{ finite aspect ratio} \tag{5-2}$$

$$E_{TT} = \frac{E_m}{V_m + \left(\frac{E_m}{E_f}\right) V_f} \tag{5-3}$$

$$G_{LT} = V_f G_f + V_m G_m, \text{ large aspect ratio} \tag{5-4}$$

$$G_{LT} = G_m \left[\frac{1 + J_g \eta_g V_f}{1 - \eta_g V_f} \right], \text{ finite aspect ratio} \tag{5-5}$$

$$G_{TT} = \frac{G_m}{V_m + (G_m/G_f) V_f} \tag{5-6}$$

$$\nu' = \frac{\left\{ \frac{E_f V_f \nu_f}{(1 + \nu_f)(1 - 2\nu_f)} + \frac{E_m V_m \nu_m}{(1 + \nu_m)(1 - 2\nu_m)} \right\}}{\left\{ \frac{E_f V_f}{(1 + \nu_f)(1 - 2\nu_f)} + \frac{E_m V_m}{(1 + \nu_m)(1 - 2\nu_m)} \right\}} \tag{5-7}$$

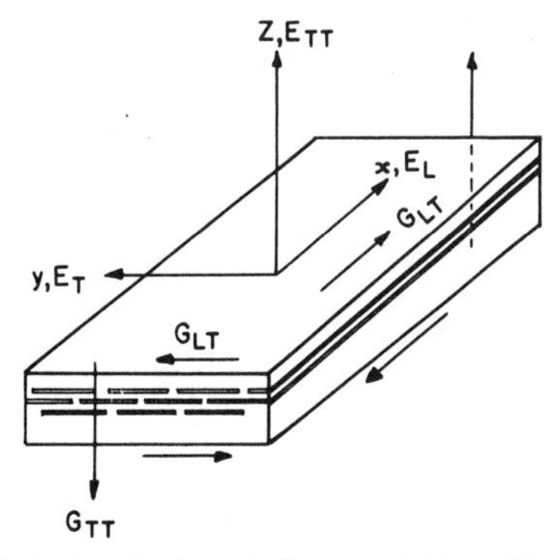

Fig. 5-6. Basic elastic and shear moduli of ribbon composites.

where the subscripts f and m refer to reinforcement and matrix respectively and a = thickness, b = width, and (b/a) = aspect ratio.

$$J_e = 2\left(\frac{a}{b}\right) \tag{5-8}$$

$$\eta_e = \left(\frac{E_f}{E_m} - 1\right) \Big/ \left(\frac{E_f}{E_m} + J_e\right) \tag{5-9}$$

$$J_g = \left(\frac{a}{b}\right)^{\sqrt{3}} \tag{5-10}$$

$$\eta_g = \left(\frac{G_f}{G_m} - 1\right) \Big/ \left(\frac{G_f}{G_m} + J_g\right) \tag{5-11}$$

The principal Poisson's ratio ν, which relates E_L and G_{LT}, is also of importance. It may be shown from eqs. 5-1 and 5-4 that:

$$\upsilon = -\frac{\left(\frac{E_f V_f \nu_f}{1 - \nu_f^2} + \frac{E_m V_m \nu_m}{1 - \nu_m^2}\right)}{\left(\frac{E_f V_f}{1 - \nu_f^2} + \frac{E_m V_m}{1 - \nu_m^2}\right)} \tag{5-12}$$

Figure 5-7 shows the variation of E_L and E_T (modulus in the ribbon width direction) with V_f. The data were obtained for Corning Code 8871 glass ribbon, 2 mils thick, having an aspect ratio of 230 embedded in SCOTCH-WELD® matrix. It is clear that both the inplane moduli, E_L and E_T, approach the rule of mixture value (solid line) providing the composite isotropic properties. The dependence of E_T on aspect ratio is shown in Fig. 5-8 for $V_f = 0.35$ and two different matrices. The solid lines are the predictions of Halpin-Tsai equations.[21] The rapid increase of E_T with aspect ratio should be

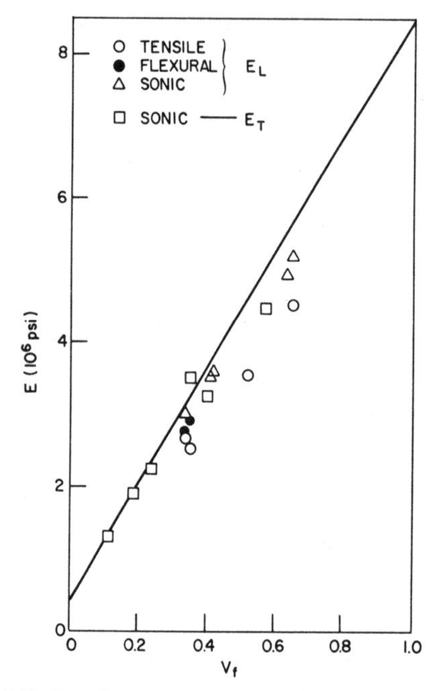

Fig. 5-7. Longitudinal (E_L) and transverse (E_T) elastic moduli as function of volume.

noted. It follows that for $(E_f/E_m) \sim 100$, an aspect ratio of ~ 100 is sufficient to achieve isotropic elastic moduli in the plane of the ribbon composite. The values for E_{TT} have not been measured owing to the difficulty of fabricating suitable specimens. Moreover, this stiffness is of little practical value in using ribbon composites with planar isotropy. The above equations, however, provide an estimate of 0.8×10^6 psi for E_{TT} for a glass-ribbon (AR = 230) – epoxy composite with $V_f = 0.40$.

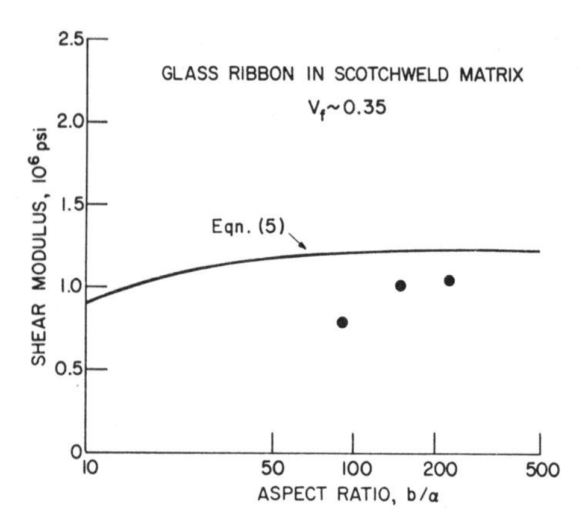

Fig. 5-9. Variation of shear modulus (G_{LT}) with ribbon aspect ratio.

The variation of shear modulus G_{LT} with ribbon aspect ratio is shown in Fig. 5-9 for $V_f = 0.35$. The data show that G_{LT} approaches the asymptotic value given by Eq. 5-4 for $(b/a) \approx 200$, and its magnitude is strongly dependent on the shear modulus of the matrix. The measurements for G_{LT} were carried out by twisting of square plates similar to the technique adapted by Whitney.[22] The values of G_{TT} were not determined due to difficulty in specimen fabrication but Eq. 5-6 gives a value of 0.3×10^6 psi for G_{TT} for a glass-ribbon (AR = 230)-epoxy composite with $V_f = 0.40$.

The variation of ν and ν' with volume fraction is given by Eqs. 5-7 and 5-12. For low values of V_f, ν' and ν are close to Poisson's ratio for the matrix; for $V_f \geq 0.3$, they approach Poisson's ratio of reinforcement.

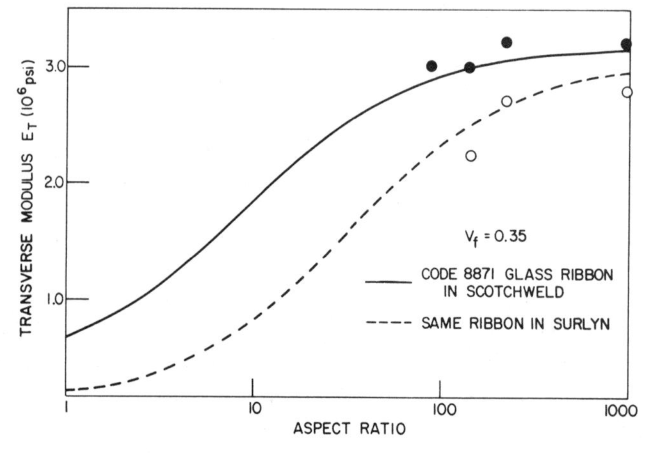

Fig. 5-8. Variation of transverse elastic modulus (E_T) with ribbon aspect ratio.

6.2 Strength Properties

The longitudinal strength of ribbon composites is, to a reasonable accuracy, given by the rule of mixtures analogous to the strength of unidirectional fiber composites.

$$\sigma_L = V_f \sigma_f + V_m \sigma'_m \tag{5-13}$$

Figure 5-10 shows the measured strength values obtained from tensile tests on single ribbon wide test bars. The longitudinal strength is independent of the aspect ratio of the ribbon.

The transverse strength, however, depends on many factors including:

- Transverse strength of ribbon
- Aspect ratio of ribbon
- Matrix properties (shear in particular)
- Stacking parameter
- Fabrication flaws

In theory, the ribbon strength should be isotropic; however, there is a finite probability of preferred orientation of surface flaws which are generally introduced during ribbon manufacturing processes. These flaws tend to be oriented in the longitudinal direction and consequently affect the transverse strength of the ribbon. The net effect of such flaws shows up in the reduced transverse strength of ribbon composites. Among the possible ways to eliminate these flaws, one could either coat the ribbon surface with a protective layer during the drawing operation or etch the ribbon in HF prior to fabrication of composites.

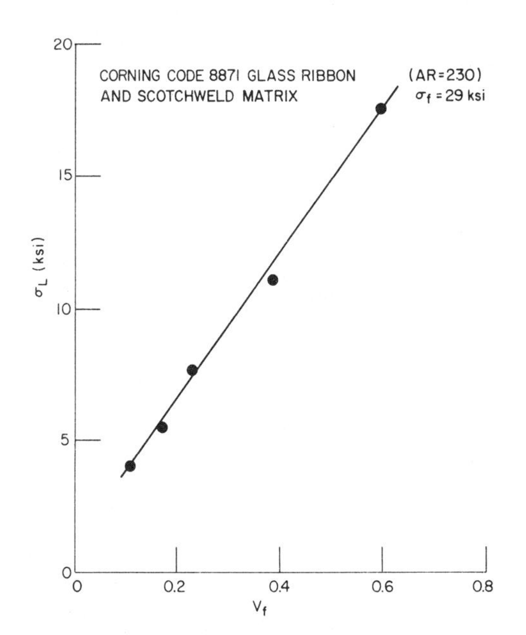

Fig. 5-10. Longitudinal composite strength as function of volume fraction.

The transverse strength of ribbon composites is analogous to the strength of unidirectional fiber composites with discontinuous short fibers. The stress transfer occurs at the edge of the ribbon via the matrix, and the normal stress builds up to its maximum value at the midwidth of the ribbon. Thus, ribbons with a large aspect ratio are more effective in providing isotropic strength than those with a small aspect ratio, assuming the ribbon itself is isotropically strong. During stress transfer, large shear stresses develop in the matrix, near the ribbon edge, and require large shear strength and interfacial bond strength of the matrix. In this connection, the ribbon cross-section also plays an important role; for example, sudden thickness changes should be avoided.

The stacking sequence, another important parameter, determines the effective reinforcement cross-section to carry the transverse load.

Burst tests and torsion tests on thin-walled tubes fabricated by winding glass ribbon through an epoxy coating bath and over a mandrel, provided the longitudinal and 45° strengths, whereas axial tension on similar tubes gave the transverse strength. The results are shown in Fig. 5-11 where strength versus orientation is plotted.

An improvement in the transverse composite strength of the tubes was observed as the aspect ratio was increased to 230, as shown in Fig. 5-12. An asymptotic value of 70% of longitudinal composite strength was obtained for an aspect ratio of 230. This compares well with the predictions of Lovelace and Tsai[23] for discontinuous fiber composites. Other investigators,[4,5,8,24] working with ductile steel ribbons in a ductile matrix, found the transverse strength to be 50% of the longitudinal strength. A simple expression relating transverse composite strength to the constituent properties is given by

$$\sigma_T = k\sigma_f V_f + \sigma'_m V_m \tag{5-14}$$

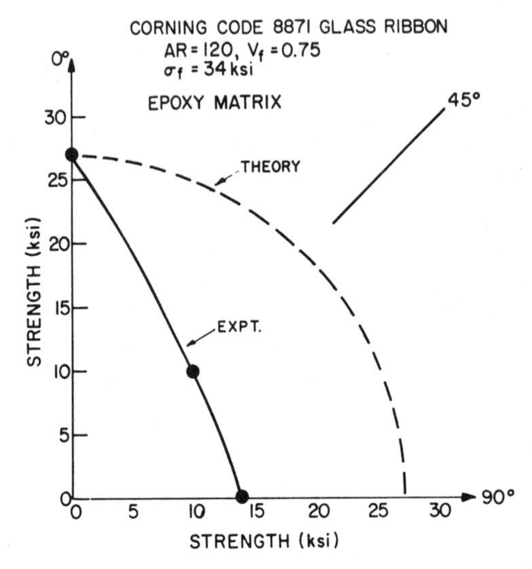

Fig. 5-11. Composite strength in different directions as determined by tube tests: theory vs. experiment.

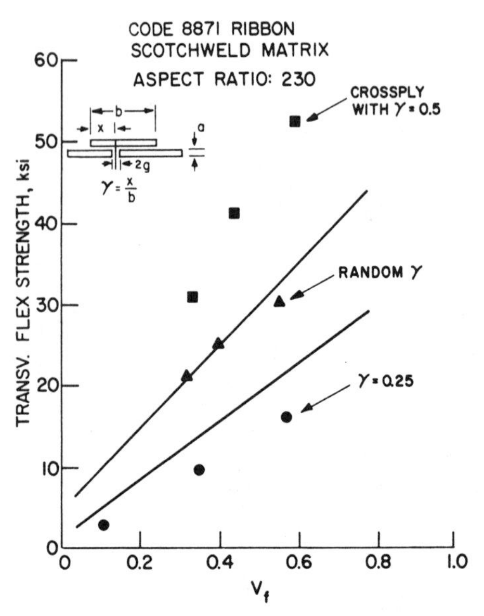

Fig. 5-13. Transverse flexural strength vs. volume fraction as determined by concentric ring tests.

where k is the efficiency factor, also called strength reduction factor. Its value ranges from 0.5 to 0.7 and depends on the aspect ratio and matrix properties.

The flexural strength of ribbon-reinforced discs was also measured in concentric ring tests, and the results are shown in Fig. 5-13. The failure was in the transverse direction for both unidirectional and crossplied ribbon composites. The effect of stacking parameter γ on the transverse strength is also shown in this figure, according to which random stacking yields higher strength values. It is also worth noting that strength anisotropy of ribbon, which is present in unidirectional ribbon composites, is masked by crossplying. Accordingly, the transverse strength is greater in the case of crossplied discs. The solid lines are the rule of mixture values using the longitudinal tensile strength of ribbon. Properly speaking, the flexural strength of ribbon, which is ~40% larger than the tensile strength, should be used in the rule of mixture equation. Crossplying combined with random stacking should yield the highest transverse strength.

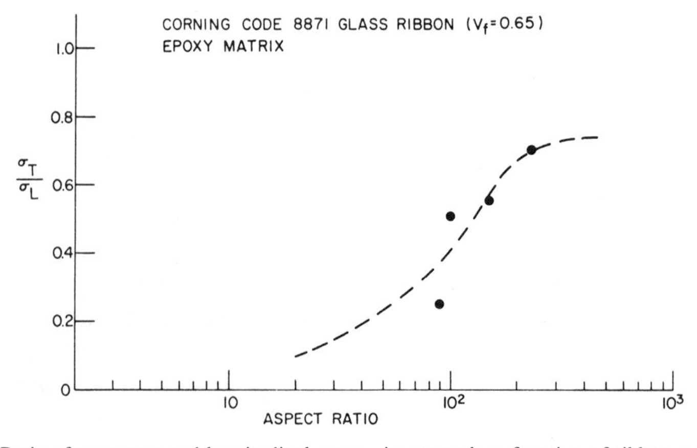

Fig. 5-12. Ratio of transverse and longitudinal composite strength as function of ribbon aspect ratio.

6.3 Notch Sensitivity

The presence of notches and holes in composite materials degrades their strength because of stress concentration. This degradation is largely a function of the aspect ratio of ribbon and the matrix properties. Secondary microcracking is generally present at the periphery of the drilled hole and may propagate through the entire width of the ribbon containing the hole. To minimize strength loss, the aspect ratio should be kept low, the case of fiber reinforcement being the limiting one. However, with a low aspect ratio, the transverse strength and stiffness are very low. Thus, a compromise is generally necessary with regard to the selection of aspect ratio, to obtain reasonably high strength and modulus values and to keep notch sensitivity down.

Notch sensitivity was also measured for transverse bars cut from a composite plate (Fig. 5-14). The cut edges where glass ribbon was exposed had microcracks that served as stress concentrators during tensile tests on the bars. One way to minimize the loss of strength due to cut edges is to etch them, thereby eliminating or rounding off the microcrack tips. This was done for various times in HF/HNO_3 solution. Indeed, as Table 5-10 shows, the transverse strength improved with etching time, reaching an asymptotic value that was two-thirds of the longitudinal strength. It is also worth noting that, in the case of no etching, the transverse strength is only 24% of the longitudinal strength, indicating a stress concentration factor (SCF) of 4, similar to that obtained for tubes with holes.

Table 5-10. Tensile strength of transverse bars with chemically etched edges.

AR = 290; SCOTCHWELD Matrix
$V_f = 0.55$, $\sigma_f = 60$ ksi

Etch time (sec)	Transverse Strength σ_T (ksi)	Strength Ratio σ_T/σ_L
0	8	0.24
2.5	15	0.45
4	21	0.64
6	21	0.64
8	21.5	0.65

The test specimens, shown in Fig. 5-14, were 0.5 in. wide, 0.06 in. thick, and 3 in. long.

Figure 5-15 demonstrates the ability of a rib-

Fig. 5-14. Failure of transverse bars in a tension test for determining notch sensitivity.

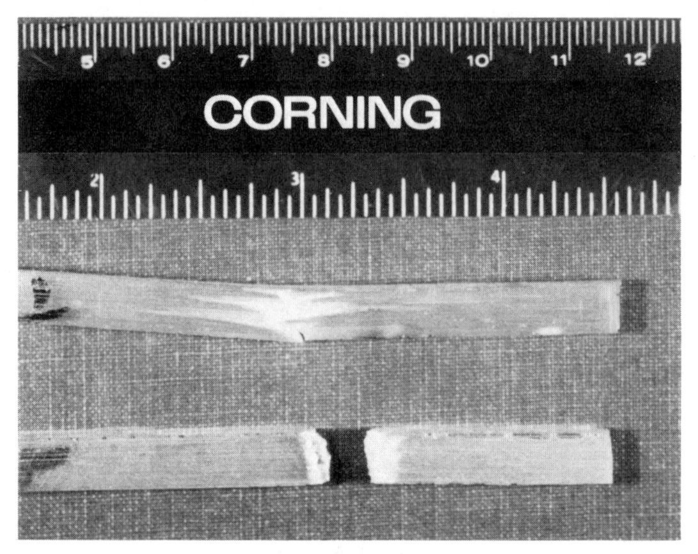

Fig. 5-15. 100-layer ribbon/epoxy bars showing effect of ribbon orientation on fracture characteristics (upper bar tested in flexure with ribbon in horizontal plane; lower bar with ribbons in vertical plane).

bon-epoxy composite to store a large amount of fracture energy. A bar containing 100 layers of Code 8871 ribbon (AR = 180) in epoxy matrix was fabricated by hand lay-up in a steel mold. The volume fraction of glass was 68%. The bar was first tested in four-point bending with ribbons in the vertical position. Failure occurred at a stress of 33,000 psi (so that the maximum stress at the bottom of the beam was shared by all of the 100 ribbon edges) and was catastrophic. The second bar, fabricated in an identical manner, was tested in four-point bending with ribbons in the horizontal position. It too failed at 32,000 psi, but the failure was not catastrophic. As can be seen in Fig. 5-15, the tensile crack did not propagate through the section but was dissipated in a series of interlaminar cracks.

6.4 Thermal Expansion

The ribbon geometry, combined with the low expansion coefficient and high elastic modulus of glass, has a strong influence on the expansion coefficient of glass-ribbon composite. Glass ribbon is isotropic in both its expansion and the elastic modulus. Consequently, plane stress analysis[13] of glass-ribbon composite subjected to thermal loading provides a simple expression for the expansion coefficient of the composite in terms of constituent properties:

$$\alpha = \left[\frac{\left(\frac{E_f}{1 - \nu_f}\right) V_f \alpha_f + \left(\frac{E_m}{1 - \upsilon_m}\right) V_m \alpha_m}{\left(\frac{E_f}{1 - \nu_f}\right) V_f + \left(\frac{E_m}{1 - \nu_m}\right) V_m} \right] \tag{5-15}$$

In Eq. 5-15, α_f and α_m are the expansion coefficients of ribbon and matrix, respectively. By virtue of $(E_f/1 - \nu_f) >> (E_m/1 - \nu_m)$ the composite expansion coefficient decreases rapidly as V_f is increased and approaches α_f for relatively small values of V_f. Thus, the large expansion coefficient of most thermoplastic and thermoset matrices may be reduced dramatically by reinforcing them with a small loading of glass ribbon. In this regard, it must be mentioned that both flakes[25] and fibers[26] are much less effective than ribbon.

Expansion measurements were carried out for glass-ribbon composites using two different matrices: SCOTCHWELD and SURLYN. The necessary constituent properties for the calculation of composite expansion coefficient are shown in Table 5-11. The glass-transition behavior of the two matrices selected was observed between 50 and 60°C. Consequently, the expansion measurements, made on a vitreous silica rod-type dilatometer in accordance with ASTM Designation E228, were restricted

Table 5-11. Constituent properties for expansion calculations.

Material	Elastic Modulus E (10^6 psi)	Poisson's Ratio ν	Expansion Coefficient α (−150C to 50C) (10^{-7} in./in./°C)
Code 8871			
Glass Ribbon	8.4	0.26	87
SCOTCHWELD	0.29	0.45	720
SURLYN	0.08	0.495	1330

to the temperature range −150–50°C. To establish the isotropy of expansion coefficient, both longitudinal and transverse bars, with SCOTCHWELD matrix, were used for expansion measurements. The results, together with calculated values (solid line) using Eq. 5-15, are shown in Fig. 5-16. It is clear that transverse and longitudinal expansion coefficients are reasonably close to the calculated value, thus establishing the isotropy of this property. In view of this, only the longitudinal specimens with SURLYN matrix were used, and again the measured values agree with the theory. It should also be noted that the large expansion of thermoplastics such as SURLYN can be reduced by an order of magnitude by reinforcing them with glass ribbon having a volume fraction as low as 30%.

For comparison purposes, a quasi-isotropic bar fabricated from SCOTCHPLY, Type 1002, containing 44% volume of E-glass fiber gave an expansion coefficient in the temperature range −100–100°C of 136 × 10^{-7} in./in./°C. It may be verified, using Eq. 5-15, that a ribbon composite containing the same amount of E-glass ribbon in epoxy matrix would have an expansion coefficient of 86 × 10^{-7} in./in./°C, which is 37% lower. Moreover, because the geometry of the ribbon permits the composite to be made with a much larger volume fraction of glass, a threefold reduction in the expansion coefficient of quasi-isotropic fiber composites is possible by using ribbons. This feature is highly desirable from a dimensional stability point of view.

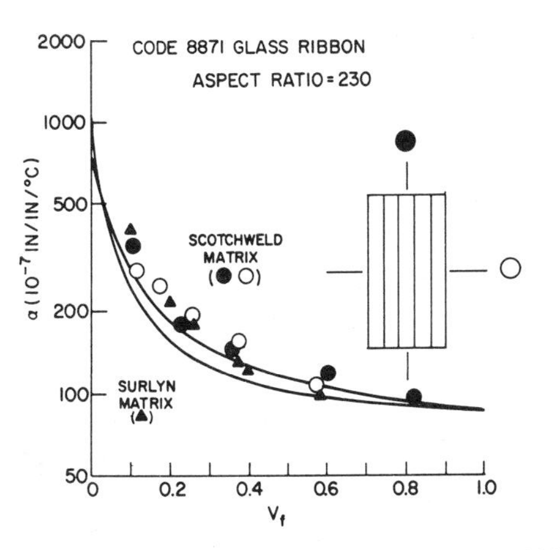

Fig. 5-16. Thermal expansion coefficient (−150–50°C) along longitudinal and transverse directions as function of volume fraction.

6.5 Chemical Durability and Corrosion Resistance

In applications such as pipes and storage tanks, chemical resistance is a highly desirable property. This property is controlled by the corrosion resistance of both the matrix and the glass. Moreover, the tortuosity inherent in a ribbon-composite construction offers a definite advantage over fiber composites with regard to corrosion resistance.

Tests were conducted on matrix specimens, fiber composites, and ribbon composites in three corrosive environments: 30% sulfuric acid, ethylene dichloride (both at room temperature), and boiling water. The matrix used for all specimens was Shell Epon 828 epoxy with curing agent Z. Two ribbon glass compositions were studied: Corning Code 8871 and Code 0211. The former ribbon had an aspect ratio of 120 with V_f = 0.74 and the latter 90 with V_f = 0.65. Fiber composite specimen (V_f = 0.33) was fabricated from fiberglass cloth impregnated with the matrix and had the same dimensions as the ribbon-composite specimen.

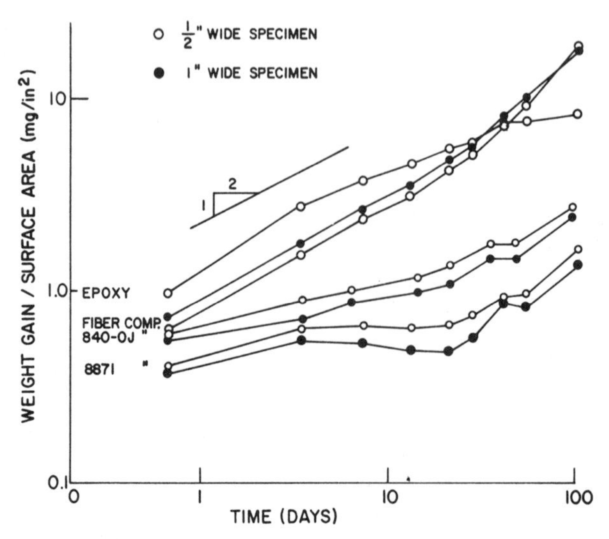

Fig. 5-17. Weight gain per unit surface area for epoxy resin, and fiber and ribbon composites as function of exposure time in 30% sulfuric acid at room temperature.

Matrix specimens were cut from bars cast in open steel molds and cured at 80°C with post-curing at 150°C. Two specimen sizes were used: 0.5 in. wide × 0.045 in. thick × 3 in. long and 1.0 in. wide × 0.045 in. thick × 3 in. long.

After immersion for various times, the specimens were removed, rinsed in tap water, dried with a paper towel, and weighed after 1 hour exposure to room conditions. Weight change per unit surface area, which is a measure of chemical durability, is shown in Fig. 5-17 for a sulfuric acid environment. The general observation is that ribbon composites gained weight more slowly than either the fiber composite or plain resin. The fiber and resin specimens gained weight at a rate nearly proportional to the square root of time, indicating a diffusion-controlled mechanism.

6.6 Permeation Resistance

Ribbon composites are less permeable than the fiber composites because of the more tortuous permeation path.[14] This advantage is a key factor in the application of ribbon composite for storage containers, pressure vessels, and pipelines. High permeability to fluids, in the case of fiber composites, has led to leakage of chemical storage tanks and loss of precious fuels. Consequently, an impervious liner is necessary when fiber composite pipe is used.

The permeation coefficient of ribbon reinforced composites has recently been deduced by Brydges et al.[15] using a simple analytical model that accounts for the ribbon geometry and the stacking parameter γ, introduced earlier. The composite permeation coefficient K_c is related to the matrix permeation coefficient K_m through:

$$\frac{K_c}{K_m} = 1 \Big/ \left[1 + \frac{b}{2g} + \left(\frac{b}{a}\right)^2 \cdot \left(\frac{V_f^2}{1 - V_f}\right) \gamma (1 - \gamma) \right] \tag{5-16}$$

in which b is the ribbon width, a the ribbon thickness, and $2g$ the matrix gap width between ribbons in the same layer. The above expression clearly shows the beneficial effect of the aspect ratio (b/a) of the ribbon. For aspect ratios >100 and $2g \sim a$, Eq. 5-16 can be approximated by:

$$\frac{K_c}{K_m} = \frac{1}{(b/a)^2} \left(\frac{1 - V_f}{V_f^2}\right) [\gamma (1 - \gamma)]^{-1} \tag{5-17}$$

In the above equations, glass was assumed to be impermeable. The dependence of K_c/K_m on the aspect ratio for various volume fractions and γ values is shown in Fig. 5-18 using the approximate relation Eq. 5-17. The curved line is the exact solution given by Eq. 5-16. Of the

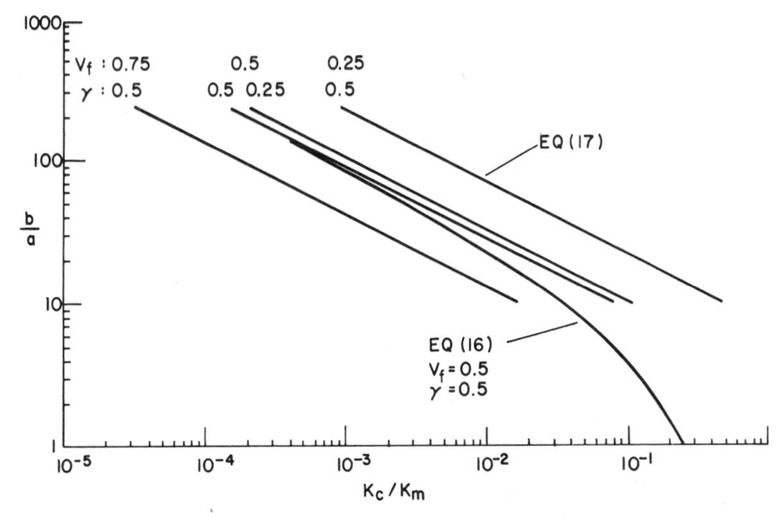

Fig. 5-18. Permeation coefficient of composites as function of ribbon aspect ratio (b/a), volume fraction (V_f) and stacking parameter (γ).

three factors that affect the permeation coefficient (V_f, b/a, γ), the stacking parameter has the least effect. The aspect ratio, by far, has the greatest effect, making ribbon composites 100 to 1000 times less permeable than fiber composites ($b/a = 1$).

The above theory was verified by mass spectrometer measurements of helium permeation through composites of various constructions. The data are summarized in Table 5-12 and confirm the permeation advantages of ribbon composites.

The predictions of Eq. 5-16 for the case of fiber composite ($b/a = 1$) agree very well with the Halpin–Tsai equations,[21] as shown in Table 5-13. It is clear that fibers, even at the highest attainable volume fraction, can lower the permeation coefficient of the composite to at most one-tenth the matrix value. In contrast to this, Table 5-12 shows that ribbons of readily attain-

Table 5-12. Helium permeation results—theory vs. experiment.

Specimen	Helium Permeation (measured) Cc (STP), mm/sec, cm^2, cmHg	K_c/K_m Measured	K_c/K_m Theory (Eq. 17)	K_c (measured)/ K_c (theory)
Epon 828/Z	1.38×10^{-9}	–	–	–
SCOTCHWELD	6.45×10^{-10}	–	–	–
Code 8871 Glass	5.60×10^{-17}	–	–	–
Code 7740 Glass	1.00×10^{-11}	–	–	–
Ribbon Composites				
Epon 828/Z, $V_f = 0.62$ $\gamma = 0.5$, $b/a = 264$	1.09×10^{-14}	7.9×10^{-6}	5.7×10^{-5}	0.14
SCOTCHWELD, $V_f = 0.22$ $\gamma = 0.5$, $b/a = 121$	5.96×10^{-14}	9.2×10^{-5}	4.6×10^{-3}	0.02
SCOTCHWELD, $V_f = 0.54$ $\gamma = 0.5$, $b/a = 121$	3.70×10^{-14}	5.7×10^{-5}	4.4×10^{-4}	0.13
Fiber Composite				
Epon 828/Z, $V_f = 0.45$ (E-Glass Fiber)	3.03×10^{-10}	2.2×10^{-1}	2.9×10^{-1} (Eq. 16)	0.76

Table 5-13. K_c/K_m values for fiber composites.

V_f	Eq. (16)	Halpin-Tsai Solution
0.11	0.63	0.79
0.25	0.46	0.59
0.44	0.30	0.38
0.50	0.26	0.33
0.66	0.17	0.20
0.83	0.08	0.09

able aspect ratios and volume fraction offer 100–1000-fold improvement over the fiber case. The assumption of neglecting permeation through the glass ribbon was also examined and found to be justifiable in those cases where $K_m K_g > 10^4$ and ribbon aspect ratio > 100. Code 8871 glass ribbon, which was used for these experiments, met these criteria.

7. APPLICATIONS—RIBBON COMPOSITE PIPE

One of the potential applications of ribbon composites is a pipe that might be fabricated by winding a resin-coated glass ribbon on the mandrel of a coil winding machine. This section will compare the performance of such a pipe with that of a conventional steel pipe. Pipe characteristics such as stiffness, strength, burst pressure, expansion, density, sagging, residual stresses, and permeation resistance are theoretically calculated as a function of the glass content and subsequently compared with those of a steel pipe. Three pipe sizes—2 in., 6 in., and 12 in. diameter—are considered.

7.1 Elastic Properties

For computational purposes, the assumed constituent properties are shown in Table 5-14. The elastic modulus of 15 × 10^6 psi and the isotropic strength of 100 Ksi for glass ribbon are considered practical within the realm of modern glass technology. The matrix properties listed in Table 5-14 correspond to those of a typical Nylon "Type 6" cast. The isotropic stiffness provided by ribbon reinforcement, assuming aspect ratio ≥ 100, follows Eq. 5-1 for E and Eq. 5-4 for G. Since $E_f < E_{\text{steel}}$, it is clear that a glass/resin composite will not outperform steel insofar as stiffness is concerned, when equal geometries and volumes are involved.

7.2 Longitudinal and Transverse Strength

Assuming continuous ribbon, the longitudinal strength of the composite pipe would be given by Eq. 5-13. The transverse strength, which is sensitive to fabrication, ribbon cross-section, and matrix properties, would fall between 50 and 100% of the longitudinal strength.

Gray[5] gives an excellent discussion of the various factors that influence the transverse strength and concludes that the matrix shear strength is the governing factor. Because of the lack of a rigorous theoretical model for predicting transverse strength, we assume that it is also given by the rule of mixtures modified by a coefficient to reflect the efficiency of strength transfer:

$$\sigma_t = k(V_f \sigma_f + V_m \sigma_m') \tag{5-18}$$

Table 5-14. Assumed constituent properties.

Property	Glass Ribbon	Matrix (Nylon, Type 6 Cast)
Elastic modulus, E	15 × 10^6 psi	0.4 × 10^6 psi
Poisson's ratio, ν	0.27	0.35
Shear modulus, G	6 × 10^6 psi	0.15 × 10^6 psi
Strength, σ	1 × 10^5 psi	12,000 psi
Density, ρ	0.097 lb/in.^{3}	0.042 lb/in.^{3}
Thermal expansion coefficient, α	43 × 10^{-7} in./in./°C	900 × 10^{-7} in./in./°C
Maximum operating temperature, T	–	250°C
Aspect ratio, b/a	100	–

The value of k will, of course, depend on many parameters. Lovelace and Tsai[23] suggest 0.6 to 0.8 as a value for k. At low volume fractions of glass, the ribbon interaction may be negligible and the matrix shear stress small so that k will approach unity. However, as the volume fraction increases, the effect of various factors becomes more pronounced, and $k < 1$. It is almost impossible to predict the transverse strength for volume fractions of the order to 0.9 to 1.0. For our purposes, we have taken $k = 0.8$ and plotted the variation of longitudinal and transverse strengths with the volume fraction in Fig. 5-19. The yield strengths of mild steel, copper, and carbon steel are also shown. To be competitive with carbon steel, a volume fraction of 70% would be necessary.

It should also be pointed out that for ribbon-wound pipes carrying pressure loading, the longitudinal strength (hoop strength) is of prime importance because the stress in the transverse direction (axial) is half of that in the longitudinal direction. If, on the other hand, the pipe is subjected to axial moment, in addition to pressure loading, the strength requirement in the axial direction becomes more critical and at times greater than that in the circumferential direction. In such cases, ribbon will have to be oriented so as to be parallel to the pipe axis, as in the pultrusion process.

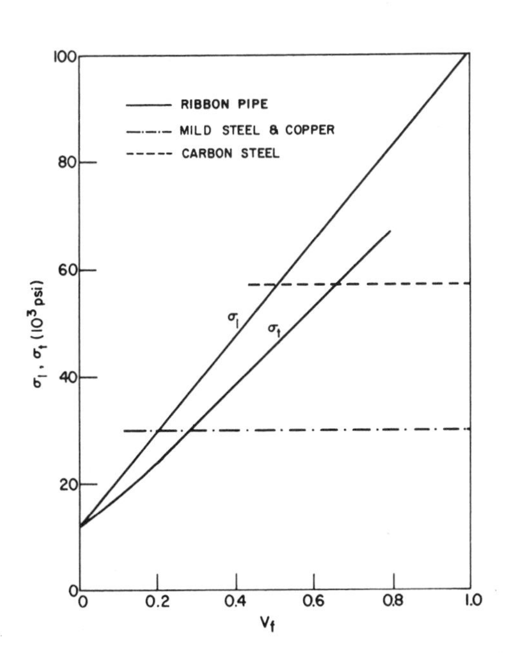

Fig. 5-19. Variation of longitudinal (σ_l) and transverse (σ_t) composite strengths with volume fraction.

7.3 Coefficient of Thermal Expansion

The composite expansion coefficient in the axial and hoop directions is given by Eq. 5-15. Assuming the constituent properties of Table 5-14, a ribbon composite pipe containing 30% glass ribbon has an expansion coefficient of 110×10^{-7} in./in./°C compared with 120×10^{-7} in./in./°C for steel. Thus, the ribbon composite pipe with $V_f > 0.3$ would have lower expansion than either steel or copper ($\alpha = 160 \times 10^{-7}$).

7.4 Burst Pressure

For calculating the burst pressure, thin-walled-tube analysis is used. Accordingly, the hoop and axial stresses in an internally pressurized tube of mean diameter d and wall thickness t are given by:

$$\left.\begin{aligned} \text{Hoop stress } \sigma_\theta &= \frac{pd}{2t} \\ \text{Axial stress } \sigma_z &= \frac{pd}{4t} \end{aligned}\right\} \tag{5-19}$$

where p is the internal pressure. Pipe strength in the hoop direction corresponds to composite longitudinal strength, σ_l, whereas the axial strength is equal to the transverse strength, σ_t. Since $\sigma_z = \frac{1}{2}\sigma_\theta$ and $\sigma_t > \sigma_l/2$, hoop stress is the more critical stress. Steel pipes are generally rated at schedule 40, which permits a factor of safety of 5 for strength. Thus, the allowable burst pressure, P_b, for ribbon composite pipe is obtained by equating σ_θ with $\sigma_l/5$, which gives:

$$P_b = 0.4 \frac{t}{d} \sigma_l \tag{5-20}$$

Since σ_l varies with volume fraction of glass, its value has been taken from Fig. 5-19. The variation of burst pressure with V_f is shown in Fig. 5-20 for three pipe sizes: 2 in., 6 in., and 12 in. diameter. Values of burst pressure are

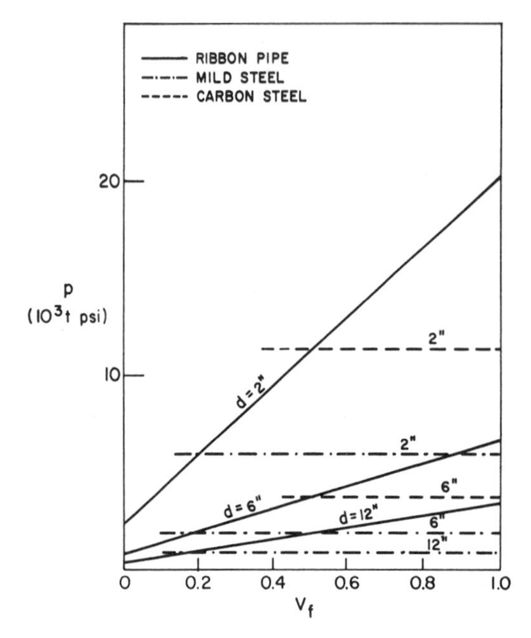

Fig. 5-20. Burst pressure for ribbon-composite tube as function of volume fraction.

shown in terms of wall thickness t. The pressure carrying capacity of mild and carbon steel pipes is also shown in this figure and is based on the tensile yield strength of 57,000 psi for carbon steel and 30,000 psi for mild steel. It follows from Fig. 5-20 that a ribbon composite pipe with volume fraction greater than 0.55 can compete effectively with the carbon steel pipe. Such a pipe with wall thickness equal to that of a steel pipe will weigh only 25% of the weight of steel pipe.

The above analysis does not account for additional axial loading that might arise from the weight of soil above the buried pipe and may require greater axial strength. In such cases, it may be necessary to reorient ribbon so as to impart adequate strength in the axial direction.

7.5 Residual Stresses

During the fabrication process, the matrix must cure at a temperature T_c. It experiences restraints against contraction during cooling from T_c to room temperature due to expansion and stiffness mismatch between the glass and matrix. Since the matrix is usually higher in expansion, it is put into tension. This residual tension weakens the matrix and reduces its effectiveness in transferring the load. As a result of tension in the matrix, the glass ribbon experiences a small amount of compressive stress.

The residual stress in the matrix reaches its maximum value for low volume fractions of glass and thereafter remains substantially constant up to a volume fraction of 0.95. The flow temperature of Nylon ''Type-6'' ranges from 205 to 315°C. Assuming a cure temperature of 250°C, a tensile stress of 12,000 to 16,000 psi can occur in the matrix. Depending on the matrix, some of this stress may be relieved through relaxation, particularly in the absence of molding pressure during cooling. Such high stresses may result in matrix failure. With thermosets, much of the stress is locked in, and there is very little relaxation during cooling.

7.6 Resistance to Fluid Permeation through Pipe Wall

Although this aspect of the performance of ribbon-composite pipe cannot be compared quantitatively with the steel pipe, it does provide a qualitative measure and points out the longer life such a pipe will experience. Also, as will be clear, comparison with plastic pipes is made possible.

The measure of enhancement of permeation resistance is provided by examining the ratio of permeation coefficient of the matrix, K_m, and that of the composite, K_c, since the permeation coefficient is linearly related to the amount of fluid permeating through the pipe wall. This ratio is given by Eq. 5-17.[15]

For a pipe wound back and forth, the dependence of K_m/K_c on the glass content is shown in Fig. 5-21 with $b/a = 100$ and $\gamma = 0.25$. It follows from this figure that a ribbon composite pipe is a thousand-fold less permeable than a pipe made from the matrix alone. Consequently, there is less chance of moisture penetration and subsequent degradation of pipe strength, leading to longer pipe life. It should also be pointed out that this important advantage is not to be found in fiber-reinforced pipes because they do not enhance the permeation resistance of plastics by more than a factor of 10.

The above analysis does not permit comparison on the basis of dynamic properties, which

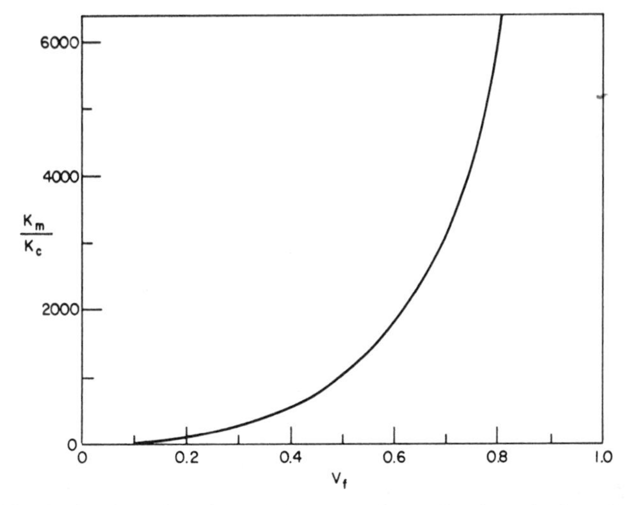

Fig. 5-21. Ratio of matrix and composite permeation as function of volume fraction.

must be handled experimentally. There is also the question of ductility offered by most metallic pipes. Undoubtedly, this property will be governed by the choice of matrix because glass must be considered brittle. Finally, the operating temperatures to which the ribbon composite pipe may be subjected will be lower than those for a steel pipe.

8. GRAPHITE AND BORON FILM COMPOSITES

Considerable attention has been given to more exotic reinforcing films, such as graphite and boron. In the early seventies, Pfizer, Inc. did some preliminary studies for the evaluation of graphite ribbon composites. This work, which was funded by The Army Materials and Mechanics Research Center, was recently reported by Froberg.[27] The following summary is based on Froberg's report.

Prepared by chemical vapor deposition technique, the pyrolytic graphite film ranges between 0.0001 and 0.0004 in. in thickness, with a nominal value of 0.0003 in. The graphite film has a tensile strength between 30 Ksi and 110 Ksi and a tensile modulus between 3×10^6 and 13×10^6 psi. It has an average strength of 60 Ksi and an average modulus of 7×10^6 psi. Its density is 1.6 g/cc. Thus, it compares well with glass ribbon in terms of specific strength and specific modulus, and has the potential of exceeding these preliminary values.

Graphite film composites enjoy the same advantages as glass ribbon composites. Thus, planar isotropy, more efficient use of reinforcement, superior specific strength and modulus, large volume loadings, and very thin laminates are distinct advantages of film composites over corresponding filamentary composites. Because they can be made as thin as 0.003 in., graphite film laminates with four to six plies are very effective as skin reinforcement. For example, the addition of 5% graphite film composite increased the flexural modulus of an acrylic beam by 100%. Another widespread use of thin film composites is in providing local reinforcement for fiber composites in those locations where stresses are concentrated or deflections are large.

Graphite film properties have been utilized effectively in a multiple-ply laminate. The composite tensile strength ranges between 30 and 60 Ksi and composite moduli are between 4×10^6 and 6×10^6 psi. For purposes of comparison, the typical room temperature properties of graphite filament composite with 52% volume loading are given in Table 5-15. Thus, graphite film composites compete favorably with quasi-isotropic filamentary composites. It should be borne in mind that in some applications where a very thin (0.003–0.005 in.) planar isotropic composite is required, graphite filaments that are 0.005 in. thick cannot be used because a minimum of five to eight plies are necessary for a quasi-isotropic structure. This

Table 5-15. Properties of HERCULES graphite filament prepreg type 2002M and graphite film composites.

	Longitudinal	Cross-Plied	Angle-Plied (±45°)	Quasi-Isotropic	Graphite Film Composites
Tensile strength (Ksi)	104	53	21	40	30–60
Elastic modulus (10^6 psi)	24	14	2.5	9	4–6

Courtesy: Pfizer, Inc.

would mean a quasi-isotropic filamentary composite with a thickness of 0.040 in., or 20 times the thickness of a planar film composite.

It is worthwhile to compare graphite film composite properties with those of boron/polyimide composites, which have been developed more extensively for a longer period of time. This comparison is made in Table 5-16, which also shows the properties of a well known aluminum alloy, 2024T-62. The table clearly shows that graphite film composites compare favorably insofar as strength is concerned. The specific modulus, however, is lower and requires additional development such as hybrid composites, in combination with fiber reinforcement.

The boron film is formed[28,29] by evaporation and deposition of pure, elemental boron in a high vacuum on polyimide film substrate. The film is passed continuously over the crucible containing molten boron in the vacuum chamber. The deposited boron forms a smooth, black, glassy film in thicknesses up to 0.0003 in. The thickness of polyimide film substrate is 0.00025 in. The sheets thus formed by Norton Research Corporation* in 1970 were about 6 in. wide and up to 60 in. long. The microcrystalline structure of boron film resembles that of boron filaments. The interest and motivation in developing boron film stem from potentially high composite efficiencies, planar isotropy, and low density—all of which are necessary in buckling—and flutter—of critical aircraft structures. The initial premise for developing these films was the inherent assumption that the films will have properties similar to boron fibers, that is, a tensile strength of 400 Ksi and an elastic modulus of 55×10^6 psi. Assuming further that these properties can be utilized efficiently in a boron film/epoxy laminate, a composite strength of 200 Ksi and a modulus of 28×10^6 psi would be realized at a volume loading of 50%. Such properties would indeed be useful in aircraft structures.

However, such strength values have not, as yet, been attained. Working backward from the measured composite properties, it was found that the film strength was 90 Ksi, instead of 400 Ksi, but the film modulus was 55×10^6 psi as expected. The drastic reduction in film strength is believed to be caused by the presence of periodic cracks that occur during the manufacturing process. The degree of cracking of the boron film is a function of the thickness of the deposit. At 40% volume loading of boron, the

*Now National Research Division of Cabot Corporation, Cambridge, Mass.

Table 5-16. Comparison of properties of graphite film/epoxy composites with other materials.

Material	Volume Fraction	Tensile Strength (Ksi)	Elastic Modulus (10^6 psi)	Density (lb/in.^{3})	Specific Strength (10^3 in.)	Specific Modulus (10^6 in.)
Boron/epoxy	–	40	20	0.06	667	333
Al2024-T62	–	62	10	0.10	620	102
Graphite film/epoxy	0.83	59	6	0.055	1060	108

Courtesy: Pfizer, Inc.

Table 5-17. Mechanical properties of boron film reinforced-epoxy composite.

Tensile strength	30–55 Ksi
Compressive strength	55–90 Ksi
Elastic modulus	20–22 $\times 10^6$ psi
Shear modulus	8–9 $\times 10^6$ psi
Interlaminar shear	12–14 Ksi
Density	0.062–0.065 lb/in.^{3}

distance between cracks varies from 0.25 to 1 in., and the film behaves like thin flakes of very large aspect ratio.

Boron film composites are fabricated by bonding multiply layers of film with a suitable adhesive. A number of resin systems including epoxy, phenoxy, and polyimide resins have been used to date. Epoxy resins have the advantage that they do not produce gaseous or liquid residues during curing. Polyimide systems provide good adhesion and high-temperature properties. Mechanical properties of composites containing 39 to 44 vol % of boron, deposited on polyimide film and laminated with epoxy resin, are given in Table 5-17.

One of the applications where boron film composites have shown good potential is the reinforcement of bolt holes in filamentary composites.[12,30] Unidirectional, bidirectional (±45°), and tridirectional (0°, ±45°) graphite fiber/epoxy resin composites were fabricated to form a series of double-lap bolted joint models. These models were tested to measure the joint strength effectiveness of boron film colaminated locally at 6 and 10 vol % levels. Results showed that the addition of film plies to the highly directional fiber laminae increases joint strength and stiffness up to 200%, the chief contribution of boron film being high bearing strength and resistance to shear distortion. The boron film was also effective in reducing the notch sensitivity of graphite fiber laminates, while at the same time substantially raising both the tensile strength and the work-to-fracture of the material.

Applications where boron and graphite films can provide optimum reinforcement include the reinforcement of bolt holes in filamentary composites, thin high-performance laminates, skins for honeycomb structures, structures requiring high planar isotropic properties, such as in aircraft where buckling and flutter are important, aircraft wings, and tubular components with high specific strength and modulus.

GLOSSARY

Quasi-isotropic—A fiber composite with 0, ±45°, 90° oriented fibers to provide isotropic properties in the plane of the composite.

ACKNOWLEDGMENT

The author is grateful to Dr. W. T. Brydges for his assistance in the preparation of this manuscript and to Mr. John F. MacDowell for constant encouragement in pursuing the important field of glass ribbon reinforcement.

REFERENCES

1. Humphrey, R. A., "Shaped Glass Fibers," in *Modern Composite Materials*, L. J. Broutman and R. H. Krock (eds.), Addison-Wesley, Reading, Mass., 1967.
2. Halpin, J. C. and Thomas, R. L., "Ribbon Reinforcement of Composites," *J. Comp. Mat.* 2, 1968.
3. Tsai, S. W. and Pagano, N. J., "Invariant Properties of Composite Materials," in *Composite Materials Workshop*, Tsai, Halpin, and Pagano (eds.), Technomic Publ. Co., Stamford, Conn., 1968.
4. Lewis, T. B., "Ribbon Reinforcement in Composite Materials," *Proc.* 25th Ann. Conf. SPI., 1970.
5. Gray, R. M., "Transverse Strength of Ribbon Reinforced Metal Matrix Composites," *Proc.* Symp. Metal–Matrix Composites, Met. Soc. AIME, 1969 (Govt. Document No. AD695-046).
6. Anderson, E., Lux, B., and Curssard, C., *Revue de Metallurgie* 2, 1972.
7. Anderson, E., "Fatigue Behavior of Ribbon-Reinforced Composites," *J. Mater. Sci.* 8, 1973.
8. Pollock, J. T. A. and Arthur, J., "Tensile Strength of Ribbon Reinforced Composites," *Mat. Science and Eng.* 18, 1975.
9. Padawer, G. E. and Beecher, N., "On the Strength and Stiffness of Planar Reinforced Plastic Resins," *Polymer Eng. and Science* 10, 1970.
10. Chen, P. E. and Nielson, L. E., "Mechanical Properties of Tape Composites," *Kolloid-Zeit. und Zeit. fur Polymere* 235, 1969.
11. Chen, P. E. and Lewis, T. B., "Stress Analysis of Ribbon Reinforced Composites," *Polymer Eng. and Science* 10, 1970.
12. Padawer, G. E., "The Strength of Bolted Connections in Graphite–Epoxy Composites Reinforced by Colaminated Boron Film," in *Composite Materials: Testing and Design (Second Conf.)*, ASTM STP497, pp. 396–414, 1972.

13. Gulati, S. T. and Plummer, W. A., "Thermal Expansion of Ribbon Reinforced Composites," *Third Thermal Expansion Symposium*, M. G. Graham and H. E. Hagy (eds.), Am. Inst. Physics, 1971.
14. Humphrey, R. A., "Glass Microtape" in *Materials Science Research*, Vol. 2, H. M. Otte and S. R. Locke (eds.), Plenum Press, New York, 1965.
15. Brydges, W. T., Gulati, S. T., and Baum, G., "Permeability of Glass Ribbon-Reinforced Composites," *J. Mat. Science*, 10: 2044–2049, 1975.
16. Hutchins, J. R., III and Harrington, R. V., "Glass," in *Encyclopedia of Chemical Technology*, ed. Kirk-Othmer, John Wiley & Sons, New York, 1966.
17. Eakins, W. J. and Humphrey, R. A., "Glass-Resin Composite Structure," U.S. Patent 3,425,454, Feb. 4, 1969.
18. Cox, H. L., "The Elasticity and Strength of Paper and Other Fibrous Materials," *Brit. J. Appl. Phys.* 3: 72, 1952.
19. Loewenstein, K. L., "Glass Systems," *Composite Materials*, L. Holliday (ed.), Elsevier, 1966.
20. Bishop, P. H. H., "An Improved Method for Predicting Mechanical Properties of Fibre Composite Materials," RAE Tech. Report No. 66245, 1966.
21. Ashton, J. E., Halpin, J. C., and Petit P. H., *Primer on Composite Materials: Analysis*, Technomic Publ. Co., Stamford, Conn., 1969.
22. Whitney, J. M., "Experimental Determination of Shear Modulus of Laminated Fiber-Reinforced Composites," *Expt. Mech.* 7, 1967.
23. Lovelace, A. M. and Tsai, S. W., "Composites Enter the Mainstream of Aerospace Vehicle Design," *Astronautics and Aeronautics*, pp. 56–61, July 1970.
24. Lewis, T. B., Nielson, L. E., and Hemmerly, D. M., "Ribbon Reinforcement in Composites," 5th St. Louis Symp. Adv. Composites, 1971.
25. Schapeny, R. A., "Thermal Expansion Coefficients of Composite Materials Based on Energy Principles," *J. Comp. Mat.* 2, 1968.
26. Halpin, J. C., "Stiffness and Expansion Estimates for Oriented Short Fiber Composites," *J. Comp. Mat.* 3, 1969.
27. Froberg, R. W., "Research Study in Evaluation of Graphite Ribbon Composites," prepared for Army Materials and Mechanics Research Center, Report No. AMMRC CTR-72-28, Dec. 1972.
28. Beecher, N., Feakes, F., and Allen, L. R., "Laminar Film Reinforcements for Structural Applications," Paper NR-3, 12th National SAMPE Symposium, Western Periodicals Co., Hollywood, Calif., 1967.
29. Crawford, R. F., "An Evaluation of Boron–Polymer Film Layer Composites for High-Performance Structures," Astro Research Corporation Report No. ARC R-276, also NASA CR-1114, Sept. 1968.
30. Padawer, G. E., "Film Reinforced Multi-Fastened Mechanical Joints in Fibrous Composites," 13th Structures, Structural Dynamics, and Materials Conference, San Antonio, Tex., Apr. 10–12, 1972; AIAA Paper No. 72-382, American Institute of Aeronautics and Astronautics, 750 Third Ave., New York, N. Y.

Section III
Short Fiber Reinforcements

6
Short Metal Fibers and Flakes

John V. Milewski
Consultant
Santa Fe, New Mexico

Donald E. Davenport
TRACOR MBA
San Ramon, California

CONTENTS

1. INTRODUCTION

The rapid growth and universal acceptance of computers, television, and microprocessor controls in many homes and businesses has accelerated the growth of the electromagnetic shielding (EMI) industry. The economics of making complex housings for these new electronic devices from plastics has created a market for fillers and fibers to add EMI shielding properties to these molded products. This chapter presents information about the kinds and types of fibers and flakes that are being marketed for these applications.

Other large markets for metallic fibers and flakes are electro-static shielding and electrical and thermal conductivity applications.

Short metallic fibers and flakes provide these properties while still retaining the mechanical properties of the base resin system. This is not possible if conductive fillers are used because they must be used in much higher loadings to develop the required conductivity. This higher loading substantially reduces the mechanical properties of the composite. Short fibers and flakes also retain the economics of high production molding found in compression, transfer, and injection molding systems, which are not available in long fiber systems.

Thus, the real advantages of using short metallic fibers and flakes are economics and flexibility in processing while still obtaining the required electrical and thermal conductivity changes.

This chapter describes the theory of how short fibers and flakes add electric and thermal properties to plastics. It also describes the forms and types of metallic fiber and flakes that are available. Suppliers are given, and typical applications and data associated with their use are described.

2. THE RATIONALE FOR USING SHORT METALLIC FIBERS

The need for a highly conductive plastic was first seen as a way of advancing the use of plastic housings in the rapidly developing electronics industry. As computers and sophisticated electronic devices began to move out of their shielded rooms and into the office, store, and home, it became highly desirable to take advantage of the light weight, low cost, and aesthetics that could be gained by substituting plastics for metal housings for this equipment. But to provide the electrostatic protection and the EMI shielding that were vital to the many solid state components that operate at 5-volt potentials, plastics needed a conductivity that would not only drain the static charge from operators dressed in plastic, but would also provide a barrier to the radiation that was generated when the static charge was dissipated as an arc to the nearest ground. There are many stories about the troubles of the first point-of-sale instruments and electronic wrist watches that first appeared with plastic cases. It quickly became clear that plastics with a resistivity less than 1 ohm-cm would be needed to serve these areas.

More recently, the importance of thermal conductivity in certain plastic applications has been better appreciated. As plastic housings became popular for portable tools, motors, and other heat-generating sources, the question of how to get the heat out of the case became important. The question has become even more critical in the automobile industry's efforts to use more and more plastic parts to reduce the weight of vehicles so that they will be more energy-efficient. But the automobile industry is keyed to making parts in a few seconds each, or a fraction of a minute at most, in order to make millions of parts per year. With conventional plastics, the problem of getting heat into the part to form it and then getting the heat out again from a poor thermal conductor meant that cycle times required minutes not seconds. This implied an increase of one to two orders of magnitude in the number of tools used to make the same number of parts. So, for the automotive industry, an order of magnitude greater thermal conductivity would be an important advance.

For these reasons much effort has been spent over the last few years in attempts to achieve large changes in the conductivities of plastics—electrical resistivities less than 1 ohm-cm and thermal conductivities at least a factor of ten higher than those of normal plastics. Significant strides have been made in creating and understanding this new class of materials, the metalloplastics, and in discovering what must be done to adapt them to conventional molding procedures and to lower their costs so they will become a practical engineering material.

3. THE THEORY OF SHORT METALLIC FIBER CONDUCTIVE PLASTICS

The basic concept that led to the creation of conductive plastics is that extremely small concentrations of additives can make plastics conductive if they are added in the form of conductive fibers with length to diameter (L/D), or aspect, ratios of 100 or more.

In the past, conductivity had always been obtained by the addition of chunky fragments of copper or silver, or, if only thermal conductivity was sought, even sand particles. The results were always the same: a weight percentage of 60 to 80 and a volume percent loading of 40 to 50 were needed to produce significant effects.

The striking difference in their effect on conductivity between chunky fragments and fibrous materials can be seen from Fig. 6-1. If one loads a plastic with 5% by volume of chunky fragments (shown as spheres for convenience), a random distribution of the spheres leads to variable gaps between them; for 6-mil spheres there will be, on the average, a 6-mil distance to the nearest sphere. Thus, heat or electrons flowing through such a matrix cross alternate paths of about equal lengths in the two media.

If one takes the same 5% of material and disperses it as 1-mil-diameter fibers that are 100 mils long, the picture is quite different. Even if the fibers were all arranged parallel to one another and at an even spacing, they would be only 3 mils apart. When they are allowed to take random orientations, it becomes inevitable that they will touch one or more of their nearest neighbors, as shown in Fig. 6-1. This provides an almost continuous path through the composite along the highly conductive fibers.

Thus, simple intuition tells us that fibers will

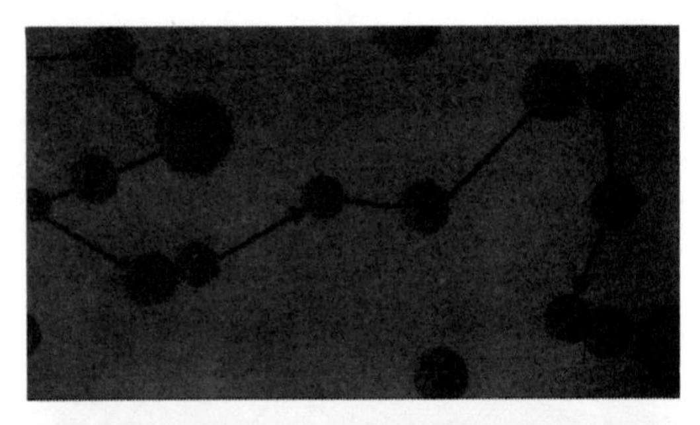

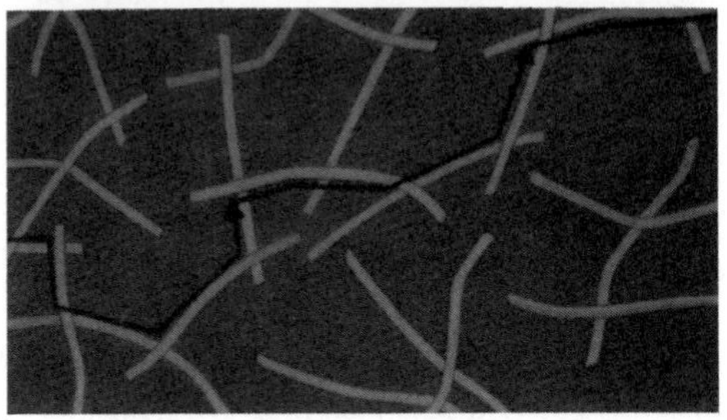

Fig. 6-1. Comparison of a typical flow path through composites using the same volume percentage of spheres and fibers ($L/D = 100$).

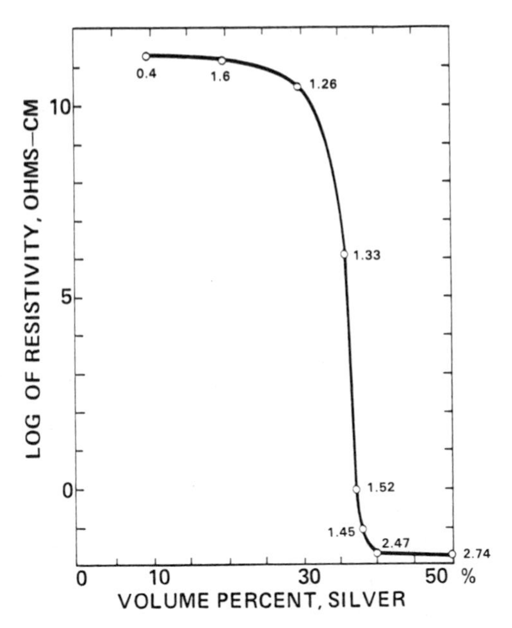

Fig. 6-2. Decrease in resistivity as a function of the volume percent of chunky silver particles added to Bakelite.

be much more effective than chunky fragments in lowering the electrical resistivity of plastics and increasing their thermal conductivity.

4. ELECTRICAL CONDUCTIVITY

We would expect the effects of low fiber concentrations on these two properties to be quite different. Whereas thermal conductivity will be increased by such low concentrations of the fibers that one fiber does not touch its neighbors, electrical resistivity will not be significantly modified until an almost continuous path is available through the conductive fibers.

This was clearly demonstrated with chunky fragments in 1966 by J. Garland.[1] His experimental data on electrical resistivity of Bakelite as a function of the volume percent of silver particles added are shown in Fig. 6-2. The resistivity changes very little until the silver particle loading approaches a critical concentration; then it drops catastrophically, and the composite becomes a good conductor.

Garland suspected the critical concentration was associated with forming continuous chains of particles in the matrix, so he devised an ingenious technique of measuring the average number of contacts each silver particle had with those around it. The data are shown on the curve. From chain forming theory, we know that the larger the average number of contacts, the longer the chains. Finally, when the average number of contacts approaches two, the entire particle population is interconnected in one long chain.

As Garland's data show, the critical concentration occurs in that narrow concentration region between the formation of significant chain lengths ($m = 1.26$) and one continuous chain ($m = 2.0$). Beyond that the resistivity falls relatively slowly because we are only increasing the number of parallel paths.

When fibers are used instead of chunky particles, the shape of the curve remains the same, but chain building starts at much lower concentrations. In fact, one can go to statistical theory of chain building and calculate the concentrations of fibers with a given L/D at which a continuous chain is formed; this should be a good approximation for the onset of conductivity. The theory, of course, assumes a random orientation of ideal fibers with the same L/D, which will lead to an optimistic estimate; that is, the lowest concentrations at which one would expect to experimentally observe significant conductivity.

If we use Milewski's[2] experimental work on fiber packing, we can obtain a good estimate of an upper limit for the concentration at which conductivity should be observed as a function of L/D. Milewski has shown that if one puts rigid fibers (he used wooden dowels in his early experiments) into a volume, their packing density decreases as their L/D increases, as shown in Fig. 6-3. This natural packing fraction results from the larger and larger voids that are spanned by the fiber before it is supported by two or more fibers. Furthermore, Milewski was able to show that fiberglass in an empty container followed the same packing fraction versus L/D curves; that is, it behaved as rigid fibers. We have extended this work to fibers with an L/D up to 375 using our metal-coated glass fibers, which we call Metafil®.

These packing fractions are then the volume fractions required to have two or more contacts per fiber in a matrix (average number of contacts/fiber greater than two) and should represent the upper limit of the volume fraction required to give conductivity. Since this is an experimental number, it contains some nonrandom fibers and inadvertent packing, depending on the care with which the experiment is carried out, and thus tends to overestimate the upper limit.

In Fig. 6-4, these two estimates of the limiting concentration are plotted along with the experimentally observed data from various plastic processing techniques. Because the experimentally observed conductivities are obtained with processed fibers that have been broken down to some extent, the data are represented as bands whose width is a best guess as to the average L/D in the matrix.

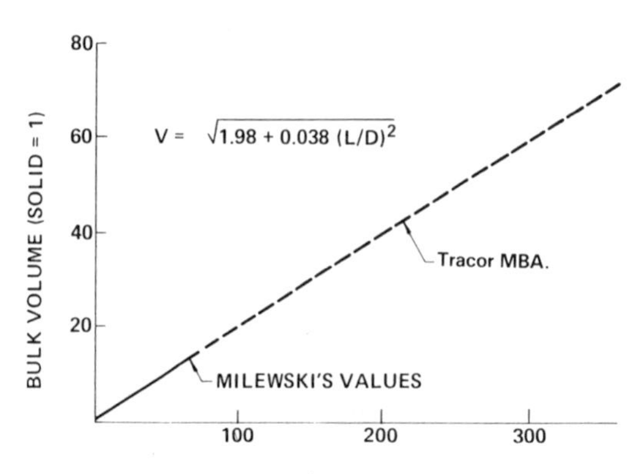

Fig. 6-3. Effect of L/D on packing fraction.

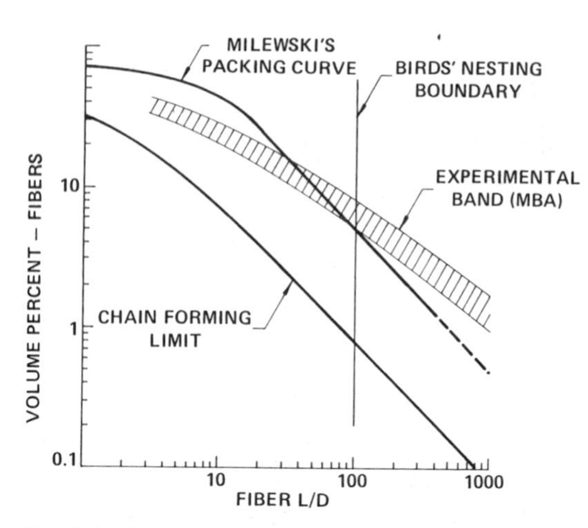

Fig. 6-4. Critical concentration for conductivity vs. L/D.

It is interesting to note that at low concentrations, the experimentally observed values fall between the two estimates, but that at large L/D, the experimental values fall above the upper limit estimate. The larger deviation at the long fiber end of the curve results from two factors that can be readily identified. The processing techniques to obtain data in this regime tend to give nonrandom orientation because the resulting parts are thinner than the fiber lengths. Also, there is a magic number for fiber tangling and matting that falls between an L/D of 50 and 100. When the L/D values are below this critical value, the fibers pour and flow readily without tangling. When they are above this value, the fibers tend to tangle and form "birds' nests" that are difficult to distribute. There are ways to avoid this tangling with long fibers, but those methods were not used in generating these data.

Thus, both theory and experiment confirm that to get good conductivity (low resistivity) at very low concentrations of fibers, one should use the maximum L/D feasible for the manufacturing process.

A puzzling aspect regarding making plastics electrically conductive with fibers concerns the way the fibers make metal-to-metal contact in the plastic matrix. One would initially suppose that there would always be a plastic film between the fibers, serving as an insulating layer.

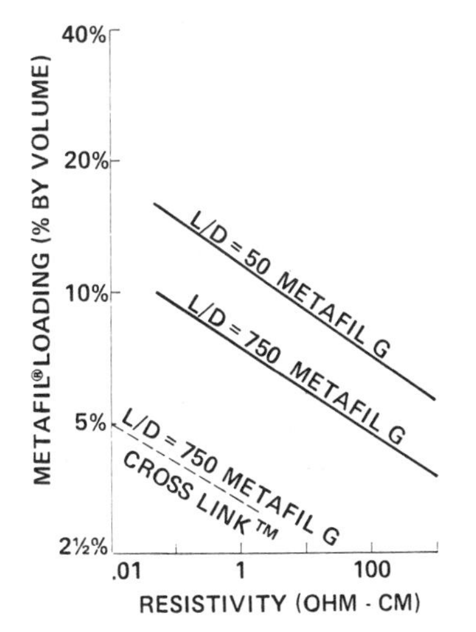

Fig. 6-5. Resistance of Metafil composites, normal and Cross-Link formulations.

This probably is true when the plastic wets the fiber well, and there are only low forces between fibers. Even above the critical concentration, it has been observed that the resistance drops markedly if the measurement voltage is greater than 10 volts, implying that a breakdown voltage is being observed. To lower this interface effect, some users have added small amounts of carbon powder. Tracor MBA developed a process (Cross-Link™), that avoids the use of any additives and still produces a very marked reduction in resistivity over that observed with simple mixing, as shown in Fig. 6-5. This figure shows the effect of increased concentrations in lowering the resistivity for two fiber lengths and the low concentrations required if the Cross-Link process is used. Note that resistivities as low as 0.01 ohm-cm can be obtained at very modest fiber concentrations, which will be important for EMI shielding applications.

5. THERMAL CONDUCTIVITY

The thermal conductivity of a composite of high-conductivity fibers in plastics can be accurately estimated by the equation of Nielson,[3] as was verified by D. Briggs[4] with experimental measurements on a wide range of fiber materials in plastics.

If one inserts the appropriate parameters for Metafil fibers into these equations, the curves shown in Fig. 6-6 are obtained. Unlike the

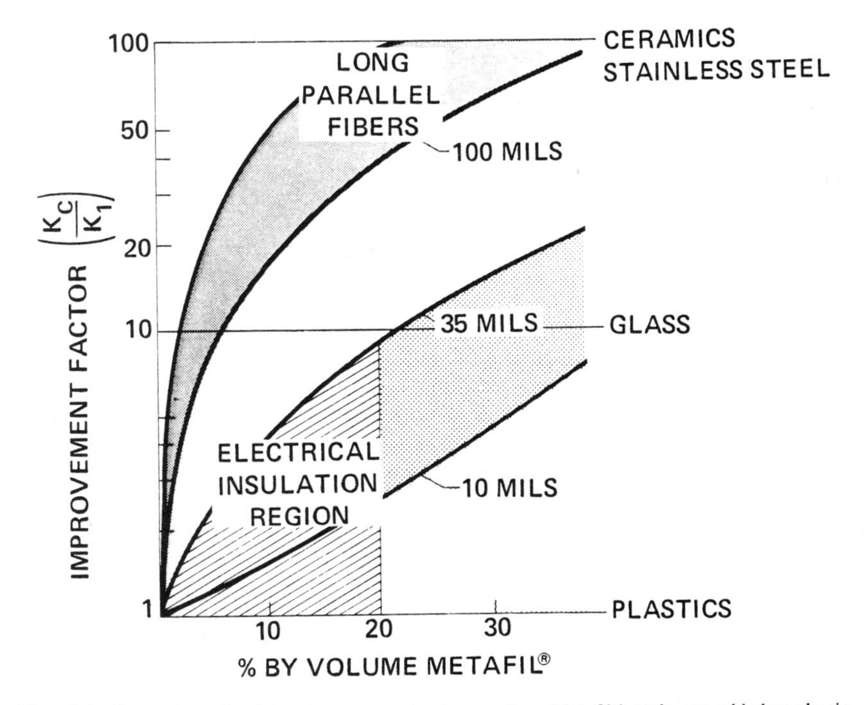

Fig. 6-6. Thermal conductivity improvement when various Metafil lengths are added to plastic.

electrical case, there is no critical concentration at which dramatic changes in properties occur; rather, the curves are all gradually changing functions of concentration. The fiber length does play a significant role up to lengths of 0.1 in. ($L/D = 100$); beyond that, the change is less dramatic.

The curves show that very great increases in thermal conductivity can be achieved with low fiber concentrations (less than 10% by volume), particularly for the longer fibers. The upper curve for long parallel fibers indicates that a two-order-of-magnitude improvement in thermal conductivity is possible in the direction parallel to the direction of the fibers, which can be very important for composites that use unidirectional lay-ups or fabrics.

It should be noted that Fig. 6-6 shows a unique region at low concentrations and short fiber lengths marked "Electrical Insulation Region." This region—which has no significant electrical conductivity, yet shows significant improvements in thermal conductivity—is to be expected because the electrical resistivity is not reduced by low concentrations of fibers with short fiber lengths. The region is not known to be as sharply or as exactly bounded as indicated in the chart because adequate experimental work has not been done. However, these boundaries are approximately correct and serve as a guide to developing plastics that have improved thermal conductivity but can also be used for electrical insulation.

6. MATERIALS AND SUPPLIERS

Three basic forms of materials are available. They are solid metallic fibers, metallic-coated glass fibers, and metallic flakes. The companies that supply these products are listed below with a description of their products.

Company Name	Product Name	Type Product
American Cyanamid Wayne, NJ		Nickel-coated graphite fiber
Bekeart Steel Wire Corporation Atlanta, GA (404) 451-6143	Bekinox	Bundle drawn S.S. filament 4–12 μm diameter, 1 mm to 3 mm length chopped fiber
Brunswick Corporation Skokie, IL (312) 470-4700	BRUNSMET	Bundle drawn S.S. filament 8–12 μm diameter
Handy & Harmon Newark, NJ		Silver flakes
Hexcel-Structural Products 11711 Dublin Blvd. Dublin, CO 94566 (415) 828-4200 *Also:* 3711 Long Beach Blvd. Suite 519 Long Beach, CA 90807 (213) 595-6311	Thorstrand	Metal-coated fabrics and prepregs
Lundy Technical Center P.O. Box 5280 Pomparno Beach, FL 33064	RO MHO Glass Conductive Fiber	Yarn roving, mat braided, milled, chopped aluminum-coated glass fiber

Company Name	Product Name	Type Product
Tracor MBA P.O. Box 1179 Lillington, NC 27546 919-893-2094 Tom Breslin	Metafil	Coated glass roving and chopped fiber, Aluminum ribbon
Meadowbrook inventions P.O. Box 360 Bernardville, NJ 07924 (201) 766-0371	MET-FLAKE MET-RIB MET-FIB	Metal and metal-coated plastic flakes, fiber, and ribbons
National Standard Company P.O. Box 1620 Corbin, KY 40701 (606) 528-2141 or 1-800-354-7844 Bob Meeks	N-S FIBREX (R) N-S MELTEX (R)	Nickel fibers, 10–40 μm dia. 50 to 100 *L/D* stainless steel fiber, 0.007 to 0.030 in chopped lengths
Novamet (An Inco Co.) 681 Lawlins Rd. Wyckott, NJ 07481 (201) 891-7976	Novamet	Nickel flakes and silver-coated nickel flakes
Transmet Corporation 4290 Perimeter Drive Columbus, OH 43228 (614) 276-5522 Robert M. Simon	Transmet	Flakes and fiber products

7. APPLICATIONS

The wide variety of applications for the use of short metallic fibers and flakes include the following:

EMI
Computers and their peripherals
Telecommunication equipment
High gain electronic equipment
Motors and relays
Automobile hoods
Cable shields

ESI
Electronic assemblies
Computers
Aircraft surfaces and interiors
Wrist watches
Ordnance manufacturing
Hospitals
Rugs
Fabrics
Clean surfaces

Electrical Conductivity
Heater elements/tapes
Bus bars (miniature)
Zebra strips
Electric junction boxes
Electrostatic spraying
Electroplating

Thermal Conductivity
Heat sinks
Solar collection panels
Heat exchanger
Electronic cabinets
Bearings and rings
Tank wheels and pads
Improved flame resistance

Production Assistance
Resistance or induction heating of plastic
Rapid heating of plastics
Rapid cooling of plastics

Miscellaneous
No-mar surfaces
Decorative effects

A number of applications are described in more detail in the following sections.

7.1 Aluminum Flake Solves Thermal Problems

In plastic applications that perform poorly because of heat degradation, an easily implemented solution is now available: thermally conductive composites made with aluminum flake modifiers.

Only flake-filled Nylon 6/6 and machined aluminum performed as effective heat sinks around the coils of one printer design, allowing the printer head to operate at top speed. Plastics filled with copper or 85% phosphorous bronze failed to disipate the heat.

Applications that require higher thermal conductivity such as electronic housings, lamp reflectors, coil bobbins, pulleys, motor housings, and other engineering or household items can be converted to use conductive composites.

The costly problems associated with graphite fiber-filled composites are eliminated with the cost-effective flake-filled materials that are currently available in the marketplace. These materials offer the following additional benefits:

- Standard plastic processing.
- Existing tooling.
- Reduced cycle time.
- Lower energy requirements.

Improved heat transfer properties coupled with cost savings are immediate benefits of using these materials in existing applications. The orientation of flakes in the plane of flow during molding provides the advantage of maximum conduction and convection heat transfer properties. A design capitalizing on the superior thermal conductivity of aluminum flake composites gives performance benefits once thought available only with metals.

Designers of heat sinks and other heat transfer systems have been limited in their choice of materials providing high thermal conductivity. Metals, particularly aluminum, were the most common materials and plastics (thought of as insulators) were seldom used. Plastic composites with a thermal effectiveness of 87% that of aluminum are opening new doors in the field of heat transfer design. Aluminum flake–modified composites provide a corrosion-free, lightweight, cost-effective alternative to metal.

Plastics with the proper loading of a high aspect ratio metal filler take on the thermal characteristics of the metal. Note the difference in aluminum flake-and powder-filled phenolic in Table 6-1.

The ability of flake-modified composites to conduct thermal energy gives designers the freedom to take advantage of cost-saving benefits in these applications:

- Pulley—improved wear.
- Electronic housings—elimination of fans.
- Lamp reflectors—reduced warpage.
- Coil bobbins—longer life.
- Motor housings—bearing stability.
- Heat exchangers—elimination of corrosion.

Designers who require the high thermal conductivity of metals and the flexibility of plastics processing are encouraged by new laboratory and field test results that demonstrate why flake-filled plastics are 87% as effective as aluminum.

Laser pulse and infrared technology have proved aluminum flake-filled conductive plastic to be a superior thermally conductive material.

A common heat-transfer situation in the electronics industry involves the mounting of a heat-producing power transistor or SCR on a thermally conductive material (heat sink) to re-

Table 6-1. Powder and flake comparisons.

	Specific gravity	Electrical resistivity (ohm·cm)	Thermal conductivity
Powder	1.85	$>10^6$	10×10^{-4}
Flake	1.85	10^{-1}	40×10^{-4}

Table 6-2. Heat sink with free convection.

Materials	Thermal conductivity, K, Btu/hr ft^2 °F/in.	Characteristics value $m = h/Kt$, in.$^{-1}$	Temp. rise ΔT, °F
Conductors:			
Aluminum	1536	0.06	20
Stainless (304)	113	0.24	27
Flake-modified resin	69	0.29	30
Semi-insulators:			
Marble	17	0.65	45
Glass	7	1.04	55
Insulators:			
ABS thermoplastic	1.30	2.1	71
Urethane foam	0.12	8.4	86

duce the peak temperature of the device. The heat is conducted away by the heat sink and is dissipated by convection. The characteristic value that describes the situation is $m = h/Kt$, where h = free convection film coefficient, K = thermal conductivity, and t = plaque thickness. Presented in Table 6-2 is a list of the K, m, and ΔT values for the case of a power transistor (dissipating 9.7 Btu/hr) mounted on a 4″ × 6″ × ⅛″ plaque of various materials in still air.

Table 6-2 can be broken down into the major categories of metal- and flake-filled plastics, mineral/inorganics, and organics/unmodified plastics. Note the small differences among the ΔT values of aluminum, stainless steel, and conductive plastics, compared with the large difference between the ΔT of these conductors and those of the unmodified plastic. The total amount of heat transfer by conduction (Q_{cd}) for aluminum was 83.8%. The corresponding flake-filled composite heat transfer was 73.2% of Q_{cd} or 87% that of aluminum. (See Table 6-3.)

When the temperature rise is plotted against time, the curves generated confirm a similar relationship for the four materials. (See Fig. 6-7.)

Table 6-3. ΔT, % of Q_{cd}.

	ΔT	% Q_{cd}	Q_{cd} normalized to Al
Aluminum	20	83.8%	100%
Flake-filled composite	30	73.2%	87%
ABS	71	21.3%	25%
Urethane foam	86	0%	0%

The improved thermal conductivity that aluminum flakes impart to any base resin is dependent not only on the flake's high aspect ratio and orientation in the plane of flow, but also on the volume loading of this filler. Figure 6-8 shows the time–temperature rise curves for aluminum, polypropylene, and composites with selected flake loadings.

Note that for a 20 C° temperature rise, the time required for the 25% flake-filled polypropylene is within 45 seconds of that of pure aluminum, while the time required for the unfilled polypropylene is seven times longer (5.5 minutes).

In addition to heat transfer, another important use of flake-filled resins is in heat exchangers, where corrosion resistance, good heat transfer, and low cost are important. Table 6-4 presents the heat exchange characteristics of a circular tube with the properties of three different materials in a forced air flow.

Heat flow rate is given by $Q = \pi \Delta T Z$, where ΔT = temperature difference between the internal temperature of the tube and forced air, and Z is a function of geometry, air flow, and conductivity of the tube. Note that at a low air velocity, the flake-modified composites are nearly as efficient as aluminum, and at an air velocity of 200 ft/sec, the composite (which has a thermal conductivity 2% of that of aluminum) still operates in excess of 81% of the efficiency of aluminum.

The above information is an example of some of the data that have been gathered during ex-

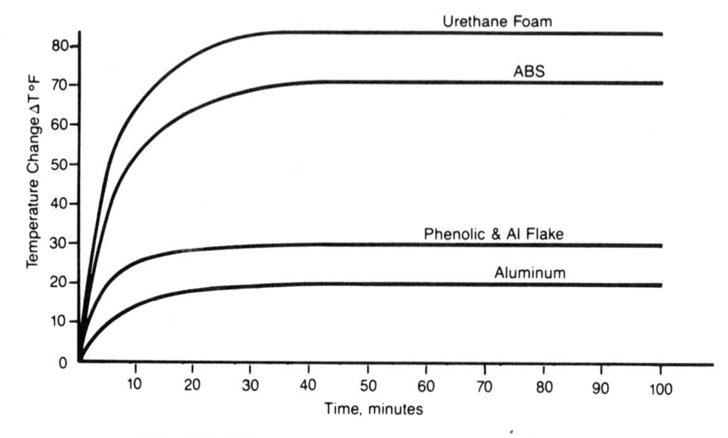

Fig. 6-7. Time-temperature rise, selected materials.

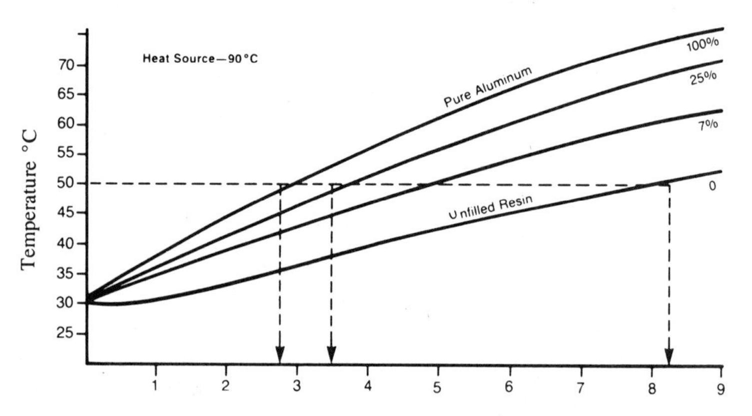

Fig. 6-8. Time-temperature rise, selected flake loadings (vol %).

Table 6-4. Heat exchanger example.

	MATERIAL		
	Aluminum	Composite	Base resin
Conductivity—(K) Btu/hr ft^2 °F/in.	1080	24.0	1.2
Tube OD (D_o) in.	0.2	0.2	0.2
Tube ID (D_i) in.	0.16	0.1	0.1
Wall thickness (*) in.	0.02	0.05	0.05
	Z VALUES:*		
Free-air velocity (ft/sec)	Aluminum	Composite	Base resin
1.2	(0.08) 1	0.99	0.78
4.8	(0.17) 1	0.98	0.63
16.4	(0.33) 1	0.95	0.47
65.0	(0.67) 1	0.89	0.30
200.0	(1.33) 1	0.81	0.18

*Z normalized to aluminum at each velocity

tensive testing of thermally conductive aluminum flake-filled composites.

7.2 Metallized Glass Fibers

The high cost of carbon/graphite fibers led to a search for less expensive conductive materials with high aspect ratios. A material already in use for radar chaff was the first to show promising results—metallized (aluminum) glass fibers. These fibers, slightly less than 0.001 in. in diameter, are available as continuous roving, chopped or milled fibers, and veil and woven fabric.

Metallized glass fibers generally handle the same way as E glass in terms of feeding and so forth. An oxide layer is formed on the aluminum surface. The effect of this layer on conductivity is a matter of debate, but it is clear that useful ESD and EMI shielding can be obtained in some applications. The metallized fibers are treated with a silane, although the silane does not appear to effect as strong an interfacial bond as silane-treated glass fibers do, and physical properties in some cases are lower. Tensile failure occurs, however, by fiber rupture rather than separation from the resin matrix. The coating apparently prevents additional oxidation. Table 6-5 gives physical properties of polycarbonate filled with 25% metallized glass fiber.

Table 6-5. Properties of polycarbonate filled with 25% metallized glass fiber.

	Solid	Foamed
Tensile strength, psi	12,000	9,000
Izod impact, ft-lb/in.	1.5	1.2
Flexural strength, psi	18,500	13,000
Flexural modulus, psi	900,000	700,000
Heat-distortion temp. at 264 psi, °F	290	285
Specific gravity	1.44	1.15
UL 95 rating solid 1/16 in.	V-O	V-O
foamed 1/4 in.	V-O	V-O

Source: Fiberfil.

Not suprisingly, metallized glass fibers are much more effective in shielding when compression-molded. Here 1- to 2-in. lengths used for SMC, BMC can be molded with minimal fiber damage, or continuous strands can be employed in such unidirectional types as XMC and UMC.

Tracor MBA reports that effective shielding in SMC can be accomplished with as little as 5% loading of 1-in. fibers.

Lundy Electronics & Systems has worked with lay-up systems for compression molding of polyesters. Using mat made from the company's metallized fibers at a 50% loading, shielding effectiveness of 70 to 80 dB over the 1- to 100-MHz frequency range was measured, and peaks to 120 dB were observed. Lundy has found that the best approach for optimum strength is a discretely layered system with glass, either sandwiching the metallized fiber mat between glass layers or putting the glass on the side that would experience tensile stress.

Injection molders of thermoplastics or BMC containing metallized glass fibers encounter three problems: breakage of fibers, difficulty in processing high loadings, and a fiber orientation that is not conducive to establishing optimum contact points between fibers. However, 2 to 5% loadings of fibers are sufficient in injection-molded parts for static bleed-off and probably for electrostatic painting. Clearly, the use of metallized glass for ESD would provide stronger composites than carbon black.

Fiberfil has applied its proprietary "long glass" process (in which continuous fiber is passed through a thermoplastic melt and the resultant strand is pelletized) to metallized glass fibers, producing pellets with continuous fibers up to 3/4 in. long. These pellets showed consistently higher shielding effectiveness in injection-molded plaques than those molded from conventionally prepared pellets with the same fiber concentration. Test panels were injection-molded in both solid and structural-foam versions and showed virtually identical shielding effectiveness for the two at 40% loading, the highest prepared. However, the level of shielding effectiveness is marginal for EDP usage. ESD performance was excellent at the lowest loading prepared—15 percent.[5]

Tracor MBA recently observed that injection molding and extrusion techniques break the fibers into lengths of 0.005 to 0.015 in. with their high shear action. Significant conductivity would require 35 to 40% loadings of such fibers. As noted in a recent paper: "Development work in progress has indicated that the critical level for conductivity can be reduced to 25 percent if steps are taken to increase the fiber lengths of 0.050 to 0.075 inch. More work is needed in this area to reduce the process to a standard manufacturing technique."[6]

8. GLOSSARY

Absorption—Amount of energy absorbed by a shield. Absorption depends on frequency of the radiation and the thickness, conductivity, and permeability of the shield.

Antistatic agents (antistats)—Additives, internally incorporated or sprayed onto a surface of plastic. Internal antistats migrate to the surface and attract moisture. This, in turn, provides a conductive layer of water to bleed off static charges.

Aspect ratio—Length to diameter ratio of a reinforcing fiber. The greater the aspect ratio is, the lower the loading required to achieve conductivity.

Attenuation—A reduction of the signal field strengths as a function of distance through materials. Commonly reported in decibel units

Conductive paints—Silver, nickel, copper, graphite, and copper graphite based paints used to coat nonconductive substrates for EMI/RFI shielding. Spray equipment or paddle guns are normally used to apply conductive paints.

Conductivity—Ability of a material to allow a current or charge to travel through it. Conductivity is the inverse of resistivity.

Critical concentration—The minimal percentage of an electrically conductive additive needed to change an insulating plastic to a conductive one. Also known as threshold concentration.

Decibel (dB)—Ten times the logarithm of the ratio of the field strengths.

Electromagnetic interference (EMI)—An electromagnetic energy that causes interference in the operation of electronic equipment.

Electrostatic discharge (ESD)—A large electrical potential (4000 volts or more) moving from one surface or substance to another. (ESD is also an abbreviation for electrostatic dissipation.)

Flame or arc spray shielding—A process of depositing molten metal (usually zinc) onto a nonconductive substrate with an arc or flame spray pistol.

Impedance—A measure of the total resistance of an object (through its width, length, and depth) to the flow of AC electricity. (A higher impedance reflects a higher resistance.)

Radio frequency interference (RFI)—The interference in electronic equipment caused by radio frequencies. These frequencies can range from 10 kHz to 1.0 GHz.

Reflection—The amount of electromagnetic energy reflected from the surface of a shield. Reflection depends on the impedance of the shield and the medium from which the signal originates.

Resistance—The ability of a material to oppose the flow of an electrical current; the reciprocal of conductance.

Shielding effectiveness—A measure of the effectiveness of an EMI/RFI shield based on the logarithmic ratio of the energy passing between a signal source and the receiver with and without the shield placed between them. Shielding effectiveness is expressed in decibels.

Static decay rate—The time required for a material to dissipate induced surface charges of static electricity. This specification is used to compare the ESD capabilities of various materials.

Surface resistivity—The ratio of DC voltage to the current that passes across the surface of a substrate. Surface resistivity, expressed in ohms / square is an indication of a material's conductivity.

Vacuum metallizing—A process in which metal, usually aluminum, is boiled in a vacuum chamber. The metal then condenses on the surface of a nonconductive substrate to form an EMI/RFI shield. The substrate surface is usually primed with a base paint.

Volume resistivity—A measurement of an object's ability to carry an electrical current through the bulk of the material. Volume resistivity is expressed in ohm-cm.

REFERENCES

1. Garland, J., "An Estimate of Contact and Continuity of Dispersion in Opaque Samples," *Trans. of Met. Soc. of AIME* 235: 642, May 1966.
2. Milewski, J. V., "Micropacking: Filling Resin More Efficiently," *Plastics Compounding* (1), May/June 1978; and "A Study of the Packing of Fibers and Spheres," Ph.D. thesis, Rutgers University, 1973.
3. Nielson, L. E., *Industrial Eng. Chem. Fund.* 13(1), 1975.
4. Briggs, D., *Polymer Eng. and Science* 15(12) Dec. 1977.
5. Woodham, G. W., "Metallized Glass: Specialized Reinforcement for Injection Molding," SPE 37th Annual *Proceedings*, 1979.
6. Davenport, D. E., "Metalloplastics—A Concept for Conductivity," SPE NATEC *Proceedings*, 1979.

7

Phosphate Fibers

M. M. Crutchfield
A. R. Henn
J. A. Hinkebein
B. F. Monzyk
Monsanto Company
St. Louis, Missouri

CONTENTS

1. INTRODUCTION

Phosphate fiber (PF) is a patented,[1] synthetic, crystalline, inorganic, polymeric, thermally stable, short fiber being developed by Monsanto Company as a new reinforcing fiber for composite materials.

The composition is approximated by the formula $2CaO \cdot Na_2O \cdot 3P_2O_5$. The empirical formula is $CaNa(PO_3)_3$. Single crystal X-ray diffraction studies[2] have shown that the crystalline unit cell contains the repeating unit

$$\left[-O-\overset{\overset{O}{\|}}{\underset{\underset{O^-}{|}}{P}}-O-\overset{\overset{O}{\|}}{\underset{\underset{O^-}{|}}{P}}-O-\overset{\overset{O}{\|}}{\underset{\underset{O^-}{|}}{P}}- \right]_{n\ (\text{large})} \quad Ca^{++}\ Na^{+}$$

which is a segment of a very long-chain, polyphosphate polyanion along with an associated Ca^{++} ion and a Na^{+} ion. These metaphosphate chains run parallel to each other in the direction of the long axis of the fiber (see Fig. 7-1). The calcium and sodium ions alternate in the interstitial spaces surrounded by four adjacent phosphate chains. Each Ca^{++} ion is associated with eight oxygens as nearest neighbors. Each Na^{+} ion is associated with six oxygens. Three of these oxygens are shared between the two cations.

Breaking the fibers in tensile requires breaking the strong covalent P–O–P bonds of the inorganic polymer. Tensile strengths measured on a large number of individual fibers using a micro tensile tester show attainable strengths of the order of 370,000 psi (2.6 GPa).

2. MANUFACTURING PROCESS

The product is obtainable from abundant, basic raw materials, such as limestone, soda ash, and

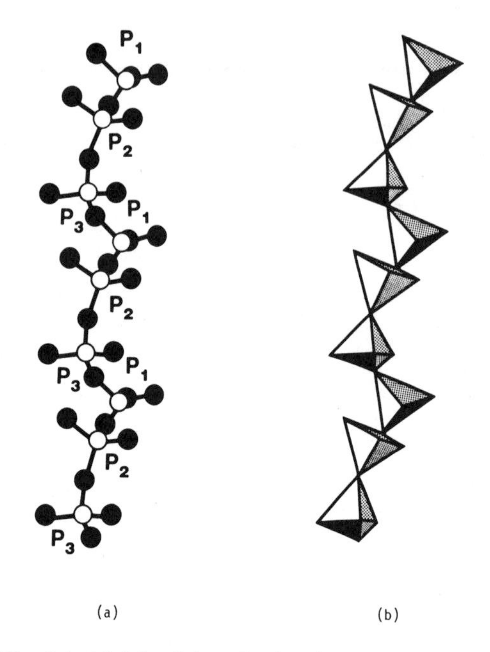

Fig. 7-1. Models of the polysphosphate backbone chain (from XRD studies[2]). (a) Ball and stick model. (b) Linked tetrahedrons.

phosphoric acid, that can supply CaO, Na_2O, and P_2O_5 on ignition. A proprietary thermal process[3] has been developed in which volatiles are driven off and crystalline $[CaNa(PO_3)_3]_n$ is obtained. The fibers are prepared by special milling and classifying techniques that preserve a high aspect ratio (length-to-diameter ratio). The fibrous crystals cleave more easily in the longitudinal direction than the transverse direction so that energetic processing can reduce fiber size without degrading aspect ratio.

3. PRODUCT FORM

3.1 Bulk Characteristics

PF is a soft, white, fibrillar material made by milling $[CaNa(PO_3)_3]_n$ crystals. The fibers can be wet-milled or dry-milled. Dry milling produces flexible fibers that are more or less rectangular prismatic (parallelepiped) rather than cylindrical in shape. Typical fibers are shown in Fig. 7-2 which is a scanning electron micro-

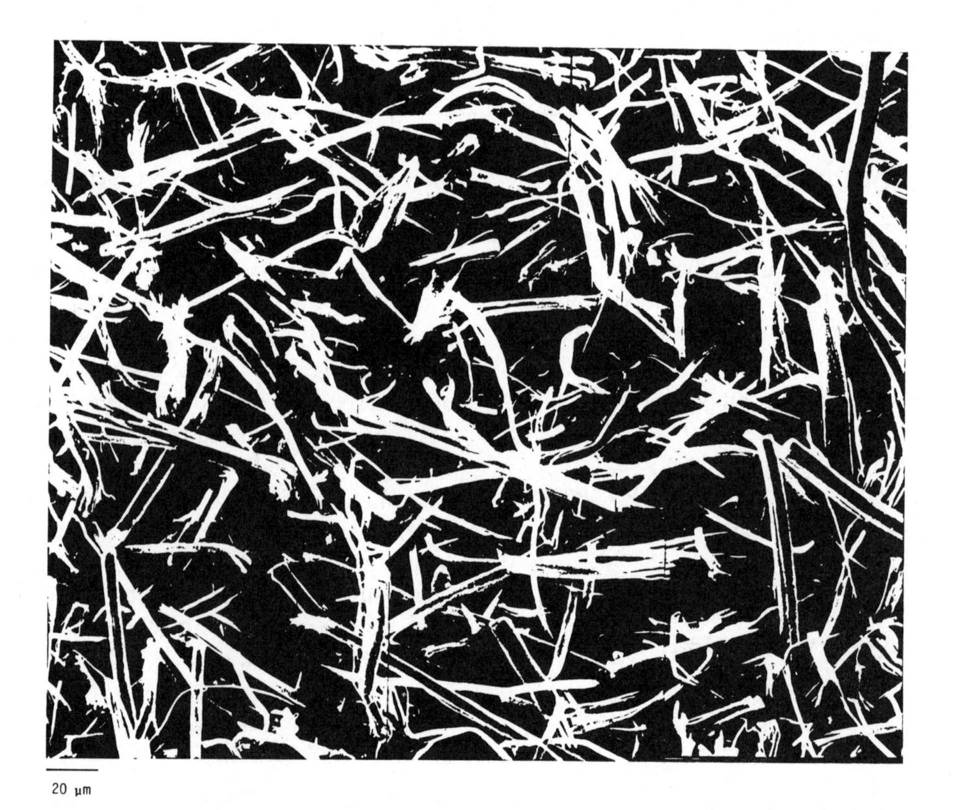

Fig. 7-2. Scanning electron micrograph of standard grade phosphate fiber (magnification—1000×).

graph of grade 001 (standard) fibers. The picture clearly shows the fibrous nature of the material. One should note the highly frayed and fibrillated ends of the fibers, which are exaggerated by wet milling. These assist the reinforcing and mat-making capabilities of PF, and permit it to be used in paper-making processes (see section 5.5). A wide range of particle sizes (see Fig. 7-3) is also evident in the picture. The smaller particles help to provide easy processing characteristics because they act as a flow-conditioning agent when blending the fibers with other solid components.

3.2 Packaging and Handling

PF is a soft, fluffy, white fibrous material. Its bulk density ranges from 5–10 lb/ft^3 loose to 15–25 lb/ft^3 packed. It is supplied loosely packed in fiber drums that contain 50 or 100 lb of fiber. Standard precautions for handling fibrous materials should be used.

3.3 Particle Size

The average fiber diameter of PF as measured by the Fisher Sub-sieve Sizer, which is based on an air permeability method, is 1 to 5 μm for grade 001 and 5 to 10 μm for grade 002. The median, mass-weighted Stokes' diameter of the smaller grade fibers falls in the range of 20 to 30 μm as determined by the Micromeritics Microsizer and the mass-weighted equivalent spherical diameter from electrozone measurements averages 11 μm. Well over 90% by weight of the small grade of PF possess an aerodynamic equivalent diameter greater than 3.5 μm. Particle size can be increased by less energetic milling if larger diameters are required.

Figure 7-3 shows the usual broad distribution of particle size typical of standard grade PF. The graph is a cumulative curve for the mass-weighted, electrozone equivalent spherical diameter. The particle size distribution is very broad because fiber diameter and length, both of which can span at least two orders of magnitude, together determine the equivalent spherical diameter.

The actual fiber dimensions have been measured from scanning electron micrographs. Fiber diameters range from less than 1 μm up to about 5 μm for the standard grade and from 3 to 30 μm for the larger grade, the averages being 2 to 3 and 13 to 15 μm, respectively. Fiber lengths vary from a few up to several

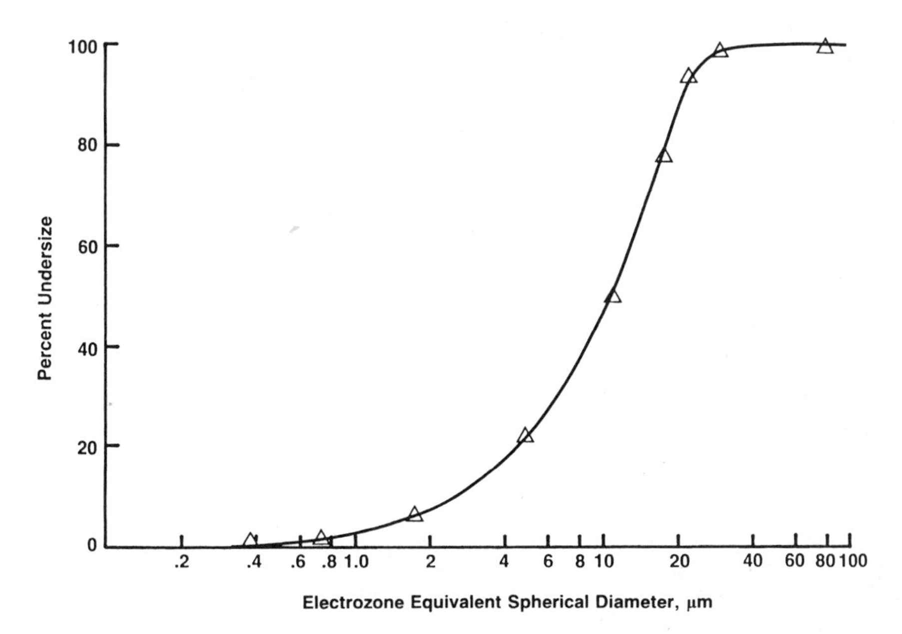

Fig. 7-3. Particle size distribution of standard grade phosphate fiber: cumulative curve for mass-weighted, electrozone equivalent spherical diameter.

hundred micrometers for both grades. Laboratory samples have been made with fiber lengths of several millimeters. The average actual aspect ratios are about 20 to 30.

3.4 Availability

Commercialization of phosphate fiber in 1987 is anticipated. For further information, contact G. D. Rawlings at Monsanto Chemical Co., 800 N. Lindbergh Blvd., St. Louis, MO 63167, or by phone at (314) 694-2925.

4. PROPERTIES OF PHOSPHATE FIBER

4.1 Chemical Properties

A typical chemical analysis of phosphate fiber is 71.39% P_2O_5, 10.74% Na_2O, 17.88% CaO versus a theoretical composition of 70.98% P_2O_5, 10.33 Na_2O, 18.69% CaO. Loss on ignition is typically 0.5 to 1.8% due to adsorbed moisture. Thermogravimetric analysis of dry fiber shows no loss of weight until about 950°C. This is well above the melting point of 749°C, which limits the thermal resistance or stability of the fiber. Figure 7-4 is a DTA/TGA curve for PF under an atmosphere of helium. Note the sharp melting point at 749°C, for which the heat of fusion is 54 ± 4 cal/g.

The dissolution rate of washed PF in water at room temperature is quite low, being attributable to slow surface hydrolysis with a half-life of many months. The initial pH of a 1 wt % slurry of PF in water is 3.6 ± 0.2. The pH will decrease slightly over time as hydrolysis slowly occurs. PF is subject to alkaline and acid hydrolysis at elevated temperatures.

PF is unreactive toward most common organic substances. Given its high melting point and lack of reactivity, PF can be used in a wide variety of matrices and environments as an essentially inert reinforcer-filler.

4.2 Physical Properties

Some relevant properties of PF are summarized in Table 7-1, in which property advantages are highlighted. Phosphate fiber possesses properties similar to those of other reinforcing fibers. It is soft yet maintains good strength and thermal stability. Phosphate fibers have a moderately high modulus but are sufficiently flexible to bend without breaking crosswise. The aver-

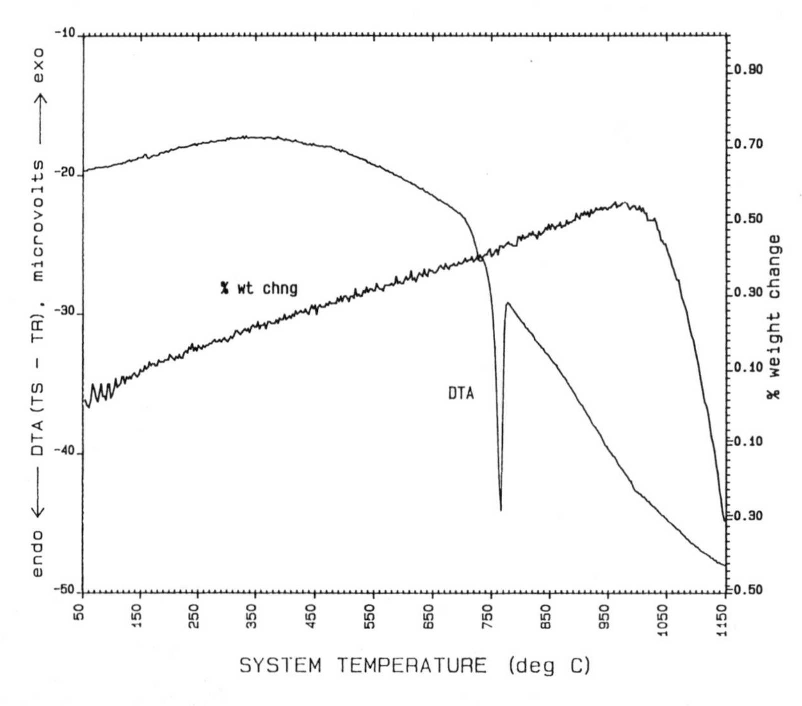

Fig. 7-4. DTA/TGA curve of phosphate fiber (He atmosphere). (Initial weight gain due to alumina crucible.)

Table 7-1. Physical properties of PF.

Property	Grade 001		Grade 002	Advantage
Dimensional				
Density, g/cm^3 (absolute)		2.86		Similar to existing mineral fibers, less than steel.
Fiber diameter, μm	0.5–5		3–30	Effective as a reinforcing fiber, without fiber entanglement problems.
Aspect ratio	~20		~20	
BET surface area, m^2/g	1–2		0.4–0.6	Allows good fiber-to-matrix contact.
Mechanical				
Tensile strength, Kpsi		370		Strong as glass.
GPa		2.55		
Tensile modulus, Mpsi		18		Higher modulus than glass.
GPa		124		
Thermal				
Stability		melts at >740° (1364°F)		Maintains integrity at higher temperature than many other fibers.
Heat capacity, cal/g − °K		0.19		
Linear expansion coeffs.				
Longitudinal, $10^{-6}/°C$		5		Similar to glass, imparts dimensional stability to composite.
Transverse, $10^{-6}/°C$		10–11		
Electrical				
Dielectric constant		8 (1 KHz)		Similar to asbestos,
Surface conductivity, 1/ohm		3×10^{-9} (18% RH)		
Optical				
Avg. refractive index		1.572		
UV/VIS/NIR reflectance	>80%	(300–2400 nm)		Not degraded by UV/VIS/NIR light.
Miscellaneous				
Oil absorption, g/100 g (ASTM D281-31)	365		175	Allows low fiber loading in liquid resins.
Mohs hardness		4.0 ± 0.3		Softer than steel (5–8) and glass (6.0), with less equipment wear.

age diameter and length of PF are large enough to limit dust problems, yet are small enough to provide a relatively large surface area for good fiber–matrix contact and bonding.

Although the actual aspect ratio of PF, as measured from scanning electron micrographs, is around 20 to 30, PF functions as if it possessed an effective aspect ratio[9] of more than twice that range. This is so because the highly fibrillated and frayed nature of the fibers reduces the bulk density relative to that of perfect cylinders of the same true density. The effective aspect ratio is high enough (>50) to provide good reinforcing characteristics without being so large that tangling and resulting processing difficulties are encountered.

In general, it is desirable for reinforcing fibers to possess small longitudinal coefficients of thermal expansion in order to impart dimensional stability to the reinforced matrix, the resin matrix usually having a coefficient of thermal expansion an order of magnitude greater than that of the fiber. The longitudinal coefficient of thermal expansion of PF is small enough ($5 \times 10^{-6}/°C$ at 25°C) to accomplish this, as its value is close to that of glass[4] and wollastonite.[5] The volume coefficient of thermal expansion at 25°C is estimated to be about $2.6 \times 10^{-5}/°C$.

The electrical conductivity of PF is very sensitive to the amount of moisture present in the sample, and can vary over a few orders of magnitude, depending on the moisture level, because the major mechanism of conduction is ionic. The value of 3×10^{-9} 1/ohm given in the table is the surface conductivity for a well-dried (18% RH) bulk crystal of $CaNa(PO_3)_3$. When the same crystal block is surface-saturated at 75% relative humidity, the surface conductivity increases by two orders of magnitude. Similar behavior is observed for ionic-type fillers[6] and fibers.

Note the relatively high oil absorption value for PF. PF absorbs more oil than many common fibers and fillers because of the numerous cracks and frayed ends on the individual phosphate fibers that are not present on other types of fibers. Oil absorption can be reduced by increasing fiber diameter or annealing the fiber. High oil absorption is an advantage when PF is used with thinner liquid resins, and it allows lower fiber loading in composites.

5. APPLICATIONS OF PF

The combination of desirable properties that PF possesses has led to the evaluation of PF in various composites and environments. Many of the physical properties of PF are quite similar to those of asbestos—in particular, its strength and fibrillated nature. Therefore, the replacement of asbestos by PF has been explored in detail. Reinforcement of phenolic resins for friction materials is an area that has been investigated to a significant extent with good promise[7]. Additional applications that appear feasible are in flooring mats and felts, both of which commonly contained asbestos in the past. Effort has also been directed at reinforcing plastics with PF. The evidence so far indicates that PF should be a novel alternative or addition to fiberglass. Moreover, PF can serve as a processing aid in conjunction with other longer fibers.

5.1 Plastics Reinforcement

An important application of PF is in the area of reinforced plastics. The presence of 15 wt % PF in a polyester thermoset clearly improves the mechanical properties of the resin, as is seen in Fig. 7-5 (S_T refers to tensile strength at break, E_B to flexural modulus, I_n to notched Izod impact strength, and $\%\epsilon$ to percent elongation at break.). PF shows utility as an alternative to fiberglass and as a processing aid for use with fiberglass. PF has processing advantages due to its ease of dispersion and relative softness. Some of the other processing advantages are good cure, warpage and shrinkage control; low minimum loading; and reduced resin runout. Furthermore, PF provides appearance benefits to the composites such as very smooth surface texture, low fiber visibility, and good translucence. Also of importance is the fact that PF tends to orient less than fiberglass under low shear fields. This results in more isotropic composite properties.

The resins for which the most data have been accumulated are unsaturated polyester/styrene thermosets (UPT). A suitable coupling agent for the PF–UPT system studies was Prosil-248 (methacrylate). All samples were cured for two hours at 175°F and then 325°F, and either cut from a sheet, molded individually, or cold-compression-molded individually, depending on the fiber loading. Figure 7-6 shows the effect of specimen preparation method on the mechanical properties of the composites. Cold compression molding gives the best results because it eliminates the problem of bubbling and the resulting voids that weaken the composite.

Figure 7-7 depicts the dependence of the flexural modulus of polyester thermoplastic composites on PF loading. All specimens were injection-molded to shape, and the PF was uncoated. The benefit of adding PF to this resin to increase the modulus is evident, and the values for the modulus compare favorably with other reinforced polyester thermoplastic com-

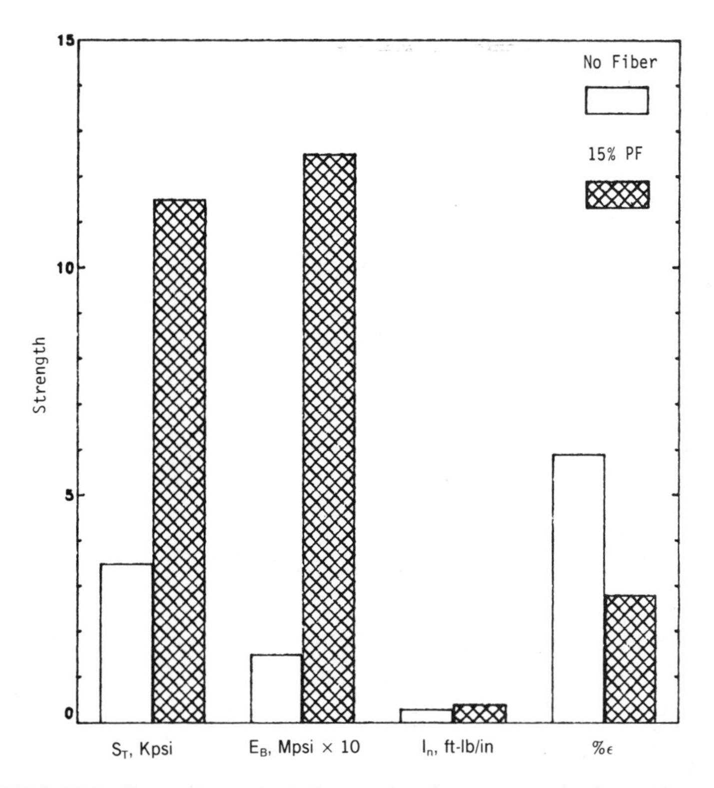

Fig. 7-5. Effect of 15% PF loading on the mechanical properties of an unsaturated polyester thermoset (compression-molded to shape).

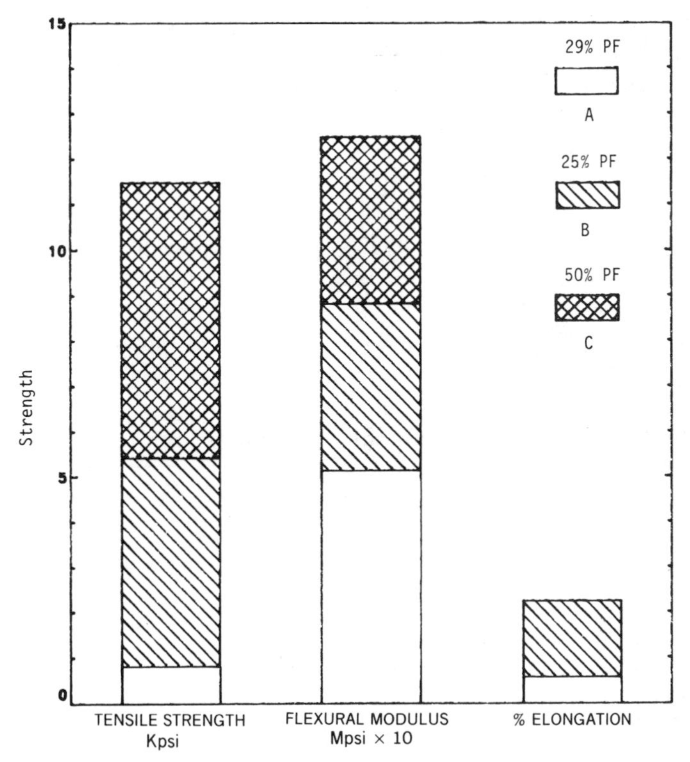

Fig. 7-6. Effect of specimen preparation method on mechanical properties of an unsaturated polyester thermoset: (A) cut from hot sheet; (B) molded to shape; (C) cold-compression-molded to shape.

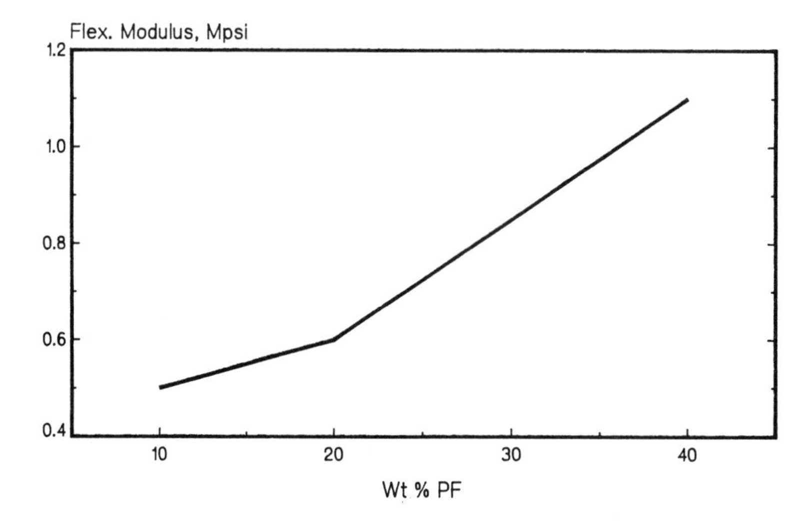

Fig. 7-7. Dependence of flexural modulus on loading of uncoated PF in polyester thermoplastic composites.

posites despite the lack of a coupling agent. PF also reinforces epoxy and phenolic thermoset resins as well as other thermoplastics such as nylon and ABS.

5.2 Phenolic Composite Hardboard

A class of resins for which PF provides excellent reinforcement is that of phenolic resins. This is notably evident in Table 7-2, which gives flexural test results and other data for phosphate fiber–reinforced, phenolic resin composites. The composites are reinforced with uncoupled fiber at the 80% level and were cured under 2650 psi and 340°F (171°C). A commercial cellulose hardboard made from wood fiber is included for comparison.

The composites containing PF are very strong and absorb significantly less water than the commercial cellulose hardboard, a characteristic that could prove important for use in moist environments.

5.3 Friction Materials

One of the major areas of expected application of PF is as a fibrous reinforcement in non-asbestos friction materials. Levels of up to 20 to 25% phosphate fiber have been used to replace 55% asbestos in traditional phenolic brake formulations. Strength, friction, and wear test results compare favorably with industry standards.[7] The high thermal stability, strength, and safety properties of the fiber are all important in this demanding application.

Table 7-3 compares the cured flexural

Table 7-2. Test data on phenolic composites reinforced with PF.

Fiber	Flexural strength, Kpsi	Flexural modulus, Mpsi	Density, g/cm^3	Specific modulus,[a] km	Water absorption, %
0% fiber, 100% resin	11.8	0.82	1.07	339	—
Wood fiberboard	4.0	0.36	1.14	222	25.9
PF (80%)	30.0	4.44	1.95	1600	—
PF (80%)	23.2	3.74	1.78	1480	6.32
PF (cured @ 50 psi)	6.2	1.34	1.26	748	8.84

[a] Specific modulus is modulus divided by density.

Table 7-3. Cured flexural strength of various phenolic brake composites.

Composition	Flexural Strength, Kpsi (ASTM D790-80)
55% Asbestos	10.0
55% PF	11.2
55% Barytes	5.8
55% Dolomite	5.4
45% Dolomite + 10% PF	7.0
30% Dolomite + 25% PF	9.4
45% Barytes + 10% PF	9.9
30% Barytes + 25% PF	11.9

strengths of various phenolic brake composites. Table 7-4 reports comparative data on brakes made with PF and with asbestos.

5.4 Sulfur Composites

As was found for phenolic composites, PF appears to possess the rare ability to reinforce sulfur quite well without coupling agents.[8] PF-reinforced composites of sulfur are superior to commonly reinforced sulfur composites with regard to flexural strength (Table 7-5). The first four entries of Table 7-5 are for fiber mats impregnated with sulfur, while the last entry is for composites made by stirring the fibers into molten sulfur. The composites are typically about 20 to 30% fiber by volume. Increasing fiber content increases strength and modulus in general. Sulfur modified by 5% cyclopentadiene is also reinforced by PF, its strength being increased by an order of magnitude.

Table 7-4. Test results on brake pieces containing PF or asbestos.

	25% PF	55% Asbestos
Cured flexural Strength, psi	12,900	10,000
Cured modulus, psi	3,900,000	5,600,000
Tensile strength in rivet direction, psi	2,200	—
Swell and Growth, % (SAE J160a)	0.44/0	0.75/0
Wear, %	11.9	10.1
Friction class (normal hot)	E E[a]	E E[a]

[a]Coefficient of friction, 0.25–0.35.

Table 7-5. Flexural data on sulfur composites reinforced with PF.

% Fiber volume	Flexural strength, Kpsi	Flexural modulus, Mpsi
16	8.6	1.40
19	7.2	1.59
21	11.0	1.56
25	16.6	1.80
23	6.6	1.51

5.5 Paper Products

The fibrillated ends of PF provide mechanical interlocking in mat form, which means that PF is a useful fiber for making thermally stable paper products. Because the phosphate fibers are relatively short, addition of a few percent of longer fibers (e.g., cellulose) is helpful in increasing the wet strength of mats and facilitates running on continuous papermaking equipment. A typical paper product containing 75% PF, 5% cellulose, and 20% styrene-butadiene rubber latex binder has shown good utility as a carrier backing for continuous sheet vinyl flooring.[10] Gasket-type materials are another possible application for PF.

6. HEALTH/SAFETY CONSIDERATIONS

Adverse health effects from respiration of durable fine fibers are well publicized. In this respect PF is expected to have a unique advantage because the polyphosphate backbone is known to be biodegradable. It is hypothesized that biodegradation by enzyme-assisted hydrolysis will facilitate dissolution of fine fibers *in vivo*, with the calcium, sodium, and inorganic phosphate being utilized in the normal biochemical mechanisms of cells. Extensive studies of fiber safety in animal experiments are under way. Preliminary results support the biodegradability hypothesis. Exposure to this fiber is not expected to produce significant adverse human health effects when recommended safety precautions are followed.

REFERENCES

1. Griffith, E. J., U.S. Patent 4,346,028, Aug. 1982.
2. Stultz, R. and Shieh, H.-S., Private communication of unpublished data, Monsanto Company.
3. Patent pending.
4. Mohr, J. G., in *Handbook of Fillers and Reinforcements for Plastics*, H. S. Katz and J. V. Milewski (eds.), Van Nostrand Reinhold Co., New York, 1978, Chapter 26.
5. Choate, L. W., in Ref. 4., Chapter 22.
6. Kummer, P. and Crow, G., in Ref. 4, Chapter 5.
7. Hinkebein, J. A., "Polyphosphate Fibers: Reinforcement of Friction Materials Composites with Calcium Sodium Metaphosphate," presented at The Society of Automotive Engineers 3rd Annual Colloquium on Brakes, Atlantic City, N.J., Oct. 21-23, 1985.
8. Hinkebein, J. A., U.S. Patent 4,484,950, Nov. 1984.
9. Milewski, J. V., "A Study of the Packing of Fibers and Spheres," Ph.D. thesis, Rutgers University, 1973.
10. Hinkebein, J. A. and Crutchfield, M. M., U.S. Patent 4,609,433, Sept. 1986.

8
Wollastonite

Joseph R. Copeland
NYCO
Division of Processed Minerals Inc.
Willsboro, New York

CONTENTS

1. INTRODUCTION

Wollastonite is an acicular particulate, naturally occurring, calcium metasilicate mineral. It provides excellent low-cost reinforcement for polymer systems. Wollastonite exhibits the general properties of low moisture adsorption, white color, excellent electrical insulation, low coefficient of thermal expansion, and good heat distortion.

The wollastonite mineral is well suited to chemical modification. Wollastonite is commercially available in surface-modified form; the prime modifiers are silane coupling agents, and wetting agents. The surface-modified grades are high-performance, cost-effective materials finding broad application in thermoplastics, thermosets, and elastomers, and as general replacements for asbestos and milled glass fibers.

2. MANUFACTURING PROCESS

2.1 Mining

In 1976, NYCO began development of a wollastonite deposit located near Lewis, New York. This deposit is proven and indicated reserves show it to be much larger than the Willsboro ore deposit. From 1976 to 1981 NYCO mined the Lewis deposit on a very small scale, primarily to assess its quality and economics. In 1982, after five years of increased activity at Lewis, NYCO closed its Willsboro underground mine and moved entirely to the open-pit Lewis operation for all its ore requirements. A primary crushing plant was built at Lewis in late 1982.

2.2 Beneficiation

Beneficiation begins with a conventional jaw crusher to reduce the run of mine ore to −5 cm. After being dried in an LP gas fired rotary kiln dryer, the rock is fed to a gyratory crusher in a closed circuit, with a 1-cm screen. Several hundred tons of 1-cm ore is stored as surge material. The ore is then fed from the surge bins to a series of 16-mesh screens.

All ore is crushed to a 16-mesh top size in order to liberate discreet wollastonite and garnet (impurity in the rock). A series of impact-type crushers is in a closed circuit with the 16-mesh screens. The entire crushing section is designed to produce a liberated feed material to the magnetic separators.

The heart of the beneficiation of wollastonite is dry, high-intensity, induced roll magnetic separators. These separators are successful in removing the magnetically susceptible garnet from the nonmagnetic wollastonite.

2.3 Milling

Beneficiated wollastonite is milled in two different systems: attrition milling to produce high aspect ratio products and pebble milling to produce products ground to a mesh size. Both systems employ air classification devices for product assurance. Finished products are packaged in bags, semi-bulk (one-ton super sacks, octainers, drums), and bulk (bulk trucks and railcars).

2.4 Chemical Modification

In 1981, NYCO constructed a large chemical modification plant to apply chemicals to the surface of its wollastonite products and other mineral products. That plant was expanded in 1982 and 1983, and now contains several dedicated and separate chemical modification circuits. Since each mineral has its own profile (i.e., particle size, surface area, chemical composition, shear sensitivity, etc.), and each coupling or wetting agent has its own profile (i.e., heat profile, reaction, viscosity, diluent requirement, etc.), a general chemical modification system cannot be used.

For the Wollastokup® product line, wollastonite is conveyed from the finished products holding tank to a pre-modification holding tank system in the chemical modification facility. The products are then conveyed to the proper circuit, coupled with chemicals, packaged, and shipped. Packaging is in 50-lb paper bags, drums, or octainers.

3. PROPERTIES

3.1 Chemical Analysis

The theoretical composition of calcium silicate is 48.3% CaO and 51.7% SiO_2. The wollastonite deposit at Willsboro and Lewis, Essex County, New York is the largest and purest in the world. The typical chemical analysis of the deposit is:

Component	Typical Weight %
CaO	47.00
SiO_2	50.00
Fe_2O_3	1.00
Al_2O_3	0.30
K_2O	0.10
MnO	0.10
MgO	0.30
TiO_2	0.05
P_2O_5	0.04
Moisture	0.20
Loss on ignition	0.20
Undermined	0.71

3.2 Physical Properties

Typical physical properties of the NYCO wollastonite deposit are:

Chemical formula	$CaSiO_3$
Appearance	White
Particle shape	Acicular
Molecular weight	116
Specific gravity	2.9
Refractive index	1.63
pH (10% slurry)	7.7
Mohs' hardness	4.5
Water solubility ($g/100\ cm^3$)	0.0095

Density (lb/solid gallon)	24.2
Bulking value (gal/lb)	0.0413
Melting point (°C)	1540
Transition point (°C)	1200
Coefficient of thermal expansion $(°C)^{-1}$	6.5×10^{-6}

3.3 Particle Size

Wollastonite has an acicular particle shape. Its aspect ratio is determined by the quality of the ore deposit and the method of processing. The NYCO material is the only wollastonite in the world that produces a high aspect ratio (20:1) product. As discussed in section 2.3, the high aspect ratio product is attrition-milled, while the mesh-sized products are pebble-milled. Figure 8-1 shows an electron photomicrograph of NYAD® G high aspect ratio. Figure 8-2 shows NYAD® 400, a 400 mesh product with an aspect ratio of 5. Figure 8-3 shows NYAD® 1250, a 10 μm top size product with an aspect ratio of 5.

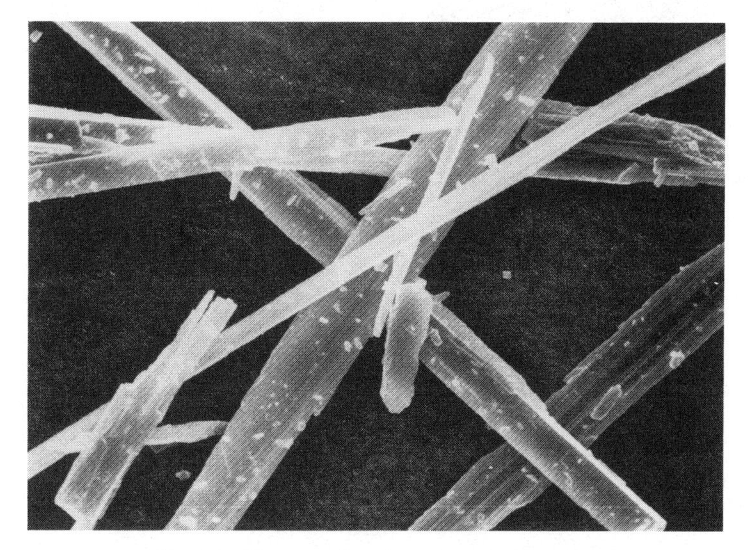

Fig. 8-1. Electron photomicrograph of NYAD-G.

Fig. 8-2. Electron photomicrograph of NYAD-400.

Fig. 8-3. Electron photomicrograph of NYAD-1250.

3.4 Toxicity, Safety and Health Data

Because wollastonite is an acicular mineral, it has come under the close scrutiny of various federal agencies as a possible health hazard. In 1976, NIOSH performed a comprehensive survey of the effects of wollastonite exposure on wollastonite mine and mill workers' health. There was no evidence to support any relationship between exposure to wollastonite dust and lung-related diseases or cancer. In 1982, NOISH did a five-year follow-up study and found essentially no change in the 1976 findings.

High aspect ratio wollastonite is an acicular particle with an aspect ratio of 20. Initial exposure to the high aspect ratio grades can produce a minor skin irritation comparable to that experienced in working with insulation materials. Continuous exposure will build an immunity; also, the initial use of a barrier cream will prevent this irritation.

Exposure to any dust is eventually damaging; so NYCO has spent considerable monies in upgrading and expanding the dust collection systems used in its plants. The use of a dust mask for work in a dusty area is highly recommended.

4. SUPPLIERS

Table 8-1A lists the three sources of domestic wollastonite. NYCO is the largest supplier in the world, with an annual capacity of 60,000 tons plus. NYCO produces a full complement of products: high aspect ratio, milled grades, and fine particle size (i.e., NYAD® 1250, a 10 μm top size, and NYAD® 475, a 100% −400 mesh product). R. T. Vanderbilt is the second largest domestic producer of wollastonite, with an annual capacity of 35,000 tons. All its products are milled grades. Pfizer's Wolcron products are local in sale, the primary outlet being in ceramics where the wollastonite is blended with Pfizer's California talcs.

Table 8-1A gives the producers, grades, current price schedule, and availability; Table 8-1B lists the distributors of wollastonite nationwide.

5. PRICE HISTORY AND AVAILABILITY

The price history of wollastonite reflects the general inflation rate of the seventies and early eighties. Pricing of NYCO products during those years was:

Grade	1970	1972	1974	1976	1980	1981	1982	1983
NYAD® 400	1.5¢	1.7¢	2.0¢	2.4¢	5.3¢	5.8¢	6.3¢	6.7¢
NYAD® 1250	—	—	—		(Introduced in 1982)		23.5¢	23.5¢
NYAD® G	1.9¢	2.1¢	2.6¢	3.2¢	6.0¢	6.7¢	7.7¢	8.2¢

Table 8-1A. Suppliers of wollastonite.

Company	Product Grades	Price	Availability
NYCO, Div. of Processed Minerals Inc. Mountain View Drive Willsboro, NY 12996 518-963-4262	NYAD® G NYAD® 400 NYAD® 475 NYAD® 1250 Wollastokups®	$214 per ton $170 per ton $360 per ton $500 per ton 25–50¢ per lb. FOB Willsboro, NY	Excellent
R. T. Vanderbilt Co. 30 Winfield St. Norwalk, CT 06855 203-853-1400	W-10 W-20 W-30	$100 per ton $122 per ton $146 per ton FOB Emeryville, NY	Excellent
Pfizer, Minerals, Pigments and Metals Div. 235 E. 42 St. New York, NY 10017 212-573-7217	Wolcron 200M Wolcron 325M	$111 per ton $128.50 per ton FOB Victorville, CA	Excellent

Table 8-1B. Distributors of wollastonite.

Cordano Chemical Co., Inc.
3322 NW 35th Ave.
Portland, OR 97210

Harwick Chemical Corp.
60 S. Seiberling St.
Akron, OH 44305-0360
216-798-9300

Harwick Chemical Corp.
725 Paramount Blvd.
Pico Rivera, CA 90660-3794
213-723-9911

Harwick Chemical Corp.
1485 Bayshore Blvd.
San Francisco, CA 94124
415-467-0505

Jensen Souders Associates Inc.
725 N. Baker Drive
Itasca, IL 60143
312-773-1830

D. H. Litter Co., Inc.
116 E. Sixteenth St.
New York, NY 10003
212-777-4410

D. H. Litter Co., Inc.
30 Lowell Junction Road
Ballardvale, MA 01810
617-475-5315

Majemac Enterprises Inc.
600 Bypass Road
Clearwater, FL 33518
813-797-8806

Peltz Rowley Chemical Co.
5700 Tacony St.
Philadelphia, PA 19135
215-537-1000

Ed Simal & Associates Inc.
3179 Maple Drive
Atlanta, GA 30305
404-231-3887

Van Waters & Rogers
P.O. Box 4579
Houston, TX 77210
713-644-1601

Walsh & Associates Inc.
1801 S. Hanley Road
St. Louis, MO 63144
314-781-2520

Walsh & Associates Inc.
500 Railroad Avenue
N. Kansas City, MO 64116
816-842-3014

Walsh & Associates Inc.
10190 Bannock St.
Suite 212
Denver, Colo. 80221

Walsh & Associates Inc.
3929 Senator St.
Memphis, TN 38118

Ore reserves for NYCO and R. T. Vanderbilt are vast, drilled reserves prove twenty five years. Processing facilities can easily be expanded if the marketplace dictates. The growth products for the plastics industry are: high aspect ratio grades, fine particle grades, and the chemically modified version of these grades. Expansion of facilities are planned for these products to meet future demand. The availability is excellent for all wollastonite products from all suppliers.

6. APPLICATIONS

6.1 General Discussion

Wollastonite products have been used in mineral-filled and mineral/glass-filled polymer systems for years. Many of these applications have been proprietary and continue to be confidential because of the unique properties contributed by wollastonite.

The addition of wollastonite and other minerals into a resin matrix is not entirely positive. Processing is affected by changes in viscosity, flow, and filler level; other than well-calculated loadings can lead to unacceptable embrittlement and loss of impact strength; the highly polar nature and high surface areas of minerals can adversely affect dispersion and wetout, the former leading to the molding of nonhomogeneous parts and the latter to attacks at the bond by moisture and other corrosives. One solution is to use surface-modified minerals.

Recent applications have been with surface-modified Wollastokup® products. The use of surface-modified minerals, in general, has several benefits. It improves wetout of the mineral in the polymer, resulting in deagglomeration, resistance to reagglomeration, and better flow; it reduces sensitivity to water (i.e., improves and retains wet electrical properties); and it improves bonding between the mineral and polymer, giving a general increase in composite strength. For many years companies have been utilizing coupling agents in situ, whereby the resin, mineral, and wetting agent are blended into the processing machine simultaneously. However, as loadings increase or multiple minerals are added, processing becomes more difficult, and end-use properties suffer. Only through pre-surface modification of the mineral(s) can the unique set of properties and requirements for each mineral be addressed.

The surface modification of wollastonite allows each product to be tailored to the individual polymer system.

6.2 Thermoplastics

6.2.1 Nylon. Wollastonite and Wollastokup® surface-modified wollastonite products are long-established reinforcement minerals for mineral-filled and mineral/glass-filled Nylon 6 and 6/6 products. G Wollastokup® 1100, a silane-modified high aspect ratio product, is most commonly used, while 10 Wollastokup® 1100, a silane-modified fine-particle-size product, is used in high impact or platable grades. Because of the excellent dispersion of 10 Wollastokup® 1100 into a nylon resin and the ability of acid to dissolve wollastonite particles, 10 Wollastokup® 1100 is used extensively in platable nylon compounds, offering excellent anchor pattern, surface, and smoothness of plating.

Table 8-2 gives properties in Nylon 6 and Nylon 6/6 resin systems of 35%, 40% Wollastokup®-filled Nylon 6, 40% Wollastokup®-filled Nylon 66, and 40% and 55% Wollastokup®/glass (25/15)-filled Nylon 66 systems. This table is based on data presented by Comalloy and Allied Chemical at SPE ANTEC meetings.

6.2.2 Polypropylene. Wollastonite has been evaluated in polypropylene over the years, and is commercially utilized with it in several compounds. The usage level is small, however, as wollastonite is added for stiffness and heat deflection temperature improvement. Recent developments in wetting and coupling agents show additional uses for wollastonite in polypropylene compound and polypropylene flake, to improve color, impact strength, and flow. This requires a trade-off of certain noncritical properties, which are exchanged for other, adequate properties while particularly desirable qualities are added. Table 8-3 is a property chart comparison of 25% 400 Wollastokup® TTS, surface-modified wollastonite; 25% Suzorite 200NP, surface-modified mica; and 20%

Table 2. Properties of WOLLASTOKUP Reinforced Nylons

Property (dry as molded)	Nylon 6 (std)	Nylon 6 40% 400 WOLLASTOKUP	Nylon 6 40% 10 WOLLASTOKUP	Nylon 6 40% 400 WOLLASTOKUP/glass
Tensile Str. at yield (psi)	11,600	13,300	13,000	21,000
Ultimate Elongation (%)	200	9	12	3
Flex Modulus (psi x 10^5)	4.0	9.5	9.0	13.0
Izod Impact, notched (ft. lb./in.)	1.0	0.7	1.2	1.0
Deflection Temp (°F) 66 psi	360	400	400	425
264 psi	155	270	270	400
Melt Point (°F)	420	420	420	420
Water absorption, 24 hrs., (%)	1.6	0.8	0.8	0.8
Specific Gravity	1.13	1.46	1.47	1.47
Mold Shrink (mils/in)	13	6	6	6
Coeff. of thermal Expansion (in/in/°F x 10^5)	4.6	2.0	2.0	1.7

Property (dry as molded)	Nylon 66 (std)	Nylon 66 40% G WOLLASTOKUP	Nylon 66 40% G WOLLASTOKUP/glass
Tensile Str. at yield (psi)	12,000	14,200	18,000
Ultimate Elongation (%)	60	3	3
Flex Modulus (psi x 10^5)	4.1	10.5	14.0
Izod Impact, notched (ft. lb./in.)	1.0	0.6	1.0
Deflection Temp. (°F) 66 psi	455	480	—
264 psi	195	445	490
Melt Point (°F)	490	490	490
Water Absorption (%)	1.2	0.6	0.6
Specific Gravity	1.14	1.50	1.50
Mold Shrink (mils/in)	15	8	—
Coeff. of thermal Expansion (in/in/°F x 10^5)	4.0	2.0	2.0

OCF 452 1/16 in. milled glass fibers. The outstanding property is the high impact strength. In these studies, 400-mesh wollastonite was used as the base. By going to a finer wollastonite base, 10 μm top size, and surface-modifying with the titanate, impact strengths of 13.3 ft-lb/in. were obtained for unnotched specimens and 0.7 ft-lb/in. for notched specimens.

Table 8-3. Comparison of surface-modified wollastonite, mica, and milled glass fibers on the properties of polypropylene flake.

Physical properties of test specimens	ASTM tests	Units	Unfilled PP flake	25% Surface-modified wollastonite	20% Glass fibers	25% Surface-modifed mica
Density	D792	g/cm^3	0.9	1.12	1.05	1.15
Tensile strength at yield	D638	psi	4400	4000	5200	4400
Elongation at yield	D638	%	12	8	3	5
Flexural modulus (secant at 0.1" deflection)	D790	psi × 10^{-5}	1.7	2.3	4.5	3.1
Flexural modulus (tangent)	D790	psi × 10^{-5}	—	3.2	5.6	4.3
Flexural strength	D790	psi	6900	6600	7400	7400
Heat deflection temperature	D648					
at 66 psi		°F	180	255	280	265
at 264 psi		°F	130	166	255	176
Izod impact strength	D256					
Notched		ft-lb/in.	0.8	0.7	1.0	0.7
Unnotched		ft-lb/in.	—	10.6	4.2	3.8
Compound cost	—	\$/in.^{3}	—	0.0170	0.0197	0.0181

The polypropylene used is Hercules 6329 Propylene Flake.

6.2.3 Polyethylene. The performance of Wollastokup® in HDPE compounds would be similar to that in polypropylene. Evaluations are under way testing Wollastokup®, surface-modified materials in these applications.

6.2.4 Engineering Plastics Group. Wollastokup®, surface-modified products are being used in polyphenylene sulfide (PPS). The Wollastokup® imparts excellent temperature resistance, compressive strength, and heat distortion temperature to PPS compounds.

Wollastokup® products are being evaluated in polyetherether ketone compounds. Table 8-4 shows work performed by Alpha Precision Plastics utilizing G Wollastokup in a PEEK compound.

Untreated, 400-mesh wollastonite is used as a filler in polysulfone compounds, providing a high heat distortion temperature and electrical insulation properties in an economical process.

Wollastokup® is being evaluated and has R&D approval for use in thermoplastic polyester compounds to control creep and warpage, give good surface appearance, and maintain general properties. No test results have been published.

6.3 Thermosets

6.3.1 Phenolic Molding Compounds. High aspect ratio wollastonite has long been used in asbestos-containing phenolic molding compounds. When asbestos was regulated, many compounders formulated a wollastonite-based phenolic compound. The high aspect ratio wollastonite product is used for:

- Low resin demand.
- Good wetout in phenolic resin.
- High loadings, good viscosity and flow.
- Fast mold cycle time.
- Reinforcement properties.
- Improved physical properties at elevated temperatures.
- Good electrical insulation properties.
- Economy.

Wollastokup® products are finding broad applications in heavy duty brake blocks, disc pads, and clutch backing compounds. Table 8-5 shows the improvement of Wollastokup® versus wollastonite in a nonmetallic rolled brake lining. In general, 12% to 15% dry weight of Wollastokup® is used. If the Wollastokup® is part of a complex fiber blend (i.e., Kevlar, glass fiber, carbon fibers, PMF® fiber, etc.), then the optimum level of Wollastokup® is 8% to 10% by weight of the blend. If metal

Table 8-4. Reinforced peek (polyetheretherketone) compound.[a]

Property	Unit	Value
Specific gravity	—	1.89
Tensile strength	psi	19,600
Tensile elongation	%	8
Flexural strength	psi	43,800
Flexural modulus	psi $\times 10^{-6}$	2.6
Impact strength	ft-lb/in.	2.0
Compressive strength	psi	27,600
Heat deflection temperature, °F at 264 psi	°F	640
Coefficient of linear expansion	in./in. · °F $\times 10^5$	0.9
Continuous use temperature	°F	600
Dielectric strength	volt/m	400

[a]Reinforcement = G Wollastokup® chemically modified wollastonite and PMF® fiber 204.

Table 8-5. Nonasbestos, nonmetallic rolled brake linings—strength retention: wollastonite vs. Wollastokup®.

Fixed Formula Components (parts by weight):

Kevlar pulp	3	Hexa	2
HRJ-1415 resin	9	HRJ-2447 (dry wt)	8
Pet coke	4	HRJ-2448 (dry wt)	8

		NYAD® G (42 parts)		G Wollastokup® 1108 (42 parts)	
Barytes		8	12	8	12
Densimix		8	—	8	—
Fluorspar		8	12	8	12
Tensile Strength:					
Room temp.	(psi)	1125	1160	1630	1960
	[MPa]	7.76	7.80	11.2	13.5
400°F	(psi)	560	540	1020	1155
	[MPa]	3.86	3.72	7.03	7.96

Note: Tensile loss in nonasbestos linings of all types is a major problem.

fibers are used, wollastonite is utilized at 2% by weight; the blending of metal fibers with inorganic minerals is formula-dependent.

6.3.2 Abrasives. Wollastonite can be blended with cryolite and aluminum oxide for use in phenolic base grinding wheels and abrasives. NYAD® 325 wollastonite, a 325 mesh grade, is being used commercially in blends with cryolite to maintain grinding wheel durability, improve flow, and increase tensile strength, and in blends with aluminum oxide to replace asbestos. Table 8-6 shows levels of Wollastokup® replacing aluminum oxide (24 grit abrasive) in a commercial grinding wheel binder system. The resin system is Schnectady Chemicals' SG 3110, HRJ1212 dry powdered phenolics and SG 3100, a liquid wetting resin.

6.3.3 Epoxy. Wollastonite and 400 Wollastokup® 187 are used in epoxy molding resins. The general property contributions are:

- High loadings, reduced resin usage.
- Improved resistance to thermal shock.
- Improved heat distortion temperature.
- Improved arc track resistance.
- Improved dielectric constant.

NYAD® 475 wollastonite, a 100% −400 mesh product and 10 μm Wollastokup® 187, a silane-modified 10 μm top size wollastonite, are used extensively in epoxy powder coatings because of their excellent dispersion, high loadings, smooth flow, hot and cold water resistance, good wet adhesion, and good gloss. Loading levels are normally 200 lb of Wollastokup® per 1000 lb of coating.

Wollastokup® is an excellent reinforcement for epoxy reaction injection molding. Information from Dow Chemical is shown in Table 8-7.

6.3.4 Unsaturated Polyester—BMC. The ability of BMC polyesters to be easily processed by injection and compression molding has made them logical competitors in many metals-dominated markets, but cost-effective formulating changes must be evaluated to heighten the competitive advantages of the

Table 8-6. Phenolic grinding wheel binder.

	Flexural Strengths, psi		
Resins Systems	Room temp.	250°C	Soak[b]
Standard system (SG3110/SG3100)	4838	2085	4110
HRJ 1212/SG3100 (190:50)	3638	2663	3158
5% G Wollastokup® 1100[a]	4200	2588	4470
10% G Wollastokup® 1100[a]	4838	2918	4013
15% G Wollastokup® 1100[a]	4665	3323	4673

[a] Substituted for 24 grit abrasive.
[b] 10 days, water, at room temperature.

Table 8-7. Epoxy RIM system.

Filler	Reinforcement loading, %	Flexural strength, psi	Flexural modulus, 10^6 psi	Tensile strength, psi	Elongation, %
None	0	17,000	1.47	—	—
Amino-modified Wollastokup®	23	18,000	0.65	8,500	5.5
Amino-modiifed Wollastokup®	50	21,300	1.35	9,800	1.7
Titanate-modified Wollastokup®	50	18,000	1.5	9,500	1.5
Wollastokup®/$\frac{1}{16}$ in. milled glass fiber	23	20,000	0.75	9,500	2.3

Formula = Dow D.E.H. 383, D.E.H. 39 hardener.

Table 8-8. BMC starting formula.

	Parts by wt	Wt %
Isophthalic resin	65	13.0
Low-profile additive	35	7.0
Catalyst	1.0	0.21
Inhibitor	0.2	0.04
Low-shrink additive	5.0	1.0
Release agent	4.25	0.85
Calcium carbonate	304.5	60.9
Clay	20	4.0

Take 87% of this composite and add 13% $\frac{1}{4}$ in. chopped glass fiber.

BMC's. Several studies evaluating partial glass replacement for glass fiber with both untreated and surface-modified wollastonite have been conducted.

A complete study for a typical general-purpose polyester BMC formulation containing 13% ¼ in. chopped glass, was performed. (See Table 8-8 for the starting formulation.) The equipment used included:

- A 300-lb. Day mixer wtih Sigma blades.
- A 425-ton Natco plunger injection press.
- A 300-ton Stokes screw injection press.
- A 50-ton Rodgers compression press.

Table 8-9. Effect of surface-modified wollastonite as a replacement for glass fibers in BMC systems.

Compound	Tensile strength, 10^3 psi	Flexural strength, 10^3 psi	Flexural modulus, 10^6 psi	Notched Izod impact, ft-lb/in.
Compression-molded				
Standard; all glass	8.16	16.9	2.23	4.9
G Wollastokup, 30%	9.46	15.9	2.19	4.02
10 Wollastokup, 30%	9.05	15.0	2.26	3.42
G Wollastokup, 50%	5.54	10.1	2.3	2.71
10 Wollastokup, 50%	6.31	12.6	2.20	2.68
Screw-injected				
Standard; all glass	6.41	11.9	2.3	1.07
G Wollastokup, 30%	5.89	10.60	2.23	1.06
10 Wollastokup, 30%	5.29	8.52	2.18	1.01
G Wollastokup, 50%	5.21	10.60	2.14	1.02
10 Wollastokup, 50%	5.20	9.06	2.2	0.89
Plunger-injected				
Standard; all glass	5.07	9.56	2.10	2.38
G Wollastokup, 30%	4.90	11.10	2.04	1.99
10 Wollastokup, 30%	6.91	13.10	2.19	2.18
G Wollastokup, 50%	4.95	11.20	2.18	1.69
10 Wollastokup, 50%	5.38	11.50	2.14	1.67

Two grades of wollastonite were used: 10 μm Wollastokup® 174, a 10 μm top size silane-modified wollastonite product, and G Wollastokup® 174, a high aspect ratio silane-modified wollastonite product. (Refer to Table 8-9 for all data.) In compression molding, the effect of glass replacement on physical properties varies with the property measured, the level of replacement, and the grade of wollastonite substituted. In general, tensile properties increased as the 30% replacement level while decreasing at the 50% level for both surface-modified wollastonite products. Flexural strength falls off slightly at 30%, drastically at 50%.

There are economic advantages in replacing chopped glass with both grades of Wollastokup®. The more glass fiber that is replaced, the lower the cost of the formulation:

Glass replacement level	Savings in price, ¢/lb
30% G Wollastokup®	1.4
30% 10 Wollastokup®	1.1
50% G Wollastokup®	2.3
50% 10 Wollastokup®	1.7

However, increasing glass replacement means more erosion of physical strength; so trade-offs must be made. In compression molding, if impact is not critical, 30% replacement of glass fiber with either wollastonite makes economic sense. In screw injection molding, 30% replacement of glass with G grade is cost-effective. In plunger injection molding, 10 grade substituted for glass at levels from 30% to 50% gives an economic incentive to making a formulation change. In these experiments, the 10 grade gave a smoother molded surface in all tests, masking resin-rich areas and fiber patterns.

6.3.5 Urethane Elastomers. Wollastokup® is used in castable urethanes to prevent hardpanning and settling. Castable urethanes have a history of settling and hardpanning when fillers are incorporated into the polyol side to increase physical properties. The incorporation of NYCAST™ U surface-modified wollastonite prevents hardpanning and shows an increase in impact strength, tensile strength, flexural

Table 8-10. G-Wollastokup in RRIM fascia.

Sample Description	% G-Wollastokup (1100-4)		Specific Gravity	Plaque Dimension (mm)		Flex. Modulus (psi × 10^3) Mean/Std. dev.		CLTE (m/m × 10^{-6}/°F)		Heat Sag (in.) 1 hr @ 250°F 6″ Overhang (⊥)/(∥)	Tensile Strength (psi × 10^3) Mean/Std. dev.		% Elongation Mean/Std. dev.	
	Calc.	Actual		(⊥)	(∥)	(⊥)	(∥)	(⊥)	(∥)		(⊥)	(∥)	(⊥)	(∥)
X	15	14.3	1.09	598.29	1053.15	32.3/.8	41.7/1.2	90	58	.76/.48	2.63/.05	2.66/.04	174/11	132/8
Y	17	16.2	1.10	598.65	1053.75	34.1/.9	46.6/.2	94	54	.57/.41	2.60/.02	2.70/.06	166/5	122/8
Z	20	19.0	1.10	599.02	1054.75	35.5/.7	46.8/.9	88	47	.70/.49	2.48/.04	2.65/.03	147/10	104/5

(⊥) = perpendicular to flow
(∥) = parallel to flow

strength, and flexural modulus. The surface is also improved.

Urethane RRIM also utilizes Wollastokup® as a reinforcement material. Typical data in a flexible RRIM urethane fascia are shown in Table 8-10.

7. FUTURE MARKETS

The surface modification of wollastonite and other minerals will open up new markets to mineral-reinforced polymer systems. Producers in the plastics industry expect to expand sales, not by developing new polymers but by modifying existing ones to improve their performance. As the price spreads between resin costs and minerals increase, there will be greater economic advantage in using fillers. New end-uses for mineral-reinforced plastics will develop where unfilled resins or other materials were previously used. Large end-markets for unfilled plastics, including toys, packaging, and housewares, could conceivably become significant applicants.

Excellent growth is seen for the surface-modified wollastonites in the thermoplastic and engineering resin areas.

9

Asbestos

John W. Axelson

Manville Corp.
Denver, Colorado

CONTENTS

1. INTRODUCTION

Asbestos—"the magic mineral," "the fiber with thousands of uses"—has been described in extravagant terms over the years, and with good reason. Chrysotile, the most common type, representing more than 95% of the world's asbestos production, has long been known for its low cost and excellent properties. Prices range from about 6¢/lb to about $1/lb for the better spinning grades. Chemically it is a hydrated magnesium silicate with the idealized formula $[Mg_6(OH)_4Si_2O_5]_2$. It is available in a variety of forms, including bulk fiber of various lengths and paper, yarn, cloth, and felt. It has excellent weathering characteristics and is resistant to most chemicals except acids and strong bases. In plastics, it improves creep resistance and heat deflection temperature, imparts flexural strength and modulus, lowers the coefficient of expansion, and controls resin flow properties. Its disadvantages, which often can be overcome, include low impact strength, dark color, some processing difficulties, and the need for better stabilization with some polymers.

However, the use of asbestos, especially the development of new uses for it, has been curtailed by the health problems associated with the mineral. These health problems are widespread and are most associated with improper use of the material. As a result of more stringent requirement in the U.S. many earlier unsafe applications are being discontinued. Special precautions are required for handling asbestos as it is being processed; so much special equipment has been developed by the asbestos industry, including bag openers, enclosed transfer units and processing equipment, and disposal facilities. Exposure of workers to asbestos can cause asbestosis, bronchial cancer, and mesothelioma. OSHA has set forth strict rules on asbestos exposure, with a threshold limit of two fibers per ml longer than 5 μm on an 8-hour time-weighted average. With proper equipment design and asbestos-handling equipment, this regulation is regularly met by the asbestos producing and using industry. So, even though new uses for it are not being developed, particularly in the plastics industry, a number of industries are using asbestos, and they probably will continue to do so. The development of replacements for asbestos has been very slow, and usually results in products that are inferior in performance and considerably higher in cost.

2. TYPES OF ASBESTOS

There are six types of asbestos, but only three have had any significant commercial use, a fourth having had limited use in a few areas. The most important fiber is chrysotile, which is the only member of the serpentine family. Comprising more than 95% of world asbestos production, it is found throughout the world and mined in many countries.

The other types of asbestos belong to the amphibole class, which contains five known fibers. The most important of these is crocidolite, the blue asbestos of commerce. It is noted for its high acid resistance and so finds use with plastics where acid resistance is required. However, its major use is in asbestos-cement products where it is useful as a drainage aid. Amosite, the second amphibole with significant tonnage, is used primarily in insulation products to give a porous structure with good insulating properties. Another type of amphibole, now used very little, is anthophyllite. It was mined in Finland and North Carolina, but these mines have ceased to operate. Anthophyllite was used extensively as a filler in polypropylene because it did not require significant changes in stabilization. The other two types of amphiboles, actinolite and tremolite, never have had any commercial use.

All types of asbestos are hydrated metal silicates, with various changes in the cationic portion. Idealized formulas are given in Table 9-1.

The classification of asbestos is by type and by fiber length. Although the Quebec Screen Classification grading system was developed for Canadian chrysotile, it, or some version of it, is the basis for the grading system of most asbestos sold commercially. The system was developed by the Quebec Asbestos Mining Association and consists of a dry screen analysis of a sample of asbestos. The classification is based on numbers running from 1 to 8 and an accompanying series of letters from A through Z. Many of possible letters are not used. In addition to these two designations, most manufacturers use a third suffix that generally denotes the type of processing the fiber has received and its degree of subdivision or openness. Each producer has a different type of designation for this suffix so it has to be identified separately.

Although asbestos fiber is produced by a screening operation, the fiber length is not discrete as it is with fibers that are produced by cutting specific lengths from continuous strands. Rather, each grade contains a spectrum of long and short fibers with a few long fibers in the shorter grades and many short fibers in the longer grades. Table 9-2 gives a Quebec Screen Analysis for a few typical grades of asbestos, and Table 9-3 shows the approximate length of the average fibers in typical grades.

3. PRODUCTION OF ASBESTOS

A naturally occurring mineral, asbestos is mined and processed by techniques common to the mining industry. Major mining is by open pit methods, but a considerable amount of asbestos is removed from the earth by underground techniques.

There are three modes of occurrence for asbestos fiber. The most common is the cross-vein where the fibers are essentially perpendic-

Table 9-1. Asbestos classes and types

Class	Type	Chemical Formula
Serpentine		
	Chrysotile	$Mg_6[(OH)_4Si_2O_5]_2$
Amphibole		
	Crocidolite	$Na_2MgFe_5[(OH)Si_4O_{11}]_2$
	Amosite	$MgFe_6[(OH)Si_4O_{11}]_2$
	Anthophyllite	$(Mg, Fe)_7[(OH)Si_4O_{11}]_2$
	Tremolite	$Ca_2Mg_5[(OH)Si_4O_{11}]_2$
	Actinolite	$Ca_2(Mg, Fe)_5[(OH)Si_4O_{11}]_2$

Table 9-2. Grading of asbestos.

Grade	QUEBEC SCREEN ANALYSIS, oz.			
	½-In.	4 Mesh	10 Mesh	Pan
3K	7.0	7.0	1.5	0.5
3T	2.0	8.0	4.0	2.0
4D	0.0	7.0	6.0	3.0
4T	0.0	2.0	10.0	4.0
5D	0.0	0.5	10.5	5.0
5R	0.0	0.0	10.0	6.0
6D	0.0	0.0	7.0	9.0
7D	0.0	0.0	5.0	11.0
7R	0.0	0.0	0.0	16.0

ular to the face rock. In the case of chrysotile, this rock is almost always serpentine and has the same chemical composition as the asbestos. However, it is not fibrous. The asbestos veins will range in width from a few that are up to 2 in., decreasing to many veins in the shorter range for group 7 fibers. The second type of occurrence is slip fiber, which also has veins interspersed within the serpentine rock structure but with the fibers deposited essentially parallel to the rock faces. This type of occurrence is not generally predominant, but it appears to some extent in most deposits and is quite predominant in a few. For a mine with cross-vein or slip fiber to be commercially attractive, it must have a minimum of 3–4% extractable fiber although a few mines may have areas with as much as 6% recoverable fiber.

The other mode of occurrence for asbestos is as a massive fiber having no specific fiber orientation. This form is quite rare for chrysotile, but is being mined commercially in several locations. The fiber is all short but occurs in a platy form as a massive deposit in which the asbestos is interspersed with the gangue rock and other material. Recovery of asbestos can be as much as 40–60% from these deposits.

Table 9-3. Approximate comparative lengths of asbestos grades.

Grade	Approximate Comparative Length, in.
2	5/8
3	1/2
4	3/16
5	1/8
6	1/16
7	1/32

Most asbestos is processed by dry milling techniques. Usually the asbestos deposit is blasted to produce boulders 6 ft or less in diameter. These are processed through primary crushers to reduce the size and then secondary crushers where the maximum size is further reduced to about 1 in. Somewhere in this size-reduction operation it is possible to separate the ore that is essentially non-asbestos-bearing from the ore that contains relatively large amounts of asbestos. As the ore comes from the mine, it contains a relatively large amount of free moisture. This is removed by dryers when the ore reaches the 1-in. maximum size stage. As it emerges from the dryer, it contains very fine ore particles, large ore particles, and a relatively large amount of the longer asbestos fiber. This ore is passed over horizontal screens where the small particles go through the screens; the large particles are collected from the ends of the screens; and the free fiber is vacuumed off the ends of the screens, having moved to the top surface of the ore bed because of its light, fluffy nature. The fiber goes into cyclones where it is collected and held for further processing. The undersize ore and oversize ore are fed to different crushing units where more fiber is released and collected by similar screening operations. Eventually, the ore has been processed so that no fiber with commercial value remains, and the tailings are transported to a refuse stockpile.

Fortunately, the fibers are removed from the ore with a somewhat automatic classification. The longer fibers tend to break loose first and appear predominantly from the first screens. As additional crushing and screening takes place, the fibers become shorter and shorter. When these fibers are first collected, they are subjected to further processing that may remove entrapped rock and other materials including shorter fibers. Eventually, the fibers from two or more lines may be blended to form the specific grade desired. And somewhere during the process, often as a final step, the fiber may be subjected to additional processing to separate the fiber bundles and produce a more open array.

Asbestos is generally packaged in 100-lb or 50-kg pressure-packed bags for health reasons, as well as ease of shipping and handling. The asbestos density will range from 50 to 60 lb/ft^3. Bags are made from paper, woven polypropylene, and polyethylene film. In the last few years some grades of asbestos have become available in 100 lb/ft^3 blocks that can be encased on a pallet and covered with a polyethylene film that is heat-shrunk.

In a few cases, wet processing of asbestos is also used, sometimes in conjunction with dry processing. One of the primary objectives of wet processing is to reduce health problems.

Both to reduce health problems and to aid in the use of these fibers, special treatments are applied at the mill. These include: surface treatments to improve drainage characteristics in the production of asbestos-cement products; treatments that will improve the thixotropic effects of asbestos in certain systems such as a polyester mix; and production of asbestos in pellet or flake form to reduce airborne fines during handling and generally to allow easier handling of asbestos.

4. CHEMICAL AND PHYSICAL PROPERTIES OF ASBESTOS

Inorganic silicates generally have good chemical, heat, and weathering resistance; and all types of asbestos, as a member of this class of mineral, have these properties. Chrysotile is a highly hydrated magnesium silicate with a structure that consists of alternate layers of silica tetrahedra and magnesium hydroxide. This structure accounts for the curved, hollow nature of asbestos fibers, and the size of the individual fibrils can be calculated at their actual size, 150 to 280 Å, based on the curvature required by the structure. The hollow interior has a diameter of 45 to 80 Å. Because of its basic nature, chrysotile reacts readily with strong acids and leaves a silica network behind. It is resistant to weak bases but is attacked by strong caustic solutions at elevated temperatures. Table 9-4 gives the solubility of the three commercial grades of asbestos in common acids and sodium hydroxide. There is no known listing of the resistance of asbestos to various solvents and chemicals, but its use in thousands of diversified products would indicate excellent resistance in most cases.

One of the most important attributes of asbestos is its resistance to weathering, which is the reason for many of its applications, from asbestos-cement products to sealing compounds and plastics. Asbestos is best known for its temperature resistance. Actually, the 1–2% absorbed water that it contains is driven off at temperatures of 130°C. This is a completely reversible reaction; the absorbed water is regained immediately on exposure to a normal atmosphere at normal temperature. Above 200°C the hydroxyl portion of asbestos is slowly evolved, a reaction that becomes quite rapid for chrysotile at 600°C. At about 810°C, a phase change begins, and chrysotile changes to the mineral fosterite, with fusion taking place at 1520°C. The effects of temperature on the weight loss of the three main types of asbestos are shown in Table 9-5, and the loss in tensile strength at various temperatures after 3 minutes for chrysotile is shown in Table 9-6.

The amount of absorbed water associated with asbestos under normal atmospheric con-

Table 9-4. Solubility of asbestos. Percent weight loss, refluxing 2 hr in 25% acid or caustic.

Fiber	ACID OR CAUSTIC				
Type	HCl	CH_3COOH	H_3PO_4	H_2SO_4	NaOH
Chrysotile	55.7	23.4	55.2	55.7	1.0
Crocidolite	4.4	0.9	4.4	3.7	1.3
Amosite	12.8	2.6	11.7	11.3	7.0
Anthophyllite	2.7	0.6	3.2	2.7	1.2
Tremolite	4.8	2.0	5.0	4.6	1.8
Actinolite	20.3	12.3	20.2	20.4	9.2

Table 9-5. Percent weight loss of asbestos after 2 hr at each temperature.

Temp. °C	Chrysotile	Crocidolite	Amosite	Anthophyllite	Tremolite
205	0.3	0.1	0.2	0.1	0.0
315	0.9	0.3	0.6	0.2	0.1
370	1.8	0.5	0.8	0.3	0.1
425	2.2	0.7	1.0	0.4	0.2
480	2.8	0.8	1.1	0.4	0.3
540	4.0	0.9	1.2	0.4	0.3
595	10.4	1.0	1.4	0.5	0.4
650	12.8	1.0	1.4	0.5	0.4
760	13.5	1.0	1.4	0.5	0.5

ditions, which is usually in the range of 1% to 3%, does not seem to have any effect on its use in plastics or other systems. When chrysotile is wet by water, a caustic solution is developed that has corrosive properties, especially with aluminum or other materials that cannot withstand caustic solutions.

Common physical properties of the three commercial types of asbestos are given in Table 9-7. Contrary to many reports, asbestos and especially chrysotile are essentially soft minerals that do not cause excessive wear on production equipment. In some cases, magnetite, quartz, or other hard materials could be present and cause wear, but usually these minerals are quite fine and act like a polishing powder similar to finely ground alumina. Both chrysotile and crocidolite have the advantage of being quite flexible so they can be processed with a minimum of fiber length deterioration. Often it appears that the fiber length has deteriorated, but actually the fiber bundles have opened to such a degree that the individual fibrils are not visible. At high magnification, as with an electron microscope, the true fiber length/diameter or aspect ratio can be seen to be more than enough for good reinforcement.

Table 9-6. Tensile strength of chrysotile after 3 minutes at each temperature.

Temperature	Tensile Strength, psi	% Loss in Strength
Room	131,000	–
315 °C	120,000	8.4
425 °C	96,000	26.7
540 °C	78,000	40.5
650 °C	42,000	68.0

There are many forms of chrysotile, and sometimes crocidolite, available for use in plastics. The strongest reinforced parts are obtained with asbestos paper or cloth that has been impregnated with resin and then formed and cured into the final product. Up to 70% asbestos reinforcement can be obtained in this way. Impregnation of asbestos paper can be somewhat difficult because of its nonporous nature, but higher-porosity papers for resin saturation are available. Instead of paper, a dry felt of asbestos can be used, which will ensure easy saturation but poorer handleability. Yarns made from asbestos are also available, but are usually utilized only after being woven into tape or cloth. Physical properties of a few asbestos laminates with various resins are given in Table 9-8.

Most asbestos used in the plastics industry is incorporated in bulk form. Grades ranged from small amounts in the relatively long group 4 to large quantities in the group 7. Many of these uses have disappeared in the last few years because of the problems associated with asbestos and health, but two markets are still strong.

One use is a true asbestos-plastic product in brake linings, brake blocks, and transmission linings. Most of the products are made by the premix process where the asbestos is premixed with the dry or wet phenolic resin plus other ingredients and shaped into the final product either in a mold at elevated temperature and pressure or after extrusion and cutting in a confined cavity at a high temperature. Even after

Table 9-7. Physical properties of asbestos.

Property	Chrysotile	Crocidolite	Amosite	Anthophyllite
Color	White to gray	Blue	Brown	Brown to gray
Tensile strength, psi	300,000	500,000	160,000	350,000
Modulus of elasticity, 10^6 psi	23.2	27.1	23.6	22.5
Hardness, Mohs	2.5–4.0	4.0	5.5–6.0	5.5–6.0
Flexibility	Good	Fair	Poor	Poor
Specific gravity	2.4–2.6	3.2–3.3	3.1–3.2	2.9–3.2
Specific heat, Btu/lb. °F	0.266	0.210	0.193	0.210
pH	10.3	9.1	9.1	9.4
Refractive index	1.50–1.55	1.70	1.64	1.61
Fibril, diameter, Å	160–300	600–900	600–900	600–900
Surface area, BET m^2/g	17–60	9–10.5	8–9	6–7
Coefficient of cubical expansion, $(°F)^{-1}$	5×10^{-5}	—	—	—
Charge in water	Positive	Negative	Negative	Negative
Isoelectric point	11.3–11.8	—	—	—

molding in a press, the parts are usually given a long-term oven bake to fully cure the resin. Then they are ground or cut to final dimensions, and any grooves or bolt holes are machined. Although some inroads have been made in this business by ceramic and metal systems, these systems are all considerably more expensive and generally do not meet the performance requirements.

Another use of asbestos with plastics is its incorporation in an asbestos paper with a rubber latex binder. This product is produced on standard paper-making equipment, usually a Foudrinier paper machine, and the product is used as the underlayment with a vinyl coating of various types as roll-type vinyl flooring. This use was very extensive, but inroads have been made by a fiberglass mat.

5. SUPPLIERS

The number of suppliers of asbestos has decreased in recent years because of the increase in mining costs and the decrease in consumption caused by health problems; so marginal mines have become nonprofitable. There have been some sales in the asbestos mining industry and some consolidation of several companies into one. A list of the major U.S. producers is given in Table 9-9.

6. PRICE RANGE AND HISTORY

Historically asbestos has been sold on an FOB basis with customers paying the shipping charges. These charges vary with location, type of shipment, size of shipment, and whether or not a commodity rate has been established between the two points of shipment. Over the years these shipping charges have increased quite regularly, and they now range from 2.0 to 3.5¢/lb.

The FOB selling price of asbestos is related to length. All the major producers sell at approximately the same price. The selling price for asbestos fiber was relatively stable through

Table 9-8. Mechanical properties of pressure laminates reinforced with asbestos.

Laminate	Tensile strength psi $\times 10^{-3}$	Flexural strength psi $\times 10^{-3}$	Tensile modulus psi $\times 10^{-6}$
Phenolic resin/crocidolite felt	28–31	57–58	3.25
Epoxy resin/crocidolite felt	20–22	42–43	2.80
Polyester resin/crocidolite felt	24–25	44–45	3.06
Melamine–formaldehyde/chrysotile fabric or paper	6–12	12–24	1.6–2.2

Table 9-9. Major asbestos producers.

Company	Address	Type of Asbestos
Asbestos Corp. Ltd.	Thetford Mines, Quebec	All grades of chrysotile
Bell Asbestos Mines, Ltd.	P.O. Box 99 Thetford Mines, Quebec	All grades of chrysotile
Johns-Manville Asbestos, Inc.	2000 Peel Street Montreal, P.O.Q. Canada H3A ZW5	All grades of chrysotile
Carey Canadian Mines, Ltd.	East Broughton, Quebec	All grades of chrysotile
Lake Asbestos of Quebec	Black Lake, Quebec	All grades of chrysotile
Cassiar Asbestos Corp., Ltd.	85 Richmond St. W Toronto, Ontario	Longer grades of chrysotile
Union Carbide Corp.	Mining and Metals Division 270 Park Ave. New York, NY 10017	Short "Calidra" chrysotile
Hedman Mines, Ltd.	P.O. Box 590 Timmins, Ontario	Short chrysotile
Special Asbestos Co.	3628 W. Pierce St. Milwaukee, WI 53215	All grades of crocidolite and amosite
Huxley Development Corp.	Time-Life Building Rockefeller Center New York, NY 10020	Short anthophyllite
Vermont Asbestos Group	Hyde Park, VT	All grades of chrysotile

the 1960s, but then prices increased as a result of higher mining costs and greater demand. By 1975, prices had essentially doubled. Prices continued to increase but at a reduced rate because of a decrease in demand. The list price in 1986 for a short grade of asbestos used as a filler for plastics was about 9¢/lb for Canadian fiber, compared to about 4¢/lb in 1974. Production of asbestos has been substantially reduced in the past few years because of the health problems associated with its use. This decline has been accompanied by price increases of about 8% per year since 1975, has practically eliminated development of new mines, and has encouraged the closing of marginal operations. Instead of the worldwide shortage anticipated a few years ago, there is now an excessive capacity with no anticipated shortage in the near future. Typical prices for various grades of asbestos in 1983 are given in Table 9-10.

Table 9-10. Asbestos fiber prices FOB mine location, January 1, 1983.

Grade	Price range, $ per metric ton
3	1450–1500
4	1080–2000
5	600–850
6	525
7	150–325

7. PROCESSING ASBESTOS WITH PLASTICS

Asbestos is a bulky fiber, and when it is dry-blended with a resin powder, the resultant mix has poor flow properties and a low bulk density. With thermoplastic resins, it is common practice to go through a preform procedure rather than use a dry mix directly. The original mix may be fed through an extruder and pel-

Table 9-11. Asbestos as a reinforcing fiber or filler.

Advantages	Disadvantages
High modulus	Lower impact strength
Good flexural strength	Increase in resin viscosity
Improved tensile strength	High density
Low thermal coefficient	Dark color
Improved heat deflection	Processing problems
Dimensional stability	PVC and PP heat stability
Heat resistance	
Low water absorption	
Easy flash removal	
Low cost	
Flow control in mold	
Minimum fiber degradation	
Good surface finish	
Chemical resistance	

letized or flaked to produce a feed for further molding or extrusion. A Banbury or other high-intensity mixer may be used for mixing to give a product that can be reduced to flakes for further processing.

Another technique, used with thermoset resins, is to mix the resin, asbestos, and solvent, if required, in any common high-shear mixer and extrude the mix into a preform. These mixes are then dried, pressure-molded, and baked to form the final product. The preform may also consist of dried flakes for molding.

When used in plastics, asbestos has some distinct advantages and disadvantages, which Table 9-11 shows in decreasing order of importance so those at the bottom of the list have relatively insignificant effects. The major ad-

Table 9-12. Physical properties of asbestos-filled plastics.

ASBESTOS WITH THERMOPLASTICS

% Asbestos	Flexural Strength psi	Flexural Modulus psi $\times 10^{-5}$	Tensile Strength psi $\times 10^{-3}$	Notched Izod ft lb/in.	Heat Distortion °F (264 psi)
		Polyethylene			
0	4170	1.54	3260	0.9	120
20	4560	2.82	2440	0.4	131
30	5840	4.21	3380	0.6	157
40	5440	6.44	3900	0.9	192
		ABS			
40–Long fiber J-M PLASTIBEST No. 20	10650	0.54	5855	1.42	190
40–Short fiber JM7D04	8765	1.11	4715	0.52	202
		Polystyrene			
0	5195	2.91	2585	1.42	179
40–Long fiber J-M.PLASTIBEST No. 20	5180	6.24	3570	1.09	201
40–Short fiber J-M7D04	7410	5.97	2500	0.56	194
		Polypropylene			
27-Short asbestos UC RG144	10400	4.60	5900		187
27-Short asbestos UC RG600[a]	12900	9.92	7300		253
		Nylon			
0	12000	2.90	8400	1.31	165
15-RG144	22000	7.54	12400	0.94	304
15-RG600[a]	24800	8.34	14000	1.01	327

Table 9-12. (*Continued*)

ASBESTOS WITH THERMOPLASTICS

% Asbestos	Flexural Strength psi	Flexural Modulus psi × 10^{-5}	Tensile Strength psi × 10^{-3}	Notched Izod ft lb/in.	Heat Distortion °F (264 psi)
		Asbestos with Thermosets **Phenolics**			
0	8900	7.8	—	0.28	—
40	13500	14.4	—	0.20	—
60	12900	20.3	—	0.39	—

[a]RG600 has a special coupling treatment.

vantages are increases in modulus and flexural strengths and heat deflection temperature. The major disadvantage, especially with thermoplastics, is a loss of impact strength and increasing process problems. Typical properties of some asbestos-resin products are given in Table 9-12.

8. SAFETY AND HEALTH AND THE FUTURE USE OF ASBESTOS IN PLASTICS

As noted, asbestos is a hazardous material, known to cause various types of cancer and asbestosis under certain conditions. Medical evidence shows that the health problems associated with exposure to airborne asbestos generally occur after long exposure to excessive amounts of asbestos. It has been shown that smokers definitely increase their susceptibility to cancer problems. This evidence comes from epidemiological studies of people exposed to asbestos many years ago and from animal studies utilizing accredited tests and exposure.

A standard for "Exposure to Asbestos Dust" was published in the *Federal Register*, Volume 37, No. 110, on June 7, 1972. It set the present OSHA threshold limit for airborne concentrations of asbestos in the workplace. This threshold limit was placed at 2 fibers/ml longer than 5 μm time-weighted average or 10 fibers/ml longer than 5 μm ceiling concentration, both measured by membrane filter/personal air sampler technique. These levels are being readily attained by producers and maintained in the asbestos industry with good housekeeping and proper dust control practices.

Although asbestos use is expected to continue at its present level in the plastics industry, no growth or new uses are expected. Other new uses are also expected to be very limited although growth in some of these areas may continue at the rate of the expanding economy. For any users or potential users of asbestos, the asbestos industry will assist in providing information and engineering assistance on proper handling for the safe use of the substance.

Papers

1. Axelson, J. W. and Piret, E. L., "Crushing of Single Particles of Crystalline Quartz," 42: 665–670, Apr. 1950.
2. Johnson, J. F., Axelson, J. W., and Piret, E. L., "Energy–New Surface Relationship in the Crushing of Solids," *Chem. Eng. Prog.* 45(12): 708–715, Dec. 1949.
3. Axelson, J. W. et al., "Basic Laboratory Studies in the Unit Operation of Crushing," *Trans. AIME Mining Eng.*, pp. 1061–1069, Dec. 1951.
4. Homiak, M. and Axelson, J. W., "Liquid Partition Chromatography for Polyhydric Alcohols," *Ind. Eng. Chem.* 52(8): 689–690, Aug. 1960.
5. Kietzman, J. H. and Axelson, J. W., "Polyolefin Composites: Comparison of Mineral Fillers," 28th Tech. Conf. 1973 Reinforced Plastics/Composite Inst., Sec. 11-C, pp. 1–16, Soc. of Plastics Ind., 1973.
6. Axelson, J. W., "Asbestos as a Reinforcement and Filler in Plastics," *Advances in Chemistry Series No. 134*, pp. 16–28, Am. Chem. Soc. 1974.
7. Kietzman, J. H. and Axelson, J. W., "Mineral–Fiber Composites That Resist Fracture and Cracking," *Plastics Eng.*, pp. 46–48, Aug. 1976.

Books

1. Axelson, J. W., "Gaskets, Packings and Seals," *Engineering Materials Handbook* 40: 1–23, McGraw-Hill Book Co., New York, 1958.

2. Axelson, J. W., "Grinding Equipment," *The Encyclopedia of Chemical Process Equipment*, pp. 485–507, Reinhold Publishing Co., New York, 1964.
3. Axelson, J. W. and Pintard, F. B., "Pressure Components (Seals)," *Mechanical Design and Systems Handbook* 24: 1–38, McGraw-Hill Book Co., New York, 1964.
4. Axelson, J. W., "Asbestos Fibers," *Modern Plastics Encyclopedia*, pp. 344–348, McGraw-Hill Book Co., New York, 1970–71.
5. Axelson, J. W., "Magnesium Silicate (Asbestos)," *Pigment Handbook*, Vol. 1, *Properties and Economics*, pp. 243–248, Wiley-Interscience, John Wiley and Sons, New York, 1973.
6. Axelson, J. W., "Asbestos in Plastics," *Polymer-Plast. Technol. Eng.* 4(1): 93–109, Marcel Dekker, Inc., New York, 1975.
7. Axelson, J. W., "Asbestos," *Handbook of Fillers and Reinforcements for Plastics*, pp. 415–427, Van Nostrand Reinhold Co., New York, 1978.
8. Axelson, J. W. and Streib, W. C., "Grinding, Blending, Screening and Classifying," *Laboratory Engineering and Manipulations, Technique of Chem.* XIII: 122–181, John Wiley and Sons, New York, 1979.

10
Inorganic and Microfibers

John V. Milewski
Consultant
Santa Fe, New Mexico

CONTENTS

---

1. GENERAL INTRODUCTION

This chapter will review the following microfibers: dawsonite from Alcoa; Fiberkal from Kaopolite Inc.; Franklin Fiber from United States Gypsum Co.; micro fiber glass from Johns-Mansville Corp.; Processed Mineral Fiber (PMF) from Jim Walter Resources Inc.; Tismo from Otsuka Chemical; and Magnesium Oxysulfate Fiber (MOS) from UBE Industries Ltd. Dawsonite is now in pilot plant production and test marketing, while Fiberkal and MOS are available only in research experimental quantities. Both Franklin Fiber and PMF recently moved out of the pilot plant stage, and they are now commercially available. Tismo, a Japanese product, is essentially equivalent to the former DuPont product Fybex, which was taken off the market. Data on these experimental and pilot plant products are included here to allow the reader to have a better understanding of the whole microfiber field. Even if these fibers never become available, new microfibers of a similar nature may be developed; so the data and processing information presented in this chapter can be expected to be useful.

1.1 Definition of Microfiber

Microfibers are discontinuous fibrous materials that range from 0.1 μm to 10.0 μm diameter, and are polycrystalline or amorphous in structure. They are generally produced by relatively low-cost processes such as fiber throwing or precipitation from solutions. These processes produce fibrous products that generally contain many structural defects (dislocation, imperfections, and vacancies in their crystalline lattice). Therefore, microfibers do not possess the high mechanical properties generally found in the more nearly perfect single crystal fibers known as whiskers. However, their properties are good, and they are useful in many reinforcement applications. Also, the cost of microfibers is significantly more competitive than the cost of whiskers.

2. DAWSONITE*

2.1 Introduction

Dawsonite, sodium aluminum hydroxycarbonate in its synthetically manufactured acicular form, shows promise as a microfiber reinforcement for a variety of thermoplastics. Significant improvements in tensile and flexural strengths and moduli, higher heat deflection temperatures, and reduction in coefficient of thermal expansion are realized for thermoplastic composites containing dawsonite.

Over and above its reinforcing properties, dawsonite, by virtue of its endothermic decomposition at elevated temperatures into sodium aluminate, carbon dioxide, and water, provides flame-retardant properties to composites.

2.2 Methods of Producing Dawsonite

The detailed process for producing dawsonite in crystalline fibers of controlled quality and size is proprietary to Alcoa. Bureau of Mines report RI-7664 describes a process for synthesis and characterization of dawsonite. This report states that dawsonite is made by a hydrothermal process in which the raw materials are digested, reacted, and precipitated. The raw materials are: sodium aluminate, sodium carbonate, and sodium hydroxide. These materials are reacted in solution at 175–200°C and 15 psig for 4 to 5 hr. After that period, the reaction vessel is cooled to 75°C, and the water-soluble carbonates are removed by three hot-water, 70°C, rinses.

It appears reasonable to assume that the Alcoa process differs mainly in process controls

*This section was written by P. V. Bonsignore (Aluminum Co. of America).

Table 10-1. Physical properties of Dawsonite microfiber crystals.

Chemical formula	$NaAl(OH)_2CO_3$
	Na_2O 21.53%
	Al_2O_3 35.40%
	CO_2 30.56%
	H_2O 12.51%
Diameter (D)	0.4–0.6 μm
Length (L)	15–20 μm
Aspect ratio (L/D)	25–40
Density	2.44 g/cm^3
Refractive index	1.53
BET N_2 surface area	15–17 m^2/g

and additives that in effect favor crystallization of acicular morphologies.

2.3 Physical and Chemical Properties of Dawsonite

Dawsonite, as recently manufactured by Aluminum Company of America (Alcoa) in pilot plant quantities has the typical properties listed in Table 10-1.

An SEM photo of dawsonite microfiber crystals is shown in Fig. 10-1.

Of the physical properties listed in Table 10-1, those of special interest are the low density and the low refractive index compared to other microfiber reinforcements such as potassium titanate fibers. Low density (2.44 g/cm^3) means higher volume loadings at lower weight percent. Low refractive index (1.53) means low tinting strength in pigmented colored composites.

Dawsonite is a relatively stable compound toward most temperatures encountered in plastic processing operation. A thermogravimetric analysis (TGA) (Fig. 10-2) reveals the onset of a sharp weight loss occurring at about 300°C. This weight loss reflects the primary decomposition reaction:[1]

$$NaAl(OH)_2CO_3 \rightarrow NaAlO_2 + CO_2 + H_2O$$

Accompanying this decomposition is a considerable absorption of heat, as shown in the differential scanning calorimetry (DSC) curve in Figure 10-3. This endothermic decomposition at temperatures above about 300°C (575°F) accounts for dawsonite's ability to impart some degree of flame retardancy to plastics in which it is incorporated (see section 2.5.2). Of additional interest is dawsonite's ability to offer some degree of smoke suppression and to act as an HCl scavenger for flame-exposed PVC compounds.

Fig. 10-1. Dawsonite.

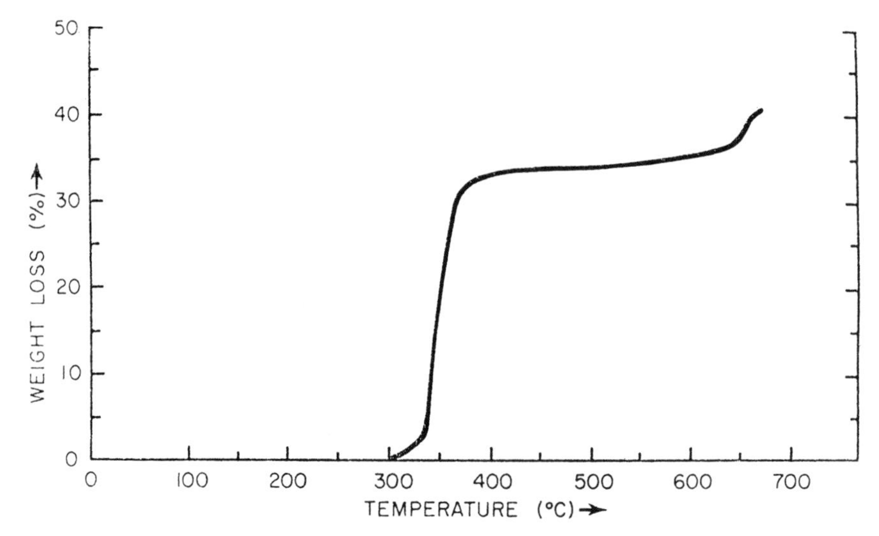

Fig. 10-2. Thermogravimetric analysis (TGA) of dawsonite, $NaAl(OH)_2CO_3$.

2.4 Material Supplier, Cost, and Availability

Synthetic acicular dawsonite has been produced in pilot plant quantities in East St. Louis, Illinois, by the:

Aluminum Company of America (Alcoa)
1501 Alcoa Building
Pittsburgh, PA 15219
Phone: 1-800-245-6333
Larry Musselman

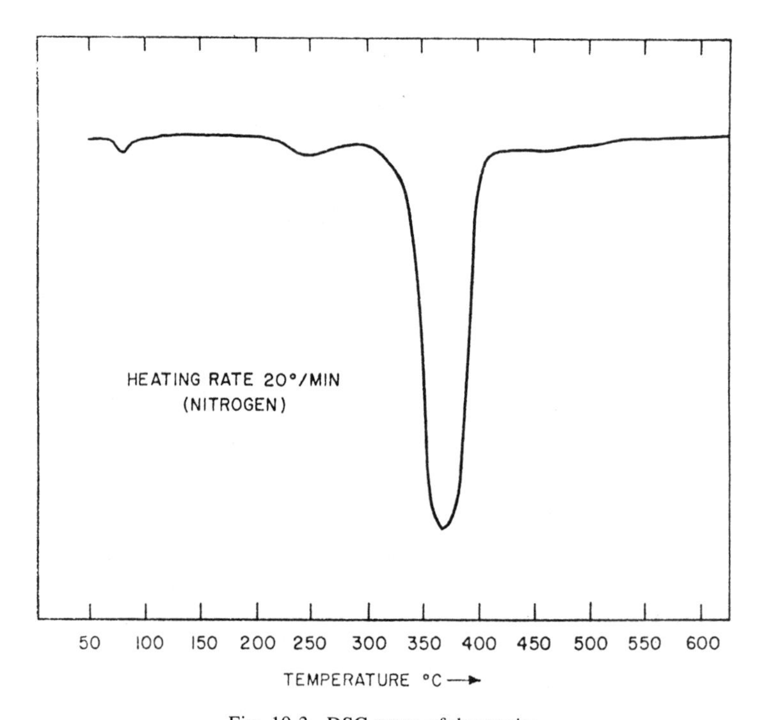

Fig. 10-3. DSC curve of dawsonite.

The availability of dawsonite is still in question because of continuing medical tests and pending regulations on allowable airborne fiber concentrations. The reader is advised to contact Alcoa directly for a report on the current availability and cost status of dawsonite.

2.5 Applications

2.5.1 Behavior in Thermoplastics

2.5.1.1 Compounding Compounding can have an important effect on the properties of a reinforced thermoplastic. The compounding operation should completely wet-out and uniformly disperse the fibers with a minimum reduction in fiber aspect ratio. The higher the compounded aspect ratio, the more efficient the reinforcement in improving resin properties. Fiber damage can be minimized by quickly melting the resin and wetting the fibers prior to applying compressive forces.[2,3]

Compounding of dawsonite in the respective thermoplastics was done by Banbury flux mixing. The thermoplastic was completely melted before the required amount of dawsonite was added. Levels of 0, 7.5, 15, and 30 wt % of dawsonite were studied in three different thermoplastics: polypropylene, rigid PVC, and ABS resin.

2.5.1.2 Physical Properties of Dawsonite/Thermoplastic Composites

Polypropylene. A typical injection-moldable grade of polypropylene (PP), Hercules Profax 6523, was compounded in a Banbury mixer with dawsonite at levels of 0, 7.5, 15, and 30% by wt. The resultant compounds were then injection-molded. Physical properties of these test samples are shown in Table 10-2.

Rigid PVC. The effect of dawsonite on the physical properties of a rigid PVC compound was likewise examined at 0, 7.5, 15, and 30% by wt loadings. The base PVC compound, Geon 87213, was supplied by BF Goodrich, who noted it to be a lead-stabilized injection-moldable compound based on Geon 80X6 PVC resin (0.65 IV). Banbury melt flux compounding was used, followed by sheeting, granulating, and injection-molding of test specimens. Physical properties of rigid PVC/dawsonite composites are given in Table 10-3.

Rigid PVC, from the data of Table 10-3, appears to respond very favorably to reinforcement by dawsonite. Flexural modulus (stiffness) is increased some 300%, while coefficient of linear thermal expansion is decreased to about one-third of its original value by incorporation of 30% dawsonite. For some applications of rigid PVC, such as vinyl siding, such

Table 10-2. Physical properties of Dawsonite/PP composites.

Property	ASTM Method	Units	DAWSONITE LEVEL (% BY WT)			
			0	7.5	15	30
Melt flow rate	D1238L	g/10 min	4.8	4.1	3.1	1.5
Deflection temperature	D648 @ 264 psi	°C	56	59	69	78
Coefficient of linear thermal expansion	D696	10^{-5} in/in / °C	7.3	6.1	5.0	4.1
Izod impact (notched)	D256A	ft-lbs / in notch	1.41	1.33	1.20	1.12
Izod impact (unnotched)	D256E	ft-lbs / in	10.3	7.9	4.9	3.5
Tensile strength	D638	psi	4290	4410	4515	4620
Tensile modulus	D638	10^5 psi	2.43	3.73	4.37	6.49
Flexural modulus	D747	10^5 psi	1.70	1.98	2.31	2.78

Table 10-3. Physical properties of Dawsonite/rigid PVC composites.

Property	ASTM Method	Units	DAWSONITE-LEVEL (% BY WT) 0	7.5	15	30
Melt flow rate	D1238-73F	g/10 min	4.7	3.5	3.2	2.7
Deflection temperature	D648 @ 264 psi	°F	152.6	155.3	159.8	163.4
Coefficient of linear thermal expansion	D696	$\frac{10^{-5} \text{ in/in}}{°C}$	6.4	3.8	3.1	2.1
Izod impact (notched)	D256A	ft lbs/in (notch)	3.39	3.00	2.62	2.36
Izod impact (unnotched)	D256E	ft lbs/in	21.0	18.1	16.5	10.4
Tensile strength	D638	psi	5,647	5,900	6,100	6,200
Tensile modulus	D638	10^6 psi	0.41	0.54	1.43	1.60
Flexural strength	D790	psi	11,760	12,420	12,900	13,160
Flexural modulus	D790	10^6 psi	0.75	1.13	1.54	2.33

increase in stiffness would mean thinner extrusion cross sections to achieve the same degree of stiffness. Similarly, the decrease in thermal expansion, and higher HDT, could permit easier mounting procedures while allowing wider latitude in dark color formulation for rigid PVC siding compounds. In common with other particulate mineral fillers, impact resistance does decrease. The decrease in this particular PVC compound, however, does not appear to be particularly severe, about 3.4 unfilled as compared with 2.4 ft lb/in. with 30% by wt dawsonite, for notched Izod impact.

ABS Resin. The efficiency of dawsonite as a reinforcing agent for ABS resin at 0, 7.5, 15, and 30 wt % is given in Table 10-4. An electroplatable grade, Cycolac EP-3510 of Borg Warner Corporation, was used. Test samples were injection-molded from composites that had been melt-fluxed in a Banbury mill.

In common with its behavior in PP and rigid

Table 10-4. Physical properties of Dawsonite/ABS composites.

Property	ASTM Method	Units	DAWSONITE LEVEL (% BY WT) 0	7.5	15	30
Melt flow rate	D1238G	g/10 min	2.18	1.43	1.17	0.83
Deflection temperature	D648 @ 264 psi	°F	178	182	189	196
Coefficient of linear thermal expansion	D696	$\frac{10^{-5} \text{ in/in}}{°C}$	7.2	4.8	3.5	2.4
Izod impact (notched)	D256A	ft lbs/in (notch)	3.62	2.00	1.61	0.66
Izod impact (unnotched)	D256E	ft lbs/in	10.9	5.7	3.6	2.5
Tensile strength	D638	psi	5,928	6,466	6,667	6,848
Tensile modulus	D638	10^5 psi	3.44	6.47	7.16	8.24
Flexural strength	D790	psi	9,168	9,767	10,341	10,888
Flexural modulus	D790	10^5 psi	3.14	4.91	6.59	9.92

PVC, dawsonite in ABS improves tensile and flexural strengths, increases HDT, and decreases thermal expansion.

Dawsonite addition should improve ABS's utility in a wide range of applications including electroplating. For electroplating durability, a plastic substrate should have a low coefficient of thermal expansion, close to that of the deposited metal surface, to minimize thermal cycling problems such as bond failure.

2.5.2 Flame Retardancy. Dawsonite decomposes at temperatures above 300°C with significant weight loss and heat absorption as previously indicated in Figs. 10-2 and 10-3.[1] This heat absorption can reduce the amount of heat available for pyrolyzing virgin plastic into combustible gaseous fuel elements. Dawsonite, therefore, behaves similarly to alumina trihydrate, $Al(OH)_3$, which has found significant utility as a flame retardant and smoke suppressive filler in reinforced polyester plastics.[4] The higher decomposition temperature of the dawsonite compared to alumina trihydrate (300–320°C vs. 230–260°C) and its fibrous nature with attendant melt viscosity increases that limit loadings, are disadvantages for use as a primary fire retardant. It can be used as an adjunct to marginally flame-retarded plastic systems, with, however, the very desirable feature of improving char reinforcement.

2.5.2.1 Polypropylene The response of PP (Hercules Profax 6523) to dawsonite was examined by the UL94 HB and ASTM D2863 oxygen index tests. The results are given in Table 10-5. Even with as much as 45% dawsonite in this grade of PP, the oxygen index did not go much above 21.5. As little as 7.5% dawsonite did, however, prevent the separation of flaming drops.

2.6 Conclusions

Dawsonite microfibers can offer significant improvements in physical properties to a wide range of thermoplastics including polypropylene, rigid PVC, and ABS. These improvements include increased tensile and flexural strengths and moduli, higher heat deflection temperature, and reduced coefficient of linear thermal expansion. In addition to physical property improvements, a degree of improvement in flame retardance is noted for thermoplastic composites containing dawsonite.

2.7 Acknowledgments

Grateful acknowledgment is extended to Mr. S. Mehta of Lowell Technological Institute who performed the extensive compounding, injection-molding, and physical property evaluation under the technical guidance of Professors A. M. Crugnola and S. A. Orroth, Jr.

2.8 References

1. Huggins, C. W. and Green, T. E., "Thermal Decomposition of Dawsonite," *American Minerologist*, 58: 548-550 (1973).
2. Speri, W. M. and Jenkins, C. F., *Polymer Eng. Sci.* 13(5): 409–414, Nov. 1973.
3. Gras, D., *Plastics Technology*, pp. 40–43, Feb. 1972.

Table 10-5. Burning behavior of Dawsonite/PP composites.

Dawsonite (% by wt)	Dripping[a]	Smoke	Flame size	Rate of Burn[b] UL94 HB in./min	Oxygen index ASTM D2863
None	Flaming drips	Heavy black	Large	1.40, 1.33	18.5
7.5	N.D.	–	–	1.20	–
15	N.D.	Less than 10	Medium	1.00, 1.03	20.0
30	N.D.	Much less than 1	Low	0.73, 0.87	20.5
45	N.D.	Almost none	Very low	0.47	21.5

[a] N.D. = Nondripping.
[b] All classified as slow burning, i.e., <1.5 in./min.

4. Bonsignore, P. V. and Manhart, J. H., paper 23C, *Proceedings* of 29th Annual Technical Conference of the Reinforced Plastics/Composites Institute of the SPI, Feb. 1974.
5. Mathis, T. C. and Morgan, R. W. (to Monsanto), U.S. Patent 3,869,420, Mar. 4, 1975.

3. FIBERKAL

3.1 Introduction

Fiberkal, a new product from Kaopolite Inc., is a fibrous form of kaolin. It is processed from kaolinite into a very fine-type thread that is calcined into a ceramic fiber with an average diameter of 3 μm. The fibers are subsequently chopped into various lengths, which give aspect ratios in the range of 6000 to 25,000. Being ceramic fibers, they maintain their physical and mechanical properties over a wide range of temperatures. Other properties include excellent chemical resistance, inertness, nontoxicity, and good color.

3.2 Summary of Properties

Physical and chemical properties are as listed:

Typical Physical Properties

Color	Off-white
Specific gravity	2.62
Density (lb/solid gallon)	21.9
Bulking value (gal/lb)	0.046
Refractive index	1.62
pH (20% solids)	6.0
Oil absorption (% Gardner-Coleman)	23

Typical Chemical Properties

	%
Aluminum oxide (Al_2O_3)	47.5
Silicon dioxide (SiO_2)	48.5
Iron oxide (Fe_2O_3)	1.0
Titanium dioxide (TiO_2)	1.8
Miscellaneous oxides (Mg, Ca, K, & Na)	1.2

3.3 Materials Supplier, Cost, and Availability

Fiberkal is supplied by:
Kaopolite Inc.
Office Park
Route 22, Box 3110
Union, NJ 07083
Telephone: (201) 789-0609
Attention: George Larson

A research sample is available on request. A 100-lb quantity is sold at \$1.50/lb. Larger quantities are offered at lower prices.

Fiberkal is available in 100- to 1000-lb quantities from stock. Several product sizes are available:

Chopped Sizes Available

Designation	Average Length, in.	Maximum Length, in.
Fiberkal 25	0.25	0.50
Fiberkal 38	0.375	0.75
Fiberkal 50	0.50	1.0
Fiberkal 100	1.0	2.0

3.4 Suggested Applications

Fiberkal may prove to be useful in a number of applications:

- Coatings: roof, asphalt, epoxy
- Polymers: plastic composites, engineering resins, elastomers
- Sealants: caulks, mastics

In many applications, this new fiber makes an excellent replacement for asbestos. It is an experimental material, so the information is based upon preliminary results of investigations by the manufacturer's laboratories. However, no warranty is made. In submitting this information, no liability is assumed or license implied with respect to any existing or pending patent.

4. FRANKLIN FIBER

4.1 Introduction

Franklin Fiber is a microfibrous crystalline form of calcium sulfate, made by the United States Gypsum Co. United States Gypsum Co. bought the product from Certain-Teed a few years ago who in turn had bought it from

Franklin Keys Corp. of Philadelphia some years earlier. The product was in the research and development stage for many years and has recently emerged for commercialization. Two products are being produced: Franklin Fiber Filler H-30, Hemi Hydrate; and Franklin Fiber Filler A-30, Anhydrous.

The FF H-30 Hemi Hydrate reached commercialization first, and is available in large quantities and at a lower price than the anhydrous form. The more chemically resistant "dead burnt" anhydrous form, FF A-30, is more chemically resistant but more expensive to make. FF A-30 is available in smaller quantities and is more expensive. Franklin Fiber has a white lustrous appearance with a silky soft feel. Microscopic examination reveals fibrous crystals with an average length of 60 μm and a very uniform diameter of about 2 μm, as shown in Fig. 10-4.

When used as a reinforcing filler in plastics, Franklin Fiber significantly improves the tensile strength and, in most cases, dramatically increases the modulus of elasticity. With improved tensile and flexural behavior, there is often a decrease in impact resistance. These strength characteristics are improved with suitable coupling and dispersing agents. Besides its proven use in thermosetting and thermoplastic resins, Franklin Fiber has potential in such diversified matrices as asphalt, mineral cements, paper, and paints. More recently, it has been used very successfully as a partial replacement for glass fibers and other more expensive reinforcing fillers.[1,2]

4.2 Methods for Producing Franklin Fiber

The formation of Franklin Fiber involves a crystal restructuring process in an aqueous environment at elevated pressure and temperature.[3] The batch process for production of Franklin Fiber is the following:

> A water slurry of gypsum is first preheated and then pumped to a steam-jacked reactor. The slurry is constantly agitated in the reactor to keep the gypsum in suspension as it is brought up to the reaction temperature and pressure. The reaction is completed within a few minutes. The reaction product is gravity-fed to a holding tank. From the holding tank, the now rather thick slurry is fed to a filter press that removes most of the water. The resulting fibrous cake from the filter press is broken up and heated in an oven to remove the remaining surface moisture. This produces the FF H-30 Hemi Hydrate product. To produce the anhydrous FF A-30, the H-30 fibers are then calcined at high temperature for sufficient time to drive off the water of hydration.

Fig. 10-4. Typical Franklin Fiber—500X.

Continuous processes for making Franklin Fiber have been developed and are now being used to increase production and reliability and maintain its low cost.

4.3 Summary of Properties of Franklin Fiber

Chemical. Franklin Fiber has the same relatively inert chemical properties as normal orthorhombic beta calcium sulfate anhydrite. Its solubility in water is less than 1200 parts/million at 22°C. Table 10-6 is a typical chemical analysis.

Physical. Franklin Fiber is an acicular unicrystal with tapered ends. The lengths of the individual fibers can be as long as 200 μm but

Table 10-6. Typical chemical analysis of Franklin Fiber.

	Anhydrous form	Hemihydrate form
$CaSO_4$	97.68%	50 to 70%
$CaSo_4{}^{1/2}H_2O$		30 to 50%
$CaSo_42H_2O$		Negligible
$CaCO_3MgCO_3$	1.27%	1.27%
SiO_2 and insolubles	0.13%	0.13%
Fe_2O_3 and Al_2O_3	0.13%	0.13%
pH, 10% slurry	10.4	8.2

Table 10-8 Weight loss data for Franklin Fiber and asbestos.

Temperature °F	PERCENTAGE OF WEIGHT LOSS (AFTER 2 HOURS)	
	Franklin Fiber	Asbestos
800	0.40	0.30
1000	0.52	3.99
1400	0.65	13.43
1800	0.75	13.77

average out to around 60 μm. The diameter or width of the fiber is a fairly constant 2 μm. With certain additives, the width of the fibers can be increased slightly, which is helpful at times for better dispersion.

The true density of Franklin Fiber is 3 g/cm^3 and its bulk density is around 10 lb/ft^3.

Franklin Fiber has a smooth, silky feel and when pressed has a tendency to mat. Table 10-7 gives typical properties.

Optical. Franklin Fiber is a lustrous white. If waste system gypsums are used in its manufacture, the impurity may impart color to the final product. The index of refraction of Franklin Fiber is 1.585.

Thermal. Temperature resistance of the Franklin Fiber is excellent, even superior to that of asbestos, as indicated by Table 10-8. Its thermal conductivity is comparable, in equal density loading, to calcium silicate. (4)

Miscellaneous. When tested on animals, under controlled laboratory conditions, Franklin Fiber showed no toxic effects. (5)

4.4 Material Supplier, Cost, and Availability

Franklin Fiber is available from the United Gypsum Co., 101 S. Wacker Drive, Chicago, Il 60606 (312-321-4000).

Both forms are available in commercial quantities, packaged and priced as follows:

Packaging: Franklin Fiber™ Filler H-30 Hemi-hydrate—25-lb bags.
Franklin Fiber™ Filler A-30 Anhydrous—100-lb drums.

Table 10-7. Typical physical properties

	Anhydrous form ($CaSO_4$)	Hemihydrate form ($CaSO_4 \cdot {}^{1/2}H_2O$)
	Crystalline fiber	Crystalline fiber
Nominal average length, μm	60	60
Uniform nominal diameter, μm	2	2
Average nominal aspect ratio	30	30
Oil absorption, $cm^3/100$ g	32.0	46.0
Index of refraction	1.585	1.56
Brightness index (Beckman scale)	92	92
True density, g/cm^3	2.96	2.45 to 2.55
Loose bulk density, lb/ft^3	16 to 20	16 to 20
Solubility (decreases with increasing temp)	2000 ppm at 22C, deionized water	Unstable in water (converts back to gypsum)
Tensile strength of individual fibers (calculated), psi	300,000	300,000
Elastic modulus (calculated), psi	26,000,000	26,000,000
Mohs hardness	2.0 to 3.0	2.0 to 3.0

Table 10-9. Dry physical properties of 30 wt % Franklin Fiber reinforced Nylon-6.

Properties	Test Method ASTM	Unit	Filled	Unfilled
Tensile strength	D638	kg/cm^2	835	635
Tensile modulus	D638	kg/cm^2	4.5×10^4	2.3×10^4
Tensile elongation at break	D638	%	2.3	182
Flexural strength	D790	kg/cm^2	1208	879
Flexural modulus	D790	kg/cm^2	4.0×10^4	2.2×10^4
Izod impact strength notched	D256	kg.cm/cm	3.4	4.72
Heat distortion temperature (18.6 kg/cm^2)	D648	°C	90.5	65.0

Price: Franklin Fiber™ Filler H-30 Hemi-hydrate—$0.40/lb FOB East Chicago, IN.
Franklin Fiber™ Filler A-30 Anhydrous—$1.50/lb FOB East Chicago, IN.

4.5 Current Applications and Data

Franklin Fiber shows great promise as a reinforcing filler in many diverse mediums. It is, however, in plastics that a definite niche has been established for Franklin Fiber. This is due in part to a projected ever-increasing scarcity of fossil-based resins; abundant supplies, now and in the future, of Franklin Fiber's raw material, namely, gypsum; the strength Franklin Fiber imparts to the resins; the excellent thermal properties Franklin Fiber endows to plastic products; and the nonabrasive nature of Franklin Fiber on plastic molding equipment and its low toxic effects as a fibrous reinforcement compared to other micro reinforcements.

Table 10-9 gives the overall performance of a thermoplastic resin with and without the inclusion of Franklin Fiber. An improvement in the impact properties should be possible with a suitable fiber coating.

A comparison study between Franklin Fiber and another mineral fiber, in both thermoplastic and thermosetting resins, was made, and the results, as shown in Table 10-10, are clearly favorable to Franklin Fiber.

To demonstrate that the beneficial effect of Franklin Fiber is due to its unique fibrous structure, a series of tests were undertaken using a calcium sulfate anhydrite in an orthorhombic powder form and Franklin Fiber as the reinforcement in acrylic films. The results, as shown in Table 10-11, dramatically indicate that, while the former adds no improvement over the unfilled control, an 18% loading with Franklin Fiber nearly doubles the tensile strength and increases the modulus fivefold.

Studies now under way with Franklin Fiber as a filler in asphalt have shown that the fiber has a decided effect on raising the asphalt's softening point (see Table 10-12). This effect of Franklin Fiber on asphalt has led to further exploratory work in its use in roof shingles, specifically those utilizing a glass fiber in place of organic-based felt.

Because Franklin Fibers are highly resistant to heat degradation, they can be used as a reinforcement in metals, especially aluminum. The resulting aluminum composites are

Table 10-10. Franklin Fiber vs. wollastonite: Nylon 6,6-PBT (50 wt % fiber).

	NYLON 6, 6		PBT	
	FF	Woll.	FF	Woll.
Flexural strength, psi	19,600	13,500	14,300	11,800
Tensile strength, psi	11,000	7,900	8,400	7,300
Heat deflection temperature, °F	469	396	385	289

Table 10-11. Properties of Franklin Fiber in acrylic films (Rohm & Haas—F-10ES).

Material	Weight (%)	Tensile (psi)	Modulus (psi)
Control	0	330	1400
$CaSO_4$ Drying powder (Orthorhombic)	18	330	1300
Franklin Fiber	18	560	6900

Table 10-12. Properties of various calcium sulfate forms in asphalt.

Fillers	Percent by Weight Added	Softening Point (°F)
Control	0	104.5
Franklin Fibers	3.0	120.0
Franklin Fibers	5.6	128.5
Franklin Fibers	7.0	128.0
Franklin Fibers	18.8	163.5
Anhydrite powder	18.8	114.0
Gypsum	(Will not mix in hot asphalt)	

stronger and stiffer, allowing them to be used more readily as a weight-bearing stuctural material.

Franklin Fiber has been used to reinforce both rigid and flexible thermoplastics and thermosets. It compounds successfully into such widely used thermoplastics as PVC, polypropylene, HDPE, and nylon, as well as unsaturated polyester systems such as SMC, BMC, and other thermosets. It offers reinforcement, improved dimensional stability, retention of dielectric properties, and other improved physical properties.

In injection-molded BMC applications, such as automotive parts, recent tests have shown that up to 25% of the automotive-grade glass fiber can be replaced by Franklin Fiber, which is priced at approximately 40¢/lb (hemihydrate form; anhydrous form will be slightly higher), while retaining impact strength and improving tensile strength.

The same series of tests (Table 10-13) showed that a 50% replacement of glass fiber with Franklin Fiber will yield a BMC composite with 15% greater flexural modulus. Various surface treatments and coupling agents may improve performance and are currently being investigated.

Table 10-13. Property comparisons.

		Franklin fiber replacing:	
	Control	25% of glass fiber	50% of glass fiber
Tensile strength, psi	5040	5225	5110
Tensile modulus, psi	411,000	398,000	408,000
Flexural modulus, psi	13,972	11,652	11,024
Flexural modulus, psi	1,600,000	1,500,000	1,840,000
Izod impact strength, notched, ft-lb/in.	3.42	3.34	2.29
Impact strength, unnotched, ft-lb/in.	3.57	3.41	2.93
Standard BMC formula			
Calcium carbonate	52.66%		
Glass fibers	15.0%		
Polyester resin	19.0%		
Polystyrene/styrene	6.5%		
Other	6.8%		

4.6 Future Applications

Applications include microwave cookware, electrical components, housings for business machines, instrument covers, automotive panels and structures, tractor components, snowmobile hoods, chemical fittings, marine fittings, shower bases, recreational equipment, furniture compounds, power-tool handles and housings, and appliance bases.

Franklin Fiber is currently available in hemihydrate form, with limited quantities of the anhydrous form also available. Initial annual production is estimated at 8 million lb. This long-awaited large-scale production brings to the plastics industry a functional filler/reinforcement with excellent physical properties and relatively low cost that offers opportunities for thermoset and thermoplastic composites.

4.7 References

1. Jennings, B. D., Watkins, K. R., and Duncan, R. E., "Calcium Sulfate: A Filler with Potential," *Plastics Compounding*, p. 70, July/Aug., 1981.
2. Milewski, J. V., "Short-Fiber Reinforcements: Where the Action Is," *Plastics Compounding*, p. 17, Nov./Dec. 1979.
3. Eberl, J. J., et al., U.S. Patent 3,822,340, pp. 423–555, 1974.
4. "Inorganic Fibers Want to Fill Gaps in Insulation," *Chemical Week* 31, Apr. 17, 1974.
5. Hazleton Laboratories, Inc., *Toxicity Study-Calcium Sulfate Anhydrite Whiskers Fibers*—Final Report, 7/27/74.

5. MICRO FIBER GLASS

5.1 Introduction

Micro fiber glass is a series of fine-diameter fiberglass products made by Johns-Manville International Corporation of Denver, Colorado. They are primarily made for and used in insulation and paper filtration products. However, more recently they have been considered for reinforcement applications. These fibers are discrete (noncontinuous) glass filaments made in eight size ranges from 0.2 μm to about 3.8 μm, and in five different chemical compositions, two general-purpose baro-silicates, one electrical grade alkaline-free baro-silicate, and two high-silica materials. They are available without binder additions in bulk form, in rolls or web form, and in water-felted sheets or blankets. These fine glass fibers generally tend to be soft and are nonirritating to the skin.

J.M. Glass Tempstran Fiber (their microglass fiber) has been used since 1950 for the production of 100% glass fiber paper. In more recent years their fibers have found an application in reinforcing and modifying the properties of various cellulose paper products.

Small additions have been found to:

- Increase pulp drainage rate.
- Increase web wet stength.
- Reduce hygroexparosivity and therefore improve dimensional stability.
- Reduce machine shrinkage.
- Increase tear resistance.
- Increase drying rate.
- Control porosity.
- Increase bulk.

5.2 Properties

5.2.1. Wide Range of Service Temperatures. Because the fibers are made of glass and contain no organic binder, they are capable of wide ranges of thermal application. The maximum service temperature will vary according to the code, type of glass, and application, but, in felted form (Code 108), 753 and 475 fibers have been successfully used to 1000°F, E to 1200°F and Q and HPS to 1800°F. Type 475 glass felts have been tested at cryogenic temperatures.

5.2.2 Chemical Durability. Based on weight loss data before and after exposure, Q and HPS fiber have comparatively excellent resistance to water and most mineral acids except hydrofluoric and phosphoric. None of the glass fibers is recommended for use with strongly basic solutions.

The general rate of chemical attack on these fibers will depend on fiber size, concentration and type of solution, temperature, and dynamics of the system.

5.3 Applications

5.3.1 Excellent Thermal Resistivity. Owing to the fine fiber sizes and ease of coating,

Tempstran fibers impart high thermal resistivities. In fact, Tempstran felts exhibit some of the lowest conductivity x density values of any commercially available product. Representative thermal conductivities for 4 and 6 pcf E, 475, and 753 felts are given below:

Mean Temp.	4.0 pcf	6.0 pcf
100°F	0.22	0.22
200	0.25	0.25
500	0.39	0.37
800	0.55	0.50
1000 (E only)	0.68	0.59
−170 (475)	0.12	—

5.3.2 Sound Absorption. Owing to fiber diameter, Tempstran fibers exhibit good sound absorption characteristics. The percent absorption depends on fiber, diameter, frequency of sound, mass, density and configuration of the Tempstran product.

5.3.3 Nuclear Insulation. Q and HPS fibers contain less than 0.01% boron, making them potentially useful for thermal control in nuclear applications.

5.3.4 Casting. Because of the short lengths and small fiber diameters, Tempstran filements are readily dispersible in water or carefully controlled acid solutions, and frequently can be cast into rigid shapes without binders.

5.3.5 Other. The high surface area attendant to fine fiber diameters paves the way for Tempstran's uses as high-efficiency filtration media and catalyst supports, for both low and high temperature applications. The inherent strength of glass suggests reinforcement for organic resins. The high surface activity of some fiber types could invite new application as additives for rubber, high temperature coatings, ceramic bodies, and other multicomponent systems.

Interstitial reinforcements are suggested because their small size will permit these fine fibers to fit and pack efficiently between larger-diameter fibers as a secondary component reinforcement.

5.4 Supplier, Sizes, and Cost

Bulk micro fiber and micro Quartz are available under the trade name of Tempstran fibers from:

Johns-Manville
Filtration and Industrial Minerals Division
Ken-Caryl Ranch
Denver, CO 80217
(303) 978-2000

Standard fiber sizes are shown in the table.

The price of bulk micro fiber (475 glass) ranges from about $1.00/lb for the larger-diameter material to about $10.00/lb for the fine-diameter fibers.

Prices of Micro Quartz range from $15.00 to $25.00/lb, depending on size and quantity.

Standard Fiber Sizes Available, Tempstran

Designation	Average fiber diameter–μm	Glass Type***
Code 100*	.2–.29	475
Code 102*	.3–.33	475
Code 104*	.34–.48	753, 475, E
Code 106*	.49–.58	All
Code 108A*	.59–.88	All
Code 108B*	.89–2.16	All
Code 110*	2.17–3.10	753, 475, Q
Code 112**	2.6–3.8	753, 475, Q

*Based on typical values obtained using Williams Freeness Test (FG-436-209, calibrated by BET surface area).
**Determined by Micronaire J-M Test No. 436202.

***Glass Fiber Types

475 and 753,	general-purpose borosilicate formulations
E	electrical grade, alkali-free borosilicate
Q	high silica (>98.5% SiO_2)
HSP	high purity silica (>99.6% SiO_2)

6. PROCESSED MINERAL FIBER

6.1 Introduction

Processed Mineral Fiber (PMF) is obtained from mineral wool (fiberized blast furnace slag)

Fig. 10-5. PMF fiber illustrating typical fiber size and aspect ratio distribution—63X.

by a unique processing and classification technique. In its base form, the single-strand fibrous material contains as much as 50 wt % nonfibrous material or shot. PMF fiber is a milled form of mineral fiber from which most of the shot is removed by a patented classification process. Its composition is basically 75% calcium silicate with the balance consisting of oxides of aluminum, magnesium, and other lightweight metals.

Processed Mineral Fiber is a white to light gray, free-flowing, short fiber. Its diameter ranges from 1 to 10 μm (average 5 μm) and length averages 275 μm. The manufacturer gives an average range for the aspect ratio of 40 to 60, as illustrated in Fig. 10-5. It is chemically stable and has a safety classification of "Inert Mineral Dust," in the same class as calcium carbonate and most clays. For this reason, it has been recommended for many applications as a replacement for asbestos, and it is useful in both thermoplastics and thermosetting plastics as a fibrous filler.

6.2 Summary of Properties

The chemical composition of PMF typically is as follows:

Typical Analysis	*% by wt*
SiO_2	43
Al_2O_3	8
CaO	39
MgO	6
Other inorganics	4
	100.0

The following physical properties can be expected:

Color	white to light gray
Fiber size	
diameter	range 1–10 μm average 2–5 μm
length	average 275 μm
aspect ratio	average 40–60
Nonfibrous material	4%
Bulk density	25 lb/ft^3
Specific gravity	2.7
Flexibility	brittle
Flammability	does not burn
Tensile strength (single fiber)	3×10^3 to 2×10^5 psi, average 7×10^4 psi
Modulus of elasticity	15×10^6 psi
Heat resistance	good to 1400°F
Glass transition temperature, T_g	1400°F
Devitrification temperature, T_{dev}	1560°F
Melting point, T_{fus}	2300–2400°F
Material safety category	"Inert Mineral Dust"

6.3 Material Supplier, Cost, and Availability

PMF is available in production truckload quantities from:

Jim Walter Resources, Inc.
P.O. Box 5327
Birmingham, AL 35207
Telephone: (205) 841-8339

The basic fiber is designated Fiber 204. This fiber is available as is; or it can be purchased with several different surface treatments to make it more compatible with different resin systems:

- Fiber 204AX is treated with an organosilane designed to provide improved wetting and fiber–polymer bonding in a variety of polymers, including: phenolic, epoxy, melamine, nylon, polyimide, PBT, PVC, polycarbonate, and urethane.
- Fiber 204BX is treated with an organosilane designed to provide the same improvement in wetting and fiber–polymer bonding as the treatment on Fiber 204AX, plus improved fiber dispersion when dry blending is used.
- Fiber 204CX is treated with a low level of a fatty-acid-type material for improved fiber dispersion during dry or low-solvent blending.
- Fiber 204EX is treated with an organosilane designed to provide improved wetting and fiber–polymer bonding in DAP, polybutadiene, urethane, thermoset polyester, and polyolefin materials.

PMF is a registered trademark of Jim Walter Resources, Inc.

The price schedule for PMF is shown in Table 10-14.

6.4 Current Applications and Data

Recent tests at the Jim Walter Resources, Inc. Laboratory and by others indicate that PMF is useful as a fibrous filler in both thermoplastics and thermosetting plastic compositions. Also, extensive evaluation of PMF is under way in rubber. Other uses are being considered for applications in adhesives, sealants, caulks, and reinforcements in thermoplastic elastomers.

PMF has a low binder demand in liquid plastics formulations and can be incorporated into plastics by methods now employed for other short fiber materials.

Information on PMF/plastic composites is contained in the following sections and was supplied by Jim Walters Resources, Inc.

6.4.1 PMF Mineral Fiber as a Filler/Reinforcement for Plastics. While its ultimate reinforcing properties do not equal those of chopped fiberglass strand, PMF Mineral Fiber does offer excellent reinforcement at a substantial cost/performance advantage. Table 10-15 shows the results of tests on various resin systems using different loadings of PMF Mineral

Table 10-14. Price Schedule for the various grades of PMF® Fiber 204, price per pound.[a, k]

Carload or truckload[c]	10,000 lb to truckload	2,000 lb to 10,000 lb	Less than 2,000 lb
$.31	$.33	$.36	$.39
	PMF Fiber 204 AX		
$.41	$.43	$.46	$.49
	PMF Fiber 204 BX		
$.41	$.43	$.46	$.49
	PMF Fiber 204 CX		
$.32	$.34	$.37	$.40
	PMF Fiber 204 EY		
$.43	$.45	$.48	$.51

[a]Standard containers:

- Bag: 23.5″ × 15.5″ × 6″; net weight: 50 lb; gross weight: 50.5 lb.
- Pallet (Nonreturnable pallet): 48″ × 42″ × 4″, 40 bags per pallet; net weight: 2000 lb; gross weight: 2070 lb.

[b]All prices FOB North Birmingham, Alabama, and subject to change without notice.
[c]Minimum truckload: to be determined by governing tariff.
Terms: net 30 days.
Freight classification: fiber, synthetic, N.O.S.
DOT classifications: N/A.

Table 10-15.

Properties of PMF Mineral Fiber-Filled Polymers						
Resin/PMF Mineral Fiber %	Tens. Str., 10^3 psi	Flex. Str., 10^3 psi	Flex Mod., 10^5 psi	Notched Izod, ft-lb/in.	Heat-Distortion Temp., F 66 psi	264 psi
G-P Phenolic						
0	–	13.1	6.62	0.13	–	–
33	–	9.2	8.23	0.25	–	–
33ST	–	13.4	11.58	0.24	–	–
50	–	8.7	9.38	0.28	–	–
50ST	–	12.4	15.55	0.27	–	–
HIPS						
0	3.9	–	2.97	1.29	181	–
33	4.4	–	7.50	0.58	196	–
50	4.1	–	12.00	0.38	200	–
Acetal						
0	7.6	–	2.80	1.32	308	–
20	6.9	–	5.50	0.55	317	–
33	6.2	–	16.00	0.56	–	278
G-P PP						
0	4.9	–	1.56	0.45	236	–
33	4.5	–	5.50	0.63	266	–
50	4.1	–	8.10	0.69	268	–
PBT						
0 Dry	7.3	9.5	3.30	0.34	–	165
0 Wet	7.0	8.3	2.83	0.46	–	–
33 Dry	7.8	11.8	8.15	0.47	–	359
33 Wet	4.6	8.7	6.45	0.31	–	–
33ST Dry	9.5	14.5	7.84	0.51	–	365
33ST Wet	8.5	13.2	6.02	0.47	–	–
50 Dry	8.2	11.9	12.80	0.57	–	394
50ST Dry	10.8	17.2	12.70	0.53	–	394

Table 10-15 (*Continued*)

Properties of PMF Mineral Fiber-Filled Polymers

Resin/PMF Mineral Fiber %	Tens. Str., 10^3 psi	Flex. Str., 10^3 psi	Flex Mod., 10^5 psi	Notched Izod, ft-lb/in.	Heat-Distortion Temp., F 66 psi	Heat-Distortion Temp., F 264 psi
Nylon 66						
0 Dry	10.3	11.4	2.04	0.88	—	360
0 Wet	6.9	5.9	1.80	1.85	—	—
33 Dry	9.3	14.7	7.80	0.57	—	394
33 Wet	5.5	8.9	4.30	0.90	—	—
33ST Dry	12.7	17.7	7.30	0.59	—	400
33ST Wet	8.7	11.4	4.30	1.20	—	—
Nylon 612						
0 Dry	8.6	9.4	2.98	0.32	—	217
0 Wet	7.3	6.3	2.12	0.45	—	200
33 Dry	7.3	12.4	6.63	0.41	386	329
33 Wet	6.0	8.7	4.33	0.59	383	324
33ST Dry	10.8	15.0	6.42	0.35	386	322
33ST Wet	9.0	11.2	4.67	0.54	383	328

Fiber. Surface treatment of the fiber generally improves properties of the composite, especially after water conditioning. Other tests have shown that PMF Mineral Fiber compares favorably with 1/32″ milled glass fibers at equal loadings in nylon, PBT, and polypropylene.

6.4.2 PMF Mineral Fiber as a Filler/Reinforcement for Nylon. Tables 10-16 through 10-19 demonstrate the physical and thermal properties of Nylon 6,6 filled with PMF Mineral Fiber. The PMF Mineral Fiber used for the majority of this work had an average fiber diameter of 5–6 μm and an aspect ratio of 46. PMF Mineral Fiber samples of various fiber lengths were specially prepared to study the effect of aspect ratio on physical and thermal properties.

Surface-treated PMF Mineral Fiber was evaluated to determine the effect of chemical coupling agents on reinforcement. The coupling agents used were applied to the fibers by two techniques. The first method entailed adding a mixture of coupling agent and solvent, followed by evaporation of the solvent. The second involved adding the coupling agent next to the PMF Mineral Fiber.

Milled glass was also incorporated into the Nylon 6,6 for comparison purposes. The 1/32″ glass fibers were heat cleaned to remove the starch coating before compounding.

The procedure for obtaining test specimens was initially to dry the resin for three hours at 175°F. The PMF Mineral Fiber and resin were well blended by using a 12″ Henschel mixer for 30 seconds prior to compounding in a Brabender 3/4″ single-screw extruder. A standard nylon screw was in the extruder. The extruder was controlled at 540°F on the rear and front sections and at 520°F on the die. The compounded material was chipped and then injection molded by using a 1-oz. Newberry Industries machine. Injection temperatures ranged from 520–580°F depending on the loading of the PMF Mineral Fiber. The die was heated to 200°F for all samples.

Two sets of test bars were produced for each compound. The first set was tested in the dry, as-molded state. Samples were tested immediately or kept in a desiccator until they could be tested. The second set of samples was water-treated prior to testing. The samples were immersed in 122°F water for 16 hours and then tested immediately. This was done to evaluate the effects of fiber/resin interactions with and without coupling agents after a water conditioning.

The testing of injection molded samples was carried out consistent with the following ASTM test procedures:

Tensile Strength and Modulus	D638
Flexural Strength and Modulus	D790
Notched Izod Impact Strength	D256
Heat Distortion Temperature	D648

The incorporation of PMF Mineral Fiber into Nylon 6,6 was trouble-free in all cases. The complete dispersion of the fibers was accomplished in a single pass through the extruder. Table 10-16 shows the test results for samples with fiber loadings of 0, 33, and 40 weight percent, as well as for a sample of 33 weight percent milled glass.

For example, at a 33% level of PMF Mineral Fiber the heat distortion temperature at 264 psi stress increased 42°F, the tensile modulus increased 61%, the flexural strength increased 33% PMF Mineral Fiber composition were virtually the same as those of the 33% milled glass compound. Because the cost per pound of PMF Mineral Fiber is approximately one-third that of milled glass, the cost/performance relation is an effective argument for the potential use of PMF Mineral Fiber.

The effect of varying the aspect ratio of PMF Mineral Fiber was determined by compounding at a constant loading of 33%. The aspect ratios varied between 17 and 88. The results are shown in Table 10-17, where it is readily seen that the larger the aspect ratio the greater the property enhancement. With the exception of impact strength, at an aspect ratio of 88 all composite property values are greater than those of the unfilled resin and hence PMF Mineral Fiber provides good reinforcement.

The physical properties for various filled resins can be enhanced when the resin and reinforcer are chemically bound by using a chemical coupling agent. PMF Mineral Fiber was treated with a coupling agent and compounded at 33 and 50 weight percent. The results are

Table 10-16.

Physical and Thermal Properties of a General Purpose Nylon 6,6 Reinforced with PMF Mineral Fiber or Milled Glass

Property	PMF Mineral Fiber Loading (wt. %) 0%	33%	50%	Glass Loading (wt. %) 33%
Tensile Strength (psi)	9,940	9,580	9,030	9,290
Tensile Modulus (psi)	294,000	473,000	548,000	414,000
Izod Impact Strength (ft.-lb./in. notch)	0.75	0.53	0.63	0.48
Flexural Strength (psi)	11,480	15,260	14,900	15,400
Flexural Modulus (psi)	181,000	793,000	1,163,000	816,000
Heat Distortion Temperature (°F @ 264 psi)	352	394	430	400

PMF Mineral Fiber: L/D = 46

Table 10-17.

Physical and Thermal Properties of a General Purpose Nylon 6,6 Reinforced with PMF Mineral Fiber of Different Aspect Ratios

Property	Unfilled	PMF Mineral Fiber Aspect Ratio (Length/Diameter) L/D = 17	L/D = 46	L/D = 88
Tensile Strength (psi)	9,940	8,280	9,580	10,100
Tensile Modulus (psi)	294,000	367,000	473,000	534,000
Izod Impact Strength (ft.-lb./in. notch)	0.75	0.44	0.53	0.57
Flexural Strength (psi)	11,480	14,335	15,260	15,440
Flexural Modulus (psi)	181,000	591,000	793,000	804,000
Heat Distortion Temperature (°F @ 264 psi)	352	375	394	427

PMF Mineral Fiber: Loading = 33 wt. %

Table 10-18.

Physical and Thermal Properties of a General Purpose Nylon 6,6 Reinforced with a Surface-Treated PMF Mineral Fiber

Property	Unfilled	33 wt. % PMF Mineral Fiber Untreated	33 wt. % PMF Mineral Fiber Treated	50 wt. % PMF Mineral Fiber Untreated	50 wt. % PMF Mineral Fiber Treated
Tensile Strength (psi)	9,940	9,580	12,700	9,030	16,590
Tensile Modulus (psi)	294,000	473,000	495,000	548,000	608,000
Izod Impact Strength (ft.-lb./in. notch)	0.75	0.53	0.59	0.63	0.77
Flexural Strength (psi)	11,480	15,260	17,700	14,900	24,000
Flexural Modulus (psi)	181,000	793,000	731,000	1,163,000	1,230,000
Heat Distortion Temperature (°F @ 264 psi)	352	394	400	430	454

PMF Mineral Fiber: L/D = 46

Table 10-19.

Effect of Water Conditioning of Nylon 6,6 Reinforced with PMF Mineral Fiber

Property		Unfilled	33 wt. % PMF Mineral Fiber Untreated	33 wt. % PMF Mineral Fiber Surface-Treated
Tensile Strength (psi)	Dry*	9,940	9,580	12,700
	Wet**	6,330	5,340	8,700
Tensile Modulus (psi)	Dry	294,000	473,000	495,000
	Wet	165,000	347,000	335,000
Izod Impact Strength (ft.-lb./in. notch)	Dry	0.75	0.53	0.59
	Wet	2.06	0.91	1.20
Flexural Strength (psi)	Dry	11,480	15,260	17,700
	Wet	6,280	8,300	11,400
Flexural Modulus (psi)	Dry	181,000	793,000	731,000
	Wet	180,000	391,000	433,000

* **Dry:** Tested as molded.
****Wet:** Tested after a 16-hour soak in 50°C distilled water.

PMF Mineral Fiber: L/D = 46

shown in Table 10-18. It can be seen that at both fiber content levels the coupling agent enhanced the properties markedly. It is especially noteworthy that at a 50% loading all composite properties were improved over those of unfilled nylon or nylon filled with untreated fibers. In short, a true reinforcement of the resin was observed at this level.

The normal usage of materials such as nylon exposes the part to atmospheric moisture. This results in a pickup of water and subsequent change in physical properties. The compounded materials were therefore water-conditioned to give some indication of the relative changes that might be expected. The method of water conditioning was chosen as a convenient means of quickly observing changes. As a result no attempt was made to equalize total water absorption. The results are shown in Table 10-19. The dry, as-molded samples were used for comparisons. As expected, the specimens with surface-treated fibers retained a higher percentage of the original, dry sample measurement.

6.4.3 PMF Mineral Fiber as a Filler/Reinforcement for PBT. Tables 10-20 through 10-22 demonstrate the physical and thermal properties of a general purpose PBT filled with PMF Mineral Fiber. The fiber utilized for most of the compounding had an average fiber diameter of 5–6 μm and an average aspect ratio of 46. A variety of other aspect ratios was specially prepared in the laboratory to determine the effect of aspect ratio on the physical properties.

Surface-treated PMF Mineral Fiber was evaluated to determine the effect of chemical coupling agents on reinforcement. Surface treatment was achieved either by adding the chemicals next to the fibers or by a chemical/solvent addition followed by solvent evaporation.

The 1/32″ milled glass, from which the starch treatment was removed by heat cleaning, was also compounded into PBT for direct comparison with untreated PMF Mineral Fiber.

Standard ASTM test specimens were molded for all materials. The resin was dried for 16 hours at 190°F prior to compounding. The PMF Mineral Fiber and dried resin were dry mixed for 30 seconds by use of a 12-inch Henschel mixer.

Compounding was accomplished using a 3/4″ Brabender single-screw extruder. The extruder was controlled at 490°F in the rear section of the extruder and 500–510°F in the forward and die sections. The compounded material was chipped and then injection molded using a 1-oz. Newberry Industries machine. Injection temperatures ranged from 500 to 510°F. The die was heated to 150°F for all samples.

All testing was carried out consistent with the following ASTM procedures:

Tensile Strength and Modulus	D638
Flexural Strength and Modulus	D790
Notched Izod Impact Strength	D256
Heat Distortion Temperature	D648

No problems were encountered in the compounding of PMF Mineral Fiber into the resin. A single pass provided excellent dispersion of the fibers. Table 10-20 shows the results of fiber loadings of 0, 33 and 50 weight percent of PMF Mineral Fiber and a 33 weight percent loading of milled glass.

For all loadings of PMF Mineral Fiber shown there is enhancement of every property. The PMF Mineral Fiber is therefore acting as a true reinforcing agent, rather than only as a filler. Specifically, at the 33% loading the heat distortion temperature increased 194°F, the tensile strength increased 6%, the impact strength rose 38%, the flexural strength increased 23%, and the flexural modulus increased 147%, respectively. Also, at the 33% loading, the PMF Mineral Fiber is equivalent or superior to the milled glass for all properties. With the cost being approximately one-third that of milled glass, PMF Mineral Fiber provides an excellent alternative to milled glass.

The effect of varying the aspect ratio was determined by compounding at a constant loading of 33%. The aspect ratios ranged between 17 and 88. The results are shown in Table 10-21. The tensile strength, tensile modulus, impact strength, and flexural modulus of the filled materials were found to be essentially independent of the aspect ratio. There was a general trend toward higher values for the flexural strength and the heat distortion temperature as the aspect ratio increased. Regardless of trends, the

Table 10-20.

Physical and Thermal Properties of a General Purpose PBT Reinforced with PMF Mineral Fiber or Milled Glass

Property	PMF Mineral Fiber Loading (wt. %) 0%	33%	50%	Glass Loading (wt. %) 33%
Tensile Strength (psi)	7,340	7,790	8,180	5,330
Tensile Modulus (psi)	259,000	441,000	589,000	358,000
Izod Impact Strength (ft.-lb./in. notch)	0.34	0.47	0.57	0.47
Flexural Strength (psi)	9,520	11,750	11,900	11,200
Flexural Modulus (psi)	330,000	815,000	1,280,000	767,000
Heat Distortion Temperature (°F @ 264 psi)	165	359	386	363

PMF Mineral Fiber: L/D = 46

Table 10-21.

Physical and Thermal Properties of a General Purpose PBT Reinforced with PMF Mineral Fiber of Different Aspect Ratios

Property	Unfilled	PMF Mineral Fiber Aspect Ratio (Length/Diameter) L/D = 17	L/D = 46	L/D = 88
Tensile Strength (psi)	7,340	7,340	7,790	6,925
Tensile Modulus (psi)	259,000	435,000	441,000	419,000
Izod Impact Strength (ft.-lb./in. notch)	0.34	0.49	0.47	0.50
Flexural Strength (psi)	9,520	10,840	11,750	12,900
Flexural Modulus (psi)	330,000	677,000	815,000	740,000
Heat Distortion Temperature (°F @ 264 psi)	165	322	359	363

PMF Mineral Fiber: Loading = 33 wt. %

Table 10-22.

Physical and Thermal Properties of a General Purpose PBT Reinforced with a Surface-Treated PMF Mineral Fiber

		33 wt. % PMF Mineral Fiber		50 wt. % PMF Mineral Fiber	
Property	**Unfilled**	**Untreated**	**Treated**	**Untreated**	**Treated**
Tensile Strength (psi)	7,340	7,790	10,220	8,180	10,800
Tensile Modulus (psi)	259,000	441,000	508,000	589,000	634,000
Izod Impact Strength (ft.-lb./in. notch)	0.34	0.47	0.48	0.57	0.53
Flexural Strength (psi)	9,520	11,750	14,440	11,900	17,190
Flexural Modulus (psi)	330,000	815,000	799,000	1,280,000	1,270,000
Heat Distortion Temperature (°F @ 264 psi)	165	359	354	386	386

PMF Mineral Fiber: L/D = 46

Table 10-23.

Physical and Thermal Properties of Polypropylene Filled with PMF Mineral Fiber

	Loading (wt. %)		
Property	**0**	**33**	**50**
Tensile Strength (psi)	4,810	4,520	4,070
1% Secant Modulus (psi)	192,000	270,000	295,000
Izod Impact Strength (ft.-lb./in. notch)	0.45	0.63	0.69
Flexural Modulus (psi)	160,000	550,000	811,000
Heat Distortion Temperature (°F @ 66 psi)	236	266	268

PMF Mineral Fiber: L/D = 58

reinforcing properties of PMF Mineral Fiber were still evident at all values of the aspect ratio.

Increased fiber–resin bonding can be achieved by providing a chemical linkage between the two. A bifunctional coupling agent, with one function capable of reacting with the resin and the other with the fiber, was applied to PMF Mineral Fiber. The treated fibers were incorporated into PBT at 33 and 50 weight percent levels. The test results are shown in Table 10-22, which compares treated versus untreated fibers. Most noticeable were the 31% increase in tensile strength, the 8–15% increase in tensile modulus and the 23–44% increase in flexural strength for the treated versus the untreated composites. The surface treatment had no significant effect on the other properties evaluated, but they were in all cases greater than the unfilled resin.

6.4.4 PMF Mineral Fiber as a Filler/Reinforcement for Polypropylene. Tables 10-23 through 10-26 demonstrate the physical and thermal properties of general-purpose and heat-sustained grade polypropylene filled with PMF Mineral Fiber. The PMF Mineral Fiber used for most of this work was a standard grade with an aspect ratio of 58. Other aspect ratios were used to show variations of properties with length. The fibers had no surface treatments or coupling agents on them and were used as produced.

Milled glass was also incorporated into the general purpose polypropylene for a comparison of properties. The 1/32″ glass fibers with a starch surface treatment were used as received. A heat cleaned, milled glass was also evaluated. The procedure followed to produce test specimens was first to flux the polypropylene material on a 13″ Black-Clausen two-roll hot mill. The PMF Mineral Fiber was added slowly and mixed well for 8 minutes on the mill. The compounded material was cut from the mill and granulated prior to injection molding. Molding was accomplished by means of a 1 oz. Newberry Industries machine.

The testing of injection molded samples was carried out consistent with the following ASTM test procedures:

Tensile Strength	D638
Flexural Strength and Modulus	D790
Notched Izod Impact	D256
Heat Distortion Temperature	D648

The long-term heat aging studies were carried out by placing molded test samples in an oven controlled at 300°F. Samples were removed at intervals of 50, 75, and 100 hours and then tested. Test results were reported as percent retention of the property of interest as compared to non-heat-treated samples.

The incorporation of the PMF Mineral Fiber into polypropylene in all cases was trouble-free. No dispersion problems were encountered with either resin or at any of the loadings used. Table 10-23 shows the results of various physical tests run on samples with PMF Mineral Fiber loadings of 0, 33, and 50 weight percent. With the exception of the tensile strength, all other physical properties increased as PMF Mineral Fiber loading increased. For example, at the 50% level the Izod impact strength increased 53%, the heat distortion temperature rose by 14%, the flexural strength improved 400%, and the 1% secant modulus increased 53%. The tensile strength decreased 15%.

Milled glass fibers were also compounded into the general-purpose resin for comparison. The milled glass fibers were 1/32″ long and contained a starch surface coating. In addition, some of the glass fibers were heat-cleaned to remove the coating and were then incorporated into PP. A comparison of physical properties is shown in Table 10-24. In general, the two materials compare quite favorably in physical properties. With the cost being approximately one-third that of milled glass, PMF Mineral Fiber would appear to be a viable alternate to milled glass as a PP filler/reinforcement.

The effect of fiber length was evaluated by taking the original PMF Mineral Fiber with an aspect ratio of 58 and breaking it down into shorter fibers by using a hydraulic press. The resulting fibers had aspect ratios of 28 and 16. Table 10-25 shows the definite trend to lower physical properties values as fiber length is decreased.

Since polypropylene is being utilized more aggressively in higher-temperature applications, the effect of PMF Mineral Fiber on the

Table 10-24.
Comparison of Milled Glass Fibers and PMF Mineral Fiber in Polypropylene

Property	PMF Mineral Fiber	Milled Glass Starch Coating	Milled Glass Heat Cleaned
Tensile Strength (psi)	4,520	4,640	4,330
1% Secant Modulus (psi)	270,000	295,000	258,000
Izod Impact Strength (ft.-lb./in. notch)	0.63	0.74	0.72
Flexural Modulus (psi)	550,000	678,000	557,000
Heat Distortion Temperature (°F @ 66 psi)	266	271	261

PMF Mineral Fiber: L/D = 58

Table 10-25.
Effect of Aspect Ratio of PMF Mineral Fiber in Polypropylene

Property	Aspect Ratio 16	26	58
Tensile Strength (psi)	4,150	4,480	4,520
1% Secant Modulus (psi)	238,000	261,000	270,000
Izod Impact Strength (ft.-lb./in. notch)	0.62	0.64	0.63
Flexural Modulus (psi)	425,000	582,000	550,000
Heat Distortion Temperature (°F @ 66 psi)	253	261	266

PMF Mineral Fiber: loading = 33 wt. %

Table 10-26.

Physical Properties of Sustained Heat Grade Polypropylene Filled with PMF Mineral Fiber after Heat Treatment at 300°F

Property	Loading Wt. %	Control	% Retention of Properties at 50 Hrs.	75 Hrs.	100 Hrs.
Tensile Strength (psi)	0	4,980	98	98	99
	30	4,400	94	94	96
Flexural Modulus (psi)	0	184,000	123	118	118
	30	407,000	112	125	109

PMF Mineral Fiber: L/D = 58

Table 10-27.

PMF Mineral Fiber vs. Asbestos in General-Purpose Phenolic

Filler	None	PMF Mineral Fiber	PMF Mineral Fiber, ST	Asbestos
Weight %	–	33	33	40
Flexural Strength, 10^3 psi	10.4	8.8	12.6	9.1
Flexural Modulus, 10^5 psi	6.75	7.98	10.87	10.55
Tensile Strength, 10^3 psi	7.52	6.62	8.78	5.01
Tensile Modulus, 10^5 psi	4.53	6.01	5.96	4.93
Notched Izod Impact Strength, ft.-lb./in.	0.18	0.23	0.26	0.25
Volume Resistivity, 10^{11} ohm-cm	1.5	0.7	1.5	0.03
UL94 Flammability (⅛ in.)	V-O	V-O	V-O	V-O

long-term heat stability of a sustained heat grade PP was investigated. The data are reported as a percentage retention of the property as compared to controls which were not heat-aged. The compounding was the same as for the G.P. grade. Table 10-26 shows the test results of unfilled and 30% PMF Mineral Fiber–filled heat grade polypropylene. The tensile strengths and flexural moduli followed quite closely the values for the unfilled grade.

6.4.5 PMF Mineral Fiber as a Filler/Reinforcement for Phenolics. Table 10-27 compares properties of phenolic compounds containing PMF Mineral Fiber and asbestos. All samples contain the same general-purpose phenolic resin. The asbestos-filled samples were a commerically available general-purpose compound.

PMF Mineral Fiber-filled compounds were prepared by initially blending all components except the fiber and milling the dryblend for 45 seconds at 212°F on a differential-speed hot mill. The fluxed resin blend was stripped from the mill, cooled, chipped, and passed through a 30-mesh screen. Then the PMF Mineral Fiber was intimately blended with the resin using a Henschel high-intensity mixer for 30 seconds. This type of mixing is important to fully disperse the fibers and prevent agglomeration. Excellent dispersion reportedly can also be obtained by dry-blending the phenolic/PMF Mineral Fiber compound and then extruding at low fluxing temperatures. In recent years, this technique reportedly has gained favor for compounding phenolics on a continuous basis.

Compounds were charged into a standard ASTM compression mold and cured at 356°F.

As shown in the table, when untreated PMF Mineral Fiber is incorporated into the general-purpose phenolic, some property values increase while others actually decrease relative to unfilled resin. However, the surface-treated PMF Mineral Fiber–filled compound clearly exhibits superior properties to the unfilled resin.

Compared with the asbestos-filled phenolic, the surface-treated PMF Mineral Fiber–filled compound equals or excels it in all the properties evaluated.

6.5 Current Status of the Health Aspects of Rock and Slag Wool Fibers*

Based on the rock and slag wool fiber studies available at this time—covering numerous workers in both the U.S. and Europe and conducted by researchers from public, private and educational institutions since the 1930's—there is no consistent evidence of malignant or non-malignant respiratory disease related to exposure of workers who manufacture and fabricate man-made rock and slag wool products.

- Completed rock and slag wool fiber animal inhalation experiments show no convincing evidence of permanent fibrosis or carcinogenesis, including mesothelioma. Recent rock and slag wool fiber animal implantation studies reaffirm previously published findings that the use of highly artificial methods of exposure can cause adverse effects in animals.
- The results of morbidity research covering rock and slag wool fiber workers continue to show no evidence of identifiable disease or changes in lung function as a result of exposure to rock and slag wool fibers.
- The results of rock and slag wool fiber mortality studies fail to demonstrate any consistent evidence of malignant or non-malignant respiratory disease which could be attributed to occupational exposure to rock and slag wool fibers. While some increases in respiratory disease have been noted, investigators have indicated that other factors—such as smoking and/or other agents—could be responsible.

7. TISMO

7.1 Introduction

Since our *Handbook of Fillers and Reinforcements for Plastics* was published (1978), Tismo

**Source*: Thermal Insulation Manufacturers Association 7 Kirby Plaza, Mt. Kisco, NY 10549. Other fact sheets are available from TIMA concerning the health aspects of man-made vitreous fibers.

has become available. Tismo is a crystalline, inorganic titanate microfiber that is essentially equivalent to the Dupont discontinued product Fybex. The manufacturer, Otsuka Chemical Co., Ltd. of Japan, claims it to have the same properties as Fybex except that some physical specifications have changed such as L/D ratio which they claim to be able to control. Therefore, much of our material on Fybex is used here, with just the name changed where appropriate. Lab data from earlier Fybex studies are identified as such. Materials specification have been updated to include data from the new manufacturer.

Tismo consists of poly-crystalline, inorganic titanate microfibers with relatively high modulus and reasonably high strength, which qualifies them as reinforcements for plastics. Evaluation of these microfibers in a variety of resins, particularly themoplastics, show significant enhancement of mechanical, thermal, and physical properties.

Tismo, as a result of its small diameter, similar to that of other microfibers, has a specific advantage in compounding and molding, as there is less fiber breakage during processing than is experienced with larger-diameter fibers such as fiberglass. Moreover, parts molded with plastic that is reinforced with these fibers show surface smoothness and gloss equivalent to parts made with unfilled plastics, suggesting aesthetic as well as engineering applications.

Data are presented for reinforced thermoplastics. Typical results for a 30 wt % loading of these fibers show a twofold increase in tensile strength and a fourfold increase in flexural modulus. Other advantages of Tismo microfiber–reinforced composites are resistance to warp and moldability in thin sections.

7.2 Manufacturing Process, Products, and Supplier

Tismo (Fybex) crystalline fibers are formed by a recrystallization process from a molten salt. They are formed by a process that blends TiO_2, KCl, and K_2CO_3 to yield $K_2Ti_4O_9$ after kilning to 900°C. Next, the $K_2Ti_4O_9$ is reprocessed by potassium leaching to form an octa-titanate, $K_2Ti_8O_{17}$. The product formed has the properties given in Table 10-28 and is illustrated in Fig. 10-6.

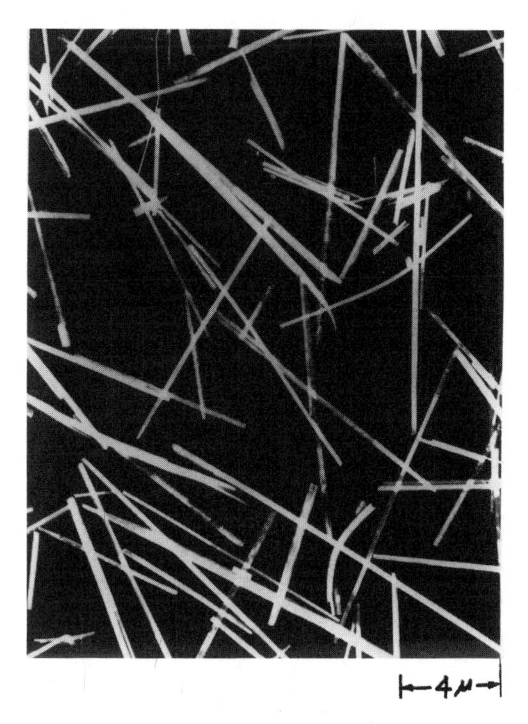

Fig. 10-6. Fybex. (Courtesy E. I. DuPont.)

Tismo is a product of Otsuka Chemical Co., Ltd. Japan. It is distributed in the United States by:

Biddle Sawyer
2 Penn Plaza
New York 10121
Phone: (212) 736-1580
Contact: Jay Nussbaum

7.3 Summary of Properties

Table 10-28 lists physical and chemical properties of Fybex.

7.4 Compounding Techniques

Most of the following information on compounding, properties, and application of Fybex has been obtained from a paper by Linsen and Regester.[1]

In compounding with Fybex and other microfibers, it is particularly important to use pro-

Table 10-28. Fybex properties.

Properties	Type L	Type D
Appearance	Needle crystal	Needle crystal
Hue	White	White
Average length, μm	20–30	20–30
Approximate diameter, μm	0.2	0.2
Bulk density, g/cm^3	0.1–0.2	0.1–0.2
Specific gravity	3.1–3.3	3.3–3.5
Chemical composition	$K_2O \cdot 6TiO_2 \cdot \frac{1}{2}H_2O$	$K_2O \cdot 6TiO_2$
Maximum percentage water content	4.0	0.7
Hardness*	—	4
PH (dispersion in water)	6.5–7.5	6.5–7.5
Dehydration temperature, °C (combined water)	350–440	—
Melting point, °C	1250–1310	1250–1310
Minimum tensile strength,* kg/mm^2	700	700
Specific heat, Cal/g. °C	0.22	0.22
Specific surface area, m^2/g	1.5–2	1.5–2
Electric resistance,* ohm-cm	—	$3.3. \times 10^{15}$
Dielectric constant	ε, 5–7, tan δ, 8–14%	ε, 3.5–5 tan δ, 6–12%
Acid resistance	Stable in 10% acid at room temperature	Stable in 10% acid at room temperature
Alkaline resistance	Stable in 30% alkali at boiling point	Stable in 30% alkali at boiling point
Affinity (water)	Good	Good
Affinity (toluene)	Average	Average

* Literature.

cessing techniques that lessen the probability of breaking the fibers. Breakage usually occurs if the fibers are subjected to high compressive or mixing forces before the polymer is in the molten state. It is therefore necessary to allow the polymer to soften before dispersion work is applied. This is best accomplished by premelting the polymer and then adding fiber, or heating the polymer/fiber blend while gradually increasing shear.

Prefluxing the polymer and then adding fiber is recommended when processing in a Banbury or a rubber (two-roll) mill. In extruders, hot, long feed sections should be used to heat the Fybex/polymer blend before high shear is applied. Another acceptable extrusion technique is to add only polymer in the feed section with fiber addition through downstream ports. This is particularly advantageous in twin-screw extruders. For additional information on processing resins with fibers and fillers see Chapter 22.

In general, molding conditions for plastics reinforced with Fybex are similar to those for unreinforced counterparts.[2,3] Slightly higher temperatures are required because of the increased viscosity of the reinforced plastic. Melt toughness is increased and setup time decreased, resulting in faster cycles than with unreinforced resins.

7.5. Properties of Fybex-Reinforced Plastics

In general, property improvements with Fybex in thermoplastics are comparable to those with glass fibers at equal weight loadings. The stiffness of plastics reinforced with Fybex is generally higher than glass fiber composites because the elastic moduli of the titanates are more than four times that of glass. Fybex has been particularly effective in reducing the coefficient of thermal expansion.

The effect of Fybex reinforcement on substrate expansion coefficient of ABS and polypropylene is shown in Figs. 10-7 and 10-8. The expansion coefficient changes most rapidly at

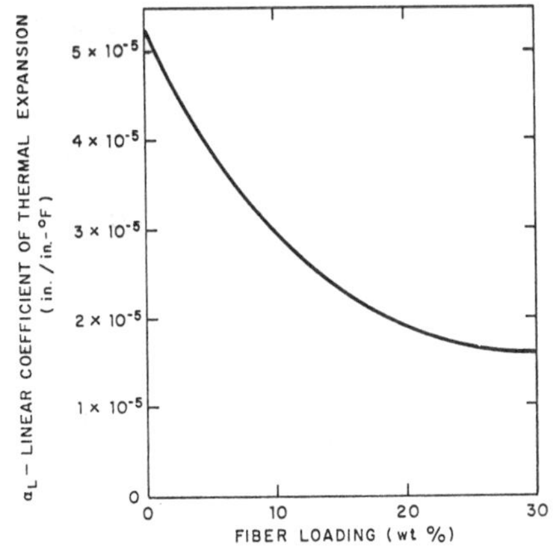

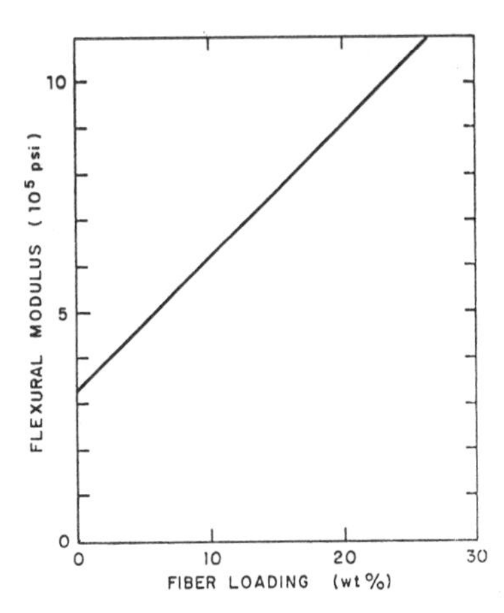

Fig. 10-7 Typical physical properties— Fybexs reinforced ABS. (Plating grade; injectin-molded bars 5'' × 1/2'' × 1/8''. Source: Ref. 3.)

low fiber loadings. High stiffness is achieved for composites with the preferred 10-15 wt % fiber loading. This is particularly important for polypropylene, where the low modulus of the unreinforced resin can cause handling problems and distortion in hot applications. Fybex loadings up to 30 wt % give parts which combine superior engineering properties with high-quality appearance and performance. Thin parts with ultra-high stiffness are thus possible.

The aspect ratio of Fybex is less than the starting aspect ratio of short glass fibers used to reinforce thermoplastics; therefore, the strength of plastics reinforced with Fybex is generally slightly less than that of glass-reinforced plastics. Plastics reinforced with Fybex can be further strengthened by using coupling agents to improve the titanate-to-plastic bond.

Impact strengths of the plastics reinforced with Fybex are generally at the level of the unfilled base plastic. Notch sensitivity appears to depend on the characteristics of the base resin, and Fybex neither improves nor degrades this effect. As stated previously, good dispersion is essential for good tensile and impact strengths. Typical property data for plastics reinforced with Fybex are given in Tables 10-29 through 10-36, and some of the unique characteristics of these composites are discussed in the following section.

Dupont's experience with Fybex (2 and 3) has shown Fybex to provide a high degree of reinforcement in addition to the other advantages.

While reinforcement of thermoplastics improves engineering properties, it often does so

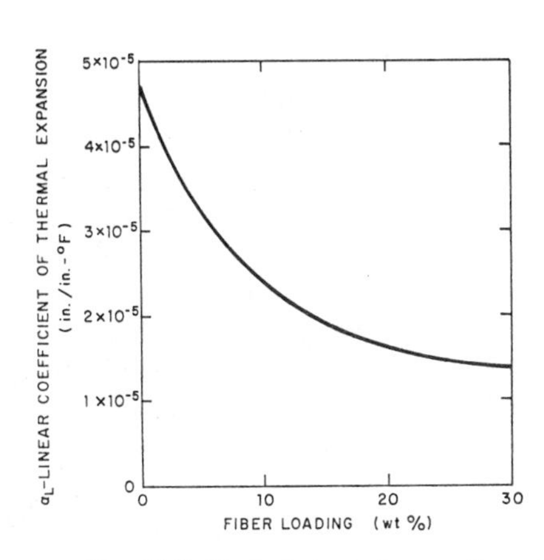

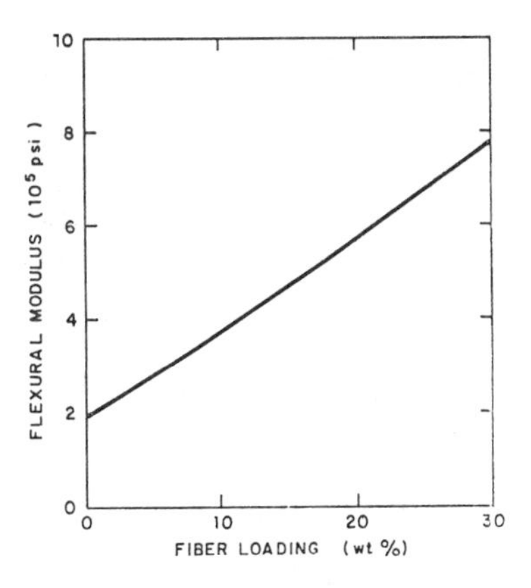

Fig. 10-8 Typical physical properties— Fybex reinforced polypropylene. (General-purpose grade, MI = 15; injection-molded bars 5'' × ½'' × 1/8''. Source: Ref. 3.)

Table 10-29. Surface roughness of reinforced and unreinforced resins.

	AVERAGE SURFACE ROUGHNESS (microinches)	
	ABS[a] Composites	Polypropylene[b] Composites
Unreinforced resin	0.7	0.8
Titanate fiber reinforcement (20 wt. %)	1.0	1.5
"E" glass fiber reinforcement (20 wt. %)	>50	>10

[a]Plating grade.
[b]General purpose grade.
Source: Ref. 2.

at the expense of other desirable features. Ideally, a reinforcing fiber should impart engineering properties with minimum sacrifice in moldability, surface finish, part isotropy, resin regrindability, machine wear, cycle time, and compatibility with pigments, flame retardants, and fillers. Failure to meet one or all of these criteria can exclude a fiber from an application, regardless of its reinforcing capability. Experiments have shown that Fybex provides a high degree of reinforcement in addition to the following advantages:

- Breakage resistance of Fybex during compounding and subsequent processing allows high regrind ratios or more regrind cycles without loss of physical properties. See Table 10-30.
- Surface smoothness of molded or extruded plastics is improved. See Table 10-29.
- In addition to aesthetics, the small uniform fiber size of Fybex provides an important engineering feature to molded thermoplastics—isotropy, as indicated in Table 10-31.

Table 10-30. Physical properties versus number of molding cycles[a]

	MOLD CYCLE			
	1	2	3	4
30/70 FYBEX/Nylon 6-6				
Tensile strength ($\times 10^3$ psi)	19.6	17.3	17.3	18.4
Tensile modulus ($\times 10^6$ psi)	1.4	1.4	1.4	1.3
Elongation (%)	2.2	1.6	1.6	2.1
Flexural strength ($\times 10^3$ psi)	31.8	29.7	30.0	29.0
Flexural modulus ($\times 10^6$ psi)	1.3	1.2	1.2	1.2
Notched Izod impact (ft-lb/in.)	1.4	1.5	0.9	0.8
Unnotched Izod impact (ft-lb/in.)	11.8	12.0	9.9	11.2

[a]100% regrind ratio.
Source: Ref. 1.

Precision Reinforced Moldings: Small gears, camera body components, electronic assemblies, and parts exposed to high thermal gradients frequently require dimensional stability beyond that attainable with unreinforced resins. Larger fiber reinforcement may not satisfy attendant requirements such as smoothness, isotropy, fast cycles, and others already mentioned. Fybex in engineering polymers provides durable, precise moldings that are often preferable to metals, as shown in Table 10-32.

The potential for the Fybex-type microfiber is good. Foam reinforcements have been investigated with Fybex, since theoretical calculations show it to be ideally sized for excellent foam reinforcement efficiency.

Vacuum-formed ABS or PVC sheet reinforced with Fybex is another promising end use. The hot strength, high elongation, and toughness of extrusion grades of ABS and rigid PVC reinforced with Fybex suggest them as promising substitutes for sheet metal (Table 10-33). Structural adhesives having high strength, low expansion, and toughness are other possible applications.

In summary, Fybex broadens the utility of

Table 10-31. Reinforcement uniformity vs. fiber size.

Material[a]	Tensile Strength (X 10³ psi)	Flexural Strength (X 10³ psi)	Flexural Modulus (X 10⁶ psi)	Notched Izod (ft-lb/in.)
30/70 FYBEX/Nylon 6-6				
Machine	18.5	27.8	1.83	1.71
Transverse	15.5	22.3	0.92	1.50
33/67 Glass Fiber/Nylon 6-6				
Machine	18.5	30.4	1.28	1.31
Transverse	11.9	17.3	0.67	0.80

[a]Molded into 3 in. X 5 in. X 1/16 in. plaques.

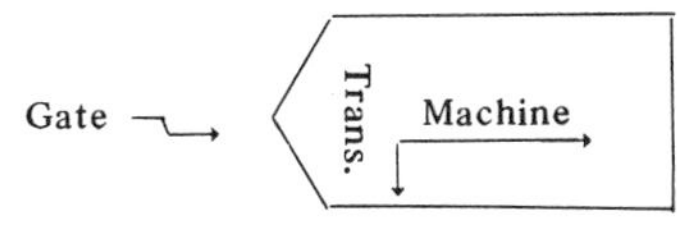

Source: Ref. 1.

Table 10-32. Fybex in engineering polymers.

Resin Fybex Loading	Polysulfone 25%	Modified Polyphenylene Oxide 20%	Nylon 40%
Tensile strength (X 10^3 psi)	14.0	9.0	26.0
Flexural modulus (X 10^5 psi)	10.0	8.8	19.0
Heat distortion temperature (°F at 264 psi)	–	270	480
α_L (in./in./°F X 10^{-5})	1.3	1.6	1.0
Notched Izod impact (ft-lb/in.)	1.8	1.3	1.8

Source: Ref. 1.

Table 10-33. Fybex in thermoformable ABS.

Sample	Tensile Strength (X 10^3 psi)	Flexural Strength (X 10^3 psi)	Flexural Modulus (X 10^5 psi)	Notched Izod (ft-lb/in.)	αL (in./in./°F X 10^{-5})	Room Temperature % Elongation
High impact ABS	4.4	7.0	2.4	7.5	6.1	>20
High impact ABS + 10% Fybex	5.0	7.8	4.2	7.0	3.1	>20
High impact ABS + 15% Fybex	5.8	9.0	6.0	3.0	2.3	>20

Source: Ref. 1.

Table 10-34. Fybex[a] in ABS extrusion grade system.

Fybex Loading (wt.%)	Tensile Strength (10^3 psi)	Flexural Modulus (10^5 psi)	Notched Izod (ft-lb/in.)	Expansion Coefficient (10^{-5} in./in./°C)
0	4.8	2.4	8.0	10.8
15	5.0	5.8	5.5	5.4
25	5.5	8.6	3.5	3.1

[a]DuPont trademark for inorganic reinforcing titanate fibers.
Source: Ref. 2.

Table 10-35. Fybex in 6,6 nylon, mechanical and thermal properties.[a]

Reinforcement Loading[b] (wt. %)	Tensile Strength (X 10^3 psi)	Flexural Modulus (X 10^5 psi)	IZOD IMPACT (ft-lb/in.)		Expansion Coefficient (10^5 in./in./°C)	HDT 264 psi (°C)
			Notched	Unnotched		
0	10.7	4.15	0.94	>60	8.4	66
30% Fybex[c]	20.0	13.25	1.10	16	3.0	225
30% Clay	12.4	6.75	1.24	48	6.1	89
20% Clay 10% Fybex[c]	14.0	7.90	1.15	28	5.0	168
10% Clay 20% Fybex[c]	16.5	9.45	1.10	22	3.7	175

[a] Dry-as-molded properties.
[b] 1% Silane added based on reinforcement loading.
[c] DuPont trademark for inorganic reinforcing titanate fibers.
Source: Ref. 2.

Table 10-36. Relative wear rates of processing equipment.

Filler	Duration of Test (hr)	Part Weight Loss (%)	Relative Rate of Weight Loss
Glass fibers	200	0.066	1.0
Silica	100	0.069	2.1
Fybex[a]	200	0.006	0.1

[a] DuPont trademark for inorganic reinforcing titanate fibers.
Source: Ref. 2.

thermoplastics. Its unique combination of properties has qualified it for certain end-uses not now captured by reinforced thermoplastics.

7.6 References

1. Linsen, P. G. and Regester, R. F., "New Inorganic Fibers for Plastic Reinforcement," 27th Annual SPI RP/CI, 1972.
2. Product Information Bulletin No. 1, "Reinforcement of Thermoplastics with Fybex Inorganic Titanate Fibers," Inorganic Fibers Division, Pigments Department, E. I. DuPont de Nemours and Co., Wilmington, DE 19898, Nov. 1971.
3. Production Information Bulletin No. 2, "Fabrication and Electroplating of Fybex Fiber Reinforced Thermoplastics," Inorganic Fibers Division, Pigments Department, E. I. DuPont de Nemours and Co., Wilmington, DE 19898, Nov. 1971.

8. MAGNESIUM OXYSULFATE FIBER (MOS)

8.1 Introduction

A new inorganic microfiber is available for reinforcing plastic. It has the composition of magnesium oxysulfate (MOS), and is synthesized by a hydrothermal reaction using magnesium sulfate and magnesium hydroxide. The new microfiber has high crystallinity and a good aspect ratio that make a promising material for reinforcement.

8.2 Fiber Properties

MOS fiber has a soft white texture, a specific gravity of 2.3, and a bulk density less than 0.1 g/cm^3 (see Table 10-37). Its aspect ratio is es-

Table 10-37. Characteristics of MOS.

Color	White	Fiber length	10–100 μm
Specific gravity	2.3	Fiber diameter	<1 μm
Specific bulk density	<0.1 g/cm^3	Aspect ratio	50–100
Moisture	<1%	Specific surface area	10 m^2/g
PH	9.5		

timated to be in the range of 50 to 100; the fibers generally are less than 1 μm in diameter with length from 10 μm to 100 μm. Tensile and modulus data for the fiber have not been obtained because it is difficult to test individual fibers that are so small. However, the reinforcing potential has been demonstrated, in composite testing that is described below.

8.3 Experimental

8.3.1 Materials. The polymer used in this work is polypropylene (PP), a chemically modified PP containing organic silane groups, UBE PP ZP 711 (MFI: 10–15 g/10 min).

MOS-reinforced PP was prepared by a twin-screw extruder with a vent at 210°C. Stearic acid of 1 wt % was added to obtain the preferable dispersion of MOS in a PP matrix. The test specimens for mechanical properties were injection-molded at 200°C.

All tests were carried out according to the corresponding ASTM standards.

8.3.2 Mechanical Properties. Some fundamental mechanical properties of a 20 wt % MOS-reinforced PP are shown in Table 10-38. The corresponding values for talc and short glass fiber-reinforced PP are also included. This table proves that MOS-reinforced PP has a high modulus, good thermal properties, and impact properties. In addition, the appearances of molded articles are good compared to glass fiber–reinforced composites.

These favorable properties can be interpreted in terms of microstructural characteristics. The whisker-like nature of MOS accounts for the high modulus of the composites. Photographs of the cross section of MOS-reinforced PP specimens, brittle-fractured in a liquid nitrogen, show that MOS fibers are uniformly dispersed and also highly oriented to the flow direction in a PP matrix. These results have a desirable effect on the modulus of composites using MOS.

Although MOS-reinforced PP has many advantages, its mechanical strength is not as good as that of the glass fiber–reinforced PP, as seen in Table 10-38. Table 10-39 shows improved

Table 10-38. Mechanical properties of MOS-reinforced PP.

Property*		Unfilled	20% MOS	20% Talc**	20% GF***
TYS	kg/cm^2	255	265	262	324
TEL	%	>500	52	86	8
FM	kg/cm^2	13,500	60,400	26,400	36,100
I_{IZ}	kg·cm/cm	6.4	6.6	5.1	7.0
I_{FW}	kg·cm	100	50	50	6
HDT	°C	64	98	80	102

*TYS; tensile yield strength; TEL: tensile enlongation to break; FM: flexural modulus; I_{IZ}: notched Izod impact strength; I_{FW}: falling weight impact strength; HDT: heat distortion temperature (fiber stress: 18.6 kg/cm^2).
**Average diameter: 8 μm.
***Surface-treated with a silane coupling agent (fiber length = 7 mm, diameter = 13 μm).

strength data with the use of a chemically modified PP. Chemical coupling between PP and MOS was obtained, and, as a result of better interfacial bonding, greater reinforcement was realized in the modified PP composites.

Relative to the thermal stability of MOS at elevated temperatures, two processes of dehydration of the crystal water that it contains are observed. The primary dehydration process appears to be initiated around 250°C. As a result, MOS processing must be carried out below this temperature. This may be a serious problem for MOS-reinforced PP composites. As an example, Table 10-40 shows the influence of injection-molding temperature on the flexural properties of MOS/PP composites. These properties decrease rapidly when the temperature exceeds 220°C. Composite appearance also becomes unsatisfactory.

Table 10-41 compares the melt flow indices

Table 10-39. Effects of a chemically modified PP on mechanical properties of MOS-reinforced polypropylene.

Type of PP Resin:			
J709HK	(wt %)	40	80
ZP711	(wt %)	40	0
Fiber:			
MOS	(wt %)	20	20
Tensile yield strength	(kg/cm^2)	339	269
Flexural strength	(kg/cm^2)	589	439

Table 10-40. Effects of injection-molding temperature on mechanical properties of MOS-reinforced PP (MOS: 40 wt %).

Temperature,* °C	200	220	240	260
Appearance	Good	Poor**	Poor**	Poor**
FS,*** kg/cm^2	469	473	446	437
FM*** kg/cm^2	112,000	115,000	108,000	102,000

*Measured at the nozzle.
**Silver streaks observed for the molded test specimens.
***FS: flexural strength; FM: flexural modulus.

Table 10-41. Melt flow indices of PP melts filled with MOS.

	MFI (g/10 min)	
Sample	230°C	200°C
Unfilled	9	4.5
20% MOS-filled	15	7.3
20% Talc-filled	9.3	—
20% Glass fiber-filled	5.3	—

(MFI) of MOS, talc, and glass fiber–filled PP. MOS is clearly superior.

8.4 Conclusions

Based upon the information presented here, it can be concluded that:

1. MOS-reinforced PP is an interesting, high-quality composite. This is a direct result of the whisker-like nature of MOS. It also is highly oriented, has good dispersion, and has good chemical affinity in a PP matrix.
2. MOS can be chemically bonded to a PP matrix using a modified polymer to increase composite strength.
3. MOS-reinforced PP has an upper limit to its processing temperature because of the dehydration of crystal water contained in MOS. However, its excellent low melt viscosity allows lower temperature processing and therefore, reduces this disadvantage.

8.5 Supplier

Magnesium oxysulfate is available from:

Hirakata Plastics Laboratory
UBE Industries, Ltd.
Hirakata, Osaka 573, Japan

Price and delivery information are obtainable from the supplier.

8.6 Credit

Most of the information used in this section came from an article by S. Hoshino, H. Tanaka, and T. Kimura of UBE Industries, Ltd.

11

Chopped and Milled Fibers and Technology of Cutting Fibers

John V. Milewski

Consultant
Santa Fe, N.M.

CONTENTS

1. INTRODUCTION

This chapter is composed of three parts. The first is a general introduction to chopped and milled fibers and their properties. Next is a section on techniques for cutting fibers, based on a presentation at the Tappi Non-Woven Fiber Seminar on March 1979 by G. B. Keith. A final section lists the suppliers of milled fibers, chopped fibers, and fiber cutting services.

2. CHOPPED AND MILLED FIBERS

Chopped and milled fibers are made from continuous fibers such as glass fibers, carbon/graphite, and aramid polymer fibers, as well as metal fibers. Chopped fibers generally run from ⅛ in. to several inches long, and are usually chopped in strand bundles; because of a binder on the strands, they usually remain in fiber bundles, resulting in relatively low L/D.

One problem with chopped fibers is failure to recognize the severe effect the number of strands per bundle has on the L_iD (aspect ratio) (Fig 11-1). This has an effect on the bulk density, which in turn has a dramatic effect on processing conditions. Also, the amount and type of binder on the strand bundle can cause two problems—too much binder keeps the strand too tight and prevents resin wetting, resulting in poor properties, while too little binder in the strand bundle causes premature blooming of the bundle during processing, severely increasing the amount of work needed for processes. If too much additional work is added, it causes excessive reduction in the fiber's aspect ratio.

Milled fibers are made by milling chopped fiber into shorter lengths and screening to sizes such as 1/32, 1/16, ⅛, and ¼ in. The milling process individualizes the fibers, making the full fiber L/D available for reinforcement (Fig. 11-2). This gives most milled fibers a greater

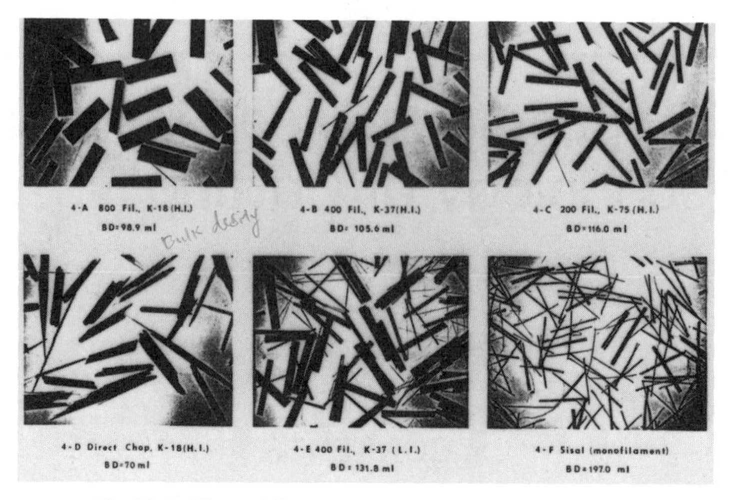

Fig. 11-1. Chopped fiber at various aspect ratios, mag. ½×.

L/D than chopped fibers have that are shorter than ½ in. in length. This result can be explained by examination of the method used to calculate the L/D of the chopped fiber bundle. It is calculated by the addition of the total cross-sectional area of all fibers into an equivalent single fiber diameter at the chopped length. The data are given in Table 11-1.

The novice compounder immediately thinks that because chopped fibers are longer than milled fibers, they have significantly greater equivalent L/D, and this greater L/D accounts for their significantly greater reinforcing ability. This is not entirely true. In the first case, the L/D's are not greatly different. But more significant is a fact with which even the more experienced compounder may not be acquainted—milled fibers have a problem that chopped fibers do not have: milled fibers carry with them a significant amount of strength-degrading debris formed in the milling process, which can be seen in Fig. 11-3. This debris represents about ¼ of the mass of milled fiber, especially the shorter grades, and is sharply edged glass pieces and fibers of very short L/D's. The weakening effect of this debris probably accounts for a 30% to 50% strength loss in milled fiber reinforcements that chopped fibers do not have, so that chopped fibers perform better in composites.

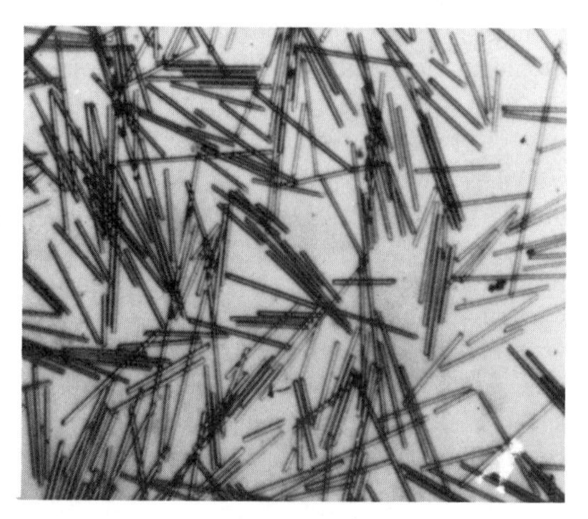

Fig. 11-2. Cleaned milled glass, mag. 50×.

Table 11-1. Comparing the *L/D* of chopped and milled fibers.

Chopped-fiber[a] length, in.	Filaments/ strand	Equivalent L/D
$\frac{1}{4}$	800	16.5
$\frac{1}{4}$	400	23.3
$\frac{1}{4}$	200	32.9
$\frac{1}{8}$	400	11.7
$\frac{1}{4}$	400	23.3
$\frac{1}{2}$	400	46.7
Milled-fiber[b] length (screened), in.		
$\frac{1}{32}$	1	10.5
$\frac{1}{16}$	1	15.3
$\frac{1}{8}$	1	25.3
$\frac{1}{4}$	1	50.5

[a]Data on chopped fiber from reference 1.
[b]Data on milled fiber from reference 2.

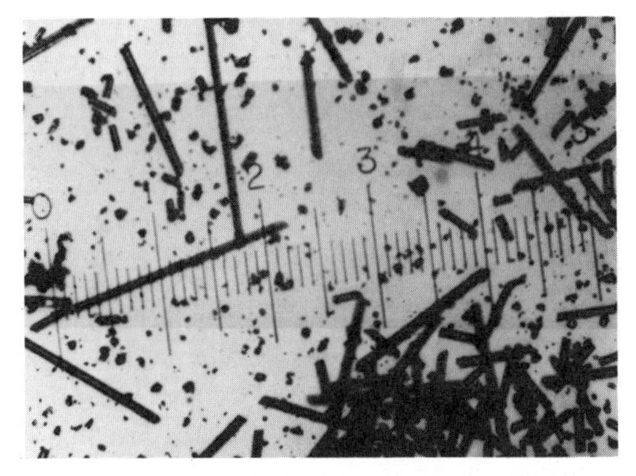

Fig. 11-3. Dirty commercial milled glass, mag. 80×.

This debris problem also helps explain the anomaly of why such weaker short fibers as wollastonite and PMF compete as well as they do against 1/32 in. milled glass, as seen in the data in Table 11-1. The reason for this is that the manufacturers of wollastonite and PMF screen out the sharper debris and particulate materials during the processing of these two fibers and sell them as powder grades, leaving a relatively cleaner, shorter-fiber product.

The properties of the chopped and milled fibers are related to structure, size, and manufacturing method used for the continuous fibers from which they are made. For example, amorphous fibers such as glass fiber and polymeric fibers such as aramid (DuPont's Kevlar) generally are low in modulus, high in tensile strength, and high in elongation; polycrystalline fibers such as carbon/graphite and boron are generally high in modulus, low in tensile strength, and low in elongation.

Toughness and nonfriability are substantially greater for polymeric fibers such as aramid than for glass fiber, although the fibers' relative tensile strength and elongation are almost the same. Thus cutting aramid is very difficult and milling it impossible. Glass fiber, on the other hand, cuts and mills very readily because the polymeric fiber's basic strength comes primarily from its internal structure, while the glass fiber's strength comes primarily from its surface. Glass fibers, then, are surface-sensitive and lose strength very quickly with scratching of the surface; polymeric fibers do not have this problem. A proposed solution to this problem is to coat glass with a thin, tough layer of resin such as urethane.

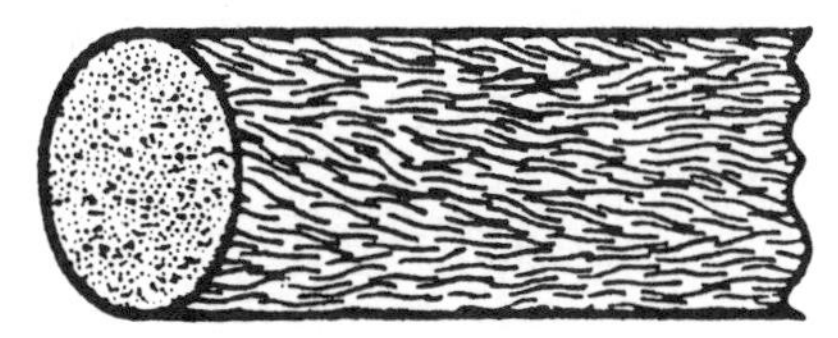

Fig. 11-4. Structure of aramid (Kevlar): a semiloose bundle of relatively long polymer chains in semiparallel array.

Aramid. The structure of Kevlar is basically a lightly bound bundle of relatively long polymer chains in a semiparallel array, as shown in Fig. 11-4. The polymer chain length alignment and degree of radial bonding determine the properties of the fiber. The total strength derives from the additive strength of each polymer chain. Scratching the surface ruptures only a few of the surface polymer chains; most of the chains remain intact, contributing their full strength to the remaining fiber.

The loosely bonded polymer-chain structure also accounts for the poor compressive strength of the Kevlar fiber. Under a compressive load, for example, the relatively long polymer chains buckle and separate, since the radial bonding of the polymer-chain bundle is not very strong.

Glass Fiber. The internal structure of glass fiber is a random network of amorphous glass, while the surface is a semi-oriented structure highly stressed in longitudinal compression, which gives it good tensile strength (see Fig. 11-5).

Almost all the strength of glass fiber is associated with this surface. Scratching the surface destroys the compressive stress and the associated strength of the fiber.

Carbon/Graphite. The structure of high-strength carbon/graphite fibers is quite different from that of either glass fiber or aramid. The matrix is amorphous carbon, filled with microcrystallite fibrils of graphite (see Fig. 11-6).

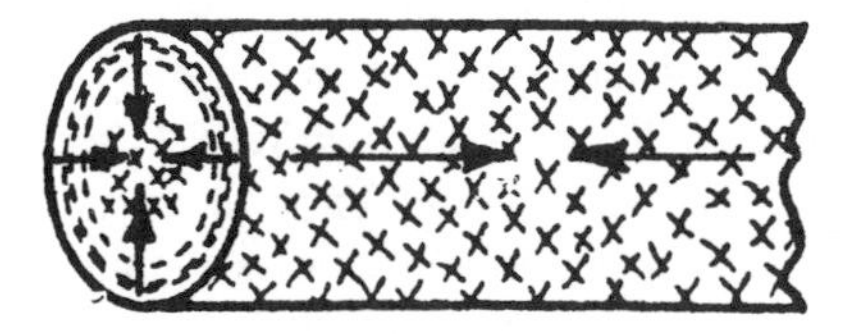

Fig. 11-5. Structure of glass fiber. Internal structure is a random network of amorphous glass. The surface is semi-oriented, highly stressed in longitudinal compression, which gives it good tensile strength but poor scratch resistance. All strength is associated with the surface.

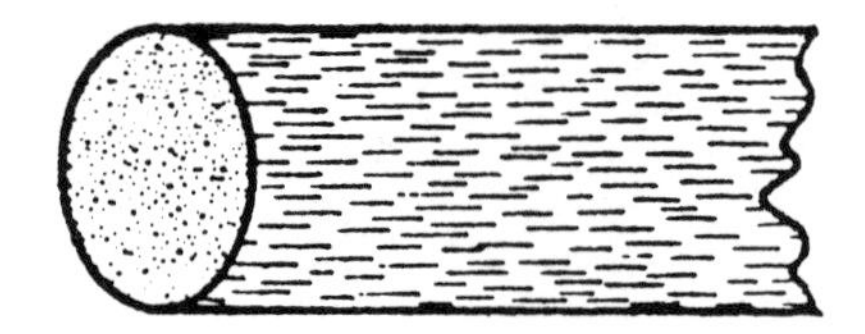

Fig. 11-6. Structure of carbon fibers. Matrix is amorphous carbon filled with acicular microcrystallites of fibrils. The amount of L/D and the degree of alignment determine both the tensile strength and modulus of the fiber.

The amount, the L/D, and the degree of longitudinal orientation of these fibrils determine both the tensile strength and the modulus of the fiber. There is also some strength associated with the surface area and surface structure, but most of the carbon-fiber strength is due to the internal structure.

Unfortunately, none of the microcrystalline fibrils of any of the carbon/graphite fiber compositions are of adequate length to utilize their full whisker-like strength and elongation. Consequently, they pull out of the amorphous carbon structure prematurely, under relatively low strain levels. This accounts for the low elongation and lack of toughness of carbon/graphite fibers in general and for the significantly lower elongation of the higher-modulus carbon/graphite fibers that contain significantly shorter reinforcing fibrils.

3. TECHNIQUES OF CUTTING FIBER

3.1 Evolution of Fiber Cutting

Before synthetic fibers, there were mainly cotton and wool. These fibers were processed on "cotton system" or "woolen system" machines. Fiber was sometimes not cut; it was "broken," especially the longer filaments. However, they didn't need to be cut, as they were about the right length already.

Synthetics did not erupt onto the fiber scene in billion-pound quantities overnight. They developed gradually. And "naturally" there were still cotton and wool. Fiber processors wanted to keep their cotton and wool equipment, and would improvise for years to keep it. Continuous synthetics were cut to the proper lengths based on the capabilities of existing cotton and wool machines. Hesitant manufacturers assumed a posture of "make do" until they could regroup, assess the situation, and catch up with the industry. They have not caught up to this day, and the nonwovens industry has suffered as a result.

3.2 The State of the Art

Improvising to manufacture nonwovens on existing equipment has worked just well enough to hinder the development of short fiber cutting. Reworked cotton and woolen machinery runs the longer lengths of fiber. However, no one knew just what exotic webs lurked in the shadows of shorter fibers. Manufacturers had none, at first. They experimented with a paper machine, reworking it for wet lay work. Where would they get fibers longer than wood pulp but shorter than cotton? They would use cotton linters. But cotton linters are random lengths and an unpredictable byproduct. How much more could they possibly accomplish with the correct length fibers accurately cut? They would have to wait, then wait some more.

The wet lay webmaker has been waiting, crying out for specially cut fibers, and specially designed machines to process them. He has yearned for a fiber supply and web manufacturing system worthy of the fine art he has developed. Few manufacturers have listened, and only now is the webmaker beginning to be heard. In the meantime, fiber has continued to be hacked by blacksmiths on the old knife and bedplate cutters, and the whole nonwovens industry has been handicapped because of the poor state of the art of short fiber cutting.

The dry lay webmaker has been somewhat more fortunate. From a starting point of cotton and woolen system machinery, he has been able to improvise for longer, crimped fibers. This gave him a standard fiber supply and some possibility of varying the loft and hand of his web. Machinery manufacturers were somewhat cooperative, desiring to develop in the friendly confines of standard length fibers. Notable contributions toward these developments, one foreign and one domestic, were made by Fehrer and Rando Machine.

However, dry lay webmaking is a much slower process and prompts two questions concerning the progress of the webmaking art in general: If short, accurate fiber with water as a carrier can be laid at the present phenomenal speeds, why can't the same short fibers be laid with air as a carrier at the self-same phenomenal speeds? If they can, then with the advent of an ample supply of short cut fiber, will one system lose its franchise for existence? It would seem so.

3.3 The Elusive Filament

There is probably no task so singularly demanding as the accurate cutting of a truckload of short cut, nonwoven fiber. The purchase order could read: "One and one-half trillion pieces of polyester, each one-half thousandth of an inch in diameter, one-fourth inch long, square cut ends, cut from drawn filaments, relaxed and heat set, no longs and no fusing!" No finish is specified; that is another story. Why complicate the issue with a chemical handicap?

By way of comparison, let us inspect a spoonful of white sugar with the naked eye. One is readily quite certain that the particle size varies by a factor of 10 to 1. It does not matter; your cup of coffee will never know the difference. However, a single filament 10 times normal length, or 2½ in. long, mixed with the above ¼ in. fibers, will produce a defect in a nonwoven web so immediately discernible that it can be detected by the naked eye, even if the web is traveling at a speed of 1000 feet per minute! Justifiably, it somewhat naturally follows that while no task is so singularly demanding as cutting fiber, no task is so singularly rewarding or valuable. The webmaker requires fiber with at least no extreme longs or fused filaments. He demands it, and is willing to pay the price. To do otherwise exacts an economic judgment too severe to contemplate. Stockholders know it as "going out of business"!

3.4 What is a Poor Cut?

Though cutting fiber may not be an exacting science, poorly cut fiber is easy to describe: it may be too long or too short, have a faulty end condition, or both.

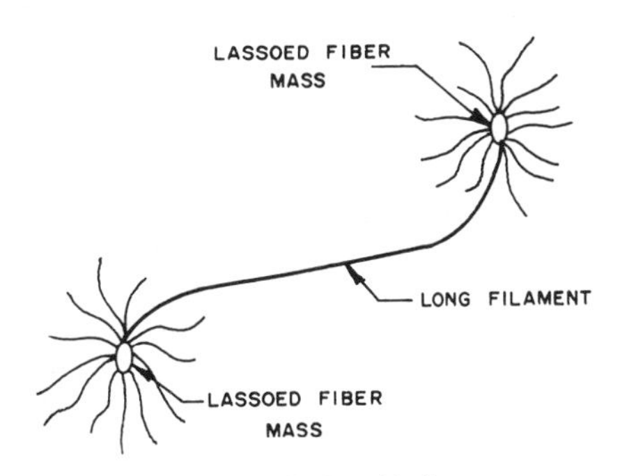

Fig. 11-7. Dumbbell.

3.4.1 Longs. Filaments that are too long lasso other fibers, gathering concentrations and entanglements that will not satisfactorily pass between downstream calendering rolls. These clumps form web defects described as dumbbells, which occur when both ends of a long fiber lasso other fibers. Such a formation is shown in Fig. 11-7.

Should only one end of a long filament lasso other shorter fibers, web defects known as scabs are produced. Such a defect is shown in Fig. 11-8.

Sometimes a concentration of fiber is so bulky that it will not pass between the rolls without splitting down the middle. These defects are called blisters. Such a defect is shown in Fig. 11-9.

3.4.2 Shorts. Little need be said about fibers that are too short. Other than their failure to carry a proportionate share of the load as tension members in the web, they cause no concern. They lie well, do not concentrate or entangle, and pose no other manufacturing problem. They only loaf on the job.

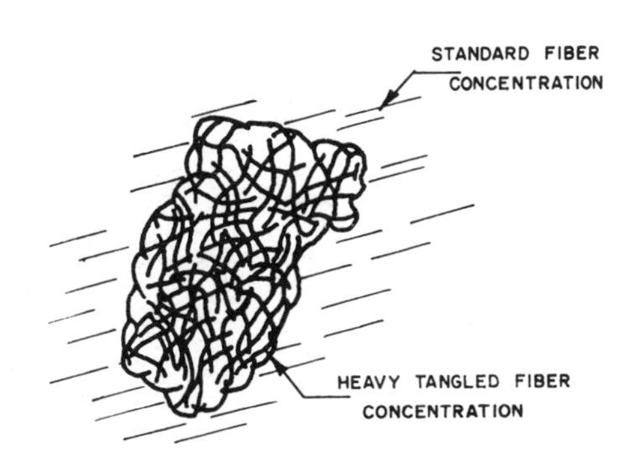

Fig. 11-8. Scab.

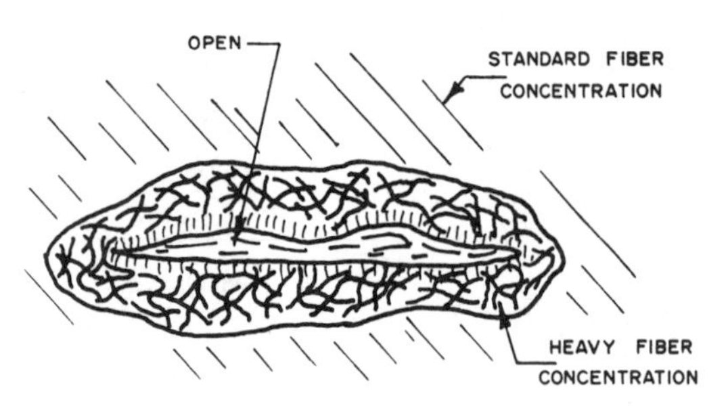

Fig. 11-9. Blister.

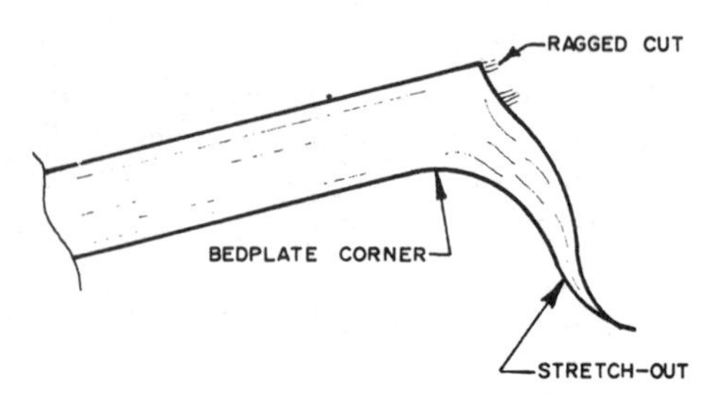

Fig. 11-12. Elf shoe end.

3.4.3 End Conditions. The ideal end condition is a smooth, square cut. Such an end, highly magnified, is shown in Fig. 11-10.

The ideal end is elusive because the hot melt synthetics, in particular, are plastic and cold-flow under pressure. The knife deforms them somewhat. Some small amount of flattening is acceptable. Any end condition that tends to entangle another filament is not so acceptable. Some such end conditions are: the finishing nail head, shown in Fig. 11-11; the elf shoe, shown in Fig. 11-12; the split end, shown in Fig. 11-13; the fused ends, shown in Fig. 11-14; the fused log, shown in Fig. 11-15; the fused daisy, shown in Fig. 11-16; and pressure married fibers, shown in Fig. 11-17.

It is heartening to note that some of these poor end conditions may go unnoticed, while running quite good web. This is especially true if they are not too prevalent and if they are not found in the company of extreme longs and/or

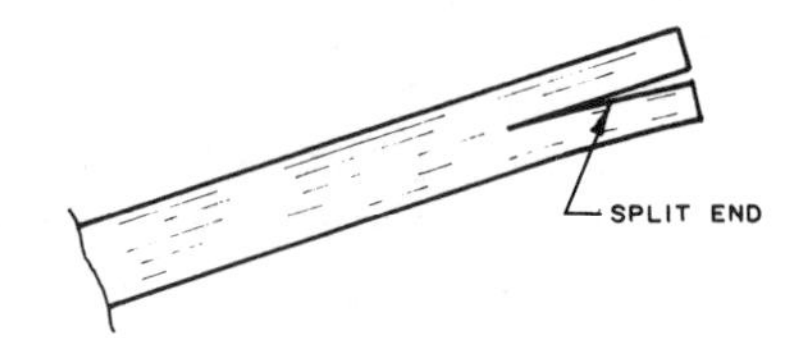

Fig. 11-13. Split end.

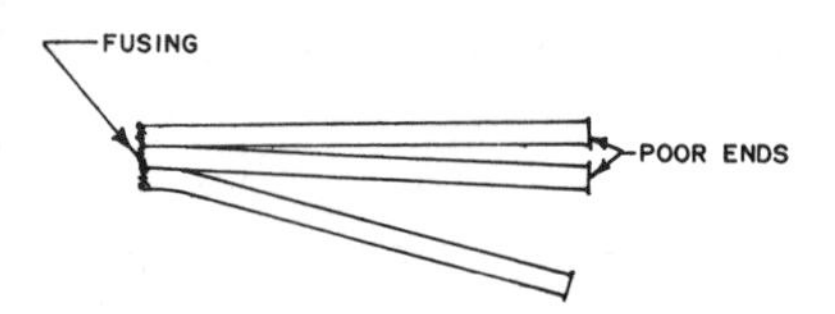

Fig. 11-14. Fused ends.

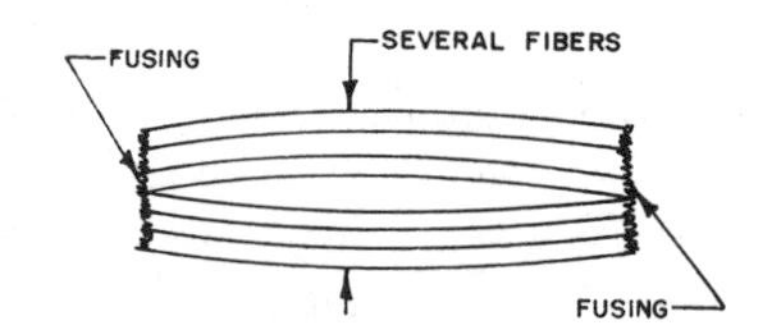

Fig. 11-15. Fused log.

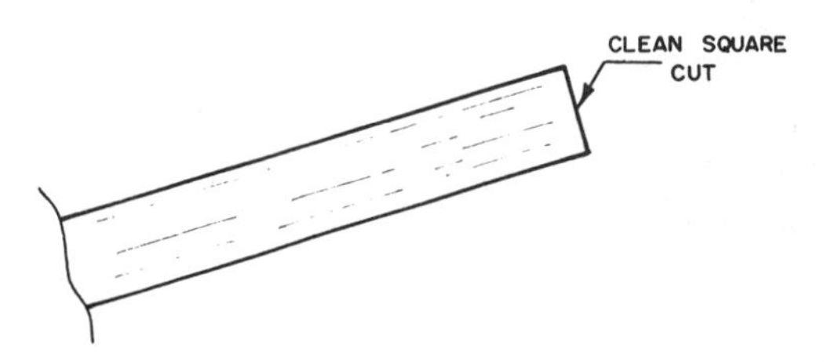

Fig. 11-10. Ideal end.

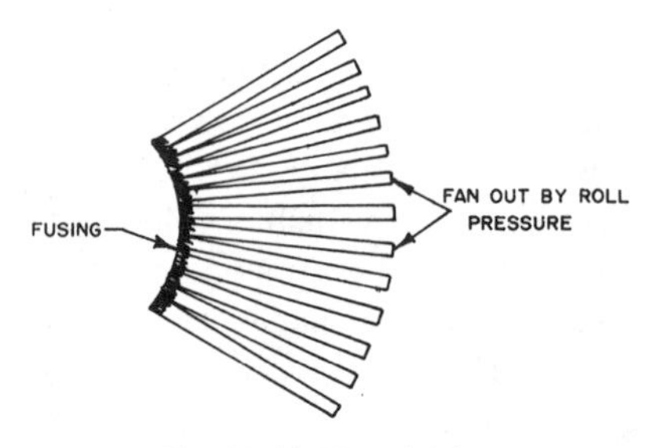

Fig. 11-16. Fused daisy.

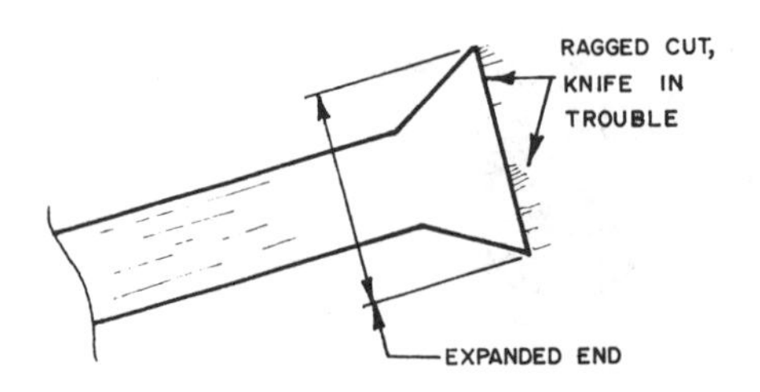

Fig. 11-11. Finishing nail head end.

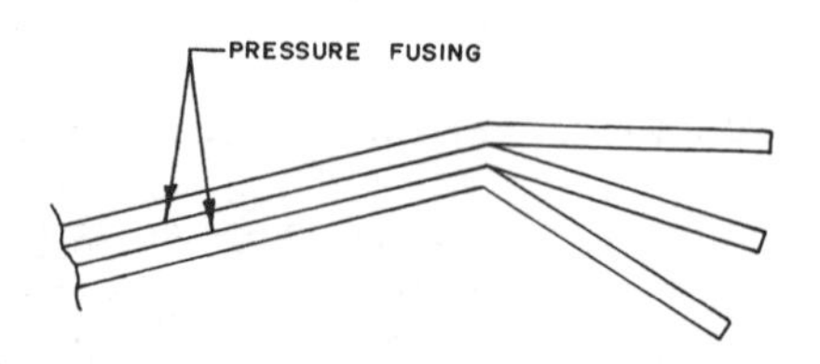

Fig. 11-17. Pressure married fibers.

fused filaments. Combinations in the same fiber stock usually spell trouble.

3.5 Fiber Cutting Theory

Many mill people use the painful term "chop" for cut. The rate of severance in the cutting of fibers can be separated into two extremes. Common reptiles have been chosen to illustrate each: Fiber is cut quickly (chopped), like an alligator's snapping jaws; or fiber is cut slowly (sliced), like a boa constrictor's squeeze. The latter, because of the absence of plastic cold flow and heat generation, is much preferred. It produces the better cut. Though developments in fiber processing are constantly urging us to cut fiber faster, the quality of such fiber is just as urgently telling us to cut fiber slower.

3.6 Methods of Cutting Fiber

With the general theory of two extremes of timing established, there are presently six methods used in cutting fibers, discussed below.

3.6.1 Knife and Bedplate (see Fig. 11-18). In this method fiber is fed between rollers, over a bedplate, and a rotating knife shears close beside the bedplate edge to cut the moving fiber. Several cutter types that espouse this oldest method of cutting are: Taylor-Stiles, Peter Brotherhood, Pekrun, Mitsuishita, and Beria. Boy Scouts start fires by striking flints together; two pieces of hardened metal are little different. Disadvantages are: impact (snapping alligator) cut, multiple lengths, fluttering feed, and fused fibers.

3.6.2 Squeeze Peels with Knife Slash Through (see Fig. 11-19). In this method, two

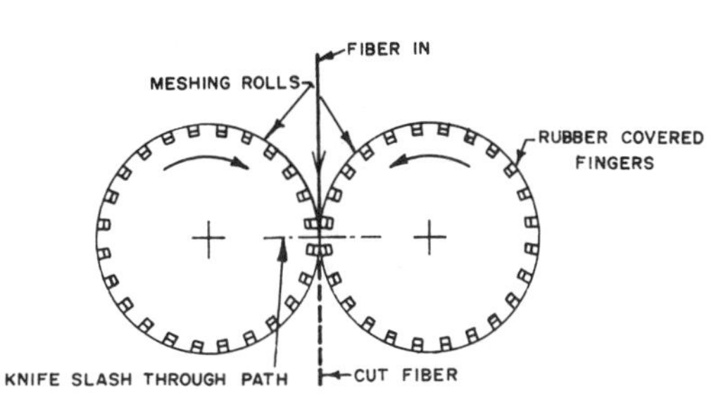

Fig. 11-19. Squeeze reels with knife slash through cutter.

meshing rolls with rubber covered fingers grip the fiber bundle on either side of the knife while the knife is timed to slash between the moving fingers and cut the fiber. The Neumag (Gru-Gru) cutter is the only machine of this type. The design, interesting and unique, is used quite widely. Disadvantages are: its impact (snapping alligator) cut; its down time to change knives (about every 30 minutes); the fact that it gets out of time and cuts rubber from the reel fingers; the fact that the knife hacks along on the filaments before cutting, damaging filaments and dulling the knife; and its inability to cut short-length staple.

3.6.3 Rubber-Covered Squeeze Rolls with Protruding Knives (see Fig. 11-20). In this method, two rolls form a pulling nip with one roll having spaced knives protruding into the rubber covering on the other. This method is widely used to cut (fracture) glass fiber and some easily cut synthetics. Disadvantages are: rapid dulling of knives, hence poor fiber end conditions; broken knives; double lengths (by embedding in rubber around edge of knife); and excessive noise and maintenance.

3.6.4 Reel and Roller (see Fig. 11-21). In this method, fiber under tension is wrapped

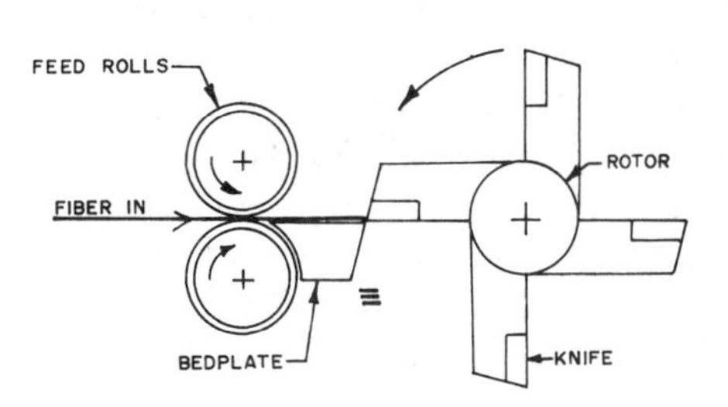

Fig. 11-18. Knife and bedplate cutter.

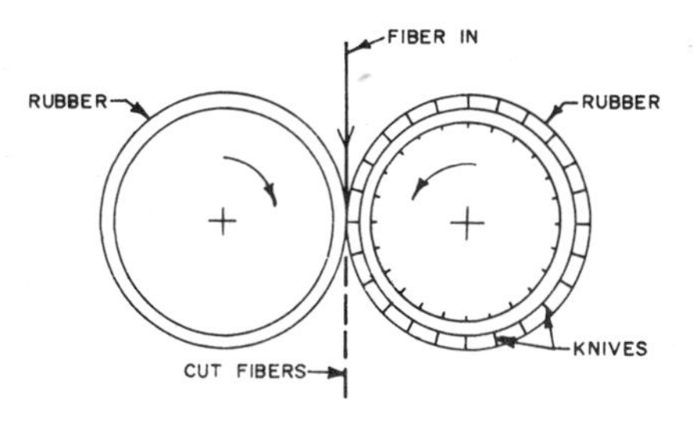

Fig. 11-20. Rubber covered squeeze rolls with protruding knives cutter.

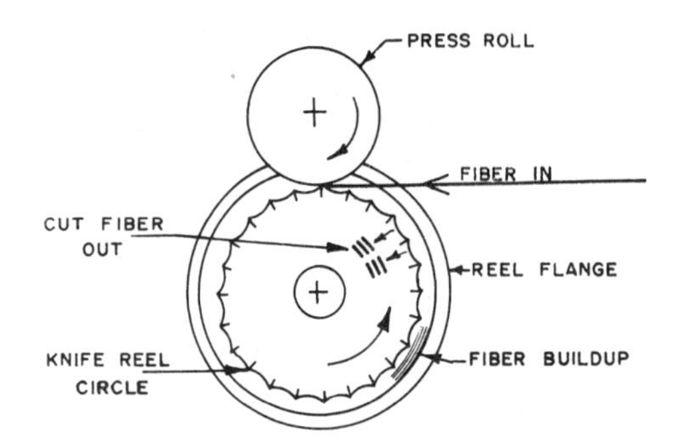

Fig. 11-21. Reel and roller cutter.

onto a reel of knives between two flanges. It is discouraged from building to a larger diameter by a rolling press wheel, thus forcing the fiber to be cut and pushed between the converging knives. The Lummus cutter is the only machine of this type. First marketed in the late 1960's, it is now considered to be the standard of the long staple industry all over the world. It is a slow squeeze (boa constrictor) type cutter with many advantages, one being that nothing touches the knife edge except fiber. Some disadvantages are: roping of output on long staple; blocking and stuffing on short staple; and because of this feature, inability to satisfactorily cut very short, wet lay staple.

3.6.5 Reel and Cam (see Fig. 11-22). In this method, fiber is fed into the hollow of a cam and then to the interior of a sunburst of inward-pointing knives. Fiber builds between the inner knife edges and the cam and is forced outward to be cut and escape between the diverging knife surfaces. The Mini Fibers, Inc. cutter is the only machine of this type. It was developed to avoid the stuffing and blocking problems found in the Lummus machine and is best suited for cutting short staple for wet lay nonwoven use. The fiber does not stuff because the fiber escape slot diverges, and the cut fiber is free to escape without pressure. The machine is a slow squeeze (boa constrictor) type cutter with many advantages. Two disadvantages are: frictional heat buildup on the fiber from the sliding cam action, and the side thrust of the fiber against the knife.

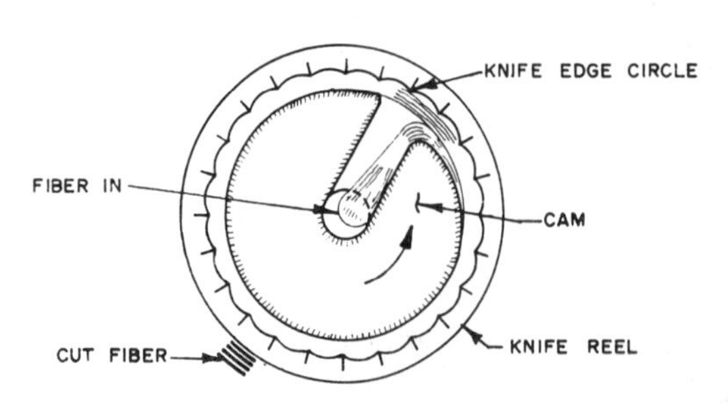

Fig. 11-22. Reel and cam cutter.

3.6.6 Variable-Length Cutting. Short fibers cut directly into variable lengths have proved advantageous over fibers cut at uniform lengths in several applications. These include a lower cutting cost and greater ease of subsequent fiber–matrix mixing. Asbestos is, of course, a good example of a variable-length reinforcing fiber.

The Buss-Condux Cutter CS 300/400-4 is a good example of this type of cutter. Suitably prepared and treated fiber tow is fed down onto a horizontally rotating drum having longitudinally mounted blades that then come into near contact with stationary blades. Fiber between these blades is cut and falls by gravity, and is then carried by airflow to a screen installed beneath the rotating drum. When the cut fiber is properly sized, it passes through the screen and is carried by air to a suitable cyclone separator. Different screens may be installed in the cutter to regulate the average size of the cut fiber, which generally has a monomodal size distribution. Proper moisture and finish levels are essential in this as in other cutting processes to assure clean cut fibers.

Further details are described in two patents assigned to Badische Corporation (Spain, Patent No. 507,568, Sept. 22, 1982; and, East German, Patent No. 208,081, Mar. 28, 1984, Brazil 8,108,011, Australia 546,861, Canada 1,178,012.)

3.6.7 Today's Cutters. Each fiber cutting machine has unique characteristics to better adapt it for production for either wet lay or dry lay work.

For wet lay fibers, most of today's production is undoubtedly cut by the knife and bedplate machines, butchering away as they have done for decades. Recently, however, the reel and roller machines, competing with a superior cut, have taken a larger share of the market. They stand to make further inroads as their cut becomes more widely known. Most short cut glass fiber is produced on the machines with rubber rolls and protruding knives, resulting in a fracture rather than a cut.

Longer, crimped, dry lay fibers, easier to cut than the short fibers, are produced most widely on the reel and roller machines. The remaining share of the dry lay nonwoven requirements is fairly well divided. The knife and bedplate and the squeeze reels with knife slash through methods appear to be nearly equal in competing for this production.

4. SUPPLIERS OF CHOPPED, MILLED FIBERS AND FIBER CUTTING SERVICES

The following is a list of the companies that are suppliers of materials and services as indicated following their address. For suppliers of chopped metal or graphite fiber, please refer to the appropriate chapter (e.g., short metal fiber: Chapter 6; carbon/graphite: Chapter 20).

Badische Corporation
P.O. Drawer D
Williamsburg, VA 23187
(804) 887-6000
Synthetic fibers—chopped

Certain Teed Corporation
Fiber Glass Reinforcement Division
P.O. Box 860
Valley Forge, PA 19482
(800) 433-0922:C
Fiberglass—chopped

Claremont Flock Corporation
169 Main Street
P.O. Box 850
Claremont, NH 03743
(603) 542-5151
Natural and synthetic fiber—chopped

Deeglass Fiber Ltd.
Queensferry
Northwest of Wreham England
Fiberglass—chopped

E. I. DuPont de Nemours & Co. Inc.
1007 Market Street
Wilmington, DE 19898
(800) 441-7515
Kevlar—chopped and pulp

Fiber Glass Industries Ltd.
Homestead Place
Amsterdam, NY 12010
(518) 842-4000
Fiberglass—chopped

Finn and Fram
13231 Louvre Street
Arleta, CA 91331
(818) 897-5209
Chopping service—all types of fibers

Henry and Frick Inc.
35 Scotland Blvd.
P.O. Box 608
Bridewater, MA 02324
(617) 697-3171
Milled glass

Johns Manville
Filtration and Industrial Minerals Div.
Ken-Caryl Ranch
Denver, CO 80217
(303) 979-1000
Chopped micro glass (Chop-Pak)

Mini Fibers Inc.
P.O. Box 186
Weber City, VA 24251
(617) 282-4242
Chopping service—organic and synthetic

McCann Manufacturers
Rt. 14A
Oneca, CT
(203) 564-4046
Chopping service—all types of fibers

Owens/Corning Fiber Glass Corp.
Fiber Glass Tower
Toledo, OH 43659
(419) 248-8000
Fiberglass—milled and chopped

References

1. Mohr, J. G., "A Bulk Density Testing Device for Quality Assurance of Fiber Glass Chopped Strands," SPI. 34th RR/C Proceeding Paper 19-D, 1979.
2. Milewski, J. V. "A Study of the Packing of Fibers and Spheres," PhD. Thesis, Rutgers University, 1973.

Acknowledgement for Section on Buss-Conduxcutter
Billy B. Hibbaed Williamsburg Va.

12

Short Organic Fibers

W. B. Hibbard
Hibbard Enterprises
Williamsburg, VA

John V. Milewski
Consultant
Santa Fe, NM

CONTENTS

1. INTRODUCTION

Reinforcement with short organic fibers grew more important during the late 1970s and early 1980s because of concerns over the potential toxicity of asbestos. Certain natural short organic fibers, such as cotton linters and wood pulp, have been used for many years; however, several of the advantages of asbestos such as nonflammability, good chemical and thermal stability, and high tenacity were difficult to match. Nonetheless, many short synthetic organic fibers have been developed, and several are finding excellent acceptance in the marketplace. These fibers include nylon, aramid, polyester, acrylic, polypropylene, polyethylene, carbon fiber, and other related materials.

Short fibers are used for the reinforcement of such products as gaskets, friction products, floor tile, roofing products, packing, bulk molding compounds, cement pipe and sheet, thermoplastic and thermoset products, and even explosives.

Short organic fibers are usually prepared by cutting long tow or strands of fiber into the desired lengths in various types of commercial cutting machines. Well-known cutting machine manufacturers are American Barmag Corp.,[1] Buss-Condux, Inc.,[2] Fi-Tech, Inc.,[3] Fleissner Inc.,[4] Lummus Industries, Inc.,[5] and many others.

An excellent source of precut short organic fibers is Mini-Fibers of Johnson City, Tennessee.[6] Mini-Fibers markets short fibers of nylon, polyester, polypropylene, rayon, acrylic, and aramid at various deniers and at lengths of ¼ in. to 1 in. Custom cutting is also available.

Uniform-length fibers, blends of different-length fibers, and short fibers cut at variable lengths have found acceptance in the marketplace. (Asbestos is a product with variable fiber lengths and fiber diameters.) Various comminuting machines are available for cutting synthetic fibers into variable lengths simultaneously. A good example is the CS 300/400-4 cutting machine marketed by Buss-Condux,[2] as described in Spain Patent No. 507,568[7] (Sept. 22, 1982), East German Patent No. 208,081[8] (Mar. 28, 1984), Brazil Patent No. 8,108,011 (Dec. 9, 1981),[9] Canada Patent No. 1,178,012 (Nov. 20, 1984),[10] and Australia Patent No. 546,861 (Feb. 20, 1986).[11] Additional details on cutting machines are found in Chapter 11.

Fibrillated short fibers are commercially available. Pulped Kevlar™ from E. I. du Pont de Nemours & Co.[12] is an excellent example of a pulped short fiber. Pulping can be accomplished in Sprout-Waldron[13] equipment. Custom fibrillation processing is available at the Herty Foundation, Savannah, Georgia.[14] Fibrillatable acrylic fibers have been available from American Cyanamid Corporation[15] and Badische Corporation.[16]

Mixing or blending of short fibers into the body material is more readily achieved when a range of fiber lengths is used although the reinforcing effectiveness may be reduced. Various mixers that are useful for mixing short fibers into a matrix are manufactured by Eirich Machines[17] and Littleford Bros.[18]

When short fibers are randomly mixed, formed, and cured, a three-dimensional reinforcement is generally achieved. Reinforcement with long fibers generally leads to reinforcement in only two dimensions.

The dimensions, physical properties, and chemical stability of a short organic fiber are critical in any given application. To achieve reinforcement of a composite body, the short fiber must have higher modulus or tenacity than the matrix. Also, the bond strength between fiber and matrix must be sufficiently great so that fracture occurs by the fiber breaking rather than slipping out of the matrix as a result of poor adhesion. The Textile Research Institute, Princeton, New Jersey, is currently conducting fiber–matrix adhesion studies.

2. SHORT CUT ORGANIC FIBERS—MINI-FIBERS, INC.

2.1 General

For several years Mini-Fibers, Inc.[6] has been a commercial supplier of several synthetic organic short fibers that are used for reinforcement as well as for nonwovens, specialty papers, and felt materials. Custom cutting is also provided for experimental materials and small-scale production quantities. The specific details of these cutting processes are closely guarded and quite specific to fiber composition, denier, properties, and end-use application.

2.2 Acrylic Fiber

Acrylic fiber is a family of synthetic fibers in which the fiber-forming substance is any long-chain polymer composed of at least 85% by weight of acrylonitrile units. Various acrylic fibers have been utilized for reinforcing gaskets, friction products, and cement sheet. Acrylic fiber containing essentially 100% acrylonitrile units, generally known as the homopolymer, has excellent stability. Table 12-1 lists technical data for a typical acrylic short fiber.

2.3 Aramid Fiber

Aramid fiber is a synthetic fiber whose fiber-forming substance is a long-chain polyamide in which at least 85% of the amide linkages are attached to two aromatic rings. High thermal stability and high tenacity are very useful properties of an aramid fiber, as shown in Table 12-2.

2.4 Nylon Fiber

Nylon fiber is a synthetic fiber whose fiber-forming substance is a long-chain polyamide in which less than 85% of the amide linkages are attached directly to two aromatic rings. Reinforcement of plastic and rubber products with nylon fiber is known. Technical data for regular and high tenacity fibers are shown in Tables 12-3 and 12-4, respectively.

Table 12-1. Acrylic fiber technical data.

Filament size	1.5 and 3.0 denier per filament.
Cut lengths—standard lengths	¼", ½", ¾", and 1". Other lengths on request.
Specific gravity	1.17.
Tensile strength	30,000-45,000 psi.
Breaking tenacity (std.)	2.0-3.0 grams per denier.
Elongation	35-45%.
Average stiffness	6.0-8.0 grams per denier.
Effect of heat	Sticks at 430-450°F.
Moisture regain	1.0-1.5 at 70°F, 65% relative humidity.
Shrinkage (boiling water)	6.7%.
Effects of acids and alkalis	Generally good resistance to mineral acids and weak alkalis; moderate resistance to strong alkalis at room temperature.
Resistance to mildew, aging, sunlight, and abrasion	Excellent.

Table 12-2. Aramid fiber technical data.

Filament size—standard deniers	5 and 2.0 denier per filament.
Cut lengths—standard lengths	¼", ½", ¾", and 1". Other lengths on request.
Specific gravity	1.38.
Tensile strength	90,000 psi.
Breaking tenacity (std.)	4.0-5.3 grams per denier.
Elongation at break	22-32%.
Effect of heat	Does not melt. Decomposes at 700°F.
Moisture regain	6.5% at 70°F, 65% relative humidity.
Shrinkage (boiling water)	2.0%.
Effects of acids and alkalis	Unaffected by most acids, except some strength loss after long exposure to hydrocholoric, nitric, and sulfuric. Generally good resistance to alkalis.
Resistance to mildew, aging, sunlight, and abrasion	Excellent resistance to mildew and aging. A 50% strength loss after 60 weeks exposure to sunlight. Good abrasion resistance.
Dispersion	Easily dispersed in aqueous and nonaqueous systems. Surface treatment to aid dispersion applied as required.

Table 12-3. Nylon fiber, regular tenacity, technical data.

Filament size	3.0 denier per filament.
Cut lengths—standard lengths	¼", ½", ¾", and 1". Other lengths on request.
Specific gravity	1.14.
Tensile strength	73,000-100,000 psi.
Breaking tenacity (std.)	4.0-7.2 grams per denier.
Elongation at break	17-45%.
Average stiffness	18-23 grams per denier.
Effect of heat	Melts at 419-430°F. Slight discoloration at 300°F when held for 5 hr. Decomposes at 600-730°F.

Table 12-3. (*Continued*)

Moisture regain	2.8–5.0% at 70°F, 65% relative humidity. 3.5–8.5% at 70°F, 95% relative humidity.
Shrinkage (boiling water)	2.9%.
Effects of acids and alkalis	Strong oxidizing agents and mineral acids cause degradation; others cause loss in tenacity and elongation. Resists weak acids. Soluble in formic and sulfuric acids. Hydrolyzed by strong acids at elevated temperatures. Substantially inert in alkalis.
Effect of bleaches and solvents	Can be bleached in most bleaching solutions. Generally insoluble in organic solvents. Soluble in some phenolic compounds.
Resistance to mildew, aging, sunlight, and abrasion	Excellent resistance to mildew, aging, and abrasion. Prolonged exposure to sunlight causes some degradation.
Dispersion	Easily dispersed in aqueous and nonaqueous systems. Surface treatment to aid dispersion applied as required.

Table 12-4. Nylon fiber, high tenacity, technical data.

Filament size	6.0 denier per filament.
Cut lengths—standard lengths	¼", ½", ¾", and 1". Other lengths on request.
Specific gravity	1.14.
Tensile strength	102,000–125,000 psi.
Breaking tenacity	6.5–9.0 grams per denier.
Elongation at break	16–20%.
Average stiffness	29–48 grams per denier.
Effect of heat	Melts at 419–430°F. Slight discoloration at 300°F when held for 5 hr. Decomposes at 600–730°F.
Moisture regain	2.8–5.0% at 70°F, 65% relative humidity. 3.5–8.5% at 70°F, 95% relative humidity.
Shrinkage (boiling water)	8.8%.
Effects of acids and alkalis	Strong oxidizing agents and mineral acids cause degradation; others cause loss in tenacity and elongation. Resists weak acids. Soluble in formic and sulfuric acids. Hydrolyzed by strong acids at elevated temperatures. Substantially inert in alkalis.
Effects of bleaches and solvents	Can be bleached in most bleaching solutions. Generally insoluble in organic solvents. Soluble in some phenolic compounds.
Resistance to mildew, aging, sunlight, and abrasion	Excellent resistance to mildew, aging, and abrasion. Prolonged exposure to sunlight causes some degradation.
Dispersion	Easily dispersed in aqueous and nonaqueous systems. Surface treatment to aid dispersion applied as required.

2.5 Polyester Fiber

Polyester fiber is a synthetic fiber in which the fiber-forming substance is a long-chain polymer composed of at least 85% by weight of an ester of a substituted aromatic carboxylic acid, including, but not restricted to, substituted terephthalate units and parasubstituted hydroxybenzoate units. Technical data for regular tenacity and fibers are shown in Table 12-5. Reinforcement of plastics, rubber, and cement products with polyester fiber is known.

2.6 Polypropylene Fiber

Polypropylene fiber is a synthetic fiber in which the fiber-forming substance is any long-chain polymer composition composed of at least 85% by weight propylene units. Short fiber reinforcement of plastics, rubber, and cement product is established. Technical data are shown in Table 12-6.

2.7 Rayon Fiber

Rayon fiber is a manufactured fiber composed of regenerated cellulose in which substitutes have replaced not more than 15% of the hydrogen of the hydroxyl groups. Technical data for rayon fiber are shown in Table 12-7. Reinforcement of plastics and rubber products is suggested.

2.8 Polyethylene Fiber

Polyethylene fiber is a manufactured fiber in which the fiber-forming substance is any long-chain polymer composition composed of at least 85% by weight ethylene units. Mini-Fibers markets a fibrillated, high density polyethylene product and recommends several reinforced formulations for "asbestos-free" asphalt roof coatings, block filler, and pavement marker systems using Short Stuff™ Polyethylene. Physical and chemical characteristics are shown in Table 12-8.

Table 12-5. Polyester fiber, regular tenacity, technical data.

Filament size	1.5 and 3.0 denier per filament.
Cut lengths—standard lengths	¼", ½", ¾", and 1". Other lengths on request.
Specific gravity	1.38.
Tensile strength	78,000–88,000 psi.
Breaking tenacity (std.)	4.0–5.0 grams per denier.
Elongation	35–38%.
Average stiffness	12–16 grams per denier.
Effect of heat	Melts at 478–490°F.
Moisture regain	0.4% at 70°F, 65% relative humidity.
Shrinkage (boiling water)	9.0–11%.
Effects of acids and alkalis	Good resistance to mineral acids. Dissolves with partial decomposition by concentrated solutions of sulfuric acid. Good resistance to weak alkalis and moderate resistance to strong alkalis at room temperature. Disintegrates by strong alkalis at boil.
Effect of bleaches and solvents	Excellent resistance to bleaches and other oxidizing agents. Generally insoluble except in some phenolic compounds.
Resistance to mildew, aging, sunlight, and abrasion	Excellent resistance to mildew, aging, and abrasion. Prolonged exposure to sunlight causes some strength loss.
Dispersion	Easily dispersed in aqueous and nonaqueous systems; surface treatment to aid dispersion applied as required.

Table 12-6. Polyropylene fiber technical data.

Filament size—standard denier	3.0, 6.0, and 15.0 denier per filament.
Cut lengths—standard lengths	¼″, ½″, ¾″, and 1″. Other lengths on request.
Specific gravity	0.90–0.91.
Tensile strength	30,000–75,000 psi.
Breaking tenacity (std.)	3.0–6.5 grams per denier.
Elongation at break	20–120%.
Average stiffness	3–40 grams per denier.
Effect of heat	Softens at 285–330°F. Melts at 320–350°F. Decomposes at 550°F.
Moisture regain	0.01–0.1 at 70°F, 65% relative humidity.
Shrinkage (boiling water)	1.4%.
Effects of acids and alkalis	Excellent resistance to most acids and alkalis with the exception of elevated temperature exposure to cholorsulfonic acid, concentrated nitric acid, and certain oxidizing agents.
Resistance to mildew, aging, sunlight, and abrasion	Not attacked by mildew. Good resistance to aging, indirect sunlight, and abrasion. Can be stabilized to give good resistance to direct sunlight.
Dispersion	Easily dispersed in aqueous and nonaqueous systems. Can be stablized to give good resistance to direct sunlight.

Table 12-7. Rayon fiber (flocking tow) technical data.

Filament size	1.5, 3.0, and 5.5 denier per filament.
Cut lengths—standard lengths	¼″, ½″, ¾″, and 1″. Other lengths on request.
Specific gravity	1.46–1.54.
Tensile strength	28,000–47,000 psi.
Breaking tenacity (std.)	2.3 grams per denier.
Elongation at break	14–19%.
Average stiffness	6.0–16.6 grams per denier.
Effect of heat	Does not melt—loses strength at 300°F. Decomposes at 350–464°F under extended periods of exposure.
Moisture regain	11–13% at 70°F, 65% relative humidity.
Shrinkage (boiling water)	2.5%.
Effects of acids and alkalis	Hot dilute or cold concentrated Acids disintegrate fiber. Strong alkaline solutions cause swelling and reduce strength.
Effect of bleaches and solvents	Attacked by strong oxidizing agents. Not damaged by hypochlorite or peroxide. Generally insoluble except in cuprammonium and a few complex compounds.
Resistance to mildew, aging, sunlight, and abrasion	Fiber attacked by mildew. Has good resistance to aging, sunlight, and abrasion.
Dispersion	Easily dispersed in aqueous and nonaqueous systems. Surface treatment to aid dispersion applied as required.

Table 12-8. Physical and chemical characteristics of Short Stuff™ Polyethylene and Fybrel™ Synthetic Fiber.

Chemical name	Polyethylene.
Trade name	Short Stuff™ Polyethylene; Fybrel™ Synthetic Fiber.
Chemical family	Polyolefin.
Formula	$(CH_2\text{-}CH_2)_n$.
Fiber type	High density, fibrillated (highly branched).
Melting point	120–130°C (248–266°F) (thermo plastic).
Surface area	8–12 m^2/g (measured by gas adsorption).
Specify gravity	(H_2O = 1.0)—0.915–0.965.
Bulking value	0.1319–0.1250 gal/lb; 7.580–7.997 lb/gal.
Molecular weight	30,000–150,000.
Effects of acids and alkalis	Excellent resistance to acids and alkalis with exception of oxidizing agents.
Effects of bleaches and solvents	Swells in chlorinated hydrocarbons at room temperature. Soluble at 160–175°F.
Resistance to mildew, aging, sunlight, and abrasion	Not attacked by mildew; good resistance to aging, sunlight, and abrasion.
Polar property	Anionic (nagative).
Appearance	Fluffy white fibers.
Odor	None.
Dispersibility	The fibers readily disperse in aqueous or solvent systems with the appropriate additives and mixing procedure.
Plasticity	Soft, flexible fibers.

3. ACRYLIC FIBER FOR CEMENT REINFORCEMENT

3.1 General

Cement pipe and sheet products have been reinforced with asbestos for a great many years. The manufacture of cement pipe consumes more than 200,000 metric tons of asbestos annually in the United States, but uses for cement pipe are primarily for potable water transmission and gravity sewer applications where there is little potential hazard.

Cement sheet is used in building construction (siding, roofing, guttering, etc.), cooling towers, flue ducts, hoods, and so on. It is manufactured by an aqueous slurry process (Hotschek-Process) and by dry-processing using a minimum amount of water where the sheet is formed under pressure. Both autoclaved and unautoclaved processes are used in the manufacture of cement sheet.

Significant effort has been directed toward the replacement of asbestos in cement sheet products even though it amounts to less than 10% of the amount of asbestos used in cement pipe.

A highly publicized material for replacing asbestos in flat and corrugated sheet products is short, alkali-resistant glass fiber. Some problems have become apparent, as glass fibers may lack adequate resistance to mechanical shock. Wood fibers have found some utility in sheet products, but insufficient long-term chemical stability has been reported. Surface treatments of wood fiber have shown some improvements. Short cut fibers of high tenacity polyvinyl alcohol have excellent alkali resistance, but may be too costly in this application.

Acrylic fiber currently appears to be the best economic replacement for asbestos in cement sheet products, as discussed in the following paragraphs.

3.2 Dolanit™ 10—Hoechst Corporation

The acrylic fiber Dolanit 10[19] is being used successfully by Eternit AG in Europe in its Eternit cement sheet products. Eternit AG

Table 12-9. Dolanit™ 10 typical properties

Tenacity	8.5 grams per denier = 128,000 psi
Elongation	9–10%
Initial modulus	170 grams per denier = 2,460,000 psi
Shrinkage	
Boiling water	7.5%
200°C hot air	9.5%
Cross section	Kidney
Luster/color	Bright yellow
Specific gravity	1.18
Alkali resistance	
pH 11, 24 hours, 80°C	No strength loss
pH 12, 24 hours, 80°C	No strength loss
pH 13, 24 hours, 80°C	6% strength loss
pH 14, 24 hours, 80°C	24% strength loss
Acid resistance	
H_2SO_4 (50%), 8 weeks, 20°C	15% strength loss
Electrical resistance	
Surface	1.3×10^{11} ohm
Insulation	2.5×10^{11} ohm

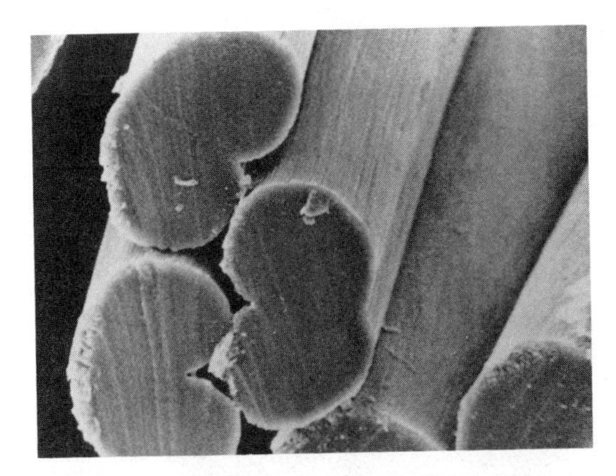

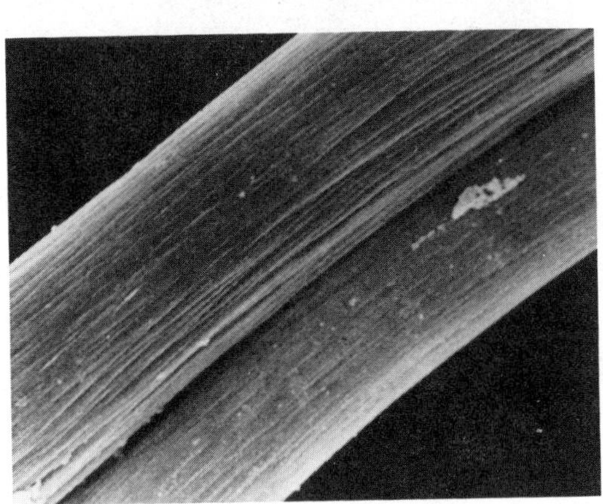

Fig. 12-1. Dolanit 10: sectional and surface view (scale 2000 : 1).

(Ametex AG) has been granted a U.S. Patent No. 4,414,031, November 8, 1983,[20] for the use of Dolanit 10 in cement. A typical application would use 2–3% Dolanit 10 plus a small amount of cellulose fiber as fill.

Dolanit 10 is commercially produced at 2.7 denier, ¼ in. uniform cut length and with no crimp. Other deniers and cut lengths may be requested. Table 12-9 describes some typical properties of Dolanit 10.

Figure 12-1 shows cross-sectional and surface views of Dolanit 10 fiber.

Figure 12-2 shows a typical surface of fracture at the fiber–cement interface.

3.3. Ministaple—Badische Corporation

Ministaple acrylic fiber of short and variable fiber lengths developed by Badische Corporation[16] is described in foreign patents issued in 1982,[7] 1984,[8] 1981,[9] 1984[10] and 1986.[11] Alkali resistance equal to or better than that of other acrylic fibers is found in these fibers. Products having high tenacity and high initial modulus give higher-bending-strength cement sheet products at equal loadings.

Badische Corporation has marketed A-551, Homopolymer Acrylic Ministaple, for use in

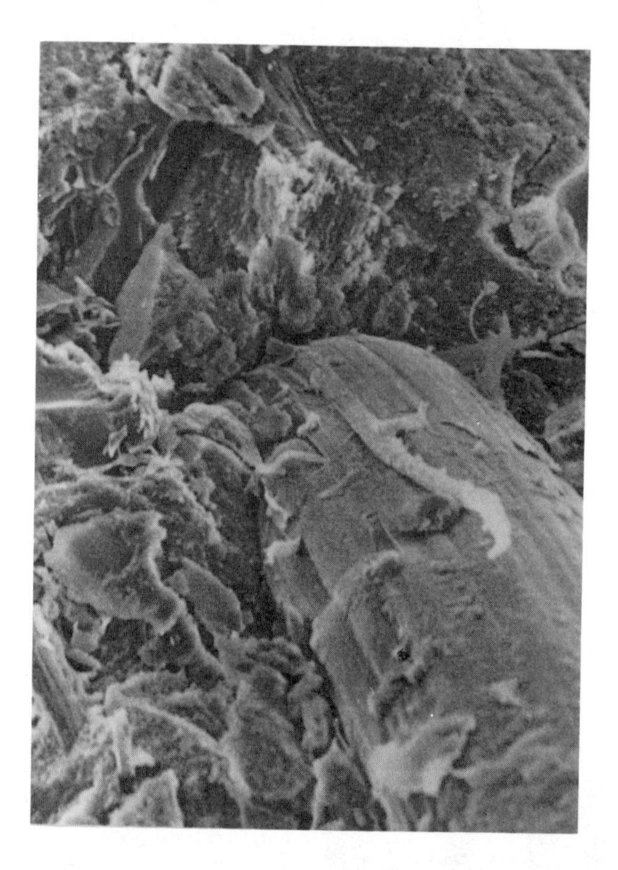

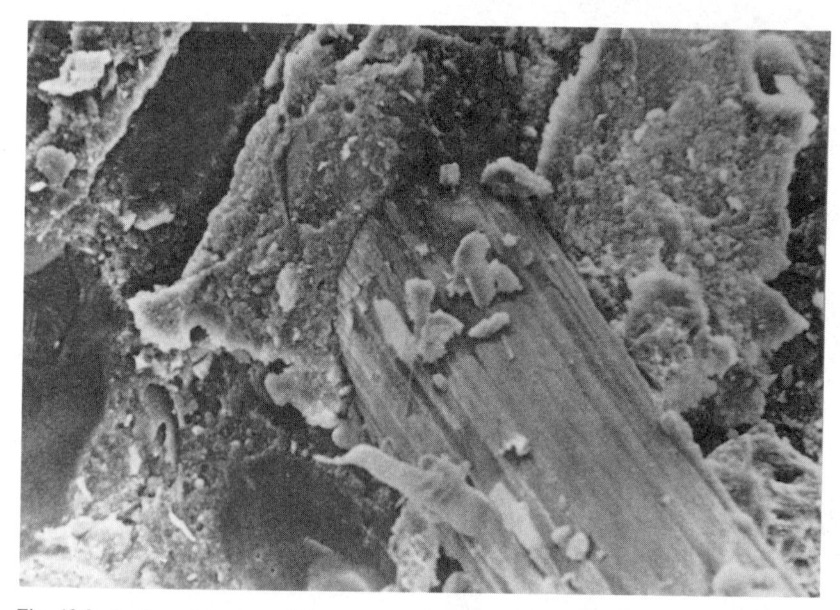

Fig. 12-2. Surface of fracture of fiber–cement shut with Dolanit 10 reinforcement (scale 2000 : 1).

Table 12-10. A-551, Homopolymer Acrylic Ministaple. Descriptive Data

I. Composition	
Polymer	Homopolymer of acrylonitrile
Zinc	100–500 ppm
Finish, average	0.1%
Finish, range	0.05–0.30%
Water, average	1.0%
Water, range	0.1–3.0%
II. Physical properties	
Density, absolute	1.18 g/cm^3
Density, apparent bulk	10 lb/ft^3 (160 kg/m^3)
Denier per filament	21
Crimp, average	3%
Crimp, crimps/in.	1
Cross section	Round
Diameter, average	50 μm
Diameter, range	30–80 μm
Length, average	1.4 mm
Length, range	0.1–8.0 mm
Tenacity	3.6 grams per denier
Elongation	52%
Modulus	37 grams per denier
III. Chemical and thermal properties[a]	
pH resistance, range	0–11
Exposure temperatures[b]	Up to 371°C (700°F)

[a]Chemical resistant to strong and weak acids, weak alkali, hydrocarbons, alcohols, ethers, ketones, etc. at ambient conditions.
[b]Confined conditions.

reinforcing gaskets, friction products, and cement products. Descriptive data are shown in Table 12-10. Experimental Homopolymer Acrylic Ministaple, X-532, is a short cut, crimped fiber. Descriptive data for X-532 are shown in Table 12-11.

Badische Corporation should be contacted directly for further technical and marketing information on short cut homopolymer acrylic fiber for reinforcement applications.

4. ARAMID FIBER—KEVLAR™ DUPONT COMPANY

4.1 General

Kevlar™ fiber is available from DuPont[12] in the form of pulp as well as the short cut fibers mentioned earlier. Pulped Kevlar has found considerable acceptance and commercial success as a reinforcing fiber in friction products and in gaskets, thus replacing significant quantities of asbestos. In the pulped form each main fiber can carry several fibrils that can increase the effective length to diameter ratio (aspect ratio) to as high as 500. Kevlar fiber is noted for high tenacity and thermal stability.

4.2 Kevlar™ Aramid Pulp

Kevlar™ aramid pulp with high tensile strength and high modulus is an excellent alternative to asbestos in gasket sheeting. This fiber offers excellent creep resistance, especially at elevated temperatures. Kevlar also has the highest strength-to-weight ratio of any commercially available man-made fiber, and this high strength enables the fiber to be used very cost-effectively as a gasket reinforcement. Very little Kevlar pulp is required (typically about 10%) to achieve a final strength equivalent to mixtures containing 85% asbestos. The 10% amount offers the best balance of cost and properties. Table 12-12 shows some of the advan-

Table 12-11. X-532, Experimental Homopolymer Acrylic Ministaple. Descriptive Data

I. Composition	
Polymer	Homopolymer of acrylonitrile
Zinc	100–500 ppm
Finish, average	0.1%
Finish, range	0.05–0.30%
Water, average	1.0%
Water, range	0.1–3.0%
II. Physical Properties	
Density, absolute	1.18 g/cm^3
Density, apparent bulk	4.0 lb/ft^3 (64 kg/m^3)
Denier per filament	3.7
Crimp, average	15%
Crimp, crimps/in.	12
Cross section	Round
Diameter, average	21 μm
Diameter, range	14–28 μm
Length, average	2.4 mm
Length, range	0.1–10.0 mm
Tenacity	3.7 grams per denier
Elongation	50%
Modulus	36 grams per denier
III. Chemical and thermal properties[a]	
pH resistance, range	0–11
Exposure temperatures[b]	Up to 371°C (700°F)

[a]Chemical resistant to strong and weak acids, weak alkali, hydrocarbons, alcohols, ethers, ketones, etc. at ambient conditions.
[b]Confined conditions.

Table 12-12. Reinforcing materials for gasket sheeting.

	Advantages	Disadvantages
Kevlar™ pulp	High strength and modulus. Thermal stability for excellent creep resistance and sealability. Acceptable tensile strength at relatively high temperatures. Low chloride. Nonbrittle character (will not break up during mixing). Excellent radiation resistance.	Mixing procedures required different from those used for asbestos. Care required to prevent static buildup.
Cellulose fibers	Sufficient strength and modulus for ambient temperature service. Ease of mixing. Low cost.	Degradation at 350–450°F. Poor dimensional stability.

Table 12-12. *(Continued)*

Glass fiber	High strength and modulus. Thermal stability.	Fiber breakage in high-shear mixing, reducing reinforcing efficiency. Irritation on handling.
Asbestos	High strength and modulus. Thermal stability. Cost low enough for its use as a filler.	Link to health problems. High chloride.

Table 12-13. Major suppliers of short organic fibers.

Company Name	Product name	Product form
Allied Corporation Fibers Division P.O. Box 166 Moncure, NC 27559 (919) 542-2200	Compet	Polyester fiber
American Enka Enka, NC 28728 Richard Wold (704) 667-7110	Encron	Polyester yarn and staple fibers
Claremont Flock Corporation Claremont, NH 03743 (603) 542-5151	Flock fibers	Polyester, nylon, cotton
Crown Zellerbach Corporation Chemical Products Div. P.O. Box 1047 Chamas, WA 98607 Bruce Mac Gowan (206) 834-4444	SWP synthetic fiber	Fibrous HDPE E-400, E-620
Danberry Chem. Co. Wallingford, CT 06493 (203) 269-3743	Flock fibers	Rayon, nylon, acrylic, polyesters, wool, cellulose
E. I. du Pont de Nemours & Co. Teflon™ Fibers Marketing Textile Fibers Department Center Road Building Wilmington, DE 19898	Teflon™ TFE	Fluorocarbon fiber
Kurray Co. Ltd. 8 Umeda Kita-KU Osaka, Japan	Kuralon	Polyvinyl alcohol fiber in yarn and chopped staple
Monsanto Industrial Chemicals Co. 260 Springside Drive Akron, OH 44313 Robert I. Leib (216) 666-4111	Santoweb	Treated cellulose fibers

tages and disadvantages of four of the common reinforcing materials.

Products reinforced with Kevlar usually meet or exceed the performance of those containing asbestos and other reinforcing materials. Although performance depends upon the formulation, manufacturers using Kevlar for both beater addition and compressed gaskets have seen:

- Lower chloride content compared to asbestos.
- Improved thermal stability versus other organics.
- Compressibility and recovery equal to that of asbestos.
- Improved creep resistance (vs. beater addition asbestos; equal to compressed asbestos).
- Tensile strength equal to that of compressed asbestos; better than that of beater addition asbestos.

E. I. du Pont de Nemours & Co., Wilmington, DE, should be contacted directly for the most current technical and marketing information.

5. SHORT FIBER PRODUCERS

5.1 Suppliers and Their Products

In addition to those already mentioned, there are several other major suppliers of short organic fibers for uses in composite reinforcement. These manufacturers are listed in Table 12-13.

REFERENCES

1. American Barmag Corporation, 1101 Westinghouse Blvd., Charlotte, NC 28217, (704) 588-0072.
2. Buss-Condux, Inc., 2411 United Lane, Elk Grove Village, IL 60007, (312) 595-7474.
3. Fi-Tech, Inc., 501 Research Road, Richmond, VA 23236, (804) 794-9615.
4. Fleissner Inc., Rt. 5, Box 1005, Charlotte, NC 28208, (704) 394-3376.
5. Lummus Industries, Inc., P. O. Box 1260, Columbus, GA 31994, (404) 322-4511.
6. Mini-Fibers, Inc., Route 14, Box 11—Boones Creek Road, Johnson City, TN 37615, (615) 282-4242.
7. Hibbard, Billy Brown et al. (to Badische Corp.), "Method of Making Reinforced Materials Having An Improved Reinforcing Material Therein," Spain Patent No. 507, 568, Sept. 22, 1982.
8. Hibbard, Billy Brown et al. (to Badische Corp.), "Method of Making Reinforced Materials Having An Improved Reinforcing Material Therein," East German Patent No. 208, 081, Mar. 28, 1984.
9. Hibbard, Billy Brown et al. (to Badische Corp.), "Method of Making Reinforced Materials Having An Improved Reinforcing Material Therein," Brazil Patent No. 8,108,011, Dec. 9, 1981.
10. Hibbard, Billy Brown et al. (to Badische Corp.), "Method of Making Reinforced Materials Having An Improved Reinforcing Material Therein," Canada Patent No. 1,178,012, Nov. 20, 1984.
11. Hibbard, Billy Brown et al. (to Badische Corp.), "Method of Making Reinforced Materials Having An Improved Reinforcing Material Therein," Australia Patent No. 546,861, Feb. 20, 1986.
12. E. I. du Pont de Nemours & Co. (Inc.), Textile Fibers Department, Kevlar Special Products, Center Road Building, Wilmington, DE, 1- (800) 4-KEVLAR.
13. Sprout-Waldron Division, The Koppers Co., Inc., Logan & Sherman Streets, Muncy, PA 17756, (717) 546-8211.
14. Herty Foundation, P.O. Box 1963, Savannah, GA 31402.
15. American Cyanamid Corporation, 1 Cyanamid Plaza, Wayne, NJ 07470, (201) 831-2000 (2532).
16. Badische Corporation (BASF Fibers), P.O. Drawer D, Williamsburg, VA 23185, (804) 887-6000.
17. Eirich Machines Ltd., P.O. Box 550, Maple, Ontario LOJ1EO, Canada, (416) 832-2241; Eirich Machines, Inc., 663 Fifth Avenue, New York, NY 10022.
18. Littleford Bros., Inc., 7451 Empire Drive, Florence, KY 41042, (606) 525-7600.
19. American Hoechst Corporation, Hoechst Fibers Industries, P.O. Box 5887, Spartanburg, SC 29304, (803) 579-5750 (5232).
20. Studinka, Josef and Meier, Peter E. (to Ametex AG, Switzerland), "Fiber-Containing Products Made with Hydraulic Binder Agents," U.S. Patent No. 4,414,031, Nov. 8, 1983.

13

Whiskers

John V. Milewski

Consultant
Santa Fe, New Mexico

Harry S. Katz

Utility Research Co.
Montclair, New Jersey

CONTENTS

1. INTRODUCTION

1.1 History

Whiskers are single crystal fibers that have a high degree of crystalline and chemical perfection giving them strength equivalent to the interatomic bonding forces of millions of psi.

The first commercial source of whiskers was Thermokinetic Fibers, Inc. (TKF), a company started in 1962 for the sole purpose of producing whiskers. From 1962 to 1968, TKF produced and offered for sale a wide variety of whisker products, including high-quality needle whiskers and wool and paper products of sapphire, silicon carbide, and silicon nitride. In 1966, TKF merged with General Technologies Inc. (GTC), which was later acquired by Citco. About 1970 Citco sold the GTC assets to the major GTC officers and Versar Inc.; GTC is now a division of Versar Inc. This company still markets old TKF stock and special grades of some whisker products. Other major whisker suppliers that have been active in the field include Horizons Inc., Carborundum Corporation, National Research Co., and Explosive Research and Development Establishment (ERDE) of England. Lonza Co. of Switzerland was still supplying silicon carbide whiskers in 1975 but no longer is. A new source of whiskers is Advanced Components Materials Corp. (ACMC) of Greer, South Carolina (formerly Silag of Exxon Enterprises, which originally was HULCO of Salt Lake City, Utah). Advanced Components Materials Corp. (ACMC) currently offers silicon carbide whisker products made from rice hulls. More details on these companies and their products are given in the section on suppliers.

One of the major drawbacks in the development of whisker composite technology has been the price and availability of good-quality whisker products. During TKF's five years of major activity, 1963–1968, the company produced about 50 lb of whiskers.

World production of whiskers during the late 1960s was less than 100 lb, over 90% of which consisted of silicon carbide whiskers. Phillips Laboratories, Findhoven, The Netherlands, developed a process for the production of alpha-silicon carbide whiskers, and licensed the process to Kempsten in Munich, Germany, and to National Research Co., Cambridge, Massachusetts, a former division of Norton Co., now owned by Cabot Corporation. During 1970, Kempsten produced from 20 to 50 lb of whiskers, mostly for internal use. Also during 1970, Norton's National Research Co. produced about 10 lb of whiskers, with the product priced at \$1600/lb. The Cabot Corporation, which has taken over this division, had hopes of offering a whisker product for sale at about \$300/lb, but this plan never materialized.

The Explosives Research and Development Establishment, Waltham Abbey, Essex, England, developed a process for the production of beta-silicon carbide. During 1970, this group made about 15 lb of these whiskers. Its process has been licensed to Lonza Co., of Switzerland, where about 20 lb of whiskers were produced during the 1970s, primarily for internal research and development work.

In the late 1960s, Carborundum Corporation introduced an alpha grade of silicon carbide whisker that they were able to produce in 100-lb quantities and sell at \$250/lb. Unfortunately, the quality of alpha-silicon carbide never approached that of beta-silicon carbide in strength and uniformity of product. Therefore, this alpha-SiC could not compete and was dropped from the company's product line.

During 1975, Silag, formerly HULCO Corporation, in cooperation with Exxon Enterprises, produced an alpha-Beta-silicon carbide product from rice hulls. In less than one year they produced about 200 lb and were test-marketing it. From the late 1970s to the mid-1980s, this source of whiskers increased, with production figures in 1986 reaching about 40,000 lbs. For the past decade, most of these whiskers have been used in government-sponsored metal matrix composites. More recently, ceramic matrix composites have attracted great attention with the introduction of silicon carbide whisker reinforced ceramic matrices. These composites are being used in a relatively high-volume application in metal cutting tools.

1.2 Methods of Manufacture

More than one hundred materials, including metals, oxides, carbides, halides, nitrides, graphite, and organic compounds have been prepared as whiskers. Whiskers may be grown from supersaturated gas phases; from melts; from solutions, by chemical decomposition or electrolysis; or from solids.

Whiskers grow by two different mechanisms—basal growth and tip growth. In basal growth, the atoms of growth migrate to the base of the whisker and extrude the whisker from the substrate. This type of growth has been observed on tin-plated structures subject to stress. On the other hand, sapphire, SiC, Si_3N_4, and AlN whiskers are formed by tip growth at temperatures high enough that the vapor pressure of the whisker or whisker-forming material becomes significant. In tip growth, the atoms attach themselves to the tip of a growing whisker by various mechanisms.

Under ideal conditions, whisker growth rates of over 1 in. in 5 minutes have been observed. (a good average is about one micron/second) The problem in quantity production is to optimize growth rate conditions in all areas of the reaction chamber by accurate control of temperature, gas flow rates, and other variables. When this is achieved, and significant end uses develop, low-price, mass production of whiskers will be economically feasible, as the basic raw materials are relatively inexpensive. Ultimately, a production process can be envisaged in which growth of whiskers is accomplished on a continuous belt passing through a furnace with the whiskers vacuumed off, air-classified, plastic-impregnated or metal-coated, and then directly processed into the final composite. Presently, whiskers are grown by various batch-process techniques, and harvested and sorted by a hand operation.

1.3 Processes for Whisker Growth

Some metallic whiskers have been produced by sublimation or evaporation of the metal, which is transported in vapor phase to a lower-temperature growth site where, under low supersaturation conditions, condensation produces whiskers. Zinc, cadmium, and other metallic whiskers have been grown in a simple chamber with a thermal gradient between the source and the growth site. An inert gas carrier may be used to help control the rate of whisker growth.

The hydrogen reduction of metal salts has been a frequent method for the production of metal whiskers such as nickel, iron, copper, silicon, and gold. The best reduction temperature is usually near or slightly above the melting point of the source material.[1-3]

Among oxides prepared as whiskers are Al_2O_3 (sapphire), MgO, MgO-Al_2O_3, Fe_2O_3, BeO, MoO_3, NiO, Cr_2O_3, and ZnO. A simple vapor transport method can be used, consisting of heating the metal in a suitable atmosphere, such as wet hydrogen, moist inert gases, or air. For example, when a stream of moist hydrogen is passed over aluminum powder, aluminum oxide, or an intermetallic compound of aluminum which has been heated to 1300–1500°C, a mass of acicular sapphire is deposited in a cooler part of the furnace.

A large part of the whisker technology efforts has been directed toward alpha-Al_2O_3 (sapphire) whisker, because this material has excellent potential as a reinforcement in metal and polymer matrix composites. Webb and Forgeng investigated the growth of α-Al_2O_3 whiskers by passing wet hydrogen over an alumina boat (containing some silica), the boat filled with molten aluminum or $TiAl_3$. At operating temperatures between 1300°C and 1450°C, large deposits of whiskers were formed around the boat after 2 to 20 hr, and the $TiAl_3$ gave a greater yield than the pure Al. Whiskers with lengths from 1 to 30 mm and hexagonal diameters of 3 to 50 were obtained.[4] Minor variations of this method have been used to produce research and development quantities of whiskers for composite fabrication. A problem that was encountered and still must be considered in current whisker technology was the lack of uniformity in each batch. Whisker type and size varied within each batch, and from batch to batch, and other growth forms such as ribbons and powders were observed. Often, a tight cocoon of whiskers forms that is difficult to disentangle and classify into discrete lengths for use in a composite. These problems spurred the development of the fiber classifer and other refining techniques, which will be discussed in a later section of this chapter.

Since about 1969, the major activity in whiskers has appeared to shift from Al_2O_3 to SiC. A main reason for this shift was the initial indication that SiC whiskers are easier to wet and bond in low-temperature metal matrices and in polymer matrices. Also, SiC whisker growth is less susceptible to branching growth and thermal degradation. Therefore, there was a greater potential for producing a uniform product and for scale-up of production.

Philips Laboratories, Hindhoven, The Netherlands, developed a process for producing alpha-SiC. Bulk SiC was vaporized by heating it under reduced pressure, and the whiskers formed upon nucleation sites containing lanthanum or another catalyst. Their process was licensed by National Research Co., Cambridge, Massachusetts, and Kempsten in Germany.

Beta-SiC whiskers have been grown by hydrogen reduction of methyltrichloro-silane at 1500°C onto carbon substrates.[5] Dr. Parrott and coworkers at ERDE, England, have explored the controlled use of combinations of chlorosilanes, CO, and CH_4 as source gases for the production of beta-SiC. Their process has produced relatively large quantities of good-quality whiskers, and has been licensed by Lonza, Switzerland.

The many different crystallographic modifications of SiC differ only in the stacking arrangement of the atom planes. The sequence formed by stacking the Si–C double layer in the sequence ABCABC results in the structure of cubic beta-SiC. All other sequences are referred to as alpha-SiC. Laboratory tests have indicated no intrinsic difference in strength between the two forms. However, in commercial products, the alpha-SiC whiskers were poorer in strength and uniformity than beta-SiC. The difference may be attributed to the fact that the beta structure is unique, and the production conditions must be more closely controlled than the conditions for producing the varying stacking arrangement of atom layers that can occur within the same batch of alpha whiskers.

A mechanism of crystal growth involving vapor, liquid, and solid (VLS) phases explains many observations of the effect of impurities in crystal growth from the vapor. The role of impurities is to form a liquid solution interface, with the crystalline material to be grown and fed from the vapor throughout the liquid interface. The solution is a preferred site for deposition of feed from the vapor, which causes the liquid to become supersaturated. Crystal growth occurs by precipitation from the supersaturated liquid at the solid–liquid interface.

A number of different whiskers have been grown by the VLS technique, including α-Al_2O_3, B, GaAs, GaP, Ge, MgO, $NiBr_2$, NiO, Se, Si, and SiC. Figures 13-1 and 13-2 illustrate the VLS growth processes, and Fig. 13-3 is a photograph of very uniform beta-silicon carbide whiskers grown by Milewski at TKF.[6] The photo clearly shows the VLS catalyst ball on the top of each whisker.

Figure 13-1 illustrates the sequence of VLS growth as follows: Solid catalyst particle on the left melts and forms the liquid catalyst ball. Carbon and silicon atoms in the vapor feed are extracted by the liquid catalyst, which soon becomes a supersaturated solution of carbon and silicon that precipitates silicon carbide on the substrates. As precipitation continues, the whisker grows, lifting the catalyst ball.

The advantage of VLS growth is that growth can be controlled and limited by the location and type of catalyst. A larger catalyst produces larger-diameter whiskers. Figure 13-2 shows that the chemistry of the catalyst ball also affects the diameter of the whiskers formed, with some catalysts wetting the initial substrate over a wider range and producing a larger-diameter whisker for the same size of catalyst.[6] Figure 13-3 illustrates the high uniformity of product obtained by the VLS process.

Advanced Composits Materials Corp. (ACMC) (formerly ARCO) is marketing a high alpha-silicon carbide whisker product made

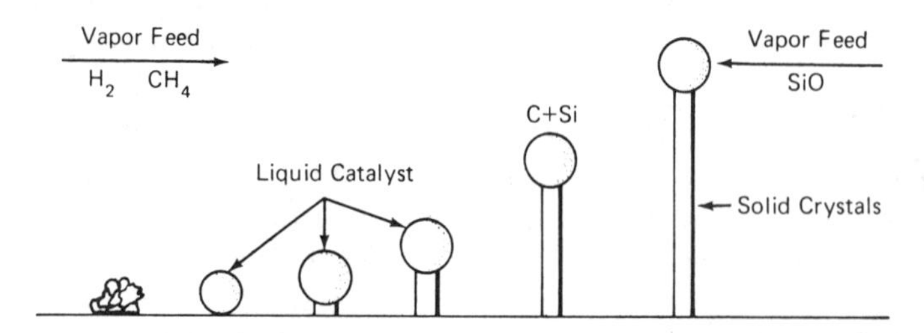

Fig. 13-1. V.L.S. (vapor feed, liquid catalyst, solid crystal) SiC whisker growth.

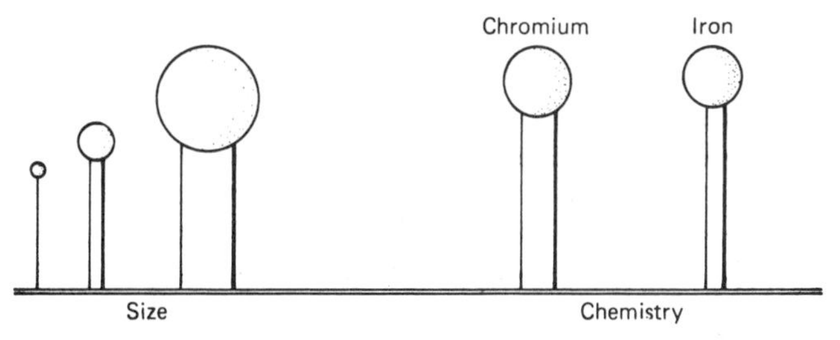

Fig. 13-2. Size and chemistry effects of the V.L.S. catalyst.

from rice hulls. The rice hulls contain a high silica content, and, upon coking, an intimate mixture of carbon and silica is formed. Upon further heating, the carbon and silica react to form a silicon carbide product, part of which is a fine-diameter beta-silicon carbide whisker. This submicron whisker is shown in Fig. 13-4.

The performance of these whiskers as a reinforcement for cutting tool bits is so good that they are expected to capture the majority of the ceramic tool bit industry by the early 1990s. This is so mainly because the whisker-reinforced bits are orders of magnitude better than any other material in performance. This is the first large-volume commercial whisker-reinforced product that is not government-sponsored and is standing on its own economically. This is a real breakthrough for whiskers and is regarded as the start of an ever expanding market for whisker reinforced metal, plastic, and ceramic matrix components.

Fig. 13-3. Illustrating V.L.S. catalyst ball on ends of uniformly grown beta-silicon carbide whiskers, 100×.

2.2. BASIC PROPERTIES OF WHISKERS

2.1 What are Whiskers?

Whiskers are a generic class of materials having mechanical strengths equivalent to the binding forces of adjacent atoms. Whiskers are strong because they are essentially perfect crystals, and their extremely small diameters allow little room for the defects that weaken larger crystals. Test data on silicon carbide whiskers reveal strengths greater than 4 million psi and

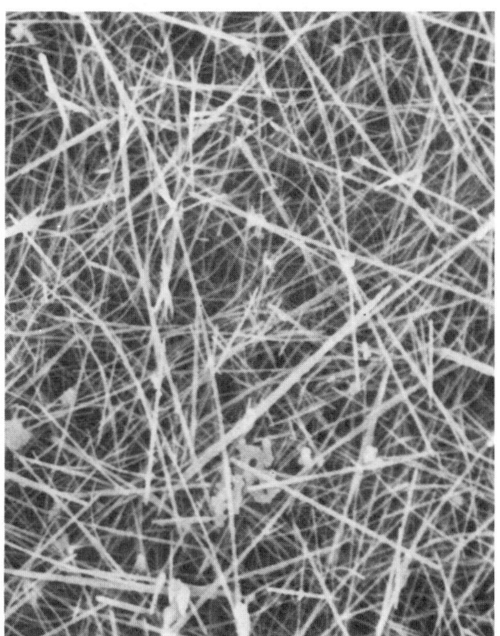

Fig. 13-4. — = One micron. Magnification = 3,000×. Submicron silicon carbide whiskers formed from rice hulls. (Courtesy: ACMC.)

a Young's modulus greater than 100 million psi. Comparing whiskers to conventional materials, the latter contain a multiplicity of grain boundaries, voids, dislocations, and imperfections, while the single crystal whisker, which approaches structural perfection, has eliminated all such defects. Because of this internal perfection, the strength of whiskers is not strictly limited by surface perfection. This property is very significant in that it gives whiskers an unusual toughness and nonfriability in handling as compared to polycrystalline fibers and fiberglass.

If the whisker's strength versus diameter is plotted, it is found that as the whisker becomes smaller (and therefore even more perfect), its strength increases rapidly. In comparing the tensile strength of whiskers with those of other reinforcing fibers, such as glass and boron, a 5–10-fold increase in basic fiber strength is evident. Boron fibers have essentially the same tensile strength as E glass fibers, but the disadvantages of boron fibers are their larger diameter and reduced elongation, which limit processing because of a greater tendency for breakage during handling.

Whiskers combine the two best properties of glass and boron. They have the elongation of glass fibers (3–4%) and the modulus of boron (60–100 million or more psi), as shown in Fig. 13-5.

A whisker may be defined as a single crystal, in a fiber or elongated form, that has a minimum length-to-diameter ratio of 10:1 and a maximum cross-sectional area of 7.9×10^{-5} in.^{2} (corresponding to a circular cross section of 0.010 in. diameter).

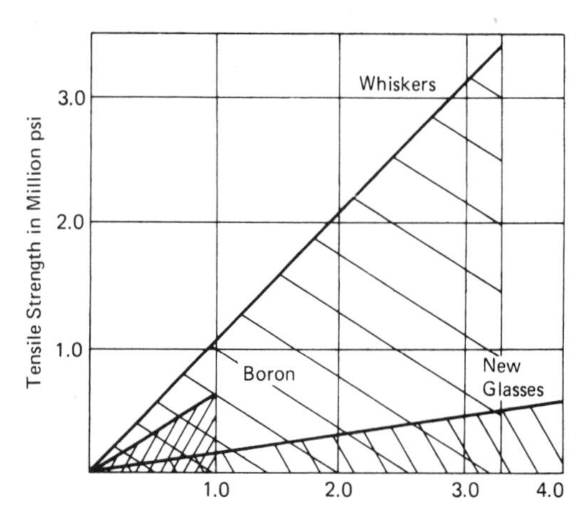

Fig. 13-5. Comparison of the strength of reinforcing fibers.

2.2 Physical Characteristics

Some of the most common whiskers and their properties are given in Table 13-1.

Whiskers are unique in that they can be strained elastically as much as 3% without permanent deformation, compared to less than 0.1% for bulk crystals. In addition, whiskers consistently exhibit much less strength deterioration with increasing temperature than the best conventional high-strength alloys. This would be expected because of the absence of imperfections which are prone to permit slip. Sapphire whiskers have been found to retain tensile strengths as high as 100,000 psi at 2750°F.

Although whiskers appear to the eye as short fibers, they have very high length/diameter ratios (aspect ratios). The aspect ratio is important in reinforced plastics because of its obvious relationship to the amount of surface available for bonding to the resin and to the amount of load that can be carried by the composite.

No appreciable fatigue effects have been observed in whiskers. They can be handled roughly and can be milled and chopped or otherwise worked in without having their strength impaired.

The geometry of fibers affects the aspect ratio. Specifically, a thin, flat, ribbon-shaped fiber has a larger perimeter than a round fiber with an equivalent cross-sectional area. Round fibers have the minimum attainable surface-area-to-volume ratio; hence any change in cross-section geometry will increase the fiber's aspect ratio. Fiberglass and boron have round cross sections, while whiskers have hexagonal, triangular, and thin ribbon cross sections. Some typical fiber cross sections and the approximate aspect ratio increase with differing geometries are given in Fig. 13-6.

It is interesting to compare the whisker diameter with the diameters of various fibers such as boron and fiberglass. A bundle of about 100

Table 13-1. Physical properties of whiskers.

Material	Density g/cm^3	Melting point °F	Tensile strength psi × 10^{-6}	Young's modulus psi × 10^{-6}	Strength density in. × 10^{-6}
Aluminum oxide	3.9	3780	2-4	70-150	14-28
Aluminum nitride	3.3	3990	2-3	50	13-21
Beryllium oxide	1.8	4620	2.0-2.8	100	31-43
Boron carbide	2.5	4440	1	65	11
Graphite	2.25	6500	3	142	37
Magnesium oxide	3.6	5070	3.5	45	27
Silicon carbide (alpha)	3.15	4200	1-5	70	9-44
Silicon carbide (beta)	3.15	4200	1-5	80-120	9-44
Silicon nitride	3.2	3450	.5-1.5	55	4.2-13

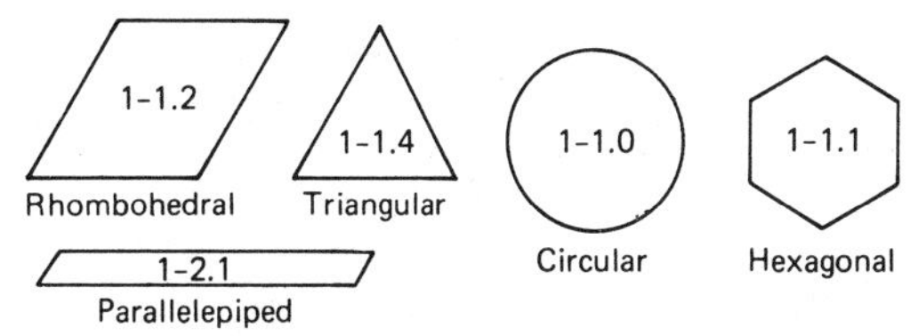

Fig. 13-6. Fiber cross section shows how increases in surface to volume ratio changes with changes in fiber geometry.

whiskers 1 μm in diameter would fit within one diameter of fiberglass—and about 10,000 whiskers within one diameter of a boron fiber. For a visual comparison of these fibers, see Fig. 13-7. In addition to the strength increase associated with small fibers such as whiskers, there is the possible added advantage of a much greater number of fibers with a more uniform distribution in the matrix because of their small size.

These new fiber cross-sectional geometries are found in whiskers as a result of the unit cell crystal structure of the whisker. The basic crystal structure for sapphire is hexagonal. A C axis growth will produce a hexagonal whisker,

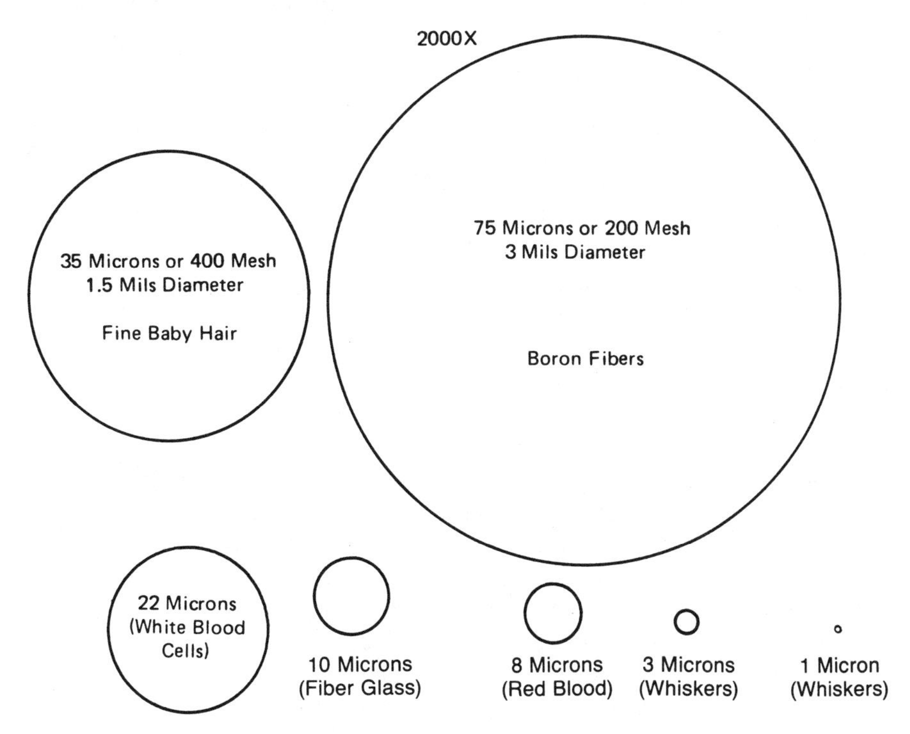

Fig. 13-7. Size comparison diameter of various fibers.

while an A_I or A_{II} axis yields a parallelepiped whisker with a ribbonlike structure. Similarly, the SiC unit cell structure is cubic, and the (111) axis growth will produce whiskers with an almost triangular cross section.

2.3 Various Types and Forms of Whiskers and Whisker Products

Whiskers have been produced in a wide range of fiber size and types. Here emphasis is placed on the types of fibers that have been or are now commercially available. (Section 3 discusses suppliers and availability of their products.)

Sapphire whiskers have been grown for some time and were commercially available in four major size categories: a submicron grade of fine ribbons with a 0.2 to 1.0 μm thickness and a length-to-equivalent-diameter ratio of about 25, a sapphire wool product of 1 to 3 μm diameter with an aspect ratio of 500 to 5000, and two sapphire needle products of 3 to 10 and 3 to 30 μm diameter with aspect ratios of 100 and 60, respectively. The sapphire wool product can be processed into a paper or felt form. In felt or paper, the whiskers are randomly oriented with fiber aspect ratios ranging from 250 to 25000. The paper, as felted, has an approximately 97% void volume and a density of 4–8 lb/ft^3, and can be compressed to about 3% of its original thickness. The paper or felt contains no binder and consequently must be handled with care, but it does have sufficient integrity for impregnation and lamination.

Silicon Nitride is available in two forms. One is a relatively large-diameter wool product with a nominal diameter of 10 μm and an average aspect ratio of 2500. The second form is a small-diameter, relatively short aspect ratio product available in small fiber clusters. The diameter ranges from 0.2 to 0.5 μm, and the aspect ratio ranges from 50 to 1000.

There are three silicon carbide products available, all very similar in size and aspect ratios. They are available in submicron fiber diameters in small clusters. The diameters range from 0.05 to 0.6 μm, with average aspect ratios in the 30 to 300 range.

An experimental 5 μm diameter beta-silicon carbide needle with a length up to several centimeters is being developed in a joint government/industry project (Los Alamos/ACMC) and is projected to be available shortly after this book is published.

A new whisker product recently completed the experimental stage and is now being produced at a rate of 400 kg per day. A cobweb whisker that has a composition close to silicon monoxide, this fine fiber product is presumed to be a core whisker of single crystal silicon with an amorphous silicon dioxide coating. Its average diameter is about 10 nm or 100 Å, and it has an aspect ratio assumed to be in the many hundreds.

3. SUPPLIERS AND THEIR PRODUCTS

3.1 ACMC Chemical

Address:

ACMC Chemical Company
Route 6—Box A
Old Buncombe Road
Greer, SC 29651
(803) 877-0123
Ceramic Product Manager: Ed Lauder

Product: SC-9, a submicron silicon carbide whisker, shown in Fig. 13-4.

Properties:

Physical:

Ave. diameter, μm	0.6
Length range (>80 wt %), μm	10–80
Surface area, m^2/g	3.0
Density, g/cm^3	3.2
Bulk density, g/cm^3	~0.2
Whisker content, %	80–90
Particle content, %	10–20

Bulk Chemical:

Crystal type	Alpha & beta
Free carbon, max. wt %	0.10
Vaporized by HF, e.g. SiO_2, max. wt %	0.75

Metals Analysis:	*Nominal ppm*
Calcium	3700
Manganese	2400
Aluminum	1300
Magnesium	800
Iron	500

Metals Analysis:	*Nominal ppm*
Chromium	<50
Nickel	<50
Potassium	<50
Sodium	<50
Copper	<25
Boron	<10
Lithium	<10
Titanium	<10

Availability and Price:

Quantity	Price/lb
5-lb minimum	$100.00
Less than 100 lb	65.00
100–999 lb	42.50
1,000–9,999 lb	29.25
10,000–99,999 lb	22.75
100,000 lb plus	19.60

3.2 J. M. Huber

Address:

J. M. Huber Corporation
P.O. Box 2831
Borger, TX 79008-2831
(806) 274-6331
Glen Conger or Bill Schofield

Product: XPV1 "Cobweb whisker," occurring as both individual fibers and as intergrown fiber bundles. Whiskers are predominantely amorphous silica. See Fig. 13-8.

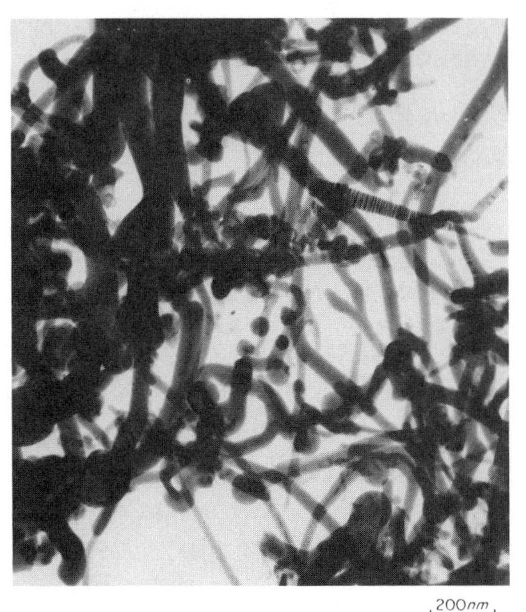

Fig. 13-8. XPV1 whisker, 100,000×.

Properties:

Typical chemical composition (wt%)	Amorphous silicon dioxide—75.6 Crystalline elemental silicon—20.2 Carbon (free and as SiC)—3.8 Nitrogen and other trace elements—0.4
Color (visual)	Tan to olive
"As-poured" bulk density	4–5 pcf (64–80 kg/m^3)
Specific gravity	2.25 g/cm^3
Individual fiber size (TEM):	Typical diameters of 2–20 nm (95%). Median diameter approximately 10 nm. Typical L/D values of 10–20.
Fiber bundle size (Coulter Counter)	Roughly spherical with median diameter of 3000–4000 nm.
Specific surface (BET)	35–50 m^2/g, 39 m^2/g typical
Surface structure (DBP)	148–184 $cm^3/100/g$, 169 $cm^3/100/g$ typical

Availability and Price: 100- to 1000-lb quantities; price range—$75. to $25./lb, depending on quantity of order.

3.3 Mitsubishi International/Tokai Carbon

Address:

Mitsubishi International Corporation
520 Madison Ave.
New York, NY 10022
(212) 605-2000
Distributor for:
Tokai Carbon Co. Ltd.
Aoyoma Bldg. 2-3 Kita—Aoyama Ichome,
Minato—Ku, Tokyo Japan
405-7211

Product: Tokama, a submicron silicon carbide whisker.

Price: $500/kilo.

Properties:

Physical:

Diameter	0.1–0.5 μm, mainly
Length	50–200 μm
Aspect ratio (*L/D*)	50–300
Density	3.19 g/cm^3
Bulk density	0.1 g/cm^3
Heat resistance	up to 1600°C (in air)
Tensile strength	3–14 GPa (about 300–1400 kg/mm^2)
Tensile modulus	400–700 GPa (about 40–70 t/mm^2)

Composition (see Fig. 13-9):

Crystal type	Beta
SiC particulate content	Less than 1 wt %
Free carbon	Negligible
SiO_2, Si_3N_4, etc.	Trace

3.4 Mitsubishi International/Tateho Chemical Industries

Address:

ICD Group Inc.
641 Lexington Ave.
N.Y., NY 10022
Stephen C. Pred
(212) 644-1500

Distributor for:
Tateho Chemical Industries Co. Ltd.
974 Koriya ARU-shi
Hyogo—Ken Japan

Products:

SCW—a submicron silicon carbide whisker.
SNW—a submicron silicon nitride whisker, Si_3N_4.

Silicon Carbide Whisker (SCW) Properties:

Color: pale green
Diameter: 0.05 to 0.2 μm
Length: 10 to 40 μm
Crystalline phase:
Beta—SiC (cubic): over 95%
Alpha—SiC (hexagonal): below 5%
Chemical analysis:

Component	% by Weight
SiC	98
Mg	0.25
Ca	0.38
Al	0.06
Fe	0.001

Silicon Nitride Whisker (SNW) Properties:

Color: white
Diameter: 0.2 to 0.5 μm
Length: 50 to 300 μm

Crystalline phase:
Alpha—Si_3N_4 (trigonal): over 97%
Beta—Si_3N_4 (Hexangonal): over 3%

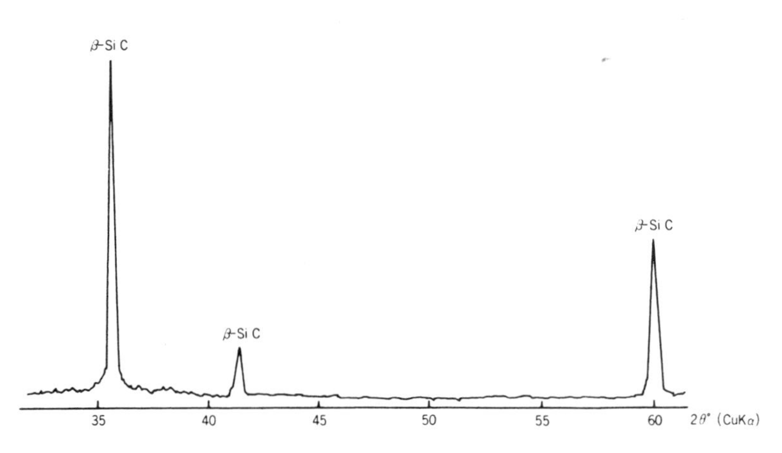

Fig. 13-9. X-Ray diffraction.

Chemical analysis:

Component	% by Weight	
	SNW #1	SNW #10
Si_3N_4	99	98
Mg	0.15	0.15
Ca	0.20	0.21
Al	0.12	0.40
Fe	0.01	0.53

3.5 Nikkiso

Address:

Nikkiso Co. Ltd.
Shizuoka—Prefect ore
Japan 421-04

Product: graphite whiskers.

Properties:

Color: black
Diameter: <1.0 μm
Length: up to 2 mm

Availability and Price: Available in multiple pound quantities, price upon request.

3.6 Versar Manufacturing

Address:

Versar Manufacturing Inc.
6850 Versar Center
P.O. Box 1549
Springfield, VA 22151
(703) 750-3000
Vice President: Dr. Robert Shaver

Products: Versite Silicon Nitride and Silicon Carbide Whisker Products.

Prices:

Powder form	
• Type VNIA Silicon Nitride	Reader must obtain quote. Price about $500/lb during 1986.
• Type VCIA Silicon Carbide	
Loose staple wool	
• Type VN2 Silicon Nitride	Reader must obtain quote. Price about $5000/lb during 1986.
• Type VC2 Silicon Carbide	

3.7 New Suppliers

American Matrix
118 Schelake Drive
Knoxville, Tenn. 37922
James A. Black
Dick NixDors
(615) 691-8021

ALCAN
1188 Sherbrook St. West
Montreal, Quebec CANADA H3A362
Richard Webb

Nippon Light Metals Co. Ltd.
#13-12 P.O. Box #5
Takanawa Post Office
Tokyo JAPAN
Phone: 03456 9347

4. PRODUCTION STATUS, PRESENT AND FUTURE

Finally, after 20 years, a number of companies are making whiskers in over 1000-lb batches per month, and one company has the capability to produce 1000/lb per day of cobweb whiskers. These changes represent a breakthrough in the technology of whisker production and availability. In the past, production and availability were expressed in grams and pounds. With several companies seriously engaged in the whisker business, the present and future availability of whiskers is promising. It is apparent that production is increasing, and prices soon will be decreasing significantly. Currently, the use of whiskers in ceramics and metals far exceeds their use in plastics, but the authors anticipate that in the near future whiskers in polymers will be an extremely large business.

5. FIBER PROCESSING, THEORY AND GENERAL APPLICATIONS

5.1 Methods of Fiber Preparation, Cleaning, and Classification

5.1.1 Cleaning. The material produced generally contains a wide range of aspect ratio fibers and a varying amount of nonfibrous debris. A general cleaning process that is applicable to most raw whisker products is wet cleaning in a hydrocyclone. The hydrocyclone separates the large particles and very low aspect ratio fiber from the medium to long fibers. For more details on this subject see the book by Parrott.[7]

5.1.2 Fiber Classification. Classification of fibers is desirable because various applications require specific aspect ratio materials. Screening of a fiber slurry does not yield a narrow range of fiber length because after the first layer of fibers is deposited on the screen, they effectively make the screen size smaller and subsequently trap shorter fiber, which again reduces the size of the screen hole. A rotary screening device that eliminates this build-up problem was developed by Glass Development, Ltd. of England. It works on the principle of depositing a monolayer of fiber on a rotating cylindrical screen and backwashes the fiber off the screen before the next layer of fiber is applied. This is done in a continuous manner. Two or more screens can be used in series or parallel arrangement, as illustrated in Fig. 13-10. The short fiber goes through the first fine screen, leaving the long and medium-size fibers on the screen. These are backwashed off later in the cycle and recycled on a larger screen, which retains only the largest fiber and permits the medium-size fiber to pass through. This combination gives three cuts of fiber lengths, a short, a medium, and a long. The actual fiber length and aspect ratio depend on the fiber diameter and screen mesh used in the experiment. A good reference to this process is given in *Composites*, 1969.[8]

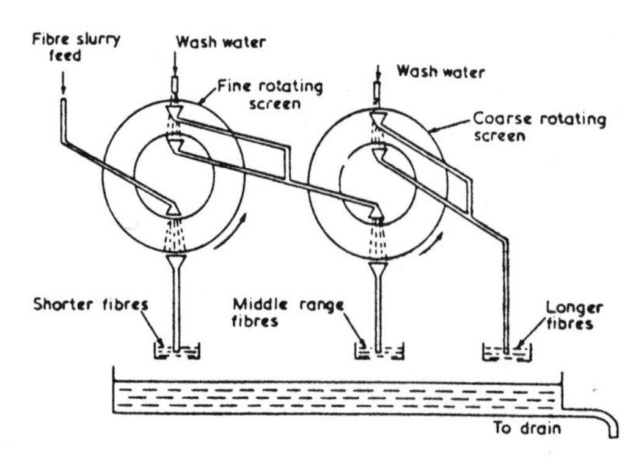

Fig. 13-10. Two wheels in series can be used to separate the fibres into three lengths, in which case the second wheel has a coarser gauze. (Courtesy IPC Science and Technology Press Ltd.)

5.2 Theory of Reinforcement

A rigorous theoretical analysis of the mechanism of discontinuous fiber reinforcements would fill a complete text in itself, so it is the authors' intent at this time to give a very generalized analysis of a discontinuous fiber reinforcement that will be helpful in understanding some of the further discussion in process development. For more detailed theory and references, see Chapter 2.

The modulus ratio of fiber to matrix affects the sphere of influence or reinforcing volume around each fiber and determines the optimum fiber length, which, as a consequence, affects both the minimum fiber concentration and the L/D required. This is a complex interaction, but the following simplified points may help the reader to visualize these factors.

The theory of reinforcement of discontinuous fibers assumes that the load or stress can be transferred through the matrix from fiber to fiber. The fiber is stronger than the matrix and has a higher modulus. When stress is applied to the composite, the fiber resists strain locally, which induces a much higher stress in the fiber than the matrix around it.

This effect has been vividly illustrated in color photos, using birefringent plastic, in some excellent work at Cornell by Schuster and Scala.[9] In Fig. 13-11A the unshaded area surrounding the whisker represents a lower stress and strain. Figure 13-11B represents the authors' conception of what composite stress/strain lobes would look like with a number of fibers unidirectionally oriented. Note that the high stress areas, always found at the ends of the fibers, are reduced when they fall within the bounds of the low stress lobes near the center of adjacent fibers. Note also that the overall composite's strain lobe will develop an aggregate shape that will represent additive strain reinforcement in the manner illustrated. A final point is that these illustrations represent the stress/strain lobe as drawn at some discrete isostrain (e.g., at 50% of maximum), and that

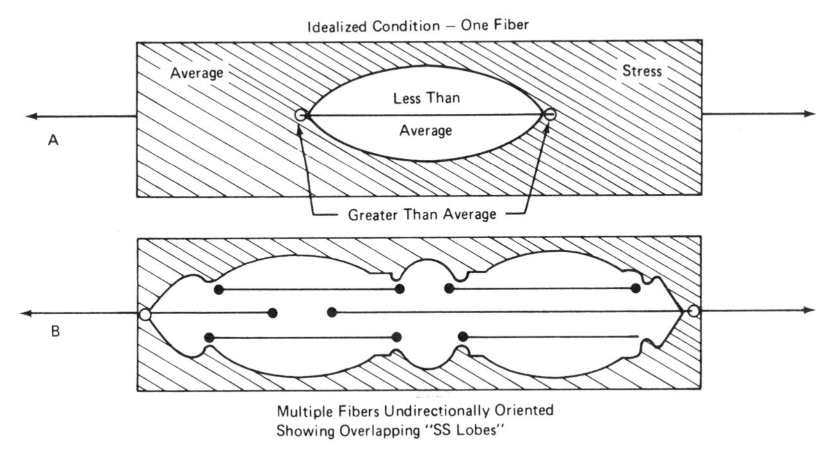

Fig. 13-11. Stress distribution around a reinforcing fiber.

in actuality it is a continuation of a gradually dissipating strain.

The ratio of the modulus of the fiber to that of the matrix has a pronounced effect on the shape of the stress/strain lobe, as is illustrated in Fig. 13-12.[10] These figures were drawn assuming equal fiber strain and the optimum fiber length required by the shear bond and fiber strength. In these figures, one can observe the major differences in strain distribution effects that come about in different matrices.

In ceramics, for example, under a given stress in the whisker, the stress/strain lobe will be broad and short because the ceramic modulus is high and the strain is not easily dissipated. Another point to consider in ceramics is that the maximum stress in the whisker can be developed in relatively short fiber lengths if a high bond strength is obtained. Fortunately, high bond strengths are possible for whiskers in ceramic matrices.

Now, considering the other extreme of whiskers in a low-modulus matrix such as plastic, it can be seen that the stress/strain lobe is thin and long. This suggests that fibers with greater L/D's and also higher concentrations would be used in order to ensure sufficient overlapping of the SS lobe. This idea, illustrated in Fig. 13-13, shows that fiber population must be over a certain minimum before significant reinforcement can be obtained. No attempt has been made at this time to calculate these minimum concentrations, but an estimated value for ceramic is less than 1%, for metals a few percent, and perhaps as high as 10% in plastics. This minimum population is affected by a number of other items not considered in detail, such as fiber diameter, fiber geometry, orientation, and

CERAMICS	*METALS*	*PLASTICS*
Ratio of the Modulus of the 2-1	Fibers to Matrix 5-1	100-1
L/D (Aspect Ratio) Required 10-20 to 1	for Optimal Reinforcement 20-50 to 1	30-100 to 1
Matrix Elongation Poor	Good	Excellent
Whisker to Matrix Bond Excellent	Poor Needs Coating	Good

Fig. 13-12. The effect of the fiber to matrix modulus ratio on the "SS" lobe.

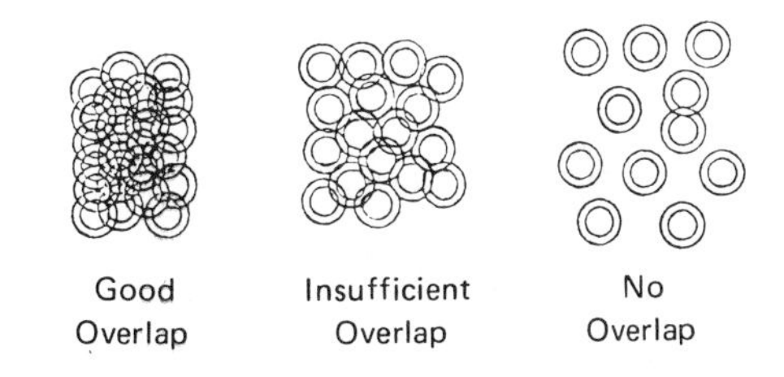

Fig. 13-13. The effect of fiber population on distribution of stresses. End view of "SS" lobe around unidirectional oriented fibers at different fiber population densities.

the addition of other non-fiber materials, all of which have a pronounced effect on the "volume strain" produced.

5.3 Type and Percent of Whiskers to Use

A usual opinion of novice engineers regarding the use of whiskers is that they must be unidirectionally aligned in order to make them useful. Actually, this is not the case. Most applications of materials require bi- and tridirectional strengths, as illustrated in Fig. 13-14.

Selection of the type and percent of whiskers for use in reinforced plastics will depend on the end product and the processing techniques to be used. For example, a tridirectional reinforced molding compound, a bidirectional reinforced laminating sheet, and unidirectionally reinforced rods or wires all require different types of whiskers for optimum performance. (See Fig. 13-14.) It is obvious, then, that engineers must become familiar with the various types and with their advantages and disadvantages, in order to realize the maximum properties available in whiskers.

The end product and processing method will dictate the preferred diameter, length, and type of whisker to be used. For example, a very fine diameter high-aspect-ratio whisker should be used for making yarn to fabricate unidirectional composites. Short, low-aspect-ratio, small diameter or submicron cobweb size whiskers will be ideal for interstitial interlaminer shear reinforcement. For producing sheet laminates with maximum planar tensile strength, whisker paper or felt should be used.

In making additions of whiskers, the following suggestions will serve as a guide:

1. For grain refinement and grain stitching in metals and ceramics, seed polymerization in plastics, and supplementary interstitial reinforcement of carbon, boron, and fiberglass composites, 1–5% addition of whiskers is recommended. This amount usually gives sizable percent increases in mechanical properties.
2. Generally, 5–50% addition of whiskers is recommended for random reinforcement of molding and casting compounds requiring three-dimensional strength. This amount usually gives a multiple increase in strength.
3. Generally 50–70% is recommended for partially oriented laminates for bidirectional strength. This amount usually gives a large multiple increase in strength.
4. Generally 70–90% is recommended for fully oriented composites such as impregnated yarn and oriented tape. This amount usually gives an order of magnitude increase in strength.
5. For a super strength, low density, bonded trusswork of whiskers, where only a sufficient amount of matrix is used to bond the whiskers together at their mutual points of contacts, 90–95% addition of whiskers is recommended.

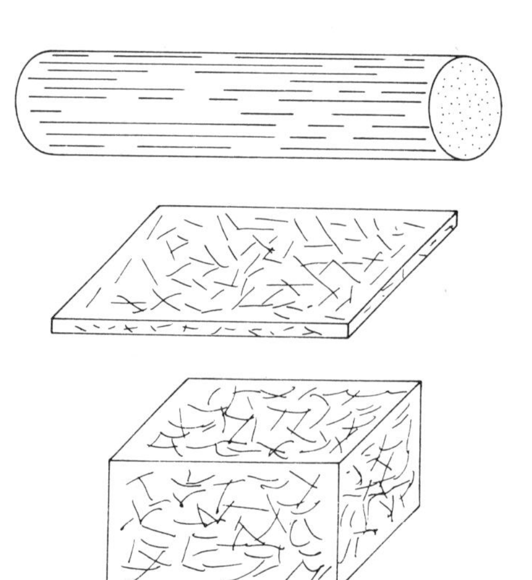

Unidirectional
in drawn wire and extruded rods axial orientation desired

Bidirectional
in laminated sheets or cylinders planer orientation desired

Tridirectional
in thick molding and casting complete random orientation desired

Fig. 13-14. Control orientation of whiskers in matrix.

6. METHODS FOR MAKING WHISKER COMPOSITES

6.1 Problems

Good composite fabrication techniques require the uniform distribution of whiskers, either in a random manner or in the orientation required for the specific application. The most important consideration is that the whiskers be "individualized" and bonded by the surrounding matrix material. The physical properties of the fiber and the matrix can often present a problem in obtaining a uniformly dispersed, bonded fibrous structure. Factors such as surface tension, wettability, density differences, excess fiber length, and matrix viscosity impede the formation of uniform dispersions. Moreover, electrostatic surface attraction and excess fiber lengths cause a clustering of the fiber into balls and parallel groups that will result in island-like clusters in the final composite. With ordinary mixing techniques, the whiskers nearly always remain in the clustered state and produce poor composites (see Fig. 13-15).

6.2 Wetting and Impregnating Techniques

One mixing technique that allows for removal of adsorbed gases and water on the whisker surfaces and avoids the clustering problems will be described. It is obvious that with different resin systems this technique will not be possible, but it is given as an example of how to approach these problems. This process involves cleaning, wetting, and coating the whisker surfaces in a dilute resin system by first forming a uniform dispersion of individualized fiber in the dilute state, and then subsequently controlling the dispersions of the coated individualized fibers.

The following steps are recommended as a lab method for preparing test specimens.

1. Dry whiskers at 250°F or higher (vacuum oven preferred).
2. Prepare the resin formulation. A typical formulation is shown below:

100 g	Shell 828 or equivalent
20 g	Shell activator Z
120 g	Resin–activator mix

3. Dilute part of above mixture in the following ratio:

10 g	Resin–activator mix
90 g	Acetone
100 g	Thinned resin-system (mix conventionally)

Balling
caused by fiber length too long or fiber mod too low, allowing fibers to ball up in non-interconnecting groups of reinforcements.

Paralleling
caused by poor processing technique or raw material forms such as chopped fiber glass.

Grouping
caused by inefficient mixing, settling out in liquid matrix.

Fig. 13-15. Problems causing nonuniform distribution of reinforcing fibers.

Coupling agents have been found helpful in improving composite properties, and may be added either to the solvent or to the resin formulation.

Impregnate whiskers by suspending them in a container of thinned resin solution. (Use approximately one cup of thinned mix to each gram of whiskers.) Place container with whiskers in a vacuum chamber, and evacuate it until the vapor pressure of the solvent is reached. Then hold at just above this pressure level to prevent excessive boiling of solvent. Vacuum-bump a few times during this period by releasing the vacuum and then pumping down again. Then pour thinned resin mix and whiskers into a food blender (Waring or Osterizer) and mix for 15 seconds at low speed and 15 seconds at high speed for average 1–10 μm diameter whiskers. For smaller-diameter whiskers, mix longer at high speeds and, conversely, shorter times for larger whiskers.

If twice the volume fraction of whiskers is used, double the mixing time. When the mixing is over, pour the slurry of fibers out into a fine metal screen, basket, or filter and let the excess resin and solvent drain through for about 5 minutes. Evaporate more of the solvent by suspending in an oven at 150°F for 15 minutes. Remove final remaining solvent in a vacuum desiccator.

At this point, the product is a relatively dry, 50–50 by volume whisker–resin, with all the whiskers individualized and their surfaces well wetted with resin. This mix now can be used directly for molding or adhesive applications or can be subsequently diluted with addition of resin for casting and potting application.

For more oriented structural application, the use of whisker paper is recommended. The process then must be modified after the impregnations by eliminating the mixing and including a vacuum paper mold in place of the steel basket to form the bidirectional felt. Impregnating of whisker yarn can be done on a spool or in a more or less continuous manner.

6.3 Casting Techniques

In a casting compound, the fiber–resin bond is not as efficient as that obtained by high-pressure molding. Consequently, to make up for this reduction in fiber reinforcement, a higher percentage of fiber must be used. In the past, it was thought that longer fibers would be the solution. However, longer fibers only lead to packing and thixotropic problems, which severely limit the volume fraction of whiskers that can be used. Further, long fibers introduce voids and uncontrolled orientation problems. A more recent understanding of packing concepts and stress/strain transfer mechanism leads one to the use of continually shorter fibers at higher concentration. For composites at 10% volume loading, one may use an aspect ratio of 50, while at 30% volume loading 25 to 30 will be more effective, and at loadings as high as 50 vol %, aspect ratios in the 10 to 15 range are more efficient. This is so because as more whiskers are added, the effective modulus of the system increases, and in turn a higher modulus matrix does not require as long a fiber for efficient stress/strain transfer. This was discussed in detail in section 5.

One other point to remember is that, in general, with larger-diameter whiskers, higher loading for the same aspect ratio will be permitted, and, conversely, with small submicron whiskers (because of the thixotropic effect) a lower percentage of whisker loading will be possible.

6.4 Transfer Molding

Transfer molding compounds are basically tridirectional stressed materials and therefore must have reinforcement in all three axes. The best way to accomplish this is by a random distribution of the reinforcing fibers. This type of reinforcement is only possible with discontinuous fibers.

A problem common to nearly every molding compound available is that it is impossible to obtain a random and uniform distribution. "Balling," "paralleling," and "grouping" of the fibers occur, as illustrated in Fig. 13-15.

In the past, it was thought that the requirement for sufficient shear loading demands a long fiber that is generally too long to allow random packing. Generally, good shear bonding theory says that fibers with an L/D of greater than 50 are required to efficiently reinforce the typical resin used in transfer molding compounds. But now, with a better understanding of packing concepts and the effect of the matrix modulus on fibers, and an understanding of required fiber aspect ratios, there are new theories: In highly loaded resin systems, much shorter aspect ratios will reinforce quite efficiently, and these much shorter fibers do not have the packing, balling, and distribution problems associated with longer fibers. Thus, good uniform isotropic reinforcement results.

Using these theories, a proposed higher-strength whisker molding compound would be composed of the following:

Component	Volume loading (%)
Resin	65
10 micron whiskers	20
0.5 submicron whiskers	10
0.02 cobweb whiskers	5

If 10 micron diameter whiskers are not available for this compound, 1/64 in. chopped graphite or 1/32 in. milled fiber glass may be substituted. For the 0.5 μm whiskers, use ACMC's SC-9 silicon carbide whiskers, and for the cobweb whisker use J. M. Huber's XPV1 200 Å diameter whiskers.

The three size ranges of whiskers used in this formulation serve to reinforce each other, allowing both good flow control and good packing and distribution because the smaller fibers will fit between the larger ones. This also permits the use of a high percentage of reinforcing fiber, up to 35 vol % of the molded composite. Besides its unusually high strength, a composite of this type would have other very desirable properties, such as the capability of molding structures with a very small cross section (less than 2 mils in thickness). Also to be expected are good dimensional stability, associated with the anticipated high modulus (greater than 5 million psi), and excellent ability to mold around and to retain micrometallic inserts for electronic applications.

It is advantageous to use whiskers in tridirectional reinforced material. A randomly distributed fiber composite will have only one-fifth the strength of the equivalent volume loaded unidirectional fiber composite. However, small-diameter whiskers, with their extremely high strength of over 2 million psi, are ideal for bulk molding compounds because they will yield a high isotropic strength. Whiskers give a high modulus to isotropic molding compounds because their initial fiber modulus is generally 7 to 10 times that of fiberglass. Therefore, even after a 1/5 dilution due to random distribution, they still make a high-modulus isotropic molding compound.

Generally, whiskers are wetted by and bond well to plastics, especially at the high pressures used for most molding. Also, whiskers that have a favorable geometry can be used effectively in shorter L/D than is generally considered necessary. Whiskers are generally much smaller in diameter than fiberglass so that, for the same volume loading of fibers, many more fibers are present, permitting a more homogeneous distribution of the fiber in the matrix.

7. UNIDIRECTIONALLY ORIENTED COMPOSITE

7.1 Various Methods for Unidirectional Orientation

Many different techniques have been studied for the production of unidirectionally reinforced plastics with whiskers. Some of these are listed below:

- Spinning of whisker wool into yarn and then processing the yarn in a manner similar to continuous filament technology.
- Making highly directional whisker paper (i.e., 90% x, to 10% y orientation), impregnating, "B" staging and cutting into strips, and then tape winding as is now done with asbestos paper.
- Electrostatic, magnetic, or vibrational alignment of loose whisker needles, followed by a spraying with a thin film of plastic, alternating the alignment and spraying techniques to produce an oriented reinforced laminate.
- Centrifugal casting of resin with loose whiskers, to force the whisker to line up in the outer hoop of the casting to produce a plastic hoop with peripheral orientation of whiskers.
- Continuous extrusion of a whisker alginate slurry into a mild acid solution that

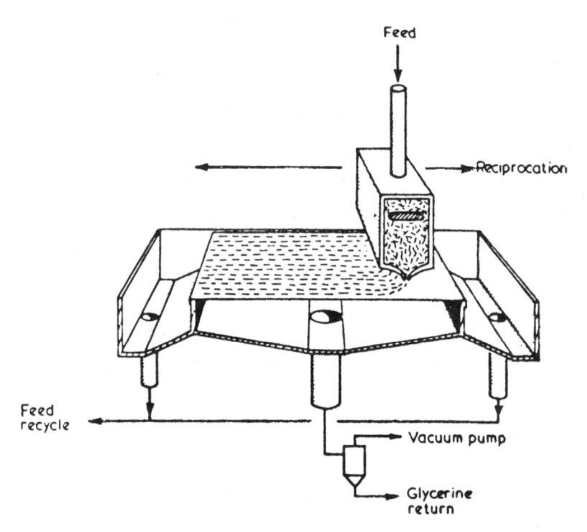

Fig. 13-16. Schematic diagram of the batch alignment plant showing deposition of aligned fibers in the glycerine onto a flat gauze filter. (Ref. 11. British Crown copyright, courtesy Controller, Her Britannic Majesty's Stationery Office.)

gels the alginate to produce a monofilament or reinforced plastic rod with oriented whiskers. The alginate is then burned off, and the unidirectionally oriented whiskers are reimpregnated with a matrix material.[10]

- A batch process in which the whiskers are suspended in a slurry with glycerine and coextruded through a convergent flow alignment slot nozzle, and are laid down simultaneously on a filtering screen. To produce a uniform layer of fiber, the nozzle travels rapidly back and forth over the screen. After sufficient fiber build-up is accomplished, the fibers are washed and can then be impregnated for compositing. See Fig. 13-16.[11]

Examples of the high strength that can be obtained in unidirectional whisker composites are shown in Table 13-2.[12] Note that the ERDE whisker composite has a flexural strength of 200,000 psi and a 20 million modulus. These are unusually high values, especially considering that this is a short fiber composite and only partially demonstrates whiskers' real potential.

7.2 Supplementary Reinforcement of Boron, Fiberglass, and Carbon Filaments with Whiskers

Many of the failings of conventional continuous reinforced plastics can be attributed to the fact that the structures are too anisotropic and lack a homogeneous distribution of strength. For example, a filament-wound structure is designed with filaments wrapped in the direction of the principal stresses. The stresses, however, do not always fall perfectly in line with the axis of the fiber. This is usually caused by such things as imperfect or unstable winding patterns and nonaxial stresses developing from torsion, bending, vibration, thermal changes, side loading, and so on. All of these stray stresses cause premature failure in filament-wound structures by overstressing the unreinforced resin-rich pockets around the fibers. This results in interlaminar shear failure. Therefore, a new application for the use of small-diameter whiskers is in supplemental interstitial reinforcement of filament-wound structures. These small fibers complete the reinforcing requirements by filling in the unsupported areas and by achieving a higher percentage theoretical

Table 13-2. Unidirectional whisker/epoxy composites.

Whisker Source	% Whiskers by Volume	Flexural Modulus GN/m^2 (10^6 lb/in^2)	Flexural Strength GN/m^2 (10^6 lb/in^2)	Effective Stress GN/m^2 (10^6 lb/in^2)
Carborundum Co.	48	180–200 (26–29)	0.82 (0.12)	1.7 (0.25)
ERDE–Explosives Research and Development Establishment, England	35	140 (20)	1.4 (0.2)	3.9 (0.57)

Source: Composites (March 1970).[12]

packing density. This idea of using whiskers to reinforce the interstitial plastic phase and at the same time to increase modulus and strength of the matrix is applicable to fiberglass, boron, and carbon reinforced structures.

Four suggested ways of using whiskers for supplementary reinforcement for boron, fiberglass, and carbon are outlined below (a more specialized case called "whiskerizing" with carbon fiber will be discussed later in the text):

- Interlaminar shear in laminates where whiskers are added to the resin between layers of boron, fiberglass, or carbon.
- Interlaminar shear in wet winding structures where whiskers are added to the resin before winding with boron, fiberglass, or carbon.
- Filament winding tape where whiskers are incorporated in especially high-stressed areas during winding.
- Multiple coating of fibers (first, with sizing; second, with resin; third, with submicron whiskers to produce a barbed-wire type fiber.)

The examples in casting, transfer molding, and supplementary reinforcement were chosen because these general molding methods are presently being used extensively in fiberglass and boron reinforced plastics and thus provide a good base to work from in the incorporation of whiskers. Under each category there are many other methods of producing similar effects, provided the proper adjustment in processing techniques is made to account for the small size and large surface area of whiskers.

7.3 Bifiber ("Whiskerized") Composites

The use of whiskers for supplementary reinforcement may play a big part in the initial commercial development of whisker composites because a small percentage addition has a major effect on the composite properties. The large reinforcing effect from a small amount of whiskers easily justifies their price. A special form of supplementary reinforcement particular to carbon fibers, called whiskerizing, is discussed below.

Carbon-graphite filaments have excellent properties. However, one area for improvement is low shear strength in resin-matrix composites, due to poor bonding. The problem is even more evident in the higher-modulus materials. A solution to this problem was the addition of whiskers to the composites, not in the loose form, as discussed earlier in the case with fiberglass and boron, but by growing the whiskers directly on the surfaces of the carbon fibers.

This process, whiskerizing, has the following effects on the carbon fibers:

1. It changes the physical surface of the carbon fibers.
2. It changes the surface chemistry of the carbon fibers.
3. It adds an integral bond between whiskers and the carbon fibers.
4. The projecting whiskers reinforce the interstitial plastic by improving its effective modulus and strength.

As a result of these physical and chemical changes to the carbon fibers, the bond strength (interlaminar shear) in general has been increased three to five times (Fig. 13-17).[11,12]

More specifically, the interlaminar shear strength increases as a function of the percent of whiskering, as illustrated in Fig. 13-18.[14] Note how important it is to make the composites with a low void content. This is accomplished by using solvent impregnation of the

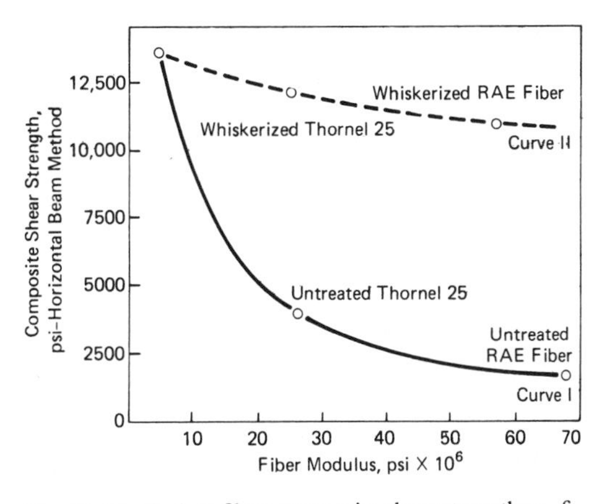

Fig. 13-17. Carbon fibers composite shear strength vs. fiber modulus (epoxy resin, amine cure). (Courtesy Naval Ordnance Laboratory.)

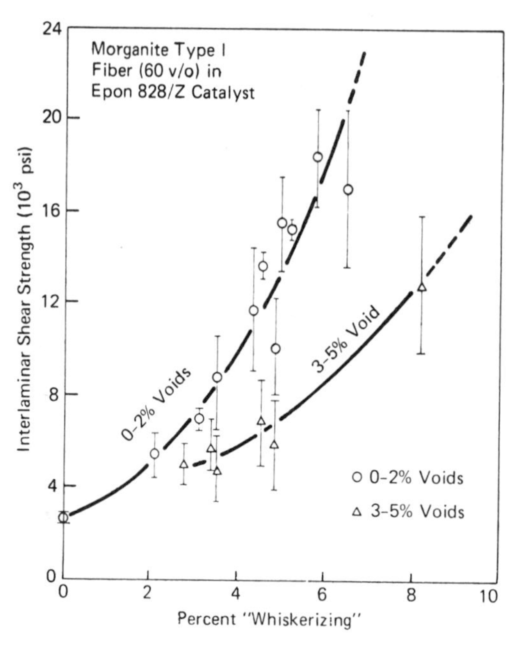

Fig. 13-18. Interlaminar shear strength vs. percent whiskering.

whiskerized carbon lay-ups, solvent evaporation, "B" staging, and then compression molding to a calculated volume using whatever pressure is required to obtain fill compaction.

The process of growing whiskers on carbon fibers is applicable to both batch and continuous methods. Figure 13-19 shows the needle growth on a continuous whiskerized graphite filament.[18] The process of whiskerizing can be applied to many other fiber systems. Whiskers have been grown on quartz, boron, SiC, and tungsten filaments. This type of bifiber compositing will no doubt have many new and special applications as its full potential is realized.

Fig. 13-19. Whiskerized RAE graphite fibers, 200×.

8. COBWEB WHISKERS, ULTRASMALL AND ULTRASTRONG

8.1 Introduction

In the last few years, cobweb whiskers have made the transition from futuristic product to commercially viable, high-volume fiber material that can be made in rates of 1000/lb a day—undoubtedly the breakthrough of the decade in new material production technology. This new submicron whiskerlike fiber, designated Fiber XPV1, has been produced by the J. M. Huber Corporation.

Cobweb whiskers are defined as single crystal fibers that have a nominal diameter of less than 0.1 μm or 1000 Å. This fiber diameter is too small to be observed clearly under normal optical instruments. They are also difficult to observe clearly using a scanning electron microscope (SEM) because of its charging effect, which causes the fiber to wave and form a blurred image. The best way to observe these fibers is with a transmission electron microscope (TEM).

Silicon carbide cobweb whiskers were first observed by Milewski at Thermokinetic Fibers Inc. in 1964. They were first discussed in the *R & D Magazine*, March 1966, and featured on the cover (see Fig. 13-23).

In general, for whiskers and other fine fibers, it has been found that as the fiber is made smaller and smaller in diameter, it becomes stronger and stronger. This effect was first noticed and explained by A. A. Griffith[16] over 60 years ago (the Griffith Crack Theory). A curve illustrating this principle is given in Fig. 13-20. Data supporting this concept were generated by Milewski at Exxon Research in 1970, where tests were run on individual fibers of beta-silicon carbide whiskers down to 1 μm in diameter, which were found to have tensile strengths up to 6 million psi. More recently, at Los Alamos, 4 μm diameter beta-silicon carbide whisker were found to have tensile strengths up to 4 million psi.

What is the ultimate strength of whiskers? What is the strength of cobweb whiskers? It may not be possible to test them as individual fibers at these sizes for some time to come. It appears from all early theories and indications

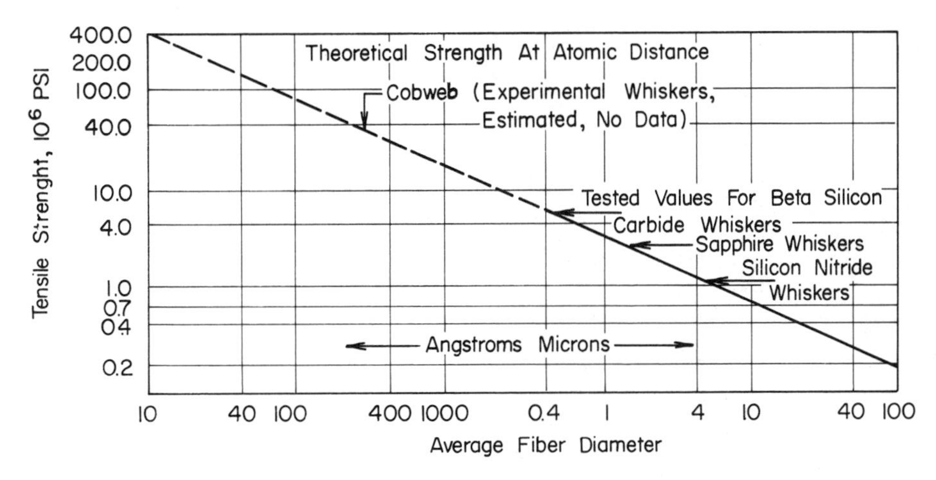

Fig. 13-20. Present and projected strength levels of whisker fibers.

that the cobweb size whisker will be stronger than its micron size counterparts of the same composition, which have been found to have tensile strengths in the millions of psi.

To prove the strength and reinforcement capability of the XPV1 cobweb whisker, a program was conducted to evaluate this fiber in a plastic matrix. This has been done, and the detailed procedure and proof of reinforcement capability are given later in this section.

Most of the fibers grow into fibrous balls while being made (Fig. 13-21). The outer edges of the fiber ball are quite loose and low in density (Fig. 13-22), with the fiber concentration increasing toward a dense growth center. The fibrous balls range from 0.5 to 10 μm in diameter. The aspect ratio of the as-grown fiber is difficult to measure because of the bundling; however, estimates range from 50 to several hundred.

8.2 Properties of Fiber XPVI Cobweb Whisker

Section 3.2 contains a detailed list of Fiber VI properties. Fiber XPV1 is a ceramic fiber that has a chemical composition very close to that of silicon monoxide. In some of the transmission electron micrographs (TEM's), it appears to be a fiber with a core of pure silicon metal surrounded by a sheath or coating of silicon dioxide; however, more often the fiber appears uniform and amorphous in structure—possibly silicon monoxide. The surfaces of such fibers are still coated with silicon dioxide, which makes them amenable to the use of silane coupling agents designed for glasslike fibers.

The fibers can be made from about 2 to 50 nm in diameter and usually range from 20 to 40 nm. They have a specific gravity of 2.3 g/cm^3 and a typical bulk volume of 0.07 g/cm^3. As produced, the material ranges in color from tan to olive.

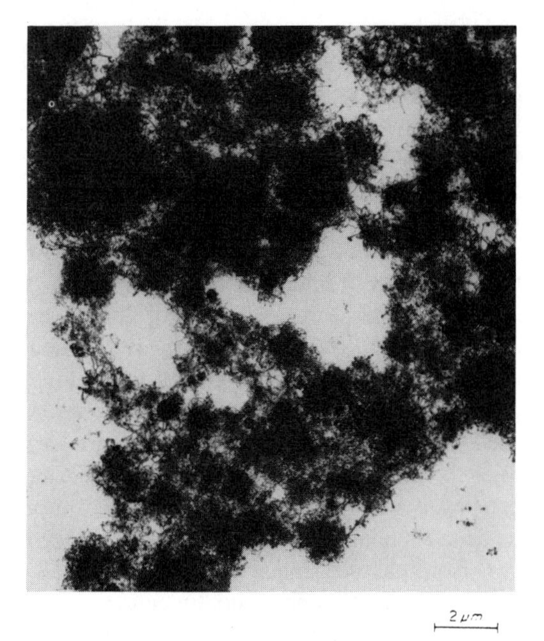

Fig. 13-21. Representative population of fiber XPV1's balled structure (X10,000).

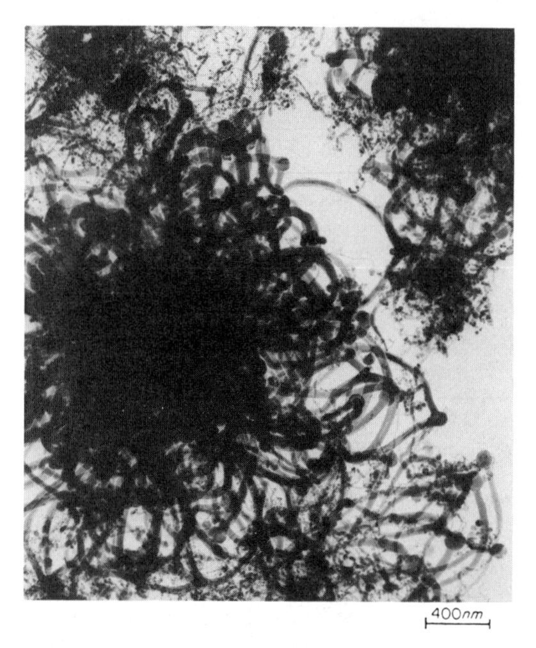

Fig. 13-22. Closeup of outer edge of a Fiber XPV1 bundle showing low density (X50,000).

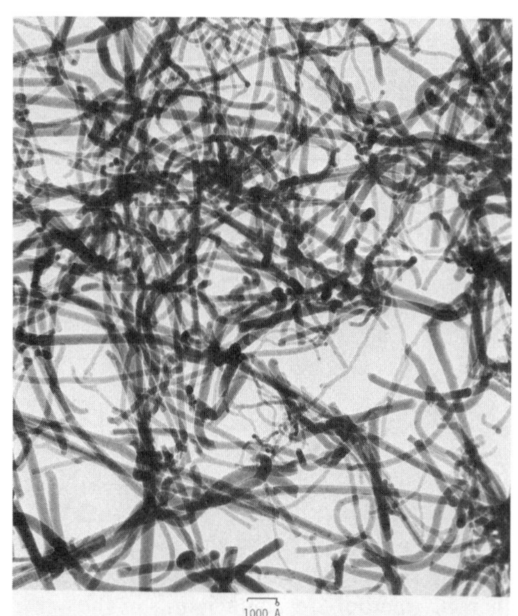

Fig. 13-23. Silicon carbide "cobweb" fiber, 200,000×. Experimental whisker from T.K.F.

8.3 Advantages/Disadvantages of Submicron Fibers

Submicron reinforcement fibers have specific advantages over the more conventional 6 to 14 μm diameter fibers such as carbon, fiberglass, and Kevlar. One advantage is the resulting smooth surface, producing a greater cosmetic appeal (no orange peel or fiber show-through). Other advantages are associated with processing. The smaller-diameter, shorter-length fibers permit the use of small gates in injection-molding with less chance of log jamming, and are not broken up in processing; they can go through screw machining and high-pressure, high-speed flow with a minimum of fiber damage. Therefore, the fiber integrity is retained after molding, preserving a greater reinforcing potential in the as-molded parts. Finally, when small-diameter fibers with low aspect ratios can be used and dispersed uniformly, the resulting composite will have more isotropic and homogeneous reinforcing. This, in turn, produces very isotropic mechanical properties. Isotropic properties are generally quite difficult to obtain with larger-diameter fibers in injection- and transfer-molded parts because of the uncontrolled, forced orientation associated with flow during the molding process.

The disadvantages of submicron fibers are basic to all materials that are very fine particles or fibers, in that they inherently have a very large surface-to-volume ratio and high surface areas. Fiber XPV1 has a large bulk volume, a characteristic that makes it a thixotropic additive. The viscosity of any liquid system increases upon its addition, which, in turn, makes large additions difficult to mix, disperse, and process.

The advantages are inherent in the fiber itself, while the disadvantages and problems can be overcome through an intelligent understanding of the problems and skillfully applied processing techniques.

The following sections are concerned with overcoming the inherent disadvantages of the submicron fibers while retaining their advantages.

8.4 Fiber Separation Techniques

Because of the tangled nature of the fibers, separate and uniform distribution is difficult. Conventional mixing and blending techniques will not suffice. This was demonstrated in early processing tests when dispersion was attempted with a Cowles blade with poor results. Slightly better results were obtained with higher shear type mixing apparatus such as the three-roll mill and the counterrotating blade in a Brabender type mixer.

It became apparent that in order to obtain the desired level of reinforcement, a much higher level of fiber separation would be necessary. Two new approaches were tried to separate the fibers from their initial growth bundles: ball milling and homogenizing. All attempts were very successful in separation of the fiber from their central growth core.

The ball milling broke off the fibers but at the same time substantially reduced their aspect ratio to 10 or less. Because of the severe fiber breakage and other processing problems, the composites made from this material were very low in strength. Next, the homogenizer was tried. It proved very effective in fiber separation, dispersion, and retention of fiber length with aspect ratios in the 10 to 50 range.

8.5 Panel Preparation Method

The resin matrix that was used to prepare the flexural strength test panels consisted of Shell Epon 828 Epoxy Resin, 90 parts by weight, and butyl glycidyl ether diluent, 10 parts by weight. This matrix was usually cured with Shell catalyst Z, 20 phr. Also, 2 wt % of Union Carbide amino silane, A1100, based on the weight of the fiber to be added, was mixed in the resin.

A multi-cavity compression mold was used to prepare most of the test panels. The molded panels were 5″ × 0.5″ × approximately 0.13″ or 0.25″ thick. All of the mixtures were sufficiently fluid to be poured into the mold. There was an initial preheat at 180°F and final cure at 310°F for 15 minutes. All panels were post-cured in an oven at 325°F for one hour.

8.6 Mechanical Testing and Results

The flexural strength and modulus were determined by a three-point loading test in accordance with ASTM D790-81, Method 1. The test span was 2 in. and the cross head speed was 0.05 in. per minute. The flexural strength, as determined by this method, is not accurate or valid when the sample deflects more than about 5% at break. Therefore, the pure matrix control samples and a small number of the test samples that had high deflections were calculated at 5% deflection. Most of the reported data was the ultimate flexural strength at break. Also, the base value for the matrix was obtained from the manufacturer of the epoxy resin as a verification of test results.

This better fiber separation and dispersion shows up as flexural strength of about 30,000 psi with an accompanying realized fiber stress in the order of 600,000 psi.

The best fiber stress to date was realized in the epoxy composite with a combination of only 14 vol % Fiber XPVI and 24 vol % microspheres. Composite flexural strength of about 25,000 psi was obtained, and fiber stress of 1,000,000 psi was realized—proving the high strength properties of the submicron whisker-like fiber. The results also support Milewski's theory that spheres improve the reinforcing efficiency of short fibers.[18,19,20]

For comparison, similar composites were made using the commercially available silicon carbide whiskers from ACMC. When their 0.6 μm diameter fiber was processed in the homogenizer, molded, and tested, the resulting composite gave a maximum flexural strength of about 31,000 psi and calculated fiber stress of about 850,000 psi. Tests to date indicate the Fiber XPV1 to be about equal in reinforcement abilities to the silicon carbide whisker when they are incorporated under identical conditions.

8.7 Conclusions

The fiber XPV1 cobweb whisker has demonstrated excellent reinforcement capabilities in levels equivalent to commercially available 0.6 μm silicon carbide whiskers. Mechanical testing gives flexural strength up to 36,000 psi at only 23 vol % loading. Calculations indicate that load-carrying fiber stress in excess of

600,000 psi has been attained in fiber-filled composites, and about 1,000,000 psi when fiber and sphere combinations were used.

To demonstrate high fiber reinforcement efficiencies, good separation of the fiber from the original bundle must be attained along with uniform dispersion. The use of the homogenizer is a good way to attain fiber separation and good fiber dispersion.

8.8 Applications

A universal reinforcement such as Fiber XPV1 can be projected for use in many applications. Preliminary tests in high-strength, high-temperature adhesives look promising, as do applications for low-shear, pressure-sensitive adhesives. Early tests in reinforced coatings and paints show apparent enhanced wear, chip resistance, and corrosion resistance.

Projected applications include interstitial reinforcement for filament winding of graphite, fiberglass, or Kevlar composites, and will be used to improve their Achilles' heel of poor transverse and shear properties, which are matrix-controlled. Reinforced thermoplastic should be a natural because this tough, fine-diameter fiber will stand the rigors of high-pressure, high-velocity molding and will pass through small gates with minimum fiber damage.

The development of high-strength molding compounds with the ability to form thin walls for small electronic components will now be possible with this new micro whisker-like fiber, and improved transverse properties in pultruded products can be projected with this new small reinforcement.

Applications in matrices other than plastic resins are also being actively pursued. Preliminary work is under way to evaluate the performance of XPV1 in glass, ceramic, elastomer, and metal composite systems. In the forseeable future, it is probable that fibers of other chemical compositions will become available (e.g., silicon carbide). Such a development will greatly expand the range of matrix materials and service conditions that can be considered for application of these whisker-like microfibers.

9. FUTURE TECHNIQUES

9.1 General

Theoretically, it is possible to grow a whisker out of any material that will crystallize—ceramics, metals, and even plastics. As a result, many different types of experimental whiskers have already been grown for specialized compatible applications. In addition to reinforcing, the specialized whiskers will bring unusual electrical, optical, magnetic, ferromagnetic, dielectric, conductive, and superconductive properties to the matrix to which they are added. As the demands of materials become more unique, so will the kinds and types of available whisker materials.

9.2 Short-Range Applications

The present high price of quality whiskers, and the projected prices in the next few years, will limit their applications to more or less specialized material problems.

The first commercial applications will probably be those that require a relatively low percentage of whiskers in relation to the total composite or those used to reinforce specific areas of weakness such as adhesive joints, weldings, hard surfacing, or interstitial reinforcements (as with carbon fibers). Next to follow will be high-priced small objects in specialized molding compounds used in electronics or medical application, such as dental and synthetic body components (bones, teeth, etc.), and other medical applications. These will then be followed by large aerospace and scientific applications, such as helicopter blades and high-strength centrifuges. Finally, applications of whisker composites in wings, tail assemblies, and space capsules will be realized.

9.3 Long-Range Applications (Commercial)

With the advent of commercial production, whiskers may be used as a universal reinforcement in all types of materials and may find their way into many commonplace products. For example, whisker-reinforced plastics will be used extensively for body and basic structures in the automotive industry where they will save weight and have twice the strength of steel, the modulus of titanium, and the density of aluminum.

The construction industry will require a whole new approach to design when using whisker-reinforced plastics. These new reinforcements will allow new freedoms in which the structural member becomes part of the architectural designs, assuming phenomenal tensile strength and crack resistance in structures having extremely thin sections.

A big step in whisker-reinforced material would be whisker-yarn-reinforced rubber. With the yarn woven into cloth and impregnated with rubber, lightweight super-pressure tires will be possible. Also, this rubber-impregnated whisker cloth would greatly accelerate the growth of the inflatable structures industry for both air and undersea applications.

Last, but not least, the use of whiskers in the human body will bring a revolution to the area of dental bridge work, reinforced plastic bones, and reinforced flexible artery tubing.

10. FUTURE ACTIVITY IN WHISKER COMPOSITES

The main criteria for determining the future utility and sales volume of whiskers will be the level of properties that are attained in the whisker composites. There were many handicaps in the early efforts in whisker technology, including a shortage of consistently high-quality whiskers, as well as a poor understanding of the subtle difficulties inherent in the handling of whiskers and subsequent composite fabrication. It is anticipated that this presentation may act as an incentive and guide to overcome these problems, and lead to the production of high-quality whiskers and ultrahigh-strength composites.

REFERENCES

1. Brenner, S. S. and Sears, G. W., *Acta Met.* 4: 268, 1956.
2. Brenner, S. S. ''Growth of Whiskers by the Reduction of Metal Salts,'' *Acta Met.* 4, Jan. 1956.
3. Brenner, S. S., ''Growth and Properties of Whiskers,'' *Science*, 128: 3324, Sept. 1958.
4. Webb, W. W. and Forgang, W. B., *J. Applied Physics* 28: 12, 1957.
5. Ryan, C. E. et al., Physical Sciences Research Papers No. 266, AFCRL-66-641, Sept. 1966.
6. Shyne, J. J., Milewski, J. V., Shaver, R. G., and A. Cunningham, ''Development of Processes for the Production of High Quality, Long Length SiC Whiskers,'' AFML-TR-67-401, Jan. 1968.
7. Parrott, N. J., *Fibre-Reinforced Materials Technology*, Van Nostrand Reinhold Co., London, 1972.
8. Parrott, N. J., ''The Rotary Fibre Classifier,'' *Composites 1*(2), Dec. 1969.
9. Schuster, D. M. and Scalla, E., ''The Mechanical Interaction of Sapphire Whiskers with a Bi-Refringent Matrix,'' Report No. 166, pp. 4 and 5, Cornell University, Dec. 31, 1963.
10. Parrott, N. J., ''Whisker Alignment, by the Alginate Process,'' *Composites* 1(1), 1969.
11. Bagg, G. E. G., Evans, M. E. N., and Pryde, A. W. H., ''The Glycerine Process for Alignment of Fibers and Whiskers,'' *Composites* 1(2), Dec. 1969.
12. Parrott, N. J., ''Silicon Carbide as a Reinforcement,'' *Composites* 1(3), 141–144, Mar. 1970.
13. Prosen, S. P. and Simon, R. A. ''Carbon Fiber Composites for Hydro and Aerospace,'' U.S. Naval Ordnance Lab, Oct. 1967.
14. Shaver, R. G., ''Silicon Carbide 'Whiskerized' Carbon Fiber,'' AIChE Materials Symposium, Philadelphia, Apr. 1–4, 1968.
15. Milewski, J. V., Shyne, J. J., and McGowan, H. C., ''Interstitial Growth of Silicon Carbide Whiskers in Carbon and Graphite Filaments,'' under contract N-60921-67-C-0293, Sept. 1968.
16. Griffith, A. A., ''The Theory of Rupture,'' *Proceedings* of the 1st International Congress on Applied Mechanics, 1924, p. 55.
17. Milewski, J. V., Katz, H. S., and Lee, K. W., ''Evaluation of a New Submicron Ceramic Reinforcing Fiber,'' 40th RP/C of SPI, Feb. 1985.
18. Milewski, J. V., ''Problems and Solutions in Using Short Fiber Reinforcements,'' 37th Annual Conference—Reinforced Plastic/Composite Institute, SPI, Jan. 1982.
19. Milewski, J. V., ''How to Use Short Fiber Reinforcement Efficiently,'' 37th Annual Conference—Reinforced Plastic/Composite Institute, SPI, Jan. 1982.
20. Katz, H. S. and Milewski, J. V., *Handbook of Fillers and Reinforcements for Plastics*, Van Nostrand Reinhold Co., New York, 1978.

Section IV
Fiberglass

14
FIBERGLASS

John F. Dockum, Jr.
PPG Industries, Inc.
Fiber Glass Products
Pittsburgh, Pennsylvania

CONTENTS

1. INTRODUCTION

The product, fiberglass, is the result of conversion by fire of earth and oxide materials into a hot-melt, vitreous state and subsequent drawing into a filamentary aspect by high-speed attenuating mechanisms. The basic forms of fiberglass for reinforcement are two: continuous filament and staple fiber. A third form, of minor application here, is comprised of blown fiber, which is the basic element of fiberglass wool. Wool products are utilized in filters, insulating products, and thin mats, one type of which forms the surfacing veils used in reinforced plastics/composites.

Reinforcing glass fibers, after being collimated into strands, are further processed into usable forms that are still only intermediates for final conversion into reinforced plastic/composite (RP/C) end products. These usable forms are designated as twisted and plied textile yarns, braided yarns, woven tapes and fabrics; continuous rovings, woven rovings, and knitted uni- and multi-axial rovings; chopped and continuous strand mats, combination mats and woven rovings; and chopped strands and milled fibers. Specialty or related glass products include such miscellaneous items as hollow fibers, flake-glass, hollow spheres, solid beads, scrim fabrics, coated yarns and bonded cordage, bulk and wound fibers for filter media, fiber-optic elements, and many others.

Continuous filament glass fibers may be produced in a wide range of filament diameters from several different fiberizable compositions. When combined with the appropriate laminating resin, this fortuitous combination provides required versatility for a myriad of end-use applications. RP/C parts are sized from a 1 in. diameter wrist watch plate of extremely fine,

*Acknowledgment is made that this is an updated chapter originally prepared by J. Gilbert Mohr, Johns Manville Fiber Glass Division, Waterville, Ohio.

precise dimensions through a 42-ft diameter × 40-ft high tank, capacity 414,600 gallons, all made possible essentially because of the technology of fiberglass.

The structural and utility RP/C products so fabricated find outlets in a steadily growing number of applications in markets such as: automotive, appliance, business equipment, aircraft, corrosion resistant equipment, construction, electrical and electronic, marine, recreational, and tooling.

As a textile and reinforcing material glass fiber may provide numerous advantages: it is a hard, incombustible, and impervious substance that is mildew- and rot-resistant and exhibits durability on exposure to many corrosive chemicals; it does not relax its rigidly held ionic bonds until a temperature of 1200–1500°F is reached; it has a high tensile strength to weight ratio, plus perfect elasticity, with an elongation of 3–4% without a yield point, characteristics desirable in spring applications; it has a low coefficient of thermal expansion (relative to resin), providing dimensional stability to RP/C; with the increased surface area per unit volume created by fiberizing, and the immediate chemical affinity of glass for (OH) groups available in the atmosphere, sites are provided for cross-linking and hence chemical bondings to specific polymers requiring reinforcement; glass fibers do not generate internal heat due to molecular friction during mechanical working as do metals; in the fibrous form, its stiffness factor, related to the minute filament diameter, permits processing on standard and applicable reinforced plastics and textile equipment without fiber breakage; and finally, numerous fiberizable glass compositions have been developed for specific applications, such as greater temperature resistance, higher electrical properties, resistance to acid or alkali, greater tensile strength or mechanical stiffness, and absorption of harmful radiation such as X rays.

On the debit side, fiberglass applications are limited to a minor degree because: the material is friable and brittle, which permits weakening from surface defects, thereby necessitating protective lubrication and/or surface treatments in processing; glass per se is a smooth-surfaced material without undulation or nodules that would provide mechanical locking sites for laminating; glass per se (and certainly no less glass fibers) is subject to static fatigue, causing the fibers to exhibit weakness when exposed to a sustained load near its innate elastic limit; and finally, glass fibers exhibit a lower-than-desired tensile modulus of elasticity.

2. HISTORY

A capsule summary of the history of fiberglass may be recorded by considering three major phases.

2.1 Invention and Development

In all probability, the original fiberglass was a creation of nature known as "Pele's Hair." This material resulted when high winds exerted a force on the leading edges of waves of vitreous volcanic dross during a lava flow. The lava was attenuated into a fine-fibered mass that was carried by the wind and deposited in trees. Strangely indicative of things to come, this fibrous mass was used by birds as reinforcement in construction of their nests. The phenomenon has been observed many places in the world, but notably in Hawaii from the Kilauea volcano, and from Mount Vesuvius in 1906.[1]

The first glass artisans in ancient Syria, Greece, and Egypt learned to draw fibers from a heated glass rod and to impress them while fluid onto the surfaces or stems of fabricated articles. This technique was used even prior to invention of the blowpipe in approximately 250 B.C. These fibers, as laid down, were either left in situ or further manipulated by reheating and tooling or scoring to form ingenious, artful decorations on the glassware of the day.

Similar and other related ways of incorporating fine-drawn glass fibers were used by Venetian glass makers in the sixteenth and seventeenth centuries (fine decorated, thin-walled "lace" glass stemware) and by the Romans (millifiori glass). English, German, and French glass producers (1650–1720) devised methods for drawing fine fibers from small melts or heated glass rods, and later (1832), glass threads from a hot glass gob onto a revolving wheel. The material so produced remained only a curiosity. Some was woven into fabric structures with other textile materials, and some was used for such decorative applications as the tails of birds in Christmas-tree ornaments.

Glass fibers drawn from heated rods were produced to make a dress and neckties and exhibited at a Chicago fair in 1893 by entrepreneur Edward Drummond Libbey and his technical genius, Michael J. Owens. A near-success by Libbey-Owens-Ford at a commercial fiberglass product occurred in the 1930s when colored glass fibers were laminated between two sheets of flat window glass to form a glare-reducing structural panel. Unfortunately, these ventures were still classed in the novelty category, and the real expansion of the glass fiber industry still waited in the wings.

2.2 Commercialization

The Owens-Corning Fiberglas Corporation was founded in 1937, following 10 years of joint research between Owens-Illinois Glass Company and Corning Glass Works. This research program was especially geared to development of viable mass production methods for glass fibers, and resulted in workable technology for all three methods: continuous filament, staple fiber, and wool, or blown glass fibers.

The research was conducted at Purdue University. The original Owens-Corning producing facility was located at Newark, Ohio.

Original products were "white wool" insulation, coarse-fibered air-filters, and continuous-filament material that was used to weave a cloth subsequently impregnated with varnish for electrical laminates. The development of polyester resins in 1935 gave a natural impetus to the combination, and reinforced plastic as a new material of construction was born.

2.3 Expansion

Circumstances of World War II spurred the glass fiber reinforcing business. Successful applications of molded and laminated products were rapidly made. Development of molding methods particularly suitable to two new materials, low-pressure curing resins and the easily handled fiberglass reinforcement, proceeded with both speed and great ingenuity. The first reinforced plastic end-products were puncture-proof, jettisonable gas tanks, radomes, flack-suit body armor, and other nonmetallic weight-reducing components for military aircraft, personnel, and land transportation.

Following the war, a concerted effort was made to adapt lessons learned on military applications to domestic production, although price–volume adjustments were necessary. In 1945 reinforced plastics pioneers, fewer than 30 companies, formed the Low Pressure Laminates Industry Division of The Society of the Plastics Industry, Inc. The RP/C industry was finally being launched.

In the ensuing decade several domestic and foreign firms received licenses to manufacture from Owens-Corning Fiberglas Corporation. Fishing rods were among the first consumer products introduced, while Project Y, an experimental all-plastics car called the Scarab, was followed by introduction of the commercially successful Corvette sports car. By 1955, 168 million pounds of RP/C end products were being produced annually. In the decade to follow, the marine industry was the predominant market, with pleasure craft increasingly being made of RP/C. Bathroom fixtures, patio panels, and the Studebaker Avanti, the first hard-top production car, were also significant applications.

Research and development, stimulated by lively competition among producers of reinforcing fiberglass products, laid the groundwork and paved the way for a great upsurge in productivity beginning in 1965. With total RP/C approaching 500 million pounds, the automotive market in the next decade was the major factor in more than doubling volume to over one billion pounds annually. Applications ranged from windshield washer gears to grille opening panels, instrument retainers, and heater/air conditioner ducts. In the decade from 1975 to 1985, the two-billion-pound mark was surpassed, and RP/C achieved the status of engineering material without equal in high pressure pipe to aircraft components and helicopter blades. In 1985, the RP/C industry with over 300 companies and now represented by the Composites Institute of the SPI and the Fiberglass Fabricators Association celebrated 40 years of innovative technology. The improved technology brought about glasses with greater chemical durability, higher modulus of elasticity, and many other desirable properties. Improvements in resins and molding technology provided fabricators the capability to produce commercial marine vessels over 150 ft in length, sophisticated beam and girder stock for structures in critical corrosive atmospheres, many elegant products for buildings and construction, automotive components produced at extremely high speed and with molded surfaces and paintability equaling their metal counterparts, and many other advanced products. Licensing of the complex technology and production methods for fiberglass is presently continuing worldwide. To turn a phrase, the golden age of fiberglass and reinforced plastic composites is with us now, but in no way near retiree status.

3. METHODS OF PRODUCING GLASS FIBERS FOR REINFORCING

There are several basic methods employed for attenuating glass fibers from hot melts to yield products that have some role, large or small, in reinforcing plastics. These include continuous filament fibers, staple fibers, blown fibers for mats, and mechanically drawn fibers for mats. Each method is described individually here, together with a listing and brief description of the characteristic products resulting from each.

3.1 Continuous-Filament Glass Fibers

The historical process of drawing glass fibers from heated rods logically resolved to drawing fibers from hot melts. The high surface tension and slow rate-of-change of viscosity with temperature of melted glass permitted formation of stable meniscuses and enabled fibers to be formed and drawn from a suitable crucible after the natural flow of molten glass downward by gravity. The unique properties of the refractory metal platinum, including its ability to perform as an electrical resistor and to become incandescent at glass melting temperatures, give rise to its use as the most favorable melting and forming vessel, or "bushing," for drawing continuous-filament fiberglass.

Marble Melt vs. Direct Melt. In the primitive era of fiberglass production, glass was melted in a separate furnace and formed into marbles, which were partially annealed and graded. Then they were fed into heated platinum bushings, where they were remelted and transformed into filaments, being drawn away rapidly at speeds approaching 12,000 ft/min. There is nothing magical about marbles except that they are easily formable mechanically, and roll by gravity through grading and feeding contrivances.

As technology improved, the marble "remelting" step was superseded by installing in-line fiber producing bushings in the appropriately designed forehearths of a glass tank furnace, and the process known as "direct melt" glass fiber production evolved. (Note that furnaces and forehearths are normally lined with brick of chromic oxide for long life.) Whereas marble bushings were originally capable of 7–9 lb/hr, production from larger direct melt bushings may now exceed 200 lb/hr.

Continuous glass filaments produced commercially range from 0.00005 to 0.00110 in. (1 to 28 μm) in individual fiber diameters. Marble bushings are still utilized in fabricating continuous-filament/strand mats, and are sometimes preferred for fabrication of finer filaments. They also permit use of glass marbles, which favor easy change to glasses of different compositions. The quality and production efficiency of glass marbles have been predetermined, and thus, poor or undesirable glass can be eliminated. However, in comparison with early direct-melt technology, current practice enables a uniform, "refined" melt that is nearly free of inhomogeneities and bubbles.

In summary, direct melt is the preferred process in most instances because it eliminates the remelting step necessary in marble melt. Figures 14-1 and 14-2, respectively, illustrate schematically the marble and direct-melt processes.*

Function of the Bushing. The stages of reduction of batch into usable glass in a melting tank or furnace are three: melting, refining, (or homogenization), and working (or processing). A fiberglass bushing may be considered as a miniature melting unit because the glass is subjected to the same three phases (eliminating actual melting from batch) when progressing from

*I acknowledge with thanks contributions made to this section by D. W. Denniston of PPG Industries—Fiber Glass Process & Quality Control—now retired.

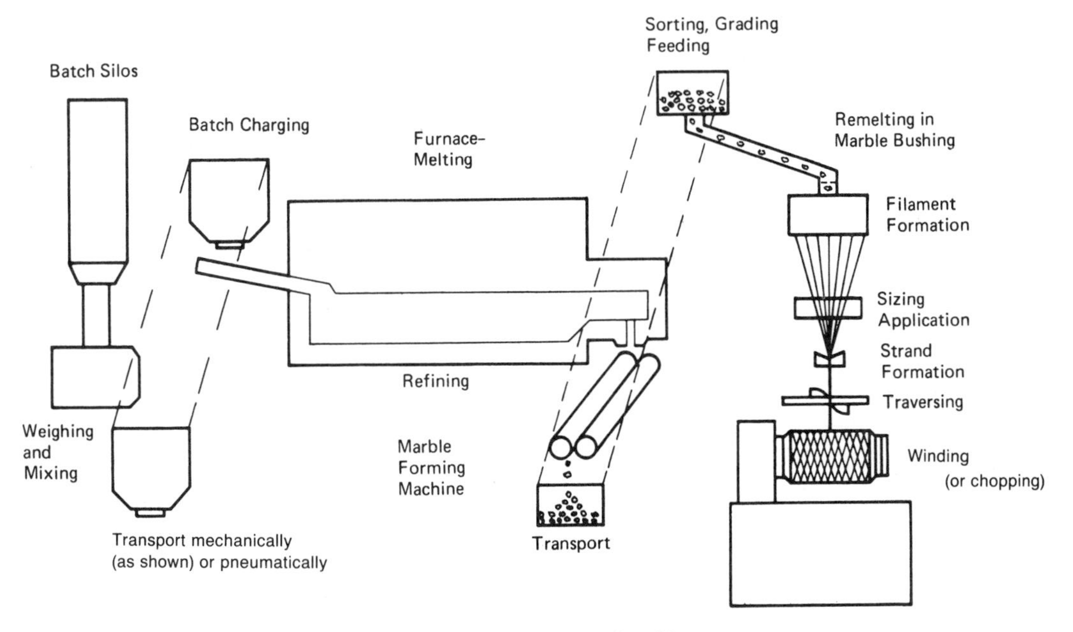

Fig. 14-1. Schematic diagram of marble process.

the point of entry to the juncture at which filaments are formed. Some differences exist in design requirements for marble versus direct-melt bushings, the latter requiring less height because the glass delivered to them is already melted and at the approximate working temperature of the furnace (approximately 2400°F). Upon entering the bushing, the glass gravitates through a perforated baffle plate, inducing homogeneity, and thence is cooled further by entering air-exposed "tips" surrounded by cooling media. These act as a further heat sink, lowering the glass temperature to approximately 2250°F (for "E" fiberglass composition, as an example). A representative marble bushing construction is shown schematically in Fig. 14-3.[2]

Strand Formation. The next step essential in

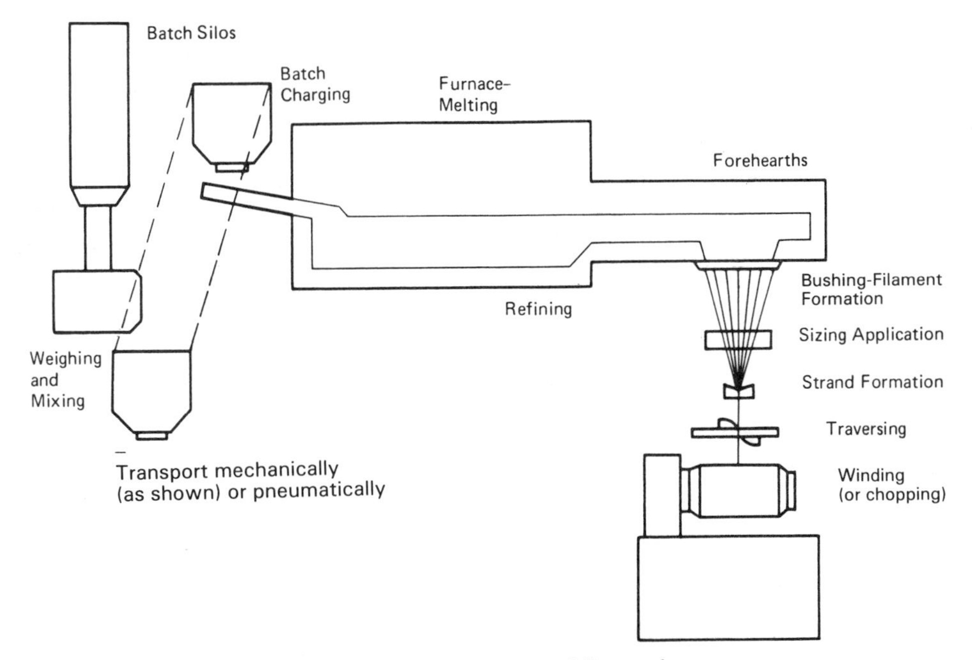

Fig. 14-2. Schematic diagram of direct melt process.

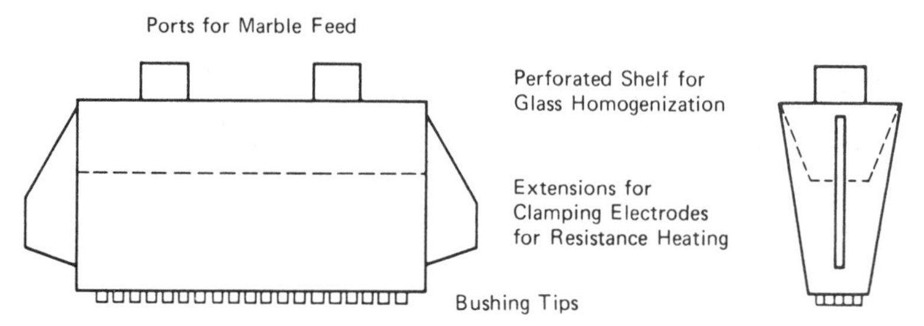

Fig. 14-3. Schematic diagram of contruction of fiberglass marble bushing.

glass fiber production is the formation of a "strand." A bushing may have a variable number of tips, depending on end-requirement, each capable of forming a filament. Originally small bushings contained 204 tips, but more recent practice, with a trend to larger bushings, has been to design with numbers of tips approximating 400, 800, 1000, 2,000, 3,000 and 4,000.

Steps in strand formation comprise the following, below the bushing: (1) formation of a cone or meniscus at each tip from which the filament is attenuated, (2) application of a water spray in a downward draft of humidity- and temperature-conditioned air, (3) application of an organic sizing composition, (4) collection into a strand using a suitable mechanical gathering device, and (5) winding onto a high-speed winder after passing over a traversing mechanism which creates a crossed, non-parallel orientation in the package. The latter can be either a precursor forming package or a direct draw roving package. In direct chopping a high speed chopper is substituted for a winder. The elements required below the bushing for strand formation and winding are detailed schematically in Figs. 14-1 and 14-2.

Though fiber production started on a "batch" basis, with each forming package requiring removal after it had reached optimum size, recent advances in winding and chopping technologies have enabled production on a semi-automatic or fully continuous basis. The strand is formed at the gathering point, but does not actually preserve its identity until the full function of the binder or sizing ingredients is brought to bear. These proprietary formulations of organic chemicals contribute three main elements to the strand: (1) lubrication, both during and after application to the glass fibers, (2) formation of a film to bind the individual filaments together when the products are removed and dried, and (3) compatibility with the resins contemplated for the end use of the fiber as plastics reinforcement.

Filament Diameters. The strand length-to-weight relationship, hence filament diameter (classified in the fiberglass industry as either yield in yards per pound or as the inverse Tex, grams per 1000 meters, rather than by denier, weight in grams of 9000 meters), is determined or influenced by the following: glass composition, and therefore its specific gravity, viscosity, tip diameter, number of tips or filaments, drawing temperature, efficacy of cooling, and speed of attenuation.

Nearly all of the filament diameters that are or might be produced commercially as textile or reinforcing fibers are shown in Table 14-1. It is necessary to understand that the basic strand designation number is determined by dividing the nominal yards per pound of the strand from a given bushing producing the particular filament size, B through W, by 100. Each filament size, designated by letter, has a small minimum to maximum definition range, noted in the tabulation.

Coincident with the development of larger bushings and filament diameters, strands have been made available for end-use RP/C processing with an increasingly larger number of filaments. As the number of filaments and their diameters increase, the strand weight also naturally increases. Hence, for strands containing multiples of the same diameter filaments, yards per pound will decrease proportionately, and the Tex value will increase proportionately.

Preparation for Processing. Glass fibers are treated with two distinct and separate classes of

Table 14-1. Description of basic continuous filament fiber designations with dimensions and corresponding length:weight relationships of strands.

	Filament			Number Per Strand →	Strand Count (yd/lb $\times 10^{-2}$)[a,b]											
Diameter Designation	Nominal Diameter Range $\times 10^{-5}$ in.	Nominal Diameter, $\times 10^{-5}$ in.	μm	50	100	200	400	600	800	1000	1200	1600	2000	2400	3000	4000
B		10.3	2.6	7200	3600	1800	900	600	450	360	300	225	180	150	120	
B	10.0–14.9	12.5	3.2	4800	2400	1200	600	400	300	240	200	150	120	100	80	
B		14.7	3.7	3600	1800	900	450	300	225	180	150	112.5	90	75	60	
BC	13.0–18.0			3000	1500	750	375	250	190	150	125	95	75	62	50	
C	15.0–19.9	17.5	4.4	2400	1200	600	300	200	150	120	100	75	60	50	40	
D	20.0–24.9	22.5	5.7	1800	900	450	225	150	112.5	90	75	61	45	37	30	
DE	23.0–27.9	25.0	6.6	1200	600	300	150	100	75	60	50	37.5	30	25	20	15
E	25.0–29.9	27.5	7.0	900	450	225	112.5	75	56.3	45	37.5	27	22.5	19	15	11.3
F	30.0–34.9	32.5	8.3	720	370	190	95	61	47	37	32	24	19	16	12	9.5
G	35.0–39.9	36.	9.1	600	300	150	75	50	37.5	30	25	19	15	12.5	10	7.5
G	35.0–39.0	37.5	9.5	540	270	135	67.5	44	34	27	22.5	17	13.5	11.5	9	6.8
H	40.0–44.9	42.5	10.8	440	220	110	55	37.5	27.5	22	18.5	14	11	9	7.4	5.5
J	45.0–49.9	47.5	12.1	350	175	92	46	30	23	17.5	15	11.5	8.7	7.5	5.8	4.4
K	50.0–54.9	52.5	13.3	300	150	75	37.5	25	18.5	15	12.5	9.5	7.5	6.3	5	3.8
L	55.0–59.0	57.5	14.6	240	120	60	30	20	15	12	10	7.5	6	5	4	3
M	60.0–64.9	62.5	15.9	200	100	50	25	17	12.5	10	8.5	6.3	5	4.2	3.3	2.5
N	65.0–69.9	67.5	17.1	175	88	44	22	15	11	8.8	7.5	5.5	4.4	3.7	2.8	2.2
P	70.0–74.9	72.5	18.4	150	75	37.5	18.8	12.5	9.4	7.5	6.3	4.7	3.8	3.2	2.5	1.9
Q	75.0–79.9	77.5	19.7	135	67	35	17	11	8.5	6.7	5.5	4.2	3.4	2.7	2.2	1.7
R	80.0–84.9	82.5	21.0	120	60	30	15	10	7.5	6.0	5.0	3.7	3.0	2.5	2.0	1.5
S	85.0–89.9	87.5	22.2	104	52	26	13	9	6.5	5.2	4.5	3.2	2.6	2.3	1.7	1.3
T	90.0–94.9	9.25	23.5	88	44	22	11	7.5	5.5	4.4	3.7	2.7	2.2	1.8	1.5	1.1
U	95.0–99.9	97.5	24.8	80	40	20	10	6.7	5.0	4.0	3.4	2.5	2.0	1.7	1.3	1.0
V	100.0–104.9	102.5	26.0				9	6	4.5	3.6	3.0	2.2	1.8	1.5	1.2	0.9
W	105.0–109.9	107.5	27.3				8.2	5.5	4.1	3.3	2.7	2.0	1.6	1.4	1.1	0.8

[a]Each number (in the row opposite a filament designation letter) in body of table under the section titled "strand count" is the approximate bare (unsized) glass yield in yards per pound $\times 10\ 0^2$ of the strand having the given number of filaments. For example, a DE bushing with 1600 tips and which is producing 1600 filaments of nominal diameter 25×10^{-5} in. may yield a strand yardage of 3750 yards/lb. The strand designation is DE 37.5.

Note: Inclusion of many of the strand counts in the table is based soley on calculated values, and their presence in the table does not indicate that manufacturers produce that particular product. The basic equations used for calculation of yields and dimaeters are as follows:

$$y = \frac{0.9789}{d^2 \rho n} \text{ and } d = \frac{0.9894}{\sqrt{y \rho n}}$$

where y = strand yield (yd/lb)
d = filament diameter (in.)
n = number of filaments per strand
ρ = filament (fiber) specific gravity or density (g/cm^3)

The specific gravity of the actual commercial (sized) fiber may differ from one manufacturer or product line to another, depending upon several factors including glass composition and the amount of sizing present. Generally, one percent application by weight will recuce specific gravity by 0.02, or density by 0.02 g/cm^3.

[b]The international TEX value, expressed as grams per 1000 meters (km), may be obtained by the formula: $\frac{496{,}055}{\text{yd/lb}}$ = TEX. For example, a strand or roving having a yield of 250 yd/lb would have a TEX of 1984.

sizing materials: (1) a starch–oil emulsion type that has humectant properties, absorbing water to provide lubrication plus resistance to abrasion in such downstream textile yarn processes as twisting, plying, weaving, cording, and so on, and (2) a resin-in-water emulsion type of sizing, generally for direct use on plastic reinforcements for RP/C. The starch–oil size must be processed in areas above 55% humidity for best performance. It is later removed or coated for specific applications. The resin or plastic emulsion type sizing must be oven-treated to drive off water, set the film, and bond the filaments into a strand.

Considerable technology has been and is continuing to be developed in making glass fibers as optimum reinforcements for matrix materials. These treatments include chrome or silane finishes for fabrics (applied after weaving and heat-cleaning); resorcinol–formaldehyde treatments of glass yarns for tire cord and industrial belting; and chrome, silane, or organic titanate ingredients of resinous sizings applied to continuous filaments specific to the particular plastic matrix for which the fiberglass product is intended as reinforcement. The full import of these treatments will become more evident following the ensuing discussion of various fiberglass reinforcing products.

3.1.1. Continuous-Filament Glass Fiber Products. Almost all glass fiber products may be classed as intermediates, since they must undergo further processing or combination to reach their full potential value as a reinforcement.

Referring to Table 14-1, the strands formed using the finer filaments in the B to F range find application in architectural and industrial fabrics and "papers," hence have little or no value as plastic reinforcements, and are not of further interest here. The bulk of plastic reinforcing fibers utilized, then, fall in the G to T range, with some small amount of fiber above and below finding specialty applications.

Generally speaking, finer fibers in reinforced plastics promote better surface appearance and also exhibit lower stiffness and hence better mold conformity. In certain thermoplastics, finer fibers contribute higher strength properties. However, the finer grade fibers are more difficult to produce, and result in lower throughput per bushing area.

Table 14-2 presents a brief description of the major forms of glass fiber products for reinforcement of plastics. Included are statements of types or grades available and whether the reinforcement is generally suitable for room temperature cure, with open lay-up (HSB—high solubility sizing), for high-temperature cure as in press molding (LSB—low solubility sizing), or for use in reinforcing thermoplastic resins (RTP). These designations will receive further clarification in a later section discussing reinforced plastics (see Table 14-1).

Before we leave the continuous filament process, mention should be made of hollow glass fiber made by a patented process.[3] Having a typical K fiber diameter and a ratio of inside-to-outside diameter of 0.65, the product is unique among glass fibers for its low specific gravity and dielectric constant. It may be fabricated into roving and yarn for preparation of tapes and prepregs, which are subsequently laminated into advanced composites.

3.1.2 Miscellaneous. Two products, different in form but related to continuous-filament fiberglass in function, deserve mention here, and are included in Table 14-2 as reinforcing products, although their contributions to composite strength are negligible.

Glass Beads. Hollow or solid glass spheres are manufactured by dropping minute glass fragments of controlled weight or screen size downward through a heated shot-chamber or equivalent. See Chapters 18 and 19 of reference 28 for detailed information on solid and hollow glass spheres.

Glass Flake. Glass flake may be produced by any of several processes. It is used as a resin filler and for shrinkage/warpage control in thermoplastic polyester and reinforced reaction injection molding of urethanes. See Chapter 4 for information on flake composites.

3.2. Staple Fibers

Fibers that are noncontinuous in derivation, but may be reduced to "strand" or yarn form are designated as staple fibers, and the individual composite strand is referred to as "sliver."

Table 14-2. Forms of continuous-filament fiber glass reinforcing products (with related forms).

Form* of Fiber Glass Reinforcement	Definition and Description	Range of Grades Available	General Types of Sizing Applied	General Usage in RP/C (and Secondary Uses if Any)
Twisted yarns	Single-end fiberglass strands twisted on standard textile tube-drive machinery.	"B" to "K" fiber, "S" or "Z" twist, 0.25–10.0 twists/in., many fiber and yardage variations.	Starch.	Into single and plied yarns for weaving; many other industrial uses (and decorative uses).
Plied yarns	Twisted yarns plied with reverse twist on standard textile ply frames.	"B" to "K" fiber, up to 4/18 ply—many fiber and yardage variations.	Starch.	Weaving industrial fabrics and tapes in many different cloth styles; also heavy cordage.
Fabrics	Yarns woven into a multiplicity of cloth styles with various thicknesses and strength orientations.	"D" to "K" fibers, 2.5–40 oz/sq yd in wt.	Starch size removed and compatible finish applied after weaving.	Wet lay-up for open molding, prepreg, high pressure lamination, and also some press molding.
Chopped strands	Filament bundles (strands) bonded by sizings, subsequently cured and cut or chopped into short lengths. Also the reverse, i.e., chopped and cured.	"G" to "M" fiber, 1/8–1/2 in. or longer lengths, various yardages.	LSB, RTP.	Compounding, compression, transfer and injection press molding.
Roving	Gathered bundle of one or more continuous strands wound in parallel and in an untwisted manner into a cylindrical package.	"G" or "T" fiber used. Roving yields 1800 to 28 yd/lb and packages 15 to 450 lb wt (up to 24 × 24 in. in size).	HSB, LSB, and RTP.	Used in all phases of RP/C. Some (HSB) are used in continuous form, e.g., filament winding, and others (LSB) are chopped, as in sheet molding compound for compression molding.
Woven roving	Coarse fabric, bidirectional reinforcement, mostly plain weave, but some twill. Uni- and multidirectional non-woven rovings are also produced.	"K" to "T" fiber, fabric weights 10–48 oz/sq yd.	HSB.	Mostly wet lay-up, but some press molding.
Chopped strand mats	Strands from forming packages chopped and collected in a random pattern with additional binder applied and cured; some "needled" mat produced with no extra binder required.	"G" to "K" fiber, weights 0.75–6.0 oz/sq ft.	HSB and LSB.	Both wet lay-up and press molding.
Mats, continuous strand (swirl)	Strands converted directly into mat form without cutting with additional binder applied and cured, or needled.	Nominally "M" to "R" fiber diameter, weights 0.75–4.5 oz/sq ft.	LSB.	Compression and press molding, resin transfer molding, and pultrusion.

Table 14-2. *(Continued)*

Form of Fiber Glass Reinforcement	Definition and Description	Range of Grades Available	General Types of Sizing Applied	General Usage in RP/C (and Secondary Uses if Any)
Mat-woven roving combinations	Chopped strand mat and woven roving combined into a drapable reinforcement by addition of binder or by stitching.	30–62 oz/sq yd	HSB.	Wet lay-up to save time in handling.
Three-dimensional reinforcements	Woven, knitted, stitched, or braided strands or yarns in bulky, continuous shapes.	—	HSB.	Molding, pultrusion.
Milled fibers	Fibers reduced by mechanical attrition to short lengths in powder or nodule form.	Screened from $1/32$ in. to $1/4$ in. Actual lengths range 0.001–$1/4$ in. Several grades.	None, HSB and RTP.	Casting, potting, injection molding, reinforced reaction injection molding (RRIM).
Related forms:				
Glass beads	Small solid or hollow spheres of glass.	Range 1–53 μm diameter, bulk densities: hollow = 0.15–0.38 g/cm^3; solid = 1.55 g/cm^3.	Usually treated with cross-linking additives.	Used as filler, flow aid, or weight reduction medium in casting, lamination, and press molding.
Glass flake	Thin glass platelets of controlled thickness and size.	0.0001 in. and up thick.	None, or trated with doupling agent.	Used as a barrier in, or to enhance abrasion resistance of, linear resins and coatings used for corrosion-resistance applications. Also used in RRIM for increased dimensional stability.

Manufacture. Staple fibers are manufactured by several different methods, almost all of which have as their primary fiberizing source a blown fiber bushing or crucible. This bushing is usually fed by marbles, and the glass compositions employed may be varied to cover a wide range of properties.

The staple fiber forming process requires collection of the blown fibers on a moving belt or other suitable surface, slightly drafting to induce alignment of the filaments, compacting or integrating the strand through a mechanical false-twister, and ultimately placing it into a suitable package on a winder or take-up machine. An oil or sizing may be applied at some suitable location to abet manufacture and enhance strand properties (see Fig. 14-4).[4] Individual filament lengths in staple fibers may vary from 5 to 15 in. The filament diameter may be varied between E and K fibers.

Usage. Staple fibers are widely used in electrical insulation applications, for battery mat separators, and as filter media. Use in reinforced plastics has, in the past, been made in roving and woven roving products. This usage in the United States is extremely limited or nonexistent because of the low laminate strength values obtained.[5]

Cardable fibers have been introduced recently for use in dry-laid nonwoven industrial fabrics. They are made cardable by virtue of proprietary polyvinyl alcohol slashing sizes that enable the naturally smooth glass fiber strands to be air and/or comb opened, blended with other textile fibers and carded to a nonwoven mat with fiber-to-fiber cohesion on conventional textile carding machinery. The mat subsequently may be drawn to yarn. To date, cardable fiber chopped from continuous strand fiberglass, is available as DE fiber (6.6 μm) in 2 inch length and as H fiber (11 μm) in 1-½ inch length. The former is appropriate for needling and most filtration applications, while other applications in insulation, facings, gasketing, packings and protective clothing are being developed. No applications in reinforced plastics are envisioned.

3.3 Blown Fiber Surfacing (Veil) Mats

Light-weight fiberglass mats have found great utility in reinforced plastics for a variety of applications. Although they contribute little to laminate strengths, the function supplied cannot be duplicated by continuous-filament reinforcing fibers alone.

Manufacture. Figure 14-5 provides a schematic representation of a typical production method for blown fiber mats. The finished products are made in thicknesses varying from 10 to 30 mils (0.6–1.8 lb/100 sq ft). A chemically resistant glass composition is sometimes

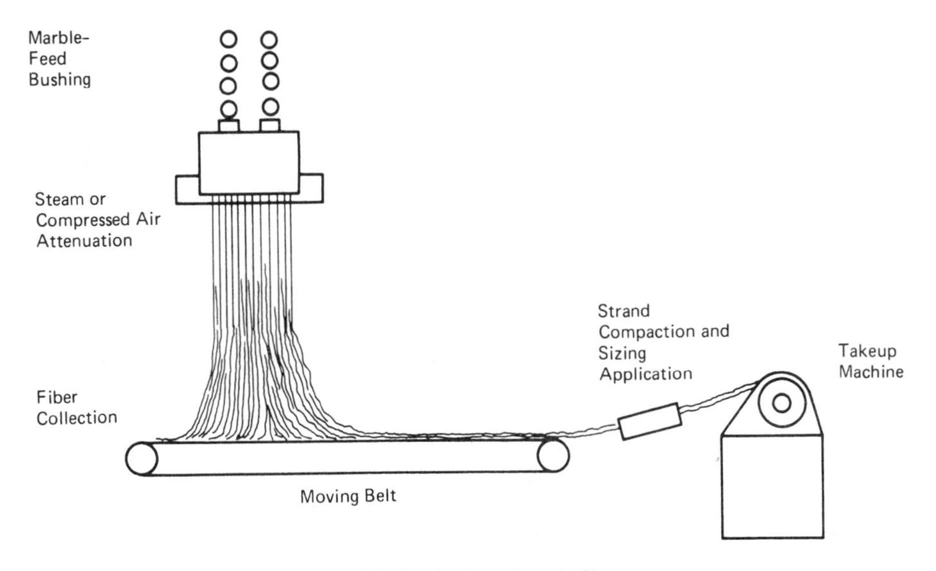

Fig. 14-4. Production of staple fiber.

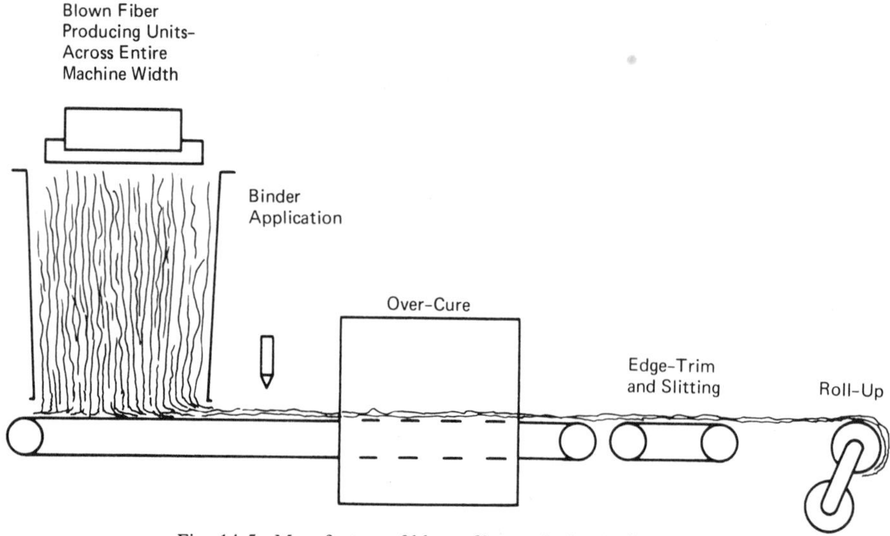

Fig. 14-5. Manufacture of blown fiber surfacing (veil) mats.

used. Filament diameters usually average 17 μm. A wide variety of binders are applied to suit end uses. Binders include none, acrylic, polyester, and others.

Usage. Surfacing or veil mats are applied to the inside of wet lay-up composites (tanks, etc.) to provide a craze-free, resin-rich, chemically resistant layer and to protect the underlying reinforcement layers, usually of "E" glass. They are also employed as a surface covering in panels and press-molded items (food trays, etc.), preventing the reinforcing fiber pattern from appearing in the prime appearance surface.

3.4 Mechanically Attenuated Mats

The Modigliani process for mechanical attenuation of glass fibers has made possible adaptation of resultant mats to many useful applications in reinforced plastics. The process is ingenious, and the resultant products are more versatile with better uniformity of fiber distribution than blown fiber mats. This process is used for production of both veil and heavier weight continuous-filament "reinforcing" mats.

Manufacture. A three-step process has evolved for conversion of the mechanically drawn fibers into mat form. These steps are illustrated schematically in Fig. 14-6. In step A, a gas-fired furnace (clay type with perforated Inconel bushing plate) is mounted on a structural frame to reciprocate parallel to and well above the main longitudinal axis of a large rotating drum 4 to 5 ft in diameter and 10 to 12 ft in length.

As the drum revolves, the fiber is drawn down from the bushing and fed so as to be forced against the drum, owing to the low-pressure condition caused at the drum surface by the high-speed of revolution of the drum. The Modigliani process is not capable of producing filaments at as high a rate or as small a diameter as those in the continuous-filament process. However, there is an advantage, in that single filaments that break out from the bushing may be restarted around the winding drum.

As the drum revolves and the furnace reciprocates, a layer of glass fiber is built up on the drum surface. The winding angle is normally 81–82 degrees from centerline, but this may be changed even within a given wind to yield products of varied characteristics.

In stage B, after winding has been completed, the fibrous mass is cut at a right angle to the fiber direction and laid flat. Binder may be sprayed onto the glass fiber during build-up on the drum, but must not be permitted to dry at this phase.

In stage C, as shown, the fibers are attached to a bait and drawn transversely. Additional binder is applied, compression means are applied if a very thin flat mat is required, and the

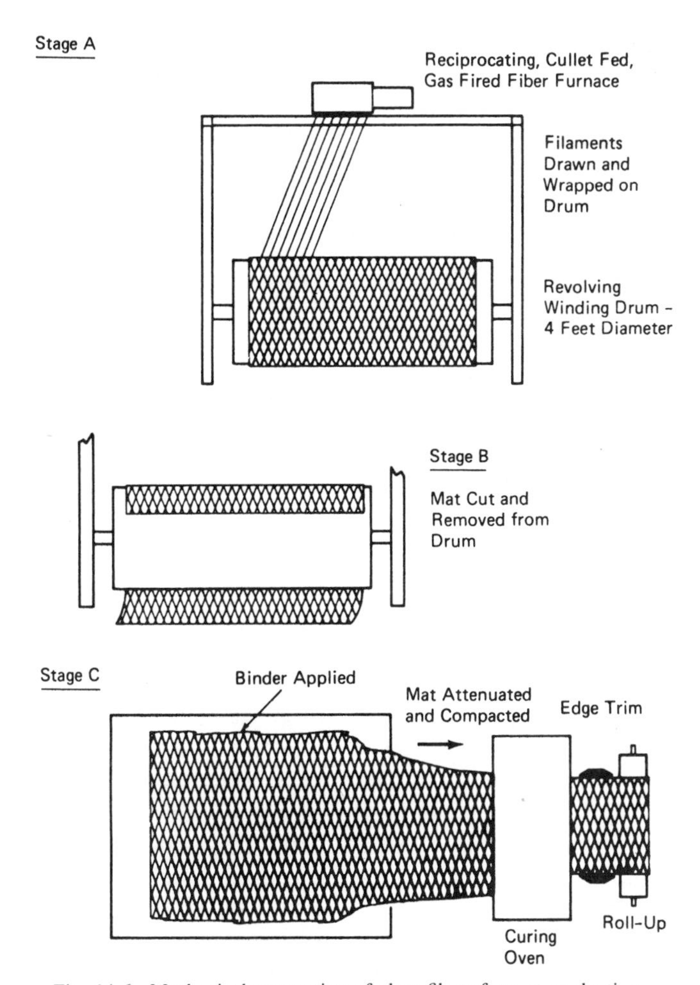

Fig. 14-6. Mechanical attenuation of glass fibers for mat production.

material is passed through a curing oven to set the binder.

Lower-melting "A" glass is usually used because the furnaces are charged with cullet (soda-lime glass). Some glasses with high modulus or other exotic compositions have been used in the past, however, and laminates, not mats, were produced directly from the drum-wound fibers. Commercial production of mats by this process, both of "A" glass and chemical resistant "C" glass, has persisted.

Mats from 10 to 100 mils thick may be fabricated by this process, and they are sometimes characterized by the typical "diamond" pattern formed between the glass filaments or residual fiber bundles. Filament diameters fall mostly in the L to P range.

In contrast, the continuous-filament process is used to make glass paper that sometimes is used as surfacing veil. It consists of chopped fibers in the low end of the reinforcing fiber diameter range that are randomly laid down from a suspension of fiber and binder in water to form a continuous sheet as in paper making.

Also, the typical methods of chopped and continuous-strand mat manufacture involving the continuous-filament process are noted in Table 14-2. In the latter case the strands are laid in a swirl pattern to build up mat thickness, and a chemical or mechanical bond is employed to achieve mat handling strength and processing integrity.

Usage. The Modigliani mechanical mats are competitive with the blown fiber mats and wet laid papers for veil or surfacing mat applications. The mechanical types are slightly better (than blown fiber) for uniformity, and may provide the persistence of the diamond-shape pattern in the laminate, enhancing the decorative aspects of the material. The thicker types, because of their superior uniformity and color, are used as reinforcing mats in low-draft, lower-strength molded and pultruded applications. However, the continuous-strand reinforcing mats, because of their "E" glass composition and sizing systems, provide significantly higher laminate strengths.

4. PROPERTIES OF GLASS FIBERS FOR REINFORCEMENT

4.1 Basic Considerations

The purpose of this brief discussion is to identify some of the basic chemical and physical changes that are brought about in glass-forming oxide materials when combined and subjected to melting temperatures; their behavior as a glass in the melted state; and the conditions that occur and prevail in processing, during which the glassy melt becomes a usable rigid solid after being cooled to room temperature. The discussion is intended as a background for a better understanding of the properties of glass fibers of the several different fiberizable compositions that will be considered.

Acknowledging that any one of the several facets of glass science touched upon here has been the subject of more than one life study,[6] we will not attempt to explain the reasons for for these fascinating occurrences. It is sufficient to say that the principles have been discovered, slowly developed, and brought to an advanced stage of beneficial utilization, where glass fibers are by no means of minimum importance. To set the stage, the word "glass" leads somewhat of a double life, referring both to the material in the molten stage and to the clear rigid solid that exists at room temperature.

Melting and Latent Compound Formation. The many definitions, long and short, large and small, that have been proposed for glass all agree that the material is an undercooled liquid with great propensity for crystal growth and compound formation. Melting is accomplished by a mutual solution of oxides which form eutectics or points of lowest melting. Addition of only 25% sodium oxide (Na_2O) to silicon dioxide (SiO_2) forms a new melting point, changing from 1730 to 793°C. Ternary, quaternary, and further multiple eutectics occur in any given oxide combination.

The compounds that may form and grow out of the melt are well-known chemical combinations of the oxides in varying proportions, and many occur elsewhere as natural minerals. Glass melted and brought directly to room temperature will usually turn out to be a clear rigid solid as intended. Crystal formation depends upon composition and is controlled by temperature—i.e., the glass composition in question must be held at a point below its liquidus temperature for the mutual existence of crystal and glass phases, and below its solidus temperature for complete crystallization. As temperature decreases, glass viscosity increases, and hence crystal formation and compound growth are inhibited. In studying compound formation out of glassy melts, and determining liquidus and solidus curves, compositions must be kept at a given temperature for sometimes as long as one year or more, to overcome the effects of viscosity and establish equilibrium. The physical chemistry of glasses (and ceramics) is well documented.[7] The result of crystals growing out of a glassy melt, termed "devitrification," is usually deleterious. These small nuclei or "stones" usually occur on the surface, interfere with production efficiency, and weaken the finished product. Most fiberizable glass compositions are subject to some type of devitrification, and must be cooled rapidly through the forming temperature range for crystallization to be avoided. Crystals induced by impurities may also grow out of a glassy melt due to unmelted batch, dissolved refractory, and other causes. As an adjunct and not a hindrance, crystals or compounds, by selective "phase separation," may be deliberately induced to grow out of a glassy melt and intermesh tightly as the glass cools to room temperature. This condition is induced by protracted, controlled post-forming heat treatment. The results are a desirable increase in strength, resistance to thermal shock, and other improved properties.[8]

Consideration of Differential Thermal Conditions. Assuming that bulk glass is successfully cooled through the critical zones without crystallization, the fact that it is a poor conductor of heat requires that care must be exercized to avoid excessive strains that would cause breakage. The method of preventing strain is "annealing," or gradually reducing temperature so that strains do not develop. In the solid form, glass has a viscosity of 10^{15} poise. At the softening point, strain will be released immediately; at the annealing point (10^{13} poise), strain will be released in 15 minutes. In controlled "chilling," "tempering," or "disannealing," the surface of a glass item, which is formed but held well above its annealing point in temperature, is rapidly cooled. By a unique process of combined contraction, flow, and contraction, the surfaces of the glass end up under a strong compressive force, held or maintained by the central portion of the cross-sectioned area under a strong tensile force. This counteraction of forces strengthens the glass by factors of from 2.5 to 10, as determined by thermal coefficient of expansion and thickness. In fibers, the identical rearrangement occurs naturally and without the need for controlled cooling because the filaments are not massive enough to set up breaking stresses. This thermal-shock treatment contributes substantially to the 40 or 50 to 1 tensile strength increase of glass in fibrous form over glass in the bulk form.

Properties vs. Composition. The next most important consideration is the effect of the various glass-forming ingredients on total properties, that is, when melted as the glassy solution, and when rigid, as the usable glass product. Silica is the most important glass-forming oxide. However, the sustained high temperatures required to melt large quantities of SiO_2 and free it from bubbles (due to its high melt viscosity) are commercially impractical. Hence, other oxides or fluxing agents are used, and this technology originally prompted the extensive investigation and use of phase diagrams. Silica or silicon dioxide (75%) amd sodium oxide (25%) combine to form a true glass that is highly susceptible to water attack, even by the moisture in the atmosphere. Addition of lime (CaO) or lime-magnesia (CaO-MgO) stabilizes this weakness for corrosion, but makes the glass susceptible to devitrification. Incorporation of a small quantity of alumina (Al_2O_3) directs the melt away from its tendency to devitrify, but adds again to its working viscosity.

Stabilization of glasses by oxide modifiers that will dissolve in the melt has been the basis for modern glass technology, and has resulted in many useful glass products. Much is known about the function of each usable oxide, and its effect in combination with other ingredients. The stages of oxidation and reduction and degree of acidity of the glass are well known. Edisonian research is no longer the rule in glass formulating. Also, many sophisticated testing systems and devices are currently available for meaningful analysis and evaluation. As an example, a controlling variable in glass furnace operation is "redox," which is the net oxidaton–reduction state of the batch measured by spectrometry.

Prior to a full-blown discussion of glass fiber compositions and their properties, Table 14-3 presents both the melt qualities and finished glass properties imparted by each of the more commonly known major glass constituents. Please keep in mind that combinations are important, and will further modify individual oxide performances.

4.2 Properties of Fiberizable Compositions

To illustrate the importance of properties of glass melts, especially viscosity, a comparison is made with different types of bulk glass and fiberglass compositions. Bulk glass intended for press molding or hand working is formulated to possess a long, slow rate-of-change of viscosity on cooling so that ample time is allowed for forcing glass to all extremities of the mold, or completing final shape manipulations before complete rigidity is reached. Conversely, glass intended for tube or rod drawing or mechanical blow-molding, is required to set fairly rapidly when cooled below the minimum working temperature, so that no distortion occurs after the final shaping takes place, and also so that high production rates may be maintained.

An important requirement of fiberglass compositions is that they possess "short" viscosity

Table 14-3. Functions of the most common glass-forming oxides.

	Glass Forming Oxide	Behavior in Glassy Melted State	Contribution to Finished Glass Properties
Silica	SiO_2	Most desirable glass-forming oxide, but requires extremely high melting temperatures, has high viscosity and slow rate of bubble release.	Very low thermal expansion.
Sodium potassium, and lithium oxides	Na_2O, K_2O, Li_2O	Low viscosity and good fluidity.	High thermal expansion, easily attacked by moisture.
Lime and magnesia	CaO, MgO	Intermediate viscosity, easily devitrified.	Improves chemical durability and resistance to attack by water, acids, alkali; high temperature resistance.
Boron oxide	B_2O_3	Intermediate viscosity; acts as fluxing agent.	Low thermal expansion; stabilizes electrical properties.
Alumina	Al_2O_3	Increases melt viscosity.	Provides high mechanical properties and assists in improving chemical durability.
Iron oxide	Fe_2O_3	Present as impurity unless used as colorant, lends some fluxing action, but absorbs infrared and interferes with transfer of heat through melt.	Discolors green in small quantities.
Zinc oxide	ZnO	Slightly higher viscosity.	Adds to chemical durability.
Lead oxide	PbO	Substantially reduces melt viscosity; acts as fluxing agent.	Increases glass density and brilliance (light transmission); induces high thermal expansion; permits identification by X ray.
Barium oxide	BaO	Intermediate glass viscosity.	High density, some benefit in increasing chemical durability.
Titanium dioxide	TiO_2	Slightly higher viscosity.	Improves chemical durability, especially alkali resistance.
Zirconium oxide	ZrO_2	Melt viscosity significantly increased; increases tendency to devitrify (crystallize).	Improves alkali resistance.

performance akin to that of the second example given. This means setting up rapidly on cooling when drawn from the meniscus, but not so rapidly as to interfere with establishment of the required filament diameters.

Other properties of fiberizable glass compositions, particularly those at room temperature conditions, will be considered here. Table 14-4 shows a listing of seven formulations representing a wide range of properties, performance, and uses of fibers for commercial and specialty applications. The various compositions, designated by an industry-assigned letter nomenclature, are defined and properties discussed as follows: chemical properties, including composition and durability, followed by physical, thermal, optical, and electrical properties. Each composition may be studied by comparison and by reference to the chemical (oxide) composition, so that the reader may establish personal, scientific "benchmarks" for the reasons the compositions perform as they do. Each of the various property classifications is discussed separately, with emphasis on the importance of the fiberglass composition in RP/C. Lastly, the safety, health, and environmental aspects of fibrous glass, which are independent of the glass composition, are discussed.

Classifications. Many different fiberglass compositions have been developed and tested, but the seven shown in Table 14-4 and included in the following list represent the principal compositions used for plastic reinforcements.

Designation	*Description*
A	A soda-lime glass similar to window or bottle glass with low thermal constants and poor water resistance and dielectric properties.
AR	An alkali-resistant glass containing zirconium oxide used for cement rather than plastic reinforcement
C	A soda-lime-borosilicate glass with excellent chemical durability
D	A low-density glass with improved electrical properties, but with reduced resistance to attack by water
E	A lime-borosilicate glass, derived from a "Pyrex®" composition, the real workhorse of general purpose reinforced plastics
E-CR	A modified E-glass to impart permanent high acid resistance, and short term alkali resistance
R	A high strength, high modulus glass with good resistance to fatigue, aging in the presence of moisture, temperature and chemicals
S, S-2 and T	High-strength and high modulus glasses developed to improve stiffness of RP/C laminates over those of E-glass

Silica glass is not a fiberizable composition, but rather is obtained by leaching out the other oxides of "E" or "S" fibers. High purity (98%+) silica fibers possess extremely high temperature capabilities and other attributes desirable in some aerospace applications.

Please note that, while the glass composition is extremely important, the functions of glass sizing, resin matrix and filler, if any, in an RP/C structure, as well as fabrication process conditions, are equally important, and are also developed for specific applications. In a critical chemical process tank, "C" (chemical) glass, or an alternative veil predetermined to be satisfactory (see ASTM or SPI's Composites Institute for criteria), is always laminated into an appropriate liner or barrier resin layer on the inside surface to protect the main body of the structure, usually fabricated using E type glass, i.e., "E" or "E-CR." On the other hand, in a DAP polyester molding compound, formulated for maximum dielectric properties, the type filler used, such as aluminum hydroxide, contributes much more to the ultimate molded part properties than does the fiberglass. Summarizing, the fiberglass industry has done its homework very well by making available many compositions for evaluation with other RP/C

Table 14-4. Fiberglass compositions and their comparative properties[a]

Fiberglass composition	"A"	"C"	"D"	"E"	"E-CR"[9]	"R"[10]	"S", "S2" & "T"
Description, definition, and characterization	Typical soda-lime silica glass, limited for reinforcement owing to poor resistance to water	Chemical glass—possesses improved durability, making it preferred composition for applications requiring corrosion resistance	Glass with improved dielectric strength and low density, developed for improved electrical performance	Borosilicate type, having a good balance of properties, used for major share of all reinforcement application	Modified "E" glass having superior long term resistance to strain crack corrosion in acid conditions	High-strength, high-modulus glass at a lower cost than "S"	Glass with high tensile strength and modulus, developed for aerospace applications
CHEMICAL PROPERTIES:							
Chemical Composition %							
SiO_2	72	65	74	52–56	58–63	60	65
CaO	10	14	0.5	16–25	21–23	9	—
Al_2O_3	0.6	4	0.3	12–16	10–13	25	25
MgO	2.5	3	—	0–5	2–4	6	10
B_2O_3	—	6	22	5–10	—	—	—
TiO_2	—	—	—	0–1.5	1.0–2.5	—	—
Na_2O	14.2	8	1.0	0–2	0–1.2	—	—
K_2O	—	—	1.5			—	—
Fe_2O_3	—	0.2	tr	0–0.8	0–0.4	—	—
ZnO	—	—	(Li_20–0.5)	—	0–3.5	—	—
SO_3	0.7	0.1	—	—	—	—	—
F_2	—	—	—	0–1.0	—	—	—
Chemical resistance-14 μ fiber % weight-loss after 1 hr boil in:							
H_2O	11.1	0.13	—	1.7	0.2	—	—
1.0 N H_2SO_4	6.2	0.10	—	48.2	0	—	—
0.1 N NaOH	15.0	2.28	—	9.7	0.2–0.6	—	—

PHYSICAL PROPERTIES							
Specific gravity (bare fiber)	2.50	2.49	2.16	2.52–2.61	2.63	2.55	2.49
Pristine tensile strength, psi	350,000	400,000	350,000	500,000	510,000	640,000	665,000
Tensile elastic modulus	9,800,000	10,000,000	7,500,000	10,500,000	10,500,000	12,475,000	12,600,000
Elongation at 72°F, %				3–4	—	—	5.4
Poisson's ratio	—	—	—	0.22	—	—	—
THERMAL PROPERTIES							
Softening point, °F	1,300	1,380	1,420	1,540–1,555	1,635	1,481	1,778
Coefficient of thermal expansion-in./in./°F × 10^{-7}	90	40	17	28–33	29	74	13–17
Thermal conductivity (k) BTU-in/hr/ft^2/°F at 72°F	—	—	—	7.2	—	6.9	—
Specific heat at 72°F BTU/lb/°F	—	—	—	0.197	—	—	—
OPTICAL PROPERTIES							
Index of refraction	1.512	1.541	1.47	1.55–1.56 (a 550 nm)	1.57	—	1.523
ELECTRICAL PROPERTIES							
Dielectric constant, 72°F, 10^6 Hz	6.90	6.24	3.56	6.1–6.7	7.01	6.2	5.34
Loss tangent, 72°F, 10^6 Hz	0.0085	0.0052	0.0005	0.001	0.002	0.0015	0.002

[a]Data taken from several sources and therefore represent approximate comparisons only.

elements to satisfy requirements and meet the demands that continually challenge the reinforced plastics industry.

Chemical Properties. Bearing in mind the earlier emphasis placed on the contribution of each oxide to glass properties, the chemical compositions may be examined to relate the importance of oxide percent to the different performance parameters. Also, please note that some new oxides have been incorporated into the "D" and "E-CR" formulations. The actual chemical durability of "A," "C," "E," and "E-CR" glass as weight loss after a boiling test on 14 μm fibers shows that "C" and "E-CR" glasses are approximately equal, although the former has superior resistance to alkali in longer-term exposure. "E" glass is particularly poor in resistance to attack by acid. Given the same glass, a decrease in filament diameter will result in an increase of the chemical attack due to the extra surface area. It should be pointed out that chemical durability of glasses may be improved by surface treatment involving ion exchange and oxidation.[11] It is doubtful if this process has yet been adapted to glass in the fiber form.

Physical Properties. Physical properties listed in Table 14-4 include specific gravity (density), pristine fiber tensile strength, and tensile modulus of elasticity. Specific gravity is an important property for controlling glass quality right at the melting operation. Its sensitivity to even minor composition variations is readily detectable. In fact, to illustrate the advanced development of modern glass technology, density factors or constants have been assigned to each glass-forming oxide. The product of the percentage of each oxide in the glass formulation and its characteristic density factor are determined, and the accumulated total of these determinations for all oxide ingredients duplicates almost exactly the actual value for the glass density determined by test after the glass has been formulated and melted.[12] Also, to illustrate further the relation of glass density to molecular orientation, also comparative and/or sequential glass sample for density determination must be annealed according to a precisely programmed decreasing temperature cycle, so that consistent results are obtained.

Because weight/volume relationships are important in applications of reinforced plastic components for aircraft, transportation, and the like, specific gravity of the sized fiber, rather than the higher value of the bulk glass, should be noted and used in design and cost estimates. (See footnote to Table 14-1.) The only fiberizable composition showing any significantly lower density is "D" glass. Hollow reinforcing glass fibers mentioned earlier have a specific gravity of 1.5–1.6 and are being evaluated for weight reduction and other contributing properties in aircraft/aerospace laminates.

Tensile strength and tensile modulus of elasticity are important physical properties of glass fibers, considering resistance to applied stresses. The more-than-adequate pristine tensile strength of E-type glasses is indicated in Table 14-4. Other glasses that are intended for similar applications, e.g., A, C, and D glass, are notably lower by comparison. In practice, from 50 to 60% of the E-type glass pristine tensile strength may be attained in a parallel-strand type laminate with 80 weight % glass loading. Only the "R" and "S" glasses, special high-performance compositions difficult to fiberize, have significantly higher virgin tensile strengths, and moderately higher tensile moduli of elasticity. Attempts to develop glasses with higher modulus and provide greater resistance to bending in laminates resulted in a class of compositions represented by YM-31A glass (modulus of 15.9 million psi). This composition has been superseded by reinforcements such as carbon and graphite (modulus 20–70 million psi), aramid (modulus 20 million psi), and/or beryllium or boron fibers (modulus 38 and 60 million psi, respectively). Carbon, graphite, and aramid fibers also have the advantage of extremely low density (1.4–1.9).

Another interesting physical property of glass, applicable to both bulk and fibrous forms, is elongation. Glass is perfectly elastic, and recovers fully to original dimensions provided the extension has not exceded 3–4%. If glass is extended beyond this limit, there is not yield point, and immediate fracture occurs. This condition is desirable for maximum strength output of the reinforcement. However, all RP/C components reinforced with fiberglass must be overdesigned by a safe margin to avoid catastrophic failure.

Thermal Properties. Despite all the comparisons made of cooling characteristics of glasses, it is well to explain that approximately the same top melting temperature (2800°F+) is reached in commercial glass tank furnaces to reduce charged batch to molten glass. Specialty glasses or exotic melting methods may require different conditions and temperatures. A viscosity versus temperature curve may be determined for glasses by any of several tests. This curve will illustrate the rate of change of viscosity for the glass below the state of pure fluidity, and indicate whether the glass has a long uniform setting rate, or whether it is "short," setting more abruptly below a specific temperature. All of the various viscosity indicators, or thermal constants, (softening point, annealing point, strain point) have precise viscosity values included in their definitions, and relate to the glass viscosity curve. The softening point of the glass is one of the main indicators of its thermal hardness. Like glass density, softening point has become an important control test to be run frequently and at close periodic intervals. As related to fiber production, maintenance of uniformity in the softening point determinations assists in maintaining filament diameters within specification.

The effect of composition charges on softening point is readily noted in the tabulation with the high alumina glasses ("E" "E-CR," "R," and "S") showing the highest values.

The thermal expansion coefficient of a fibrous glass composition is especially important in parallel (unidirectional) and woven fabric reinforced laminates. The characteristics of the other ingredients, resin and filler, often override the contribution of the glass to thermal expansion of the resultant laminate. The influence of the combined effect of boron and silicon oxides on thermal expansion is readily noted in the tabulation.

Optical Properties. When required, the ultimate clarity of reinforced plastic components is directly determined by matching the index of refraction of the glass with that of the resin, and, of course, eliminating opaque fillers. A glass with index of refraction of 1.47, such as the "D" composition, will never show near perfect clarity and high laminate translucency in a polyester resin system having an index of 1.55, as will "E" glass with an index at the lower end of the 1.55–1.56 range. Of course, a compatible or highly soluble sizing system must be used on the glass.

Additional purity of color in RP/C for high translucency applications may be obtained by maintaining the iron oxide (Fe_2O_3) content as low as possible. This usually requires the chemical treatment with HCl of silica and alumina batch ingredients to remove Fe_2O_3. This treatment together with special brick-lined tank furnace walls, results in production of a color-clear fiber and frees the laminate from a characteristic transmission-lowering greenish cast.

Like glass density and softening point, redox, mentioned earlier, is a third important control test in the production of fiberglass. It measures the net oxidation states of iron oxide, if any, and thus controls optical properties.

It is well known in the glass industry that lead oxide, PbO, and potassium oxide, K_2O, combine in a melt to provide brilliance and clarity characteristic of art glass. An illustration related to, but not necessarily involving, reinforcing fibers is found in fiber optics. Using a glass of highest optical clarity, not only can the innermost cavities of the human anatomy be probed and photographed, but also light signals converted from electrical impulses may be transmitted for miles without need of transforming or boosting.[13]

Electrical Properties. Glass is a good electrical insulating material by nature. The electrical properties of fiberglass contribute markedly to the high performance of molded RP/C parts in electrical applications. These applications include standoff insulators, arc-chutes, distributor caps, switchgear, printed circuit boards, and so on. "E" glass was originally developed for enhancement of electrical resistance properties of molded parts. The improvement in electrical performance of the "D" composition over that for "E" glass may be readily noted in the tabulation. Because of their low dielectric constants, "D" glass and hollow "E" glass fiber (dielectric constant equals 2.9) are or have been used as reinforcements in composites for transparency to radar (hence their use in protective radomes) and more recently for transparency to electromagnetic radiation. (On the other hand, the opacity to radiation of carbon/graphite, boron, and other electrically conductive fibers is desirable in electromagnetic interference shielding of sensitive electrical/electronic devices.) As stated previously, other RP/C ingredients, particularly fillers, contribute substantially to desirable electrical properties in finished molded parts.

One difficulty encountered in designing fiberglass-reinforced plastic parts for electrical applications is in the sometimes interrupted nature of the glass–resin bond.[14] This condition results in substantial loss of electrical insulation resistance in high temperature and humidity atmospheres, while the wet laminate retains a high proportion (70–75%) of its original dry strength. It is postulated that this condition is improved somewhat by press-molding, filament winding, and pultrusion where greater compaction is induced, and a higher percentage of complete glass–resin bonding occurs.

Health, Safety and Environmental Aspects. Commercial fiberglass consists of the basic glass composition plus surface sizings (binders) that vary depending on end use. These binders comprise a small percentage of the weight of the product as indicated on typical material safety data sheets of suppliers.

Health effects on the skin and upper respiratory system (short term): Exposure to glass fibers may cause irritation of the skin and, less frequently, irritation of the eyes, nose, and throat. Typically, such irritation is experienced by individuals who are newly exposed to fibrous glass. The irritation usually diminishes after several days of exposure. Good personal and industrial hygiene practices coupled with the use of loose-fitting, long-sleeved clothing and eye protection will minimize the amount of discomfort experienced. If upper respiratory irritation or very dusty conditions are encountered and ventilation cannot be used to control the exposure, a nuisance dust respirator will minimize the transitory effects.

Health effects on the respiratory system (long-term): Only fibers with diameters of less than approximately 3.5 μm are considered respirable. Typically, the average diameter of the insulation-type fibrous glass is above 5 μm, while the average diameter of textile-type fibrous glass, including plastic reinforcements is above 6 μm. Insulation-type fibrous glass always has contained a percentage of fibers with diameters below the 3.5 μm level, including some less than 1 μm. Workers have been exposed to these smaller-diameter glass fibers for over 50 years. Typical textile-type fibers are used in a more controlled manufacturing process and have only a minute number of fibers with diameters below the 3.5 μm level.

Studies of workers with up to 35 years of exposure to fibrous glass, conducted to determine the health effects of fibrous glass exposure on humans, demonstrate no credible evidence of long-term health effects attributable to textile glass. There have been no significant excesses in deaths from respiratory cancer, nor has there been progressive malignant disease of the digestive organs, mesothelioma, or nonmalignant respiratory disease associated with exposure to fibrous glass. Recent reports have shed some new light on this subject which warrents continued study. Refer to suppliers' Material Safety Data Sheets for current information.

OSHA considers fibrous glass to be a "nuisance dust" and as such would enforce a permissible occupational exposure limit of 15 mg/m^3 (milligrams per cubic meter of airborne fibrous glass contaminated air).

The American Conference of Governmental Industrial Hygienists (ACGIH) also considers fibrous glass as a nuisance material, but their recommended limit for nuisance materials is more stringent than OSHA's. A Threshold Limit Value (TLV^R) of 10 mg/m^3 of airborne fibrous glass is the maximum concentration recommended for continuous exposure. This TLV^R does not have legal enforcement status under OSHA.

Safety: Fiberglass is a nonburning material. The binder chemicals, which constitute a small percentage of total weight as indicated on the MSDS, may burn. While these binders will not support combustion, in a sustained fire, proper protection against products of combustion from the fuel and potential binder involvement must be provided. Fiberglass is not a highly reactive or unstable product and does not require special safety storage procedures. It carries no Chemical Abstract number.

Environmental: Fiberglass is generally considered to be an inert solid waste not requiring hazardous waste disposal procedures. State, local, and federal regulations should be consulted to ensure proper disposal procedures for your location. Fiberglass products that have become part of a reinforced plastic or uncured resin system must be disposed of in accordance with applicable requirements for those plastics or resins where they exist.

5. MATERIAL SUPPLIERS AND PRODUCT TYPES

Having discussed methods of manufacturing glass fiber products, and the basic properties of the material, we shall now provide information regarding sources and availability. In this section, details are provided to register both the manufacturers of fiberglass and related reinforcing materials plus actual product classifications and designations.

5.1 Manufacturing of Reinforcing Products

Nearly all of the production of glass fiber reinforcements is from E-type glass compositions. However, glass fibers from the specialty compositions discussed in the previous section are, or recently have been, commercially available in the world market in the following forms: (Their relationship to E-type glasses in volume and price is given in the next section.)

''A'' Glass: Surfacing veil or mat of low weight (1/8 to 1/4 oz)

''C'' Glass: Chopped strands
Rovings for corrosion-resistant applications
Surfacing veil or mat of low weight
Yarns for fabrics, braids, and sleeving

''D'' Glass: Yarn for fabrics, for electrical applications

''R'' Glass: Rovings for filament winding, sheet molding compounds, and weaving
Uni- and bidirectional fabrics
Yarns for weaving, braiding, and sleeving

S-Glass®: Military grade yarn for existing applications only

S-2 Glass®: Rovings for filament winding, prepreg, pultrusion, and sheet molding compound
Yarns for weaving, braiding, and knitting

®S-Glass and S-2 Glass are registered trademarks of Owens-Corning Fiberglas Corp.

Table 14-5 lists the basic U.S. manufacturers of glass fiber reinforcements for plastics and the weavers that also supply this industry. The corporate or sales headquarters of each firm is listed together with its telephone number as of the year 1986. The product forms offered by each firm are represented by a number in the right-hand column and explained in the legend.

5.2 Product Designations

Tables 14-6a through 14-6e present basic reinforcing products and their type numbers (which change frequently) made by all manufacturers according to application area. Fabrics woven from fiberglass can be engineered to meet the design requirements of many plastic reinforcement and related applications. Tables 14-6f through 14-6i list some of the more common styles used in the various applications.

6. PRICES AND VOLUMES

The relationship of continuous-filament E-type glasses to the specialty glasses in volume and

Table 14-5a. Domestic merchant manufacturers of fiberglass reinforcements for plastics (basic producers of glass fibers in capital letters).

Company	Address	Phone Number	Products[a]
Advanced Textiles, Inc.	1580 McLaughlin Run Rd. Suite 203 Pittsburgh, PA 15241	(412) 221-6110	2,5,8,9,10
Bean Fiber Glass, Inc.	Union Street Jaffrey, NH 03452	(603) 532-7765	8, 9
Brunswick Technologies, Inc.	39 Apple Wood Lane Avon, CT 06001	(203) 673-3634	8
CERTAINTEED CORP. FIBER GLASS REINFORCEMENTS	P.O. Box 860 Valley Forge, PA 19482	(215) 341-7770	1,3,5,9
Composite Reinforcements Business Div. of Gulf States Paper Corp.	P.O. Box 3199 Tuscaloosa, AL 35404	(205) 553-6200	8
The Cumagna Corp.	26 S. Wakefield Rd. Norristown, PA 19403	(215) 539-4317	10
C.H. Dexter Div, The Dexter Corp.	Two Elm St Windsor Locks, CT 06096	(203) 623-9801	4
FIBER GLASS INDUSTRIES, INC.	Amsterdam, NY 12010	(518) 842-4000	3,8,9
Henry & Frick, Inc.	P.O. Box 608 35 Scotland Blvd. Bridgewater, MA 02324	(617) 697-3171	6
Hexel Corp. (includes HiTech subsidiary, Gastonia, NC)	11711 Dublin Blvd. P.O. Box 2312 Dublin, CA 94568	(415) 828-4200	2,5,8,9
International Paper Co. Formed Fabrics Div.	77 West 45th St New York, N.Y. 10036	(212) 536-7965	4,5
King Fiber Glass Corp. (Marine Fiber Glass)	P.O. Box 25 Arlington, WA 98223	(206) 435-5501	5,8,9
Knytex Inc.	2301 Highway 46 North Seguin, TX 78155	(512) 372-4001	5,8,9
MANVILLE SALES CORP	P.O. Box 517 Toledo, OH 43693	(419) 878-8111	1,3,4,9
Mastech, div of C. H. Masland & Sons	3907 N. Front St Harrisburg, PA 17110	(717) 233-4841	2,10
NICOFIBERS, INC.	Ironpoint Rd. Shawnee, OH 43782	(614) 394-2491	4,5
OWENS-CORNING FIBERGLASS CORP.	Fiberglas Tower Toledo, OH 43659	(419) 248-8000	1,3,5,6,7,9
PPG INDUSTRIES, INC. FIBER GLASS PRODUCTS	One PPG Place Pittsburgh, PA 15272	(412) 434-3131	1,3,5,6,7,9
SUPERIOR GLASS FIBERS, INC.	499 N. Broad St. Bremen, OH 43107	(614) 569-4175	4
Xerkon Co	1005 Berkesshire Lane Minneapolis, MN 55441	(612) 593-9444	8,10

[a] Legend:
1—chopped strands
2—cloth and fabrics including tapes (woven and knitted from textile yarn)—Also see Table 14-5b
3—continuous rovings
4—finishing mats/veils
5—reinforcing mats and combinations
6—milled fibers
7—textile yarns (filaments and staple)
8—nonwoven roving reinforcements including unidirectional, double bias, biaxial and triaxial (knitted or stitch-bonded)
9—woven rovings
10—three-dimensional wovens, braids, sleeving, and multiple fabric preforms

Table 14-5b. Domestic weavers and fabricators of fiberglass yarn products

Company	Address	Phone Number	Products[a]
Amatex Corporation	1032 Stanbridge St. Norristown, PA 19404	(215) 277-6100	2c
Auburn Manufacturing Co.	P.O. Box 201 Mechanic Falls, ME 04256	(201) 345-8771	2e
Atkins & Pearce Company	3865 Madison Pike Covington, KY 41017	(606) 356-2001	2d
Bedford Weaving Mills	119 W. 40th Street New York, NY 10018	(212) 840-2877	2a
Bentley-Harris Mfg. Co., Thermal Design Group	241 Welsh Pool Road Lionville, PA 19353	(215) 363-2600	2d,e
Burlington Glass Fabrics Company A Division of Burlington Industries	P.O. Box 21207 3330 Friendly Avenue Greensboro, NC 27420	(919) 379-2000	2a,b
Carolina Narrow Fabric Co.	P.O. Box 1400 Winson-Salem, NC 27101	(919) 724-5381	2b,d,e
Chemical Fabrics Corp.	P.O. Box 1137 Merrimack, NH 03054	(603) 622-3758	2c
Clark-Schwebel Fiber Glass Corp.	5 Corporate Park Drive White Plains, NY 10604	(914) 694-9090	2a
Darco-Southern, Inc.	P.O. Box 454 Independence, VA 24348	(703) 773-2711	2e
Davylyn Manufacturing Co., Inc.	P.O. Box 626 Chester Springs, PA 19452	(215) 363-7615	2e
Fil-Tec, Inc.	1800 Woodburn Drive Hagerstown, MD 21740	(301) 797-6700	2b,e
Gividi Glass Fabrics, Inc.	Ward Hill Rd., Box 233 Ward Hill Industrial Park Haverhill, MA 01830	(617) 373-2500	2a
Garlock, Inc.	1666 Division Street Palmyra, NY 14522	(315) 597-4811	2e
Hi Temp Textiles	P.O. Box 693 Greensboro, NC 27402	(919) 582-7918	2e
Hesgon Company	P.O. Box 4182 Brownsville, TX 78520	(512) 542-5491	2b,d
Hexcel Corporation-Trevano Div.	11711 Dublin Blvd. Dublin, CA 94566	(415) 828-4200	2c
Intec, Inc.	1801 N. Orangethorpe Park Anaheim, CA 92801	(714) 525-3895	2e
Mutual Industries, Inc.	Washington Road Red Hill, PA 18076	(215) 679-7682	2b,d,e
Newtex Industries, Inc.	P.O. Box 25 Victor, NY 14562	(712) 924-9135	2e
North American Textiles	19679 John R. Street Detroit, MI 48203	(313) 369-2250	2a,c
Quinco Fabrics, Inc.	P.O. Box 392 Auburn, ME 04210	(207) 783-2261	2c

Table 14-5b. Domestic weavers and fabricators of fiberglass yarn products

Company	Address	Phone Number	Products[a]
Raybestos Industrial Products	P.O. Box 5205 N. Charleston, SC 29406	(803) 744-2940	2c,e
I. Sommers Narrow Tape Corporation	435 King St. E. Stroudsburg, PA 18301	(717) 421-4452	2b,d
Southern Manufacturing Corp.	P.O. Box 32427 Charlotte, NC 28232	(704) 372-2880	2b,c,e
J. P. Stevens & Company, Glass Fabrics Dept.	33 Stevens St. Greenville, SC 29501	(803) 239-4000	2a
Jonathan Temple, Inc.	P.O. Box 219 Hackensack, NJ 07602	(201) 487-8000	2d
Uniglass Industries, Div. of United Merchants, Inc.	1407 Broadway New York, NY 10018	(212) 564-6000	2a,b
Warwick Mills, Greenville Mills Div.	New Ipswich, NH 03071	(603) 878-1565	2c

[a]Legend: 2a—general industrial fabrics
2b—industrial tapes
2c—specialty industrial fabrics
2d—electrical sleeving, tapes, cordage
2e—tapes and tubing (asbestos replacement)

Table 14-6. Basic fiberglass manufacturers' reinforcing products by application and type.

Table 14-6a. Continuous rovings for chopping operations.

Process	Sizing/Solubility	PPG	OCF	CSG	Other
Spray-up	Chrome/Silane	Hybon® 600HTX	447BA	292B2	NEG G-20
		526	P356	21A	
		Hybon® 650HTX	P367	234B3	Fuji 2310
	Silane	535 HT	P352	290	NEG G-21
			881		
Mat forming and preforming	Silane	9000	P352	220A4,B4&C4	NEG H-14
		9500	852AA-Spun	231B1	
				247B1	
				277	
				22A	
Centrifugal casting	Chrome/Silane	Hybon® 600HTX	447BA		
	Silane	535HT	495	243	
Sheet mldg. compound	Silane/Low	523			
		521	951AC	235B4	
		1125	P980A	220B4	
		1150	964	232	
	Silane/Medium	522	956AC		JM
			P217 X5 (P)		
	Silane/High	524	433AD & AE	243B4	
			471, S-2 Glass®		
Panel	Silane	65WF	992AA(W-2)	277	NEG H-14
		7802	995AA(W-2)	27A	
		7803	P381H		
		7850	495AE & ME		
			P314		

Table 14-6. *(Continued)*

Table 14-6b. Continuous rovings for continuous reinforcement applications.

Process	Resin	PPG	OCF	CSG	Other
Filament winding	Polyester	1064	475BA, CA & KA		NEG F-70 F-31 R-10
	Vinyl ester	2002	379 475 402	55A 66A	F-183
	Epoxy	1062 2006	410AA, DA EA & FA	622A5, A8 63A 611A8	NEG F-60 NEG F-163 NEG F-164
			346 449, S-2 Glass® 463, S-2 Glass®	66A	FGI 510 JM S-1
	Multi-compatible:	2079	432BA, BC & CC	625A5, A8	NEG F-181
	Polyester, vinyl ester, epoxy	2080	366 P365A, S-2 Glass®	660	FGI 521 JM 501
Pultrusion and continuous impregnating	Polyester	764	424 AA, BA	65A 670	NEG (see FW)
	Vinyl ester	764	424 (as above)	62C 670	
	Epoxy	2086	346 (see FW) 449, S-2 Glass® 463, S-2 Glass®	66B	NEG (see FW) FGI (see FW)
	Multi-compatible	712 713 2079	425 AA 432 (see FW) 366 P311A 456 BC	625	NEG F-183 FGI (see FW) JM 501
Weaving	Polyester	Hybon® II	475BA, CA & KA	612A4,A5, A8,A9	NEG (see FW)
	Multi-compatible	2079 2080	836CA 432X10	625A8	JM 501

price are approximated for the world market in Table 14-7a.

Prices of fiberglass reinforcing products were high ($1.75–$5.00/lb, average) during the periods of development and preliminary usage in classified and strategic applications during World War II. Immediately following the war, the prices broke to the lowest possible level that the conditions and times would permit and then continued slowly downward during the 1950s and held steady until the early 1970s. Prices then eased gradually upward because of inflationary pressures and some material procurement difficulties, and then receded with the 1980–1982 recessions. The 1985 prices were little more than double those of 1967 in inflated dollars, but less than those of 1967 when compared on a real-dollar basis using the GNP and cost component deflators. Figure 14-7 shows that from 1967 to 1982, real "E" glass roving prices using a fiberglass cost component deflator declined by 11% with each doubling of industry cumulative experience.

Any price trends that might be prognosti-

Table 14-6. *(Continued)*

Table 14-6c. Chopped strands (and milled fibers) (MF) for thermosets.

Processes: Molding compounds or mixes by premix/BMC, extrusion compounding, putty compounding, roll milling and wet (solvent) methods for encapsulation, casting, potting; compression, transfer, and injection molding; and reinforced reaction injection molding (RRIM).

Resin (abbr.)	Sizing (if other than silane)	PPG	OCF	CSG	Other
Alkyd,		3029	847HE	See	JM716
Diallyl phthalate (DAP) and		3075	405AA,BA & AC	below	JM 758E
Epoxy incl. encapsulation		1156	832BB & BE		
Phenolic		3075	178AA		
	Cationic	1440 (MF)	731BB (MF)		
	Unsized	1420 (MF)	739AB (MF)		H&F MF
		1463 (MF)			
Fluorocarbon (PTFE)	Unsized	1420 (MF)	739DB (MF)		
Polyester (UP) and		3075	405BA,BB & BC	927A4	JM758E
Vinyl ester (VE)		3029	832AA,BB & BE	919A4 & A5	JM716
		1460 (MF) series	847HE		NEG 154H
Urethane (RRIM)		1460 (MF) series	737BB & CB (MF)		JM755E
Others: Furan, polyimide		1156	841		
silicone, melamine, and	Cationic	1440 (MF)	731FB (MF)		
urea formaldehyde					
Slurries		G-Fiber	453AB		

Table 14-6d. Chopped strands (and rovings) for thermoplastics.

Processes: Extrusion compounding (chopped strand or roving) and direct injection molding (chopped strand).

Resin	Abbr.	PPG	OCF	CSG	Other
Acetal	POM	3420	473AA		
			408AA		
			731FB(MF)		
High temperature resins,		3540	497AA,DB	91B	
e.g., poly (amide-imide),		3090	415		
(ether imide), (phenylene					
sulfide), (sulfone), (ether					
sulfone), (ether ether					
ketone) and (arylates)					
Nylon	PA	3540	491AA	93A	
		3540 (roving)	492 AA	931A1	
		3541	408AA	931A4	
			P353B (roving)		
Polycarbonate	PC	3090	473AA	930A4	
		3540			
		3541	408AA	963A4	JM718
		3830	415CA		
			471AA (roving)		
Polyolefins:					
Polyethylene	HDPE		415CA		
Polypropylene	PP	3830	457AA	967	JM713
		3831	P340	965A4	JM729
		3130	885BD	919A5	
Polyphenylene	PPO		497AA,DB	91B	JM715
oxide, modified			497CC (roving)		
Polystyrene and	PS	3130	885BB,BD	919A5	JM717
Co/terpolymers	ABS,SAN,	3541	414CA	931A1	
	SMA,SMM	3090	408AA		
			P490		
			P386		
Polyurethane	PU	3540	473AA		
Polyvinylchloride	PVC	3130	497DB		
Thermoplastic polyesters	PBT, PET	3540	491AA	931A1	
		3090	408AA	930A4	JM704
			473AA	963A4	
			445BA (roving)		

Table 14-6. *(Continued)*

Table 14-6e. Reinforcing mat, woven roving, and woven roving/mat.

Process	Strands/Weights/Widths	PPG	OCF	CSG	Other
			Chopped Strand Mat		
Hand lay-up	Fine 3/4–3 oz/ft^2	ABM	M-722	M113	
Cont. lam.	Medium 1.5 oz/ft^2		M-723		
Pres./vacuum bag, centri-fugal casting	Coarse (full) 1.5–3 oz/ft^2 widths: 12–120 in. plus slits	AKM	M-741	M127	
			Continuous Strand Mat		
	0.75–4.5 oz/ft^2; widths to 72 in.			Unifilo®	NICOFIBERS[a] Conformat®
Press mldg.	Fine		M8605	U720	N-700
	Coarse		M8610	U816	N-701
Cold mldg.					N-758
Pultrusion	Standard		M8641		N-921
	Conforming		M8643		N-952
Electrical sheet	Medium		M8680		N-700
	Coarse		M8681		N-751
Resin transfer mldg.			M8608		

		Woven Roving			
Hand lay-up pultrusion and other low pressure processes	Nomenclature indicates weight in oz/yd^2	Hybon®			FGI Rovecloth®
	Weave is plain except where otherwise noted	HWR 169		315 Twill	1054
	widths: 4–120 in.	HWR-180	OC18-54P-475T	318	1654
		HWR-186	OC18-43P-475T	319 Open	1854
		HWR-213	OC22-54P-475T	322	2254
		HWR-240	OC24-54P-475T	324&324(B)	2454&
		HWR-360	OC36	325 Open	2454C
		Woven Roving Mat			
Hand lay-up pultrusion, RTM filament winding and other low pressure processes	Nomenclature indicates weights of woven roving faced with chopped strand mat in oz/yd^2 and oz/ft^2, respectively	Combomat®	Bi-Ply®	Stitchmat®	FGI Fabmat®
	widths: 6–72 in.				745 Light Wght
					785 Light Wght
					1670
					16-15
			1810	518-10	
		1815	1815	518-15	
					22-10
		2215	2215	522-15	22-15
		2415	2415	524-15	24-15

NOTE: Combomat and Hybon are registered trademarks of PPG Industries, Inc. (PPG); S-2 Glass and Bi-Ply are registered trademarks of Owens-Corning Fiberglas Corp (OCF); Unifilo and Stitchmat are registered trademarks of CertainTeed Corp. (CSG); Conformat is a registered trademark of Nicofibers, Inc.; Rovecloth and Fabmat are registered trademarks of Fiber Glass Industries, Inc. (FGI). Other suppliers' abbreviations used in the above tables are: Henry and Frick, Inc (HF); Manville Corp. (JM); and Nippon Electric Glass (NEG).

[a]All products "A" glass, silane treatments.

Table 14-6. (*Continued*)

Table 14-6f. Electrical and high pressure laminating fabrics.

Fabric Style No.	Count Yarns/in. Yarns/5 cm	Warp Yarn Yarn Count TEX	Fill Yarn Yarn Count TEX	Weave	Mass oz/yd² g/m²	Thickness in. mm	Breaking Strength lb/in. N/(5 cm)
104	60 × 52	ECD 900 1/0	ECD 1800 1/0	Plain	0.58	0.0012	40 × 15
	118 × 102	EC5 5.5 1×0	EC5 2.75 1×0	Plain	19.7	0.030	350 × 131
108	60 × 47	ECD 900 1/2	ECD 900 1/2	Plain	1.43	0.0020	70 × 40
	118 × 93	EC5 5.5 1×2	EC5 5.5 1×2	Plain	48.5	0.051	613 × 350
112	40 × 39	ECD 450 1/2	ECD 450 1/2	Plain	2.10	0.0032	90 × 80
	79 × 77	EC5 11 1×2	EC5 11 1×2	Plain	71.2	0.081	788 × 701
116	60 × 58	ECD 450 1/2	ECD 450 1/2	Plain	3.16	0.0040	125 × 120
	118 × 114	EC5 11 1×2	EC5 11 1×2	Plain	107.0	0.102	1095 × 1051
2112	40 × 39	ECD 225 1/0	ECD 225 1/0	Plain	2.10	0.0034	82 × 80
	79 × 77	EC5 22 1×0	EC5 22 1×0	Plain	71.6	0.076	720 × 700
2116	60 × 58	ECD 225 1/0	ECD 225 1/0	Plain	3.16	0.0040	125 × 120
	118 × 114	EC5 22 1×0	EC5 22 1×0	Plain	107.0	0.102	1095 × 1050
7628	44 × 32	ECG 75 1/0	ECG 75 1/0	Plain	6.00	0.0068	250 × 200
	87 × 63	EC9 66 1×0	EC9 66 1×0	Plain	203.0	0.173	2189 × 1751
7642	44 × 20	ECG 75 1/0	ETG 37 1/0	Plain	6.87	0.0110	250 × 120
	87 × 39	EC9 66 1×0	ET9 134 1×0	Plain	232.0	0.279	2190 × 1050
1080	60 × 47	ECD 450 1/0	ECD 450 1/0	Plain	1.44	0.002	70 × 40
	118 × 93	EC5 11 1×0	EC5 11 1×0	Plain	48.8	0.051	610 × 350

Table 14-6g. Coating and reinforced plastic fabrics.

Fabric Style No.	Count Yarns/in. Yarns/5 cm	Warp Yarn Yarn Count TEX	Fill Yarn Yarn Count TEX	Weave	Mass oz/yd² g/m²	Thickness in. mm	Breaking Strength, lb/in. (N/(5 cm))
1581	57 × 54	ECG 150 1/2	ECG 150 1/2	8-H Satin	8.90	0.0090	350 × 340
	112 × 106	EC9 33 1×2	EC9 33 1×2	8-H Satin	302.0	0.228	3065 × 2977
1582	60 × 56	ECG 150 1/3	ECG 150 1/3	8-H Satin	13.90	0.0140	490 × 450
	118 × 110	EC9 33 1×3	EC9 33 1×3	8-H Satin	471.0	0.355	4291 × 3940
1583	54 × 48	ECG 150 2/2	ECG 150 2/2	8-H Satin	16.10	0.0160	650 × 590
	106 × 94	EC9 33 2×2	EC9 33 2×2	8-H Satin	545.0	0.406	5692 × 5166
1584	44 × 35	ECG 150 4/2	ECG 150 4/2	8-H Satin	26.00	0.0260	950 × 800
	87 × 69	EC9 33 4×2	EC9 33 4×2	8-H Satin	880.0	0.670	8318 × 7005
3706	12 × 6	ECG 37 1/0	ECG 37 1/2	Leno	3.70	0.0086	140 × 120
	24 × 12	EC9 134 1×0	EC9 134 1×2	Leno	125.0	0.218	125 × 1050
7781	57 × 54	ECDE 75 1/0	ECDE 75 1/0	8-H Satin	8.95	0.0090	350 × 340
	112 × 106	EC6 66 1×0	EC6 66 1×0	8-H Satin	304.0	0.228	3065 × 2977
7626	34 × 32	ECG 75 1/0	ECG 75 1/0	Plain	5.40	0.0066	225 × 200
	67 × 63	EC9 66 1×0	EC9 66 1×0	Plain	183.0	0.168	1970 × 1751
181	57 × 54	ECD 225 1/3	ECD 225 1/3	Satin	8.9	0.009	350 × 340
	22 × 21	EC5 22 1×3	EC5 22 1×3	Satin	302	0.229	3065 × 2971

Table 14-6h. Scrim fabrics.

Fabric Style No.	Count Yarns/in. Yarns/5 cm	Warp Yarn Yarn Count TEX		Fill Yarn Yarn Count TEX		Weave	Mass oz/yd² g/m²	Thickness in. mm	Breaking Strength lb/in. N/(5 cm)
1610	32 × 28	ECG	150 1/0	ECG	150 1/0	Plain	2.41	0.0040	115 × 100
	63 × 55	EC9	33 1×0	EC9	33 1×0	Plain	84.1	0.102	1010 × 965
1650	20 × 10	ECG	150 1/0	ECG	75 1/0	Plain	1.60	0.0040	80 × 70
	39 × 20	EC9	33 1×0	EC9	66 1×0	Plain	54.3	0.102	700 × 615

Table 14-6i. Boating and tooling fabrics.

Fabric Style No.	Count Yarns/in. Yarns/5 cm	Warp Yarn Yarn Count TEX		Fill Yarn Yarn Count TEX		Weave	Mass oz/yd² g/m²	Thickness in. mm	Breaking Strength lb/in. N/(5 cm)
1800	16 × 14	ECK	18 1/0	ECK	18 1/0	Plain	9.60	0.0130	450 × 350
	31 × 28	EC13	275 1×0	EC13	275 1×0	Plain	326.0	0.330	3940 × 3065
2532	16 × 14	ECH	25 1/0	ECH	25 1/0	Plain	7.25	0.0100	300 × 280
	31 × 28	EC10	200 1×0	EC10	200 1×0	Plain	246.0	0.254	2627 × 2452
7500	16 × 14	ECG	75 2/2	ECG	75 2/2	Plain	9.66	0.0140	450 × 410
	32 × 28	EC9	66 2×2	EC9	66 2×2	Plain	327.0	0.356	3940 × 3590
7533	18 × 18	ECG	75 1/2	ECG	75 1/2	Plain	5.80	0.0080	250 × 220
	35 × 35	EC9	66 1×2	EC9	66 1×2	Plain	197.0	0.203	2189 × 1926
7544	28 × 14	ECG	75 2/2	ECG	75 2/4	Basket	18.0	0.0220	750 × 750
	55 × 28	EC9	66 2×2	EC9	66 2×4	Basket	610.0	0.559	6567 × 6567
7587	40 × 21	ECG	75 2/2	ECG	75 2/2	Mock Leno	20.5	0.0300	750 × 450
	79 × 41	EC9	66 2×2	EC9	66 2×2	Mock Leno	695.0	0.761	6567 × 3940

Note: Data given are for guidance only and should not be used in specifications. Weight, thickness, and breaking strengths shown are, in many instances, averages of slightly different values published by two or more weavers. Many additional fabric styles are available from industrial weavers. Various widths of woven tapes (1½–12 in.) and fabrics (standard 38, 44, 50 and 60 in., with special fabrics to 72 in.) are offered for plastic reinforcements.

The nomenclature of fiber yarn in the "Glass System" consists of two parts—one alphabetical and the other numerical. The letters describe glass composition, filament type, and filament diameter. The numbers identify basic strand weight and yarn construction.

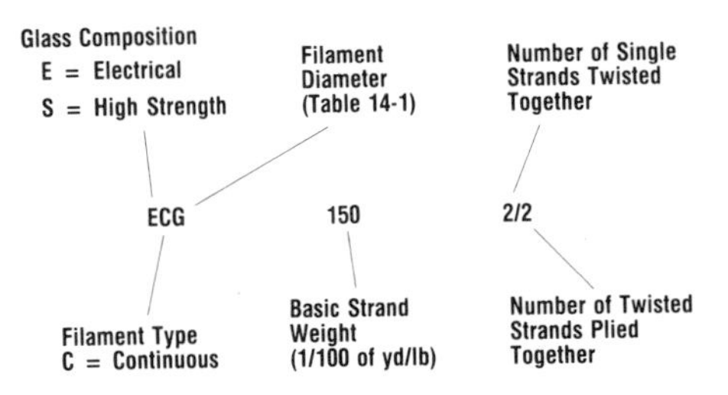

In the above illustration, the yarn is made of "E"-glass in continuous filaments of G diameter with a basic strand yield of 15,000 yards per pound. Two single strands are twisted together and two of these twisted strands are plied together. The yield of the plied yarn is obtained by dividing the yield of the basic single strand by the total number of single strands in the plied yarn (multiplying the last two digits with 0 being multiplied as 1).

Single strand yarns used as such in weaving have low twist (turns per inch), usually less than 1.0, which adds strand integrity for further processing. Single or multiple twisted strands for plying have higher twist, approximately 4 turns per inch; plying these yarns by twisting in the opposite direction with an equivalent number of turns per inch yields a "balanced" yarn that will prevent unraveling or "lively" yarn.
(Nomenclature courtesy Uniglass Industries, Div. of United Merchants and Manufacturers, Inc., New York, NY.)

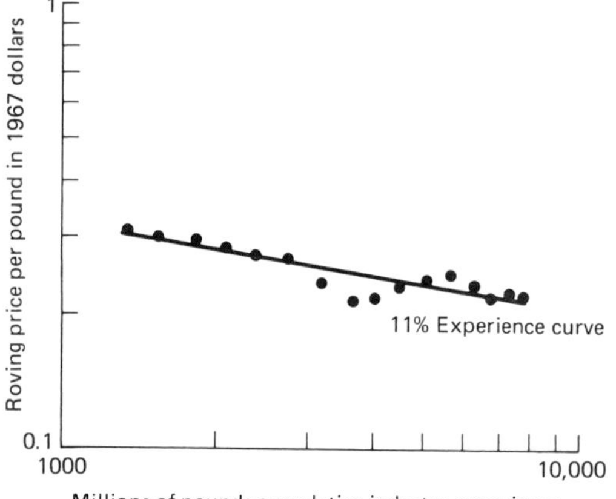

Fig. 14-7. Fiberglass roving prices in 1967 dollars, 1967 to 1982.

Table 14-7a. Comparison of volume and price for the glass fiber compositions.

	E Types	A&C Types	D Glass	R&S Types	Silica
Volume (millions of pounds)	2,500	2.5	0.02	5	0.2
Price scale (base = 1)	1	1.5 to 5–10	16–18	5–12	35–60

cated for the future would have to relate to probable slow continued inflation rates of the domestic and world economies.

Historically, the only serious shortage that occurred in glass-making raw materials was with the boron oxide contributors. Domestic supply of boron oxide became critical for a period in the mid-1970s due to fire-retardant insulation requirements during the energy shortage. Currently and for the foreseeable future, boric acid and colemanite from domestic and foreign sources are expected to be in ample supply.

Table 14-7b presents (1985) prices of fiberglass reinforcing products. All prices are tabulated as a range of values from low to high. The ranges represent approximate prices in truckload quantities for the several specific variations within each product group (thickness, weight, yardage, etc.)

Table 14-7b. Prices of fiberglass reinforcing materials (''E'' glass except where noted).

Product	Approximate range (1985)
Chopped strands for:	
Thermoplastics	$0.67–0.70/lb
Thermosets	0.75–0.80/lb
Continuous rovings for:	
Filament winding	$0.74–0.79/lb
Panel	0.85/lb
Preforming	0.85/lb
Prepreg (S: $12.00/lb) S2	4.00–5.00/lb
Pultrusion	0.74–0.83/lb
Sheet molding compound	0.76–0.80/lb
Spray-up (gun)	0.75–0.85/lb
Mats:	
Chopped strand	$1.02–1.14/lb
Combination (woven roving backed)	1.13–1.19/lb
Continuous strand	1.30–1.50/lb
Veil (C glass)	32.10–64.50/M sq. ft.
Milled fibers	$0.80–1.20/lb
Woven rovings and specialty directional reinforcements	$0.90–1.25+/lb
Woven yarn fabrics	$1.00–5.00/lineal yd

7. FIBERGLASS-REINFORCED PLASTICS/COMPOSITES

Having considered in detail the contribution glass fibers make to reinforced plastic/composites, based on their fundamental properties, we shall proceed with actual applications of glass fibers and utilization in combination with the various plastic and filler ingredients available.

Glass is reputed to have been discovered or first noted by a Phoenecian sailor.[15] Natural plastics such as rosin have been known and used for centuries. Techniques for reinforcing have been used since the days of early Egyptian culture (straw in bricks).

The science of reinforced plastics/composites, however, is strictly a modern, up-to-date twentieth-century affair. The amalgamation of a strong but friable reinforcing material with a weak but adhesive resinous matrix was conceived in antiquity, but brought abruptly into today's focus by modern materials and processing technology, and by the demand for improved performance of modern manufactured products.

This section succinctly describes the markets for RP/C products and growth patterns, materials and equipment necessary for processing, types of fabricated RP/C parts, classified generally according to their various molding methods, and an analysis of where the industry is headed. An inside look at potential future products and molding methods is also included. Brevity is essential when one attempts to include a large amount of information in a limited space. Hence, tabulations are used liberally, and ample documentation is provided.

7.1 Markets

Reinforced plastics/composites realized steady growth in annual volume through 1979 reaching 2.05 billion pounds. More recently, setbacks in the general economy in 1980 and 1982 caused temporary reductions in RP/C volumes. Shipments in 1984 of 2.15 billion pounds and in 1985 of 2.22 billion pounds evidenced resumption of the growth pattern, as seen in the accompanying Table 14-8. Shipments in 1986 are estimated to have set another record at 2.28 billion pounds.

Table 14-8. RP/C growth rate.

Year	Annual Volume, lb
1945	7,500,000
1955	110,000,000
1965	340,000,000
1975	1,180,000,000
1980	1,694,000,000
1985	2,216,000,000

Table 14-9 presents a percentage breakdown of the total poundage sold in each separate market for RP/C products in 1985.

Table 14-10 gives a matrix of markets versus fabrication processes (see section 7.3) for composites for 1984. The SPI does not include printed circuit board laminates (press laminated from epoxy impregnated glass fabric) in its RP/C data; so this large market application is not included in any of the three tables in this section.

7.2 General View of Materials and Equipment

Reinforcements. The descriptions and definitions of fiberglass reinforcements from section 3 are applicable here, and the uses for the different forms of fiberglass will become evident in the next discussion.

Resins. The resins used include the thermosets, principally polyesters, vinyl esters, epoxies, and phenolics, and the complete fam-

Table 14-9. Markets for RP/C and distribution (SPI classification).

Market	Millions of pounds	% of Total (1985)
Aircraft and aerospace	33	1.5
Appliances/business equipment	133	6.0
Construction	444	20.0
Consumer goods	144	6.5
Corrosion-resistant equipment	295	13.3
Electrical/electronic	190	8.6
Marine	330	14.9
Transportation/land	568	25.6
Specialties/other (tooling, etc.)	79	3.6
Total	2,216	100.0

Table 14-10. Estimates of 1984 reinforced plastic shipments by process and market (in millions of pounds of composite).

		Injection Molding		Compression Molding							Open Molding						
						Resin Applied At Press											
	Total reinforced plastics shipments (1)	Thermoplastics	Thermosets	BMC	SMC	Hot mold	Cold mold	Pressure bag	Resin transfer molding	Hot stamping (Thermoplastic)	Spray-up	Hand lay-up	Vacuum bag	Continuous laminating	Pultrusion	Centrifugal casting	Filament winding
Aircraft and aerospace	29	3		3	3							1	7		3		9
Appliance and bus. eqpt.	123	58	12	11	38			1	2			1					
Construction	430	8		7	27		8		2		210	6		157	2	3	
Corrosion-resistant	310			11	6				8		53	29		5	24	3	171
Electrical and electronics	189	62	23	29	24	17		2			9	1		1	15		6
Land transportation	540	189	25	68	183	8				7	8	9		39	1	1	2
Consumer	143	6		20	15	7			7	1	39	23		5	16		4
Miscellaneous	80	6	1	9	5	5	5	1	2		30	9	3		2	1	2
TOTAL	2153	334	61	162	302	43	13	4	24	8	532	188	10	207	63	8	194

Sources: (1) SPI Composites Institute, NY., NY and Composite Services Corp., Demarest, NJ: includes both thermoset and thermoplastic composites.

ily of thermoplastics. Included with the thermosetting resins are all types of cross-linking monomers, catalysts, promoters, modifiers, and other agents for performance improvement.

Equipment. Equipment for the preparation and processing of RP/C includes refrigerated storage compartments for catalysts, temperature-controlled areas or tanks for resins, dry storage for glass reinforcements, weighing and mixing equipment for batch preparation, pumping and metering equipment for continuous processes, mold-making facilities (whether commercial or in-house), and mold handling or support means, whether for wet lay-up, spray-up, filament winding, pultrusion, compression, transfer, injection, or other types of press equipment for converting the ingredients into finished usable parts. Many of the methods were newly devised to accommodate the new materials, and more recently to improve quality and productivity. An appreciable battery of test equipment for evaluating materials, processing parameters, and finished products and their performance properties have also been developed, or adapted from similar industries. These are too numerous for detailed description, but can be found in ASTM, ANSI, ASME, SPI and other industry organization standards and methods.*

7.3 Fabricaton Methods**

Because the types of reinforced plastic parts produced vary widely in size, complexity, function, and other parameters, it is most feasible to categorize the entire industry by fabrication methods.[16,17] These resolve neatly into five or six major groups and include most of the finished RP/C products, although there is some minor duplication, that is, product type (piping) being fabricated by two or more different methods.

Prior to individual method descriptions, fabrication methods and concomitant products produced are summarized in Table 14-11, and a correlation of fiberglass reinforcement forms with applicable fabrication methods is presented in Table 14-12.

Hand Lay-up. This process was one of the first developed following the perfection and utilization of polyester resins. An open-type mold with no unrelievable undercuts is taken off a master mold, polished, release agent applied, and otherwise made ready. A catalyzed polyester gel coat is applied and cured.

Lay-up involves precutting and tailoring cloth, mat, or woven roving reinforcement, placing it in the mold and applying the catalyzed resin, effecting thorough wet-out, eliminating gaps, bubbles, and voids by rolling out, and building up a thickness of at least 0.10 in. The ratio is nominally 30% glass to 70% resin, but glass contents of 35–45% are often used. Fillers are not a major ingredient except in the gel coats. Thicknesses may be built up to 1 in. or more with successive applications, and extra supports may be molded in.

As noted in Table 14-11, the main large-volume products are boats and corrosion equipment. Large chemical processing tanks and other components up to 3,000 sq ft in area have been fabricated.

Spray-up. Spray-up evolved from efforts to create greater production speeds and more consistent product uniformity over the hand lay-up process. An especially-built mechanism pumps resin and catalyst, and chops fiberglass roving through a hand-held gun suspended at the end of a boom. The boom is rotatable and extensible so that the operator holding the gun may move around to perform the work on a mold or successively on a battery of molds. Roll-out is performed as in hand lay-up.

Gel coating is likewise employed as surfacing for spray-up. However, an alternative, "rigidizing," results in improved surfacing when vacuum-formed acrylic sheeting is used as a surface, with glass and resin sprayed into it. Mold costs are reduced, and improved product properties and performance result. Special decorative effects may be incorporated into the vacuum-formable plastic sheet. The spray-up process is employed for the same type applications, and is frequently used in conjunction with hand lay-up.

Vacuum Bag Molding. Usually cloth, but

*American Society of Testing and Materials, American National Standards Institute, American Society of Mechanical Engineers, The Society of the Plastics Industry.

**Acknowledgment with thanks is made for the assistance of Dr. N. Raghupathi, of PPG Industries, Fiber Glass Research, in preparation of this section.

Table 14-11. RP/C fabrication methods and typical products.

Processes	Typical products
A. Room temperature cure	
Hand lay-up	Corrosion equipment, boats, and machine guards
Spray-up, including rigidizing	Tube and shower units
Vacuum bag molding	Radomes
Tooling	Master molds, checking fixtures
Cold-press molding	Motor mounts, hollow ware, finished two sides
Resin transfer molding	Microwave dish antenna, truck air drag foilers, chemical pumps, finished two sides
Casting, potting, encapsulation	Sinks, synthetic marble, electrical components
B. Intermediate cure	
Continuous laminating	Translucent and opaque, flat and corrugated architectural panels, truck trailer panels, sandwich panels, foam structures
Centrifugal casting	Pipe and tubing
Pressure bag molding	Pressure cylinders and tanks
C. High temperature cure (includes compression, transfer, and injection)	
Premix and BMC molding	Distributor caps, electrical gear, shower bases
Mat, preform, and SMC molding	Automotive front and rearend panels, truck cabs, tub–shower units, appliance and business machine housings
Phenolic molding	Automotive transmission and engine parts
Press lamination	Printed circuit boards
Reinforced reaction injection molding	Tractor grille housings, auto fascia, quarter panels
Foam reservoir molding	Military armor/equipment
D. Throughput processes	
Pultrusion	Rods, tubes, structural beams and shapes
Filament winding (including pre-impregnation)	Pipe, tubing, and tanks
E. Reinforced thermoplastics	
Injection molding	Gears, auto radiator tanks
Rotational molding	Hollow structures, tanks
Hot forming (stamping)	Auto seat backs, load floors; instrument cases
F. Advanced composites	
Autoclave molding and many combinations of other methods	Aircraft and aerospace parts

also chopped mat or woven roving may be laid up in a shell or paraboloidal type mold, the resin applied and uniformly squeezed out. A flexible plastic film then covers the entire lay-up, is clamped onto the mold flange and a vacuum applied. This introduces the equivalent of 13–14 psi pressure being applied to the lay-up while cure takes place. Autoclave and hydroclave cures may also be employed. Greater dimensional control and higher laminate density and quality may be obtained in the finished parts than for contact lay-up.

Machine-made prepregs of fabrics, tapes, and rovings (resin-impregnated and "B"-staged) are commonly used for autoclave molding of aircraft and aerospace components including sandwich structures with foam or honeycomb cores between two RP/C skins. The only limits to sizes of parts are the ability to construct and use molds, and the size of available pressure apparatus, if required. Radomes 12 or 15 ft in diameter and 4 ft deep are commonplace.

Tooling. Many different techniques and combinations of laminating, bonding, casting, and so on, are employed in tooling procedures. The main objectives are forming master molds, fabricating molds to be used for production of parts, and making checking fixtures and many other RP/C elements for dimensional control.

Cold Press Molding. Parts are produced in this process by catalyzing the resin for room-temperature curing and laying-up in a mold that has a matching half. Both halves may be gel

Table 14-12. Areas of usage of fiber glass reinforcing products.

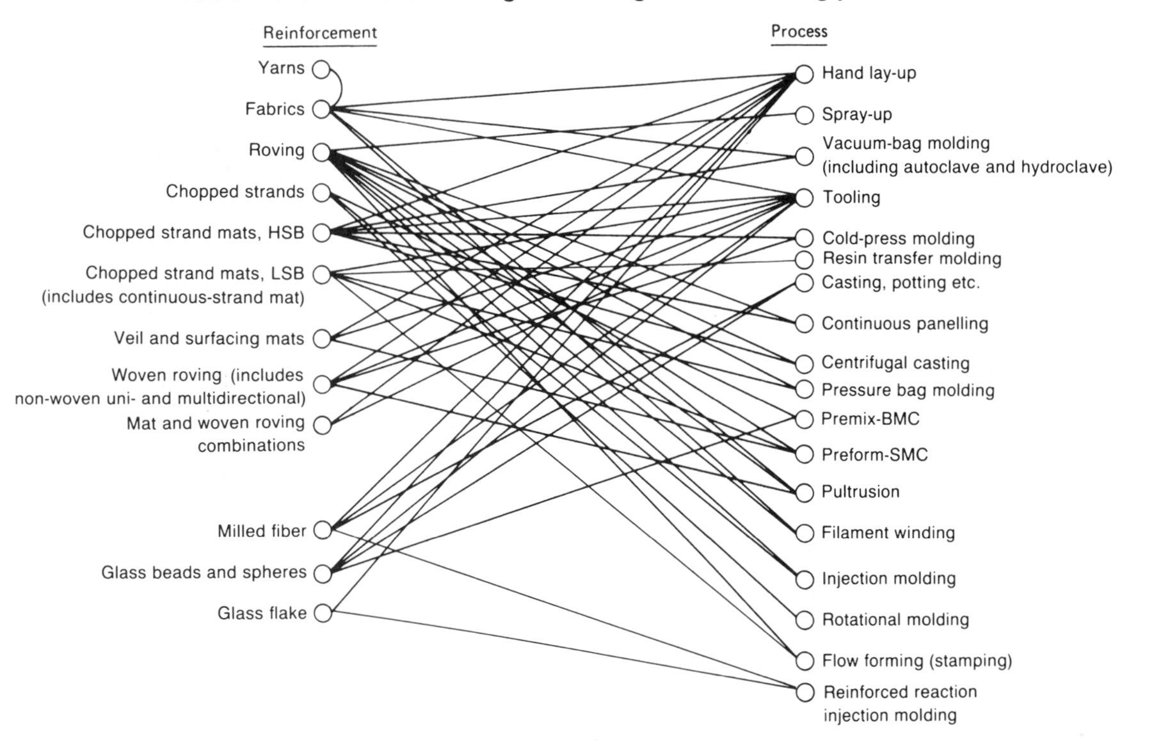

coated. The mold halves are mounted in a supporting frame or a press (no pressure required), and the molds closed to stops while cure takes place. Advantages are that the process provides two finished surfaces and is more rapid and consistent than open-mold, hand lay-up fabrication.

Resin Transfer Molding (RTM): This method is an alternative to cold molding wherein mat or preform reinforcement is prepositioned in a cold mold, mold halves are clamped together (press not required), and resin is pumped in through self-sealing entry ports. Resin may be precatalyzed and mixed with resin/accelerator in a mixing head at the entry port, or catalyst may be metered in at the entry port. Resin fills the mold, developing 40 psi internal pressure while exhausting air through peripheral release risers. Cure is accomplished without external heat. Molds may be gel-coated, or thermoformed acrylic may be used.[18]

Casting, Potting, and Encapsulation. In casting, resin and filler and sometimes reinforcement are mixed together and poured into open or closed molds to form lavatory sinks and related items. Mottled color effects may be incorporated into the resin to simulate marble and other finishes. Casting is also used in producing electrical components. Potting and encapsulation are similar processes.

Continuous Laminating. Production using a continuous machine lay-up, cure, and trim process yields several mainstays among RP/C products for construction—near-transparent, flat or corrugated architectural panelling, flame-retardant industrial roof and curtain-wall panelling, interior liner panels for coolers, dairies, etc., and panelling for highway and commercial signs. Either chopped roving or chopped strand mat is deposited uniformly into a controlled thickness of resin or film on a moving belt. Wetout devices complete the impregnation, a second film is added on top, and the sandwich passes through shapers or corrugators prior to gelling and curing in a continuous, throughput oven. Lineal speeds may reach 80 ft/min. The trimmed sheeting may be from 0.040 to 0.065 in. thick, and special resin modifiers are incorporated to resist weathering.

Centrifugal Casting. Either open-end pipe

and tubing or closed-end containers are fabricated by placing combined reinforcement and resin in a rotating cylinder in a heated area. Advantages are many, and include good densification and elimination of bubbles, easy release by shrinkage away from the mold surfaces, and producibility of a wide range of diameters and lengths.

Pressure Bag Molding. A mold is fabricated to define the outside shape of the part, usually a cylindrical form. The lay-ups (preformed end caps and L.S.B. mat plus resin) are placed into the mold, ends bolted and an elastomeric bladder shaped to fit inside is inserted. Pressure is applied, forcing the bladder against the lay-up, inducing wetout, densification, and bubble release.

The mold, with pressure maintained, is passed through a hot water bath or oven, and the part cures. Higher productivity is obtained by employing preforms and resin transfer (injection) instead of hand lay-ups. Water-softening tanks, RP/C fire extinguisher containers, and filtration housings are made by this method. Higher pressure requirements in the end products are fulfilled by changing to filament winding methods.

Premix and BMC Molding. A mixture of chopped glass fiber (lengths 1/16–1/2 in. or more), filler, and catalyzed resin is blended in a suitable mixer or compounding machine. The result is a dough-like material suitably termed "dough molding compound" (DMC) by the British, but known better as "premix" and bulk molding compound (BMC) in the U.S.

The compound is extruded, cut, weighed, and molded in heated matched metal dies, in a compression, transfer, or injection press. Typical parts range in size from 1 sq in. to 16 sq ft, and applications range in aprt size from small electrical components with molded-in inserts to large items such as industrial wash basins.

Shrinkage and the finished surface out of the mold of premix molded parts were substantially improved upon by BMC formulation technology. That involved modifications to the resin system affecting these parameters. Additives were developed to prevent resin shrinkage out of the mold, and hence also the exposure of fiber and filler surface patterns. These additives also induce a synthetic gel or thickening of the resin prior to molding. This improvement results in a more uniform distribution of reinforcement and filler in the molded parts, as well as better physical strength due to freedom from voids, and higher densification.

Preform and SMC Molding. A page of technology was borrowed from felt hat manufacturing in the early days of RP/C fabrication. This yielded a method for collecting glass fibers (chopped from roving) onto a screen prepared in the shape of a part to be molded. Molding was carried out in large metal dies in a compression press. The chopped glass preform, bonded using a spray binder, was combined with a polyester resin and filler mix at the press and molded at 235–265°F, 250–500 psi pressure.

This process was used to fabricate all the large-sized, higher-production-rate RP/C parts such as washtubs, hard hats, car bodies, and even boats up to 17 ft long.

The improved low-shrink-and-thickened-type-resin technology indicated in the premix–BMC discussion has enjoyed wide application in the preform molding field. In fact, the changes in technology made it possible to prefabricate the fiberglass, resin, and filler into a processible form designated "sheet molding compound." SMC is made on a continuous machine not unlike the continuous panel process previously described, except there is no oven for heating and curing. Instead the sandwich is maturated (i.e., thickened) via resin viscosity build in a controlled temperature/humidity environment over several days to give sheet that is tack-free, yet pliable and flowable in the mold.

For the first time, the RP/C industry had a material that could be taken off a roll, like sheet steel, fed directly to a press, and molded according to known methods. No longer was it necessary to pour fluid resin–filler mix laboriously over a shaped fiberglass preform at the press and hope for 100% successful molding results.

The difficult surface-finishing tasks originally associated with preforming were also eliminated. Parts 60–70 sq ft in area are about the largest possible using SMC, not quite as large as those for the process of preform molding.

Though SMC has not replaced preform (and mat) entirely, its ability to be molded into complex shapes with ribs, bosses, and cut-outs has made it very attractive for high-volume semi-structural applications in the automotive, business machine, and other industries. External automotive parts such as hoods and deck-lids are manufactured at competitive production levels obtaining a metal-like "Class-A" surface using techniques such as in-mold coating and vacuum applied to the mold.

Press Lamination. Flat laminates are prepared for electrical circuit boards by laying up plies of resin-impregnated mat or cloth, such as prepregs, on caul plates that are stacked in multi-opening presses. The plies are pressed together and cured by introducing heat into the caul plates, or by placing the compressed stack in an oven. Copper surface sheets may be included, which are later partially etched away to leave the circuit pattern desired.

Foam or Elastic Reservoir Molding (FRM or ERM). This is a specialized, limited shape process in which a sandwich laminate is prepared by first laying down in the mold the desired plies of mat, then a sheet of urethane foam. A predetermined amount of catalyzed epoxy resin is distributed uniformly over the foam, on top of which are laid a second sheet of foam and top layers of mat. The heated mold is closed under low pressure, compacting the laminate to about a 10/1 ratio. The resin is forced through the layers of foam and mat to the outside surface and subsequently cured. Very high rigidity is obtained.

Reinforced Reaction Injection Molding (RRIM). Plastic end-users have long been intrigued with the concept of combining, into one, the steps of polymerization of the resin monomers and molding of the part. Early work was done with caprolactam to form nylon 6 shapes, but large-scale commercial production by this method awaited the successful development of urethane chemistry. Later, the addition of fiberglass reinforcement improved the stiffness and temperature resistance of these otherwise low modulus RIM polyurethanes.

The process involves preblending short fiber and/or flake reinforcement in one of the two streams of chemicals, and catalyst in the other, and pumping the two streams to a mechanical or high-impingement mixing device mounted at or near the injection port of the heated matched metal mold. The liquid mix fills the closed mold under low pressure, and polymerization of the monomers takes place in situ (in the mold). A self-skinning reinforced part is formed, the density, strength and rigidity of which can be varied within limits to suit the application. Parts must be post-cured to relieve outgassing, and power-washed before painting. Large and fairly complex shapes such as soft front and rear fascia and "friendly" fenders (quarter panels) of autos are molded by this technique.

Pultrusion. The process for producing RP/C rod stock was re-dubbed "pultrusion" after technology improved to the point that solid, hollow, round, square, and rectangular cross-sectioned and even girder-type structural members could be continuously fabricated by impregnating roving and drawing it through shaping and curing dies. This method of fabrication of structural shapes from composites on a continuous basis is analogous to aluminum or thermoplastic extrusion, in that all these methods produce constant cross-sectional profiles from their respective materials. With the advance in pultrusion technology, an infinite number of solid and hollow profiles are now fabricated using mat and fabrics in combination with continuous roving to tailor the properties of the finished part by fiber orientation and glass content. Furthermore, by appropriate choice of the resin system along with reinforcement, chemical resistance, thermal resistance, and mechanical properties can be enhanced.

Though traditionally pultrusion has been a slow process, at running rates of 2–3 ft/min., recent advances in resins and curing technology combined with radio-frequency heating in conjunction with conventional heating methods have made pull rates up to 20 ft/min. possible, depending upon cross-sectional shape. The process provides for controlled fiber tensioning and orientation, lower void content, and uniform fiber distribution in the composite, resulting in consistent properties compared to some of the other RP/C molding processes. While initial cost in the heavier sections is slightly more than that of their metal counterparts, pultruded structural shapes are indispensable in areas where corrosive chemicals are produced and handled. Ells, tees, and I-beams up to 12 in. × 6 in. and more than ½ in. in thickness may be manufactured by this process.

Filament Winding. Ancient methods of fortifying many types of structures using wire or cloth overwraps (e.g., wire-wrapped gun barrels and cloth-wrapped Egyptian mummies) suggested the application of strong glass fibers circumferentially or axially wrapped. However, glass fibers can go one stage further. After impregnation with resin and curing, the filament-wound composite structure supports itself and needs no permanent substrate.

Fundamentally, filament winding is a relatively simple process in which the continuous fiber reinforcements in the form of strands or rovings are wound over a rotating mandrel. Control of wind angles and the placement of reinforcements are accomplished by using specially designed machines in which the carriage traversing speed and mandrel rotation are synchronized. Successive layers can be added at the same or different wind angles. The wind angle can be varied from "near longitudinal" to "near hoop" relative to the mandrel axis to avoid unidirectional orientation and provide complete coverage. Impregnation of the fibers can be done at the time of winding with the aid of a resin impregnation system, or preimpregnated B-staged rovings can be used. Normal curing is performed at elevated temperature without external pressure and the mandrel is removed after curing.

The main advantage of filament winding is that glass contents of from 60 to 85% may be attained, making possible exceptionally high part strengths. The process can be designed to give a broad variety of structural types, design features, and material combinations. Structures from simple cylinders to high pressure pipes (6 in. I.D., 5000 psi working pressure) to spherical tanks and complex windmill blades and helicopter rotor blades can be fabricated. Though thermosetting resins such as polyesters, vinyl esters, and epoxies have been conventionally used because of their low initial viscosity and ease of handling, other high-performance resins such as polyimides and certain thermoplastics are also being used in select applications. Filament winding also provides mobility in the sense that on-site fabrication of large structures is made possible by setting up the unit at the place of erection. Tanks up to 1500 sq ft in area and power plant stack liners ("cans") 40 ft in diameter × 15 ft high are wrapped on a single mold.

The first piece of RP/C filament winding equipment was obviously some type of machine tool or lathe. However, exceptionally sophisticated six-axis winding equipment, all programmed automatically, has evolved for fabrication of the many types of complex parts now produced by this method.

Injection Molding of Thermoplastics. Thermoplastic resins for injection molding were first reinforced by forming an emulsion of the resin and coating fiberglass roving. The roving, with resin fused and set, was chopped to form moldable pellets that could be fed to an injection press in the conventional manner.

In later developments, melted thermoplastics were used to impregnate the roving, which was die-wiped, cooled, and cut into pellet size for molding. Both reinforced pellets of high glass content and unreinforced pellets could be combined at the injection machine to vary glass content of the moldings as desired. Still later, developments led to the extrusion-compounding of dry blends of chopped strand and unreinforced resin, resulting in reinforced pellets in which the two ingredients were thoroughly mixed (melt-blended). A subsequent alternative method of extrusion compounding involves the continuous feeding of fiberglass roving to an opening in the barrel of a twin-screw extruder. The roving is cut by kneading block components of the screws. During the same period, advancing fiberglass technology resulted in the development of chopped strands having both strand integrity and dry flowability. This enabled the direct feeding of single-screw extruders with chopped strand, thereby eliminating the preblending step. Today, glass reinforced thermoplastics (RTP) represent over 25% of all thermoplastic compounds offered.

In injection molding, RTP pellets or a dry

blend of resin powder and chopped strand (i.e., direct injection molding) are drawn from the machine hopper into a heated barrel, the resin is melted and the mix "plasticated" on the rotating screw ram, and then the melt blend is "shot" (rammed) under high pressure into a cooled, clamped mold. The resin sets up rapdily and the part is ejected as the mold opens. Parts up to 100 sq ft in projected area (300 oz press) may be molded.

The incorporation of glass fibers into thermoplastic resins for injection molding substantially improves all properties, particularly tensile and flexural strengths, heat deflection temperature, and elongation. However, the process would be an even greater factor in RP/C (currently 16% of total composite volume—see Table 14-10) if it were less damaging to the glass fiber lengths. Dry glass strands ¼ in. long, starting out of the hopper and into an injection machine or pelletizer, are soon separated and end up in the molded part completely dispersed into the desired individual filaments, but with a range in lengths somewhere between 0.001 in. and 0.030 in. Attrition occurs primarily in nonstreamlined runners and small gates of the mold, and sharp turns in the cavity. If final fiber lengths exceed the critical lengths, which are about 0.020–0.025 in. for several engineering thermoplastics (i.e., equivalent to aspect ratios of 40/1 to 75/1), then the maximum strengths of the RTP composites will have been realized.

Rotational Molding. Chopped glass fiber lengths are directly charged together with a 35-mesh thermoplastic molding powder into the mold cavities of a rotational molding machine. In the standard heating cycle, the glass fibers are assimilated into the resin as it fuses, and they become firmly bonded into the part as the resin is cooled. Average glass contents up to 25% are possible, but loadings of from 1 to 10% are more frequently used. The presence of glass reinforcing improves all of the roto-molded physical strengths except impact resistance.

Parts molded using reinforcement include large containers for food storage and processing, housings, electrical components, and statuary.

Flow Forming. This interesting process for molding reinforced thermoplastics sheet is a sort of SMC in reverse. Fiberglass as chopped strand or mat is combined with the thermoplastic resin melt, at elevated temperatures, into a moldable sheet product. In order to convert the sheet into a molded product, precut, flat blanks are reheated in a rapid-temperature-rise oven, such as infrared, on wires. They are then transferred directly to relatively colder, matched molds in a high-speed compression molding or stamping press, where they are immediately shaped and rapidly cooled. The molds are telescoping, and neither shearoff nor trimming to size after molding is required.

Small housings to station wagon load floors and bumper beams are produced very practically by this process, on rapid cycles of 10 to 60 seconds, depending upon part thickness. Comparable steel stampings require several "strikes" in multiple dies and presses. Excellent molded physical strengths are obtainable using this method. Melt bonding of rigidizing hat sections to stamped overlays is done in assembly-line fashion by electromagnetic welding.

7.4 RP/C Parts Design and Molded Properties

Both parts design and molded properties are of great importance to potential purchasers and users of fiberglass-reinforced plastic products. Unfortunately, each subject, if treated properly and to the fullest extent, would command the space of an individual volume in its own right. The reason is that each separate composite material and molding method yields a different type product with different glass content and with an individual area of application. This versatility within the family of RP/C, one of its principal virtues, requires completely separate treatment for full representation and understanding. Therefore, these two subjects will be treated briefly, and information presented so that the reader will be able to consult other, more complete sources, if necessary.

Design Factors. Design parameters include such information as practical thickness and variations, recommended sidewall draft for proper release from molds, permissibility of undercuts, allowable diameter and placement of holes, extent or area of joining plates and location of screws or bolts, permissibility of ribs, bosses, molded-in labels or numbers, type of surface finish possible, and many other facets. These have been given a thorough review and are available for reference in a 1978 publication.[19]

Fiberglass-reinforced plastics may be fabricated into almost any shape desired, with comparatively great savings in tooling costs. They also have the versatility of drastically reducing the number of components required for a given assembly, that is, a cost benefit of parts consolidation.

Molded Properties. Table 14-13 lists representative property ranges of a cross section of fiberglass-reinforced plastic (FGRP) molded types. FGRP laminates are totally engineered structures with physical, mechanical, electrical, chemical, and corrosion-resistance characteristics that can be adjusted with remarkable flexibility to meet specific application requirements. The properties (room temperature) show wide ranges due to the various resin and filler formulations, and fiberglass forms and types, amounts (ASTM D-2584), and orientations available. Their lightweight nature is indicated by the specific gravities, and it may be noted that strength increases proportionately with increase in glass content, length, and orientation in the laminate. Properties shown for directional processes and compounds are those parallel to the direction of primary reinforcement. For XMC® composite, the ranges listed encompass XMC-3 and 2 from minimum to maximum, respectively.[20] For Azdel® composite and injection-molded RTP, ranges are for reinforced thermoplastics rather than for reinforced thermoset polyesters as in the case of all other categories. Deflection temperature and continuous resistance to heat are functions of the resin type used, but are also influenced slightly by the glass fiber length and content.

The strength properties of FGRP are adversely affected by continuous exposure to elevated temperatures and by the various types of loading. Substantial data are available from this and other sources.[21-24] Design factors of safety (ultimate strength divided by allowable working stress) may range from 2 to 10 for both conditions, depending upon the many variables that are usually involved in an end use. Their selection should be the responsibility of the designer. Because of the continuing growth potential for fiberglass-reinforced thermoplastic molding compounds, Table 14-14 is presented to correlate and compare properties in a typical glass content level of reinforced thermoplastic resins. Other reinforcement contents are available, generally from 10% to as high as 55% in a few resins. A qualitative scan is made of the base properties of all polymers considered. This is shown as a supplement to Table 14-14. Quick and easy reference has been made possible by abbreviated letter designations.

7.5 New Developments and New Directions

In this concluding section, the most notable of a series of unique new developments in RP/C products and processing methods is presented. Also a few observations are made on directions that the industry may take to assure continuation of its growth and expansion.

High-Strength Composites. Automotive and truck manufacturers are moving closer to broad use of high-strength composites, as the RP/C industry has provided materials, equipment, and fabrication technologies conducive to high-volume production. Specialty high-strength sheet molding compounds are beginning to be used in structural applications after many years of application development effort. SMC containing various proportions of random chopped and continuous fibers essentially aligned in the machine direction and HMC® sheet with up to 65% random chopped fiberglass and 35% resin (filler not required but is helpful in mold flow) are made on an SMC machine using conventional polyester/vinyl ester SMC thickening technology. Parts matched-die compression-molded of directional SMC exhibiting 3–3.5 million psi flexural modulus include rigidizing elements of RP/C truck cab doors and station wagon tailgate doors. Although they have a lower flexural modulus (2.2 million psi), prototype parts such as radiator supports molded of HMC composite exhibit more isotropic properties, including tensile and flexural strengths of 30,000 psi and 50,000 psi respectively in the flow direction. High-temperature

Table 14-13. Representative properties of RP/C molded parts.

	ASTM Method	Open Molding: Lay-up	Open Molding: Spray-up	Matched Die Compression Molding: Bulk Mldg cpd	Matched Die Compression Molding: Preform	Matched Die Compression Molding: Sheet mldg cpd	Matched Die Compression Molding: HMC® comp.	Matched Die Compression Molding: XMC® comp.	Matched Die Compression Molding: AZDEL® comp.	Injection molding RTP	Filament Winding	Pultrusion (profile-rod)
MECHANICAL PROPERTIES												
Tensile strength[a,b] psi × 10^3	D638	9–50	5–18	3–10	10–30	8–20	23–35	70–90	11	6–30	80–200 [a]	40–180
MPa		62–345	35–124	21–69	69–207	55–138	159–240	480–620	76	41–207	550–1380[a]	275–1240
Tensile modulus, psi × 10^6	D638	0.6–4.5	0.8–1.8	1–2.5	0.8–2.0	1–2.5	1.9–2.8	6.0–6.5	0.8	0.5–1.8	4–8	3–6
GPa		4–31	6–12	7–17	6–14	7–17	12–15	41–45	5.5	3–12	28–55	21–41
Flexural strength, psi × 10^3	D790	16–80	12–28	7–20	24–40	15–40	45–55	125–160	22	8–45	100–250	75–200
MPa		110–550	83–190	48–138	165–276	103–276	310–379	862–1103	152	55–310	690–1725	517–1380
Flexural modulus, psi × 10^6	D790	0.9–4.0	0.7–1.3	1–2.5	1–2.5	1–2.2	2.2–2.4	5.2–6.5	0.7	0.4–2.2	5–7	3–6
GPa		6–28	5–9	7–17	7–17	7–15	15–16	36–45	4.8	3–15	34–48	21–41
Compressive strength, psi × 10^3	D695	18–50	15–30	15–30	15–30	15–30	20–35	79/28		6–26	50–80	40–100
(edgewise) MPa		124–345	103–207	103–207	103–207	103–207	138–240	545/193		41–179	345–550	276–690
Shear strength, psi × 10^3	c	4–6					3–5	3.5–9			7–10	5–10
MPa		28–41					21–34	24–62			48–69	35–69
Impact strength,[d] Izod, ft-lb/in. notch	D256	5–30	5–15	1.5–10	8–25	8–22	14–22	65	12	1–5	40–60	
J/m notch		270–1600	270–800	80–535	427–1335	427–1170	750–1170	3470	640	54–270	2135–3200	
Deflection temp °F @ 264 psi	D648	>400	>350	>400	>350	>375	>500	>500	310	220–485	>350	>350
°C @ 1820 kPa		>205	>177	>205	>177	>191	>260	>260	155	105–252	>177	>177
PHYSICAL PROPERTIES												
Specific gravity[e]	D792	1.4–2.1	1.4–1.6	1.6–2.3	1.4–2.3	1.6–2.3	1.8–1.9	2.0	1.2	1.1–1.7	1.7–2.2	1.7–2.1
Density, lb/in^3 × 10^{-3}		51–76	51–58	58–83	51–83	58–94	65–69	73	43	40–61	61–77	61–76
Hardness, Barcol units	D2583	40–70	45–65	50–80	45–60	50–72	65–80	65–80			35–65	35–65
Poisson's ratio		0.25–0.29						0.31–0.32				
Coeff. of thermal exp., °F × 10^{-5}	D696	1.0–1.8	1.2–2.0	0.8–1.2	1–1.8	0.8–1.2	2.9–4.0	1.6/3.4	1.5	1.0–3.5	0.4–0.6	0.4–0.6
°C × 10^{-5}		1.8–3.2	2.2–3.6	1.4–2.2	1.8–3.2	1.4–2.2	1.6–2.2	0.9/1.9	2.7	1.8–6.3	0.7–1.1	0.7–1.1
Thermal conductivity (k factor)	C236											
@ 70°F BTU · in/(h · ft^2 · °F)		1.40–1.55										
@ 21°C cal · cm/(s · cm^2 · °C) × 10^{-4}		4.8–5.3										
Water absorption, 24 hr %	D570	0.2–1.0	0.2–2.0	0.06–0.28	0.01–1	0.1–0.25	0.1–0.25	0.1–0.23		0.01–1.2	0.1–0.5	0.1–0.6
Flammability[f], in./min	D635	0.8–2.0	0.8–2.0	0.7–2.0	0.7–2.0	0.7–2.0				0.7–1.1	0.7–2.0	0.7–2.0
(0.50 in.) mm/min		20–50	20–50	18–50	18–50	18–50				18–28	18–50	18–50

Table 14-13. *(Continued)*

ELECTRICAL PROPERTIES	ASTM Method	Open Molding: Lay-up	Open Molding: Spray-up	Matched Die Compression Molding: Bulk Mldg cpd	Matched Die Compression Molding: Preform	Matched Die Compression Molding: Sheet mldg cpd	Matched Die Compression Molding: HMC® comp.	Matched Die Compression Molding: XMC® comp.	Matched Die Compression Molding: AZDEL® comp.	Injection molding RTP	Filament Winding	Pultrusion (profile-rod)
Dielectric constant, 60 cps	D150	3.8–6.0	3.7–6.0	5.3–7.3	3.8–6.0	4.4–6.3	5.1			2.4–4.2	4.2–5.3	4.0–6.0
10^3 cps		4.0–6.0	4.0–6.0	4.0–6.8	4.0–6.0	4.4–6.1	5.1			2.4–4.0		4.0–6.0
10^6 cps		3.5–5.5	3.6–6.0	5.2–6.4	3.5–5.5	4.2–6.0	5.0			2.4–3.9	4.0–5.2	4.0–6.0
Dissipation factor, 60 cps	D150	0.01–0.05	0.01–0.04	0.01–0.2	0.01–0.04	0.007–0.2	0.008			0.001–0.03		
10^3 cps		0.01–0.06	0.01–0.05	0.01–0.2	0.01–0.05	0.007–0.2				0.001–0.025	0.018–0.05	
10^6 cps		0.01–0.03	0.01–0.03	0.01–0.02	0.01–0.03	0.01–0.02	0.005			0.0015–0.026		
Dielectric strength, volts/mil (short time)	D149	300–600	200–450	300–450	350–500	300–450				400–600	200–400	300–350[h]

Footnotes:

[a]For XMC® Comp values are parallel to continuous reinforcement, except compressive strength and coefficient of thermal expansion for which directional properties of XMC-3 (parallel/transverse) are given. XMC-3 provides 9-15 × 10^3 psi and 1.4-1.8 × 10^6 psi transverse tensile strength and flexural modulus respectively. For filament wound structures values are by ASTM D-2290 and D2343 for filament and strand tensile strengths, respectively.

[b]Ultimate tensile elongation ranges from 0.3 to 2.7% for FGRP thermoset polyesters, and from 1 to 5% for FGRTP thermoplastics.

[c]ASTM D3846, twin notch, in-plane shear in compression, except values for filament winding which are apparent horizontal shear, ASTM D2344, by the short beam method.

[d]Charpy values for HMC® and XMC® composites are 16 and 41 ft-lb/in of notch respectively.

[e]May be calculated as follows: Laminate sp. gr. equals the reciprocal of the sum of the ratios of weight fraction to sp. gr. for each component in the laminate.

[f]Not intended to reflect actual fire conditions. Flame-retardant types are available.

[g]Dielectric constants and dissipation factors are substantially higher for nylons conditioned to 50% R.H.

[h]Range applies to sheet. Parallel dielectric strength for rods is 50 KV/in.

epoxy SMC has the strengths at 300°F that conventional polyester SMC has at room temperature.

The ultra-high-strength material among sheet molding compounds is XMC® composite, originally developed to better utilize the strengths and modulus of fiberglass in semicomplex parts that must be compression-molded to shape. (These properties are more fully realized in uniform shapes that are filament-wound and pultruded from continuous, directionally oriented reinforcement.) XMC sheet composite is a directional material prefabricated with a combination of oriented chopped (for transverse strength) and continuous fibers nested at predetermined wind angles using a filament winding machine. The maturation, charge cutting, molding, and post-fabrication operations are similar to SMC. The exception is molding flow that is confined to the transverse direction as the "lattice" of continuous longitudinal glass expands. With glass contents as high as 75%, tensile and flexural strengths of 70,000 psi and 130,000 psi, as well as a flexural modulus of 5.5 million psi, are achieved in the direction of the continuous fibers. Applications include truck cab roof rigidizing elements, firemen's safety "hard hats," and a light aircraft nose wheel.

Commercial acceptance of these directional, high-strength thermoset sheet molding compounds has been slow owing to an extensive learning curve in controlling fiber orientation and distribution during compression molding, and owing to sophisticated on-line quality control measures needed for assurance of part performance and reliability. Nevertheless, industry is well up on the curve as composite monoleaf springs, having elastic strain energy capacity ten times greater than steel and made by a combination of filament winding and compression molding or pulforming,* are in mass production for the Chevrolet Corvette and General Motors' minivan and top-of-the-line Buick Riviera, Oldsmobile Toronado, and Cadillac Eldorado, Seville and Allante. Compression-molded composite road wheels weighing 40% less than steel and 17% less than cast aluminum have been thoroughly tested and are ready to go into production. And the newest among directional composites are the thermoplastic prepregs of continuous and woven rovings that are destined for pultrusion and filament winding, perhaps of automotive space frames and high-productivity aircraft/aerospace components.

*An adaptation of the pultrusion process has been developed by Goldsworthy Engineering Inc., Torrance, CA, for fabricating longitudinal shapes of nonuniform cross section. An example is an automotive leaf spring, which has a narrow thick center and flared thin ends. Continuous fibers impregnated with appropriate resin are pulled through heated rectangular dies. As it comes out of the die, the partially cured composite is shaped into the form of a spring on a continuous basis in a multicavity revolving mold system under heat and moderate pressures.

For many large chassis and underbody applications, automotive and truck manufacturers will prefer high-strength isotropic (nondirectional) composites for ease of part design and molding. In this category are two families of fiberglass materials: continuous strand mat reinforced thermoplastic sheet for high-speed compression molding, or "stamping" (Azdel Inc composites), and continuous strand mat or glass preform reinforced thermoset hybrid esters and urethanes molded by high speed resin transfer molding (RTM) or reinforced reaction injection molding (RRIM).

The sheet materials, supplied as rectangular blanks, are in-line preheated to above the melting point, and charged to the mold. Under rapid mold closing, isotropic flow results in mold fill-out, followed by cooling and ejection in cycles usually of less than 1 minute. Tensile strengths of 12,000 to 14,000 psi, flexural strengths of 24,000 to 28,500 psi, flexural moduli of 800,000 to 1,100,000 psi, Izod impact strengths of 13 to 14 ft-lb/in., and deflection temperatures under load (264 psi) of 300 to 478°F—for 40% reinforced polypropylene to 35% reinforced thermoplastic polyester Azdel® sheet composites,⊛ respectively—provide high strengths, toughness, and chemical and heat resistance in demanding structural body and under-the-hood applications.

The upgraded RTM and RIM structural applications of mat or preform reinforced methylmethacrylate-based vinyl ester (ICI Americas' Atlac® resin), acrylesterol-isocyanate (Ashland Chemicals' Arimax™ acrylamates),

⊛Azdel is a registered trademark of Azdel, Inc.

Table 14-14. Properties of fiber glass reinforced thermoplastic resins.

	Nylon-6	Nylon-6/6	Acetal	High Density Polyethylene	Polypropylene (chemically coupled)	Polybutylene Terephthalate	Polycarbonate
Properties of the Basic Polymers	Same as Nylon 6/6	TUF-E ABR-G MST-H CHR-G MBR-G WA-H PRO-E CO-M	MST-H CR-G ST-L DST-G FE-H CHR-G WR-G TS-P	CO-L CHR-G WR-G MST-L ST-L EPR-G SPG-L PRO-G	PRO-G CR-G TUF-G ST-M WR-G CHR-G	PRO-G HDT-L MST-M FE-G TUF-G WA-L CHR-G IR-L SP-L FL-M to H	CL-H TUF-G IR-E SC-H SP-H TS-H SP-P ABR-P
Molding parameters							
Injection pressure, psi	10-20000	12-20000	15-20000	10-20000	10-20000	10-15000	15-20000
Injection cylinder temp., °F.	450-550	525-550	375-410	350-550	450-525	450-520	550-650
Mold temp., °F.	150-200	150-200	200-250	75-125	90-150	100-250	150-250
Physical data							
Glass fiber, %	30	30	30	30	30	30	30
Specific gravity	1.36	1.36	1.63	1.18	1.13	1.53	1.43
Molding shrinkage, in./in.,							
1/8 in. section	0.003	0.004	0.004	0.003	0.004	0.003	0.001
1/4 in. section	0.005	0.006	0.005	0.0035	0.005	0.004	0.002
Water absorption, %, 24 hr @ 23°C	1.2	0.7	0.9	0.017	0.04	0.070	0.08
Mechanical properties							
Impact strength, Izod,							
notched 1/4 in.	2.0	2.1	0.9	1.3	1.9	1.4	1.7
unnotched 1/4 in.	12	13	5.0	4.5	12	11	11
Tensile strength, psi	23000	26000	12500	9000	12000	18000	20000
Tensile elongation, %	3.0	2.0	1.6	1.5	2.5	2.5	2.0
Tensile modulus, psi × 10^6	1.3	1.5	1.4	0.9	1.0	1.4	1.4
Flexural strength, psi	28000	36000	16800	11000	18000	27000	23000
Flexural modulus, psi × 10^6	1.1	1.3	1.1	0.8	0.7	1.2	1.1
Compressive strength, psi	22000	22500	11500	7000	12000	18000	20000
Hardness, Rockwell R	119	120	112	75	98	120	119
Electrical properties							
Dielectric strength, VPM, S/T	500	500	500+	500	500	500+	470
Dielectric constant, 1 MC, dry	3.8	3.8	4.0	2.7	2.7	3.8	3.4
Dissipation factor, 1 MC, dry	0.017	0.016	0.0055	0.008	0.001	0.015	0.0075
Arc resistance, sec.	120	120	182	140	125	130	120
Volume resistivity, ohm CM	10^{14}	10^{14}	10^{14}	10^{16}	10^{15}	10^{16}	10^{16}
Thermal properties							
Deflection temp., °F., @ 264 psi	400	480	325	250	295	415	300
@ 66 psi	420	500	330	265	315	420	300
Flammability	B	B	B	B	B	B	SE
Flammability, UL Sub. 94, 1/8 in.	SB	HB	HB	HB	HB	HB	VE1
Coefficient of linear thermal expansion, in./in./°F × 10^{-5}	1.75	2.0	2.3	2.7	2.0	1.4	1.3
Thermal conductivity, Btu/hr/ft²/°F/in.	3.5	3.5	2.2	2.5	2.3	1.4	2.2

and other new resins by Dow, Union Carbide, Mobay, Hercules and others, are in development. The first application is a spare tire cover. Processing is similar to the previous descriptions for RTM and RIM except that here the reinforcement must be cut or preformed to shape and placed in the mold. Low injection pressure (50 psi) permits lightweight tooling and low clamping force. Because of low viscosities permitting fast fills, and rapid gel times at elevated temperatures (140 to 205°F), in-mold times can be as short as 1 minute and overall cycle times up to 3 minutes, including loading of reinforcement and demolding. Properties for the acrylamates with 35 to 50% glass fiber reinforcement, respectively, are 14,000 to

Polysulfone	Polyphenylene Oxide (modified)	Polystyrene	Styrene Acrylonitrile	Acrylonitrile Butadiene Styrene	Modified ETFE Fluoro Plastic	Polyvinyl Chloride (rigid)	Polyurethane	Polyphenylene Sulfide	ASTM Test Method
CL-H SP-H TS-E SR-P ABR-L PRO-M	PRO-G to E WR-E SP-H FL-L SR-L ABR-P	CO-L PRO-E CL-E SP-L SR-L MST-M to L IR-L (brittle)	CO-L PRO-E CL-G SR-M to L SP-L MST-L	CO-L PRO-E SR-M TUF-E MST-M	PRO-G HAR-M ST-M ABR-M CHR-G TS-M EPR-G	HDT-L CHR-E WR-G EPR-G FL-L CR-G SP-L TS-L PRO-M to P	TUF-G IR-G AR-G ST-L MST-L	MP-H CHR-G TS-G FL-L SR-E ST-H MST-G (at hi-temp) HAR-G PRO-G SP-H MP-H TS-G (no oxidation)	
15-20000	15-20000	10-20000	10-20000	10-20000	3-20000	10-20000	10-15000	15-20000	
600-660	500-650	400-550	450-550	450-550	570-650	300-450	360-410	575-650	
200-300	450-600	100-160	100-160	150-200	–	300-415	60-150	100-350	
30	30	30	30	30	25	20	30	30	
1.46	1.36	1.28	1.31	1.28	1.80	1.54	1.45	1.53	D-792
0.001	0.004	0.001	0.001	0.001	.002-.030	0.001	0.003	0.002	D-955
0.002	–	0.001	0.001	0.0015	–	–	0.004	0.004	
0.23	0.06	0.06	0.15	0.16	0.022	0.008	0.15	0.02	D-570
1.3	1.0	1.1	1.1	1.3	9.0	1.6	2.0	1.4	D-256
10	–	2.1	4.0	5.0	–	–	15	6.0	
17000	17000	12000	19000	16000	12000	14000	17000	17000	D-638
2.1	6-0	1.0	1.6	1.8	8.0	4.0	3.5	1.3	D-638
1.2	1.2	1.3	1.6	1.2	1.2	1.8	1.05	1.6	D-638
23000	20000	16800	22000	19000	–	21000	23000	28000	D-790
1.0	1.1	1.2	1.4	1.0	0.95	1.0	0.80	1.4	D-790
22000	17900	17000	21000	15000	10000	12000	12000	24000	D-695
123	–	121	123	110	–	–	117 86 (Shore D)	122	D-785
430	600	450	500	400	425	–	450	350	D-149
3.65	2.93	3.2	3.6	3.5	3.4	–	4.5	3.5	D-150
0.005	0.0009	0.002	0.008	0.009	0.015	–	0.014	0.001	D-150
100	120	50	60	75	–	–	90	120	D-495
10^{15}	10^{17}	10^{16}	10^{16}	10^{15}	10^{15}	–	10^{11}	10^{16}	D-257
362	310	210	212	230	410	180	185	500	D-648
370	310	225	225	240	510	180	270	500+	
SE	SE	B	B	B	–	SE	B	SE	D-635
VEO	–	HB	HB	SB	–	–	HB	VEO	
1.4	2.2	2.0	1.9	1.7	3.2	1.7	2.4	1.4	D-696
2.2	1.1	2.1	2.0	1.5	–	–	2.8	2.1	C-177

29,000 psi tensile strength, 24,000 to over 45,000 psi flexural strength, 890,000 to 1,500,000 psi flexural modulus (to over 2,000,000 psi at 55% glass content but at a significant sacrifice in flexural strength), 13 to 19 ft-lb/in. Izod impact strength, and 390°F deflection temperature under load (264 psi).

High-strength, lightweight, corrosion-resistant RP/C parts for the transportation, industrial, and farm equipment markets are a long-term growth opportunity. Consolidation of parts into one molding will be the additional cost benefit for structural RP. Liberal use is being made of the application of robot-arm, automatic devices for loading and unloading charges and cured parts at the molding operation. The automatic "arm" is capable of reaching out as far as 88 in.

Class A Panels. Another opportunity being pursued in the automotive industry is manufacture of "hang-on" body panels replacing sheet steel. All elements of the RP/C industry (and the plastics industry in general) are continuing to strive toward Class A (metal-like) painted surfaces without rework. Following the lead of a car producer, compression press manufacturers recently perfected force velocity and parallelism control to enable thin (under 100 mil), large-area SMC moldings. Concurrently, urethane in-mold coating systems were developed to fill in the pores of the SMC part to prevent paint popping, but with an extension of the molding cycle. Now an in-mold coating system has been developed that can be pressure-injected without parting the mold and cured in one-half the normal cycle extension time. A primer spray coating system—developed by Ashland Chemical Co., called Vapor Injection Cure (VIC), and based upon rapid room-temperature tertiary amine cure of urethane chemicals—is reported to greatly reduce, if not eliminate, paint blowouts by filling and bridging the pores in SMC surfaces.[25] Deaeration of SMC resin paste and vacuum applied to the mold has helped remove most, but not all, of the porosity. Paralleling the effort by fiberglass producers to eliminate "fiber prominence," intensive development efforts are in progress by polyester resin manufacturers to evolve SMC systems with little or no shrinkage, that will not telegraph rib and sink marks onto the exterior surface of the molded part. Then, instead of the present two-piece, inner and outer bonded hood, a one-piece RP/C design with surface appearance equivalent to steel will become a reality, with attractive cost- and weight-saving potentials over steel. A similar need exists relative to RRIM quarter panels, in which the addition of reinforcement (milled fiber and glass flake) to increase rigidity causes surface roughness as the urethane resin shrinks following high-temperature painting.

Aircraft and Marine. The highly desirable light weight and high-strength properties of fiberglass-reinforced plastics are paying off in these fields. Also, the many innovations made by the military are reaching commercial status. For example, the Navy has contracted for mine sweeper–hunter ships that are either fiberglass/epoxy clad wood or all fiberglass/polyester. The advantages, in addition to strength, light weight, and corrosion resistance, are that RP/C is nonmagnetic and radar-transparent. Several companies have produced and marketed all-RP/C gliders, and at least four all-RP/C family-style commercial aircraft (two–four places) have been introduced since 1965. Liberal use is made of pultruded material for landing gear springs. Also, RP/C, properly designed, may be fabricated into components that, in flight, create a minimum of air drag, making them highly efficient and desirable. (The motor home industry has also made much use of this design benefit of RP/C, which is not possible in steel, for both improved appearance and gas mileage.)

RP/C in Construction. Although performing this duty well and most aesthetically, the larger RP/C parts used in general building construction have, to date, filled the function of fascia and curtain wall only. Unfortunately, increased product costs and ability to support maximum loads without cold-flow are directly related. However, by ingenious use of self-supporting designs using domes, arches, and paraboloids with inside and outside finished sandwich construction, many practical and useful designs have been successfully marketed for commercial and agricultural buildings. Ease of maintenance and cleanability are major advantages.

In industrial buildings, RP/C finds specialty uses as roof trusses and ceilings in corrosive chemical environments, and as the complete structure in test buildings for computers where electromagnetic interference (from the test equipment) cannot be tolerated.

Industrial and residential doors simulating their traditional counterparts have recently been introduced. Window frames and sash now on the market give double-pane units the thermal insulation of wood-framed windows without maintenance problems, sweating, and loss of insulating capability over the years. Energy conservation is further enhanced by RP paneling used as insulated energy collector covers in solar heating units.

Corrosion-Resistant Applications. Expanded

usage of RP/C components continues in the chemical process fields. Parts are made by almost all processing methods. The largest parts are, of course, made by either hand lay-up or filament winding. Pultruded structural members, spray-up, and compression-molded parts are widely applicable. The chief criterion is selection of the correct resin and proper laminate construction to provide the required corrosion-resistant service. A reference volume, published by the nonmetallic section of NACE,[26] presents data in graph form on the resistance of RP/C components to some 1425 different chemicals. Comparison with many other materials is included.

Notable applications include several types of corrosion-resistant pumps with both rotor and stator parts of RP/C, as well as assemblies that float in a waste-water treatment pond, supporting both delivery and out-take piping so as not to damage a special film lining the pond. High-pressure oil-field pipe is to be made available with up to 6 in. I.D., having a 5000-psi pressure rating. The Society of the Plastics Industry's Fiberglass Pipe Institute represents the RP pipe industry's mutual interests in the chemical process industry, oil and gas industry, and municipal water/waste and general mild industrial applications area.

Reinforced Foams. Almost any plastic material (thermosets or thermoplastics, but usually thermoplastics) can be foamed, either by induced gas or by chemical blowing agents. Foaming agents also may be combined with ingredients in the relatively new reaction–injection molding process.

Fortuitously, it has been determined that incorporation of glass fibers into the foamable resins from 2 to 30% by weight provides reaction sites for more rapid and thorough generation of the foaming agents. This results in faster cures, smaller and more uniform foam cells, and better cell distribution for improved finished part properties.

The great advantage of one-cycle foaming operations is that they produce directly, in one stroke and without a two- or three-stage operation, a structure with a greater cross section and hence higher modulus of elasticity. Tennis racket frames and car body parts are examples, as are one-shot panels for containerized truck trailers. The panels may measure up to 45 ft in length, and weigh up to 2000 lb each.

For large, contoured, polyester–foam–polyester structures, such as boat decks, an interesting woven fiberglass reinforcement has been developed. Mölnycke Co. of Göteborg, Sweden, produces a woven roving fabric with loops 1 in. in diameter, on 1 in. centers, extending from one surface. Cut shapes of this fabric are laminated to the top and bottom RP/C panels of a boat deck or other structure, separated by a distance approximately the diameter of the loops. Foamable polyurethane resin is injected into the cavity and allowed to foam and cure. The gap is filled, and the foam becomes substantially reinforced by surrounding and encapsulating the loops.

Another polyurethane foam core material, called Corex^{TM}, is available from Cytonics, Inc., Ambridge, PA. Foam strips are aligned in a tongue-and-groove fashion and encapsulated by layers of chopped strand mat, as both skins and webs. The "articulated" core material bends to the shape of the mold as it is laid up and wet-out in normal fashion, saving time and labor. The shear strength is higher than that for standard foam cores because of the vertical webs of mat/resin between the foam strips.

Lastly, mention should be made of reinforced urethane spray-up. The attraction of urethane instead of polyester is the fast wet-out of the fiberglass and the elimination of the roll-out step. Special spray-up equipment has been developed by Thermo-Cell Southeast, Inc., Lithonia, GA, to adapt to the requirements of urethane. Mold release formulated for RIM-urethane molding and moisture-free gel coats are suggested. By spraying a layer of foamed urethane or by using foamed core materials with the dense reinforced skin, parts as rigid as sprayed-up polyester can be fabricated.

Cure Systems and Related Mechanisms. Many chemical systems have been researched that were intended to trigger more effectively the reactions between catalysts and thermosetting resins, the basis for 84% of all RP/C production. However, two significant recent advances, using any of the most effective chemical curing systems available, are concerned with the mechanics of applications of a heating means more effective than the present standard near-infrared energy sources.

Radio-frequency preheating and curing systems are being used on pultrusion processing lines. This type of excitation generates heat that develops simultaneously throughout the entire laminate cross section, instead of merely moving from the surface inward. Faster pulling speeds and higher product quality are thus made possible.

The use of a pulsating ultraviolet energy source, reputed to cure laminates close to 1 in. thick in approximately 4 minutes was developed by the Xenon Corporation. Possibilities include production rates for pultrusion of from 10 to 20 ft/min. instead of the current 3 to 6 ft/min., and curing wet lay-up, spray-up, cold press, and bag molded parts after UV exposure and without any additional heating required. The overall ultimate effect, combined with UV curable resins that are available, would be a welcome speed-up of the more labor-intensive RP/C processes, lowering finished product costs and enhancing the competitive position.

Combinations. The first commercially important sheathing or cladding of an RP/C structure with a more chemically resistant or protective surfacing layer was developed in 1960. Using a small quantity of resin additive or modifier with a solvent action, a polyester–fiberglass lay-up was bonded to PVC to form chemical hoods, stacks, and ducting. The PVC skin protected the stronger fiberglass structural portion from corrosion.

More recently, the process was refined by vacuum-forming a decorative acrylic or other thermoplastic sheeting into a desired shape (hand lavatory, shower and bath unit, or edged wall panel), and spraying-up fiberglass–polyester on the unfinished side. Mold costs and cleanup problems were reduced, and the characteristic strength and lightness of weight of RP/C were used to advantage in creating unique decorative and utilitarian objects.

Wood, plywood, masonite, rigid foam, and other porous or semiporous materials are used as substrates or cores for RP/C skins. Glass and PVC piping have been overwrapped with layers of filament-wound RP/C for protection against shattering and to strengthen the core pipes. Metal tanks for service above and below ground level are coated with fiberglass to protect them from deterioration due to chemical and galvanic action. Underground gasoline storage tanks of 100% RP/C have been in service for 20 years, and are currently replacing corroded and leaking steel tanks. Aluminum pressure cylinders for holding natural gas fuel are overwrapped with filament-wound RP/C to increase their service capability for fleet cars and light trucks.

Hence, it should be stressed that this type of combination of RP/C with other materials should be diligently pursued where cost and performance show a true advantage.

Injection Molding. In addition to fiberglass-reinforced thermoplastics, the injection molding process has also been successfully adapted to the molding of fiberglass-reinforced thermoset plastics. These materials include premix and BMC, SMC, TMC® (Thick Molding Compound*), pelletized polyesters, and glass-reinforced, asbestos-free diallyl phthalate (DAP) and phenolic molding compounds, which have the excellent creep resistance and dimensional stability at high temperatures required for several electrical connectors and transmission parts. In the case of thermosets, fiber length reduction may be less than that observed with thermoplastics, especially in the barrel and screw zones. However, in the injection molding of all fiber-reinforced composites, the preservation of the fiber length is a major consideration. Although much progress has been made toward this goal, there will be a continued need for cooperation between molders, compounders, and glass manufacturers in order to investigate means for preserving the original fiber lengths throughout the compounding and molding processes. For a discussion and information on equipment for processing fiber-reinforced plastics with minimum fiber breakage, see Chapter 22.

Metallized Glass Fibers. Aluminum and nickel plated glass fibers are offered as rovings, chopped strands and yarns, for use in electromagnetic interference (EMI) and radio fre-

*TMC® is a registered trademark of Aristech Chemical Corp.

quency interference (RFI) shielding and for static dissipation. These high-cost specialized reinforcements, when combined with the resin matrix and molded into RP/C parts, provide reasonably good conductivity (especially the nickel) because of their fibrous (continuous) nature. However, strength contributions are reduced compared to the traditional glass–resin bond, and the systems' cost/performance should be carefully compared to alternatives, including inside surface painting of, for example, housings for business machines and electronic components.

Inorganic Polymers. The disparity in thermal end-points that exists between glass and organic polymers for continuous elevated temperature service truly limits the usage of RP/C. The presently used organic polymers are the restricting factor. Even the present high demand for reinforced plastics would increase considerably if the restrictive temperature barriers were removed.

Even ablative structures accomplish their high-temperature resistance by decomposition of the organic fraction, leaving a char layer that remains intact only because of the high tensile strength of the reinforcement that was originally incorporated.

The answer is undoubtedly in inorganic polymers. Several have been proposed and unfortunately are still curiosities of the laboratory bench. Aluminum phosphate was used to encapsulate fiberglass reinforcing fibers and fabricate radomes. Other phosphate compounds, reportedly resilient enough to be substituted for polymeric resins, have been known for quite some time, but are not yet commercially viable.[27] Metal and carbon-filber reinforced ceramics have high temperature resistance but are friable. Use of magnesium-oxychloride cements, made reinforceable and resilient, as substitutes for organic polymers, was proposed, but found no applications because of their brittleness after dehydration.

Expansion. While many users of fiberglass and other reinforced plastics solicit quotations from molders for parts they need, a continuing, energetic, and creative sales effort is required to sustain industry growth. "Bread and butter" sales for existing items are required to maintain steady incomes, but innovation and looking diligently for new applications must be regarded as a concomitant function. RP/C can be made faster, better, and usually at lower cost than any other material, and it will certainly perform better.

REFERENCES

1. Oleesky, S. S. and Mohr, J. G. *SPI Handbook of Reinforced Plastics*, 1st ed., Van Nostrand Reinhold, New York, 1964, p. 117.
2. Loewenstein, K. L., *The Manufacturing Technology of Continuous Filament Glass Fibres*, Elsevier Scientific Publishing, 1973, p. 104.
3. U.S. Patents No. 3,268,313 and No. 3,510,393.
4. Carroll-Porczynski, C. Z., "Inorganic Fibres," National Trade Press, Ltd., 1958, p. 13.
5. Beach, N. E., *Plastic Laminate Materials*, Foster Publishing, 1967, p. 33.
6. Weyl, W. A. and Marboe, E. C., *The Constitution of Glasses*, Vols. I, II, and III, Interscience Publishers, New York, 1962.
7. Levin, E. M., Robbins, C. R., and McMurdie, H. F., *Phase Diagrams for Ceramists*, American Ceramic Society, 2 vols, 1964 and 1969.
8. U.S. Patents No. 3,819,387, No. 3,785,833, and No. 3,785,834.
9. U.S. Patent No. 4,542,106.
10. French Patent No. 1,435,073.
11. U.S. Patent No. 3,816,107.
12. Morey, G. W., *The Properties of Glass*, Van Nostrand Reinhold Co., New York, 1938, p. 224.
13. Di Domenico, M., Jr., "Wires of Glass," *Industrial Research Magazine*, p. 50, Aug. 1974.
14. Mohr, J. G., "Use of the Differential Expansion Technique for Study of Internal Strain in Reinforced Plastic Structures," 15th Annual SPI RP/C Institute *Proceedings*, 1960, Section 10-E.
15. Scholes, S. R., *Modern Glass Practice*, Industrial Publications, 1941, see Frontmatter.
16. Mohr, J. G. et al., *SPI Handbook of Technology and Engineering of Reinforced Plastics/Composites*, Van Nostrand Reinhold Co., New York, 1973.
17. Richardson, M. O. W., *Polymer Engineering Composites*, Applied Science Publications Ltd., London, 1977.
18. PPG Fiber Glass Reinforcements Market Series: Reinforced Plastics . . . By Design.
19. Bralla, J. G., *Handbook of Production Design*, McGraw-Hill, New York, 1978, see section on Design of PR/C Products.
20. U.S. Patent No. 4,220,496.
21. Lubin, George, *Handbook of Fiberglass and Advanced Plastics Composites*, Polymer Technology Series of the Society of Plastics Engineers, Inc., Van Nostrand Reinhold Co., New York, 1969.

22. Lubin, George, *Handbook of Composites*, sponsored by Society of Plastics Engineers, Inc., Van Nostrand Reinhold Co., New York, 1982.
23. Simpson Gumpertz & Heger Inc., *Structural Plastics Design and Selection Manuals*, monitored by Task Committee on Design, Structural Plastics Research Council, American Society of Civil Engineers, and sponsored by U.S. Department of Transportation, U.S. Department of Housing and Urban Development, and Participating Industrial Corporations. Superintendent of Documents, U.S. Government Printing Office, Washington, D.C. 20402.
24. Grayson, M. and Kroschwitz, J. I., *Encyclopedia of Polymer Science and Engineering*, John Wiley & Sons, Inc., New York, 1970.
25. Cobb, M. G. and Mormile, P. J., "The Development of Vapor Permeation Cure (VPC) and Vapor Injection Cure (VIC) Coatings for RIM and SMC Substrates," 40th Annual Conference Preprint, Reinforced Plastics/Composites Institute, The Society of the Plastics Industry, Inc. Jan. 28–Feb. 1, 1985.
26. Hamner, N. E., "Non-Metals Sections, Corrosion Data Survey," National Association of Corrosion Engineers, 1975.
27. Stone, F. G. A. and Graham, W. A. G., *Inorganic Polymers*, Academic Press, New York, 1962, pp. 1, 68, and 189. See also Holliday, L., *Ionic Polymers*, Halsted Press, 1975. Also Buckley, J. D. (editor and compiler), "Advanced Materials: Composites and Carbon, American Ceramic Society Symposium," American Ceramic Society, 1973.
28. Katz, H. S. and Milewski, J. V., *Handbook of Fillers and Reinforcement for Plastics*, Van Nostrand Reinhold Co., New York, 1978.

15

BASALT FIBERS

R. V. Subramanian

Washington State University
Department of Mechanical and Materials Engineering
Pullman, Washington

CONTENTS

1. INTRODUCTION

Basalt fibers possess the desirable properties of glass fiber reinforcements and have some additional advantages. These include better temperature resistance, better adhesion to epoxy resins, and superior resistance to alkali solutions. The chemical composition of the fibers is determined by the native basalt rock that is used as raw material, which can show small variations from the typical composition represented by: SiO_2 : 50%, Al_2O_3 : 15%, TiO_2 : 3%, FeO : 11%, Fe_2O_3 : 2%, MnO : 0.2%, CaO : 9%, MgO : 5%, K_2O : 1%, Na_2O : 3%, $P_2O_5 < 1\%$. The strength and moduli of virgin basalt fibers have been found to be comparable to those of E-glass fibers produced and tested under the same conditions in the laboratory. The fibers are brown in color because of their iron content, which may prevent their use in some applications. At the present time, they are not being produced commercially in the United States, but a prototype unit for their production has been set up at Washington State University. It is expected that, based on this study, a pilot plant for the production of larger quantities will soon be in operation. Present estimates indicate that the cost of basalt fibers will be somewhat less than that of glass fibers.

2. MANUFACTURE

Basalt fibers are readily made by the same techniques and equipment used for the production of glass fibers. Both continuous fibers and staple fibers can be produced from basalt rock. Staple fibers that average 30 mm in length and are useful for mats and polymer reinforcements are produced by a "nozzle-blow" process. (8.2.1).* This consists of a melting tank furnace that feeds an electrically heated platinum–rhodium bushing containing a large number of nozzles. A steam jet (or compressed air jet) is located below the nozzles and creates an air stream that moves parallel to the direction of extrusion. This pulls the basalt fibers from the nozzles and tears them apart into different lengths. The fibers fall onto a porous drum that is under vacuum. Plants for the production of 5–10 tons/day of staple basalt fibers by this process have been built in Germany. Superfine fibers, 0.2–4 μm diameter, which are excellent

*Numbers in parentheses refer to abstracts in section 8.

as thermal insulation, have reportedly been manufactured in Russia.

Continuous basalt fibers are produced in standard glass fiber production equipment. The rock is melted and fed to electrically heated platinum–rhodium bushings containing 200 to 400 openings. The fibers are pulled from the apertures by winding on a drum. Size of the fiber is controlled by the temperature of the melt (1250–1400°C) and speed of the collecting drum. The usual size is 10–15 μm. Desired sizes and coatings of various kinds can be put on the fiber before it is wound on the drum. The type of coating is determined by the intended application of the fiber. Silanes and conventional sizes can be applied by the usual methods. (8.2.4, 8.3.4, and 8.3.5).

The use of basalt rock as a raw material has several advantages. The rock is available in large quantities in each of a number of locations in the United States. Even though basalts from different locations vary somewhat in composition, each location has uniform composition, and almost all will produce fiber of comparable properties. No mixing of ingredients prior to melting is necessary, and only crushing and washing is needed to prepare the raw material. The selection of the appropriate basalt rock will enable the production of fibers of specific properties (e.g., alkali resistance).

3. PROPERTIES

Chemical compositions and strengths of fibers from basalts from various locations are shown in Table 15-1. A skein of basalt fibers with a close-up view revealing the uniformity of fine fibers is shown in Fig. 15-1. Figure 15-2 shows the tensile strength of basalt fibers drawn at various temperatures.

The strength of basalt fibers has been improved by appropriate process modifications (8.3.5). Virgin fiber strengths in excess of 3.5 GPa (500,000 psi) have been achieved so that the measured strengths are close to that of E-glass fibers. Research on further improvements in strength is continuing.

The fiber fracture stress is bimodally distributed and can be modelled by Weibull statistics. The gage length dependence of strength, the severity of flaws and their distribution are altered by conditions of fiber drawing. The inert medium provided by gases, such as, nitrogen, argon or helium, minimizes the oxidation of ferrous to ferric oxides, thus reducing nucleation, by the latter, of crystallites in the otherwise glassy fiber, which can act as stress raisers. (8.3.9)

The Young's modulus of the fibers from different basalt rocks, measured by thin-line ultrasonics, is presented in Table 15-1. The measured values of moduli varying between 78 and 90 GPa are higher than the modulus of E-glass fibers. These values are comparable to the 90 GPa reported for fibers made from Berestovets basalt in the Soviet Union, but the strength shown above is higher than that reported by the Soviet scientists (8.3.1).

The softening temperature of basalt fibers is above 1000°C. The temperature resistance of the fibers is thus higher than that of glass fibers.

Tests of alkali resistance have shown that basalt fibers have better resistance than E-glass (8.3.3). After treatment with saturated $Ca(OH)_2$ at 100°C for 4 hours, E-glass fibers were reduced in diameter by about 5%, while basalt fibers suffered less than 1% reduction. Similar indications were obtained when fibers were immersed for 2 months in a "cement effluent" solution containing NaOH (88 g/l), KOH (3.45 g/l), and $Ca(OH)_2$ (0.48 g/l) that had a pH of 12.5. Scanning electron micrographs of the fibers showed the E-glass fibers to be heavily etched, with the formation of many holes; basalt fibers were devoid of any such evidence of severe alkali attack (8.3.3). The reinforcement of concrete by basalt fibers is presently being investigated to evaluate long-term resistance of basalt fibers in cement. The fiber fracture surface, examined by SEM, shows a characteristic mirror zone, whose radius bears an inverse square root relationship to the fracture stress. The measurement of mirror zone radius on the exposed fiber fracture surfaces in failed cement composite specimens enables the measurement of fiber-fracture stress *within* the composite. The durability of fibers and their strength retention when embedded in the corrosive, alkaline cement are thus evaluated. (8.3.10)

Table 15-1. Chemical composition and strengths of fibers from basalts obtained from various locations.

Sample number, rock flow, source	X-6, Lolo flow, Whitman County, Wash.	RC-11, Imnaha, Rocky Canyon, Idaho	K-9068, Elephant Mt., Saddle Mt., Wash.	K-9064, Middle Yakima, Saddle Mt., Wash.	0-2, Sweet Home, Ore.	K-9048, Pomona, Saddle Mt., Wash.	RC-3, Imnaha, Rocky Canyon Idaho	K-9017, Lower Yakima, Saddle Mt., Wash.	BCR-P, Yakima Flow	E-glass[b] Owens-Corning
Analysis, %										
SiO_2	49.10	49.41	50.02	50.49	50.50	50.86	51.41	53.61	54.50	52.20
Al_2O_3	13.80	17.92	13.29	13.62	16.00	15.18	15.14	15.14	13.60	14.80
TiO_2	3.16	2.41	3.48	3.06	2.17	1.69	2.22	1.84	2.20	0.00
Fe_2O_3	2.00	2.00	2.00	2.00	–	2.00	2.00	2.00	2.00	0.30
FeO	11.98	9.66	13.27	12.62	13.80	9.21	11.32	9.60	10.50	0.00
MnO	0.21	0.17	0.21	0.22	–	0.19	0.21	0.18	0.18	0.00
CaO	9.43	9.06	8.59	8.48	10.00	10.62	9.33	8.43	6.92	18.70
MgO	5.25	5.64	4.28	4.45	4.30	6.49	5.05	4.98	3.46	3.30
K_2O	1.26	0.81	1.35	1.54	0.35	0.80	0.68	1.14	1.70	0.00
Na_2O	3.09	2.57	2.95	2.93	3.20	2.62	2.28	2.73	3.27	0.30
P_2O_5	0.68	0.35	0.55	0.58	–	0.33	0.36	0.35	0.36	0.20
B_2O_3	0.00	0.00	0.00	0.00	0.00	0.00	0.00	0.00	0.00	10.20
Temperature of drawing, °C	1325	1300[a]	1250	1250	1250	1250	1250	1250	1360[a]	1250[a]
Speed of drum (rpm)	515	250	370	515	370	250	370	515	250	250
Number of samples	40	29	30	18	25	20	20	20	31	23
Average diameter of fiber (μ)	13.0	12.2	13.5	9.0	11.4	11.8	11.3	10.2	12.1	12.2
Average tensile strength										
(GPa)	1.97	1.99	2.13	2.23	2.08	2.08	2.25	2.45	1.97	2.52
psi	286,000	288,000	309,000	323,000	302,000	301,000	326,000	355,000	285,000	365,000
Young's modulus										
(GPa)	82.76	77.93	77.93	87.59	90.34	82.76	87.59	87.59	71.03	81.38
(millions psi)	12.0	11.3	11.3	12.7	13.1	12.0	12.7	12.7	10.3	11.8

[a] Air jet used.
[b] Produced in-house from Owens-Corning E glass marbles.

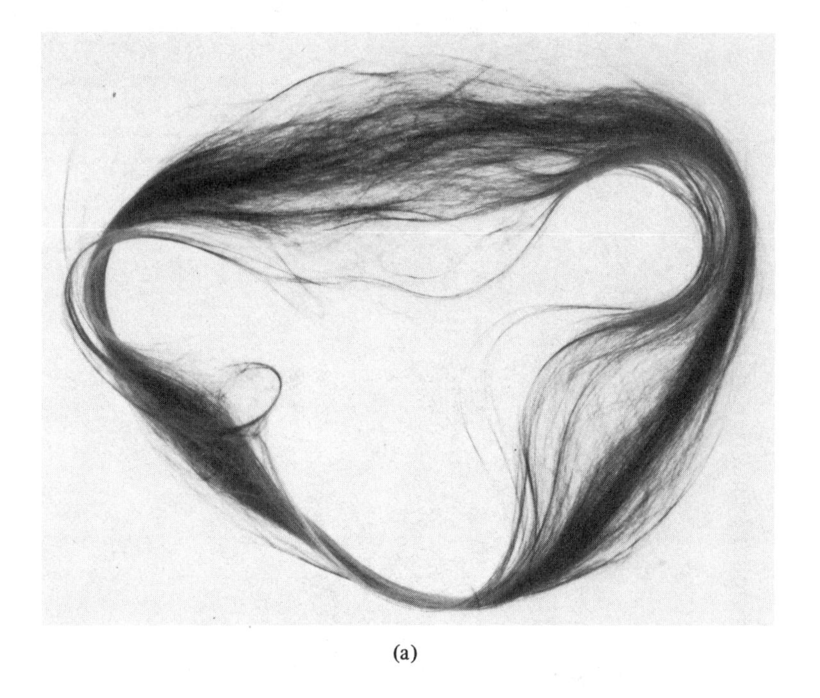

(a)

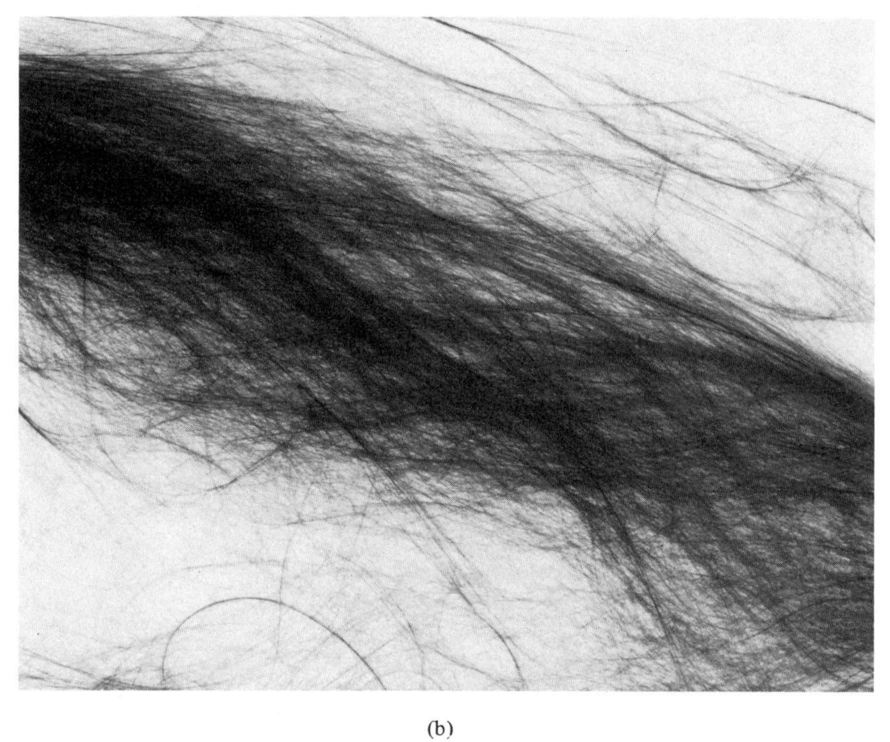

(b)

Fig. 15-1. A skein of basalt fibers (a) and an enlarged section showing uniformity of the fibers (b).

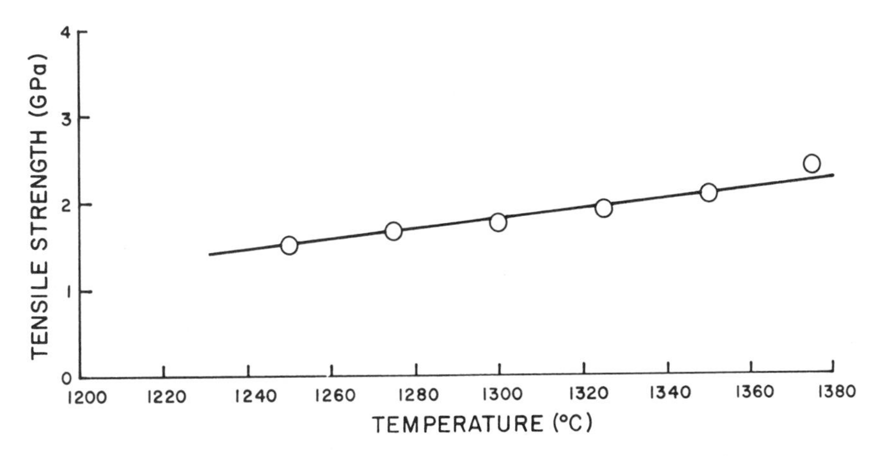

Fig. 15-2. The variation of tensile strength of virgin basalt fibers as a function of drawing temperature (8.3.4).

4. SUPPLIERS

There are now no commercial sources of basalt fibers in the United States, but it is hoped that a pilot plant will soon be in operation for their production. Further information may be obtained from the Department of Mechanical & Materials Engineering, Washington State University, Pullman, WA 99164. Present indications are that the cost of basalt fibers should be less than that of comparable glass fibers.

Elsewhere in the world, manufacture of basalt fibers for use in reinforcement in composites seems to have been achieved, as noted above, only in the Soviet Union, though basalt wool insulation, similar to rock wool, is produced in Germany and the United Kingdom by a few manufacturers:

Deutsche Basaltsteinwolle GmbH
Rodetal 40
3406 Bovenden 1, Germany

Basalan-Isolierwolle GmbH
KG Industriestr. 4
4403 Hiltrup, Germany

Cape Insulation Ltd.
Kerse Road
Stirling, Scotland FK77RW

5. HISTORY AND FUTURE

As noted, basalt fibers have been and are being produced in the Soviet Union and Germany. As long ago as 1959, the Germans reported on installations for the production of mineral fibers from basalt (8.2.1). More recently, in 1973, the official Russian news agency, Novosti, reported that "use of basalt fibers, made from basalt rocks, is being developed extensively in the Soviet Union. Several large plants for fiber output are being built."

Experimental production of basalt fibers from basalt of the Pacific Northwest at Washington State University has shown the feasibility of commercial production of quality fibers from basalt resources occurring in the United States. The ready availability of abundant supplies of raw material, and the resultant decrease in cost of production compared to glass, should provide manufacturers the incentive to initiate a basalt fiber facility. This cost decrease should arise from not having to purchase, store, and mix many individual ingredients, as is necessary to produce glass.

6. APPLICATIONS

The effectiveness of basalt fibers as reinforcement for polymers has been studied with epoxy resins (8.3.3 and 8.3.4). The results show that basalt fibers can be substituted for glass fibers in all composites where their brown color is not objectionable. This conclusion is based upon tests of adhesion of basalt and glass fibers to an epoxy matrix and upon determination of the relative improvement in adhesion and compos-

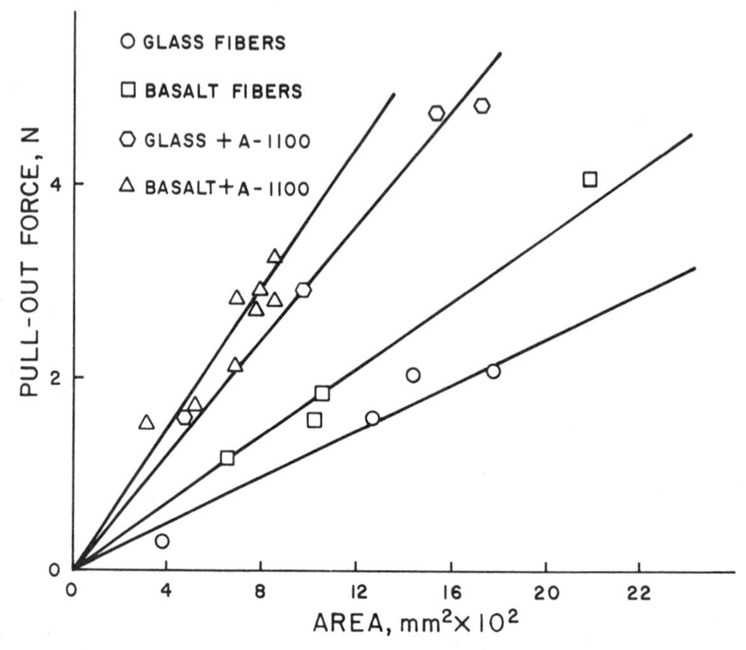

Fig. 15-3. Pull-out force for fibers embedded in epoxy matrix plotted against area of contact (8.3.4).

ite properties as a result of treatment with silane coupling agents.

The force required for pull-out of fibers embedded in an epoxy matrix in single-fiber adhesion tests is plotted in Fig. 15-3. It is seen that basalt fiber adhesion to the epoxy resin is somewhat superior to the adhesion of E-glass fibers. This superior adhesion is true for both silane-coated and uncoated fibers. The addition of silane adhesion-promoting coatings increases the tensile strength and tensile modulus of basalt fiber–epoxy composites. The data in Figs. 15-4 and 15-5 illustrate these increases when Union Carbide Silane A-1100 (γ-aminopropyl trimethoxy silane) was used. The matrix resin was a DGEBA resin (EPON 828) cured with metaphenylene diamine (MPDA).

The similarities are thus much more striking than differences between the two fibers in reinforcement of thermoset polymers. Tensile strengths of the fibers are almost identical, and the elastic modulus of the basalt fibers is a little higher, which could indicate its preferential use in some applications.

The effects of silane treatment in improving polymer adhesion to basalt fibers are seen directly in fiber pull-out tests (8.3.6, 8.3.7). In this study, the fibers were treated with a num-

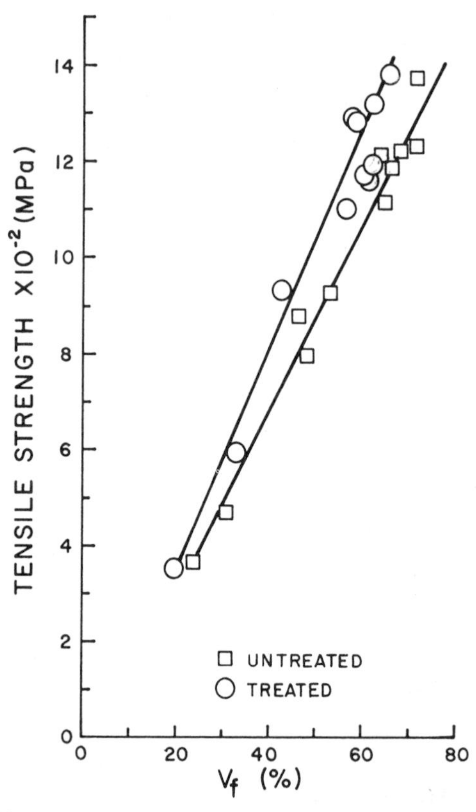

Fig. 15-4. Tensile strength of epoxy resin composite (DGEBA cured by metaphenylenediamine) reinforced by untreated and treated (Silane A-1100) basalt fibers (8.3.4).

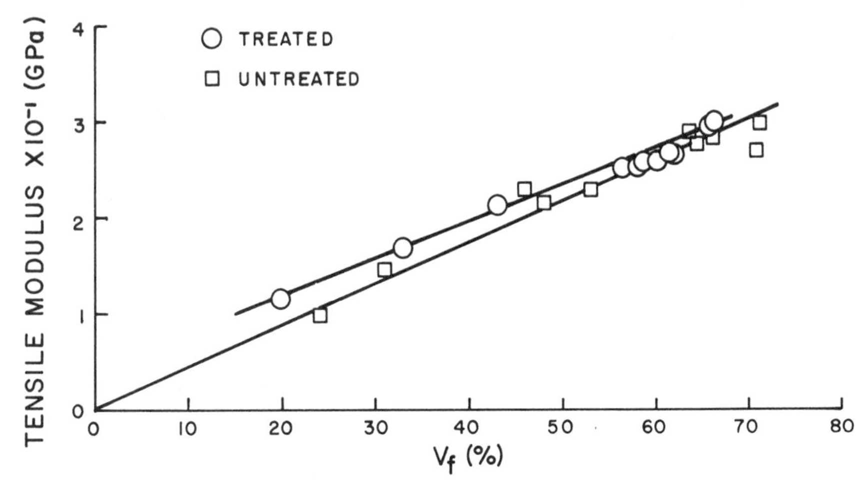

Fig. 15-5. Tensile modulus of epoxy resin composite (DGEBA cured by metaphenylene-diamine) reinforced by untreated and treated (Silane A-1100) basalt fibers.

ber of ionic and nonionic silane coupling agents under a variety of experimental conditions, changing solution pH and concentration, aging time, and fiber treatment time. The values of pull-out stresses were measured for treated and untreated fibers embedded in both epoxy and polyester matrix resins. The surfaces of treated and untreated fibers and those of pulled-out specimens were examined by scanning electron microscopy. Details of debonding and modes of failure revealed in scanning electron micrographs were also noted.

The pull-out tests showed that silane coupling agents are effective in improving interfacial bond strength in basalt fiber systems, and confirmed that basalt fiber has excellent potential as reinforcement for polymer composites. The controlling effects of silane hydrolysis, condensation, orientation on the fiber surface, and chemical bonding of silane to the fiber and polymer were revealed in the trends of interfacial bond strengths with experimental variables.

An interesting observation was the effect of the presence of iron oxides in basalt fiber. This effect was manifested in the presence of two maxima pull-out stress values when ionic coupling agents such as Dow Corning's Z-6031 were applied from silane solutions at different pH conditions. The first maximum, around pH 2, is attributed to proper orientation of the silane coupling agent for effective bonding with the surfaces at SiO_2 sites whose isoelectric point is about 2. The second maximum, at pH 9 to 10, is similarly due to bonding at the iron oxide sites on the surface whose isoelectric point is around 10 (8.3.7).

Basalt powder can be used as filler in polymer composites, as shown in a study with a polyester matrix (8.3.8). It is observed that good strength retention under wet conditions is achieved by using basalt reinforcement. Here again, the effectiveness of cationic silane coupling agents in improving composite properties is seen to be modified by the pH of silane application. As in the case of fiber pull-out stresses, wet-strength retention is high in the alkaline range because of the surface concentration of iron oxides on basalt.

7. UTILIZATION

The possible uses of basalt fibers are numerous: in air filters, heat insulation at high temperatures, insulation of vibrating surfaces, sound insulation, reinforcement in concrete, mineral paper or cardboard, reinforcement of thermoset resins with better adhesion than glass fiber, reinforcement of polyester resins, and as a building material with emphasis on the replacement of asbestos. The Russians have reported the production of microfine fiber (0.4–2 μm diameter), in a size range that will enable it to compete successfully with asbestos fibers.

The current search for substitutes for asbestos, to alleviate the health problems connected with its use, should lead to extensive testing of basalt fibers. If basalt fibers are found suitable as a substitute for asbestos fibers, the potential uses are very large. The superior alkali resistance of basalt fibers, indicated in laboratory tests so far, makes this highly probable.

The alkali resistance of basalt fibers promises an excellent potential for use in the reinforcement of concrete. These fibers are very much cheaper to produce than the special alkali-resistant glasses that have been formulated and tested.

8. ABSTRACTS OF LITERATURE

8.1 Review

Raff, R. A. V., *Engineering/Mining Journal*, pp. 76–77 (Feb. 1974).

An excellent review article that lists 50 pertinent references to the literature on basalt fibers and presents a clear picture of the technology and uses throughout the world and the potential for its production in the United States.

8.2 Production of Fibers

8.2.1 Kaswant in Kastellaun, "Modern Installations for the Production of Mineral Fibers using the Pressure Nozzle Process," *Sprechsaal für Keramic, Glass Email* 91: 577–586, (1958); *CA* **53,** 7464 (1959).

This is a most detailed description of a unit for the commercial production of 10 tons of basalt fiber per day. The unit is fed with basalt chips and produces staple fiber from a platinum–rhodium nozzle. Detailed cost figures including wages, materials, and power costs are given but will require revision in order to be useful today.

8.2.2 Dubroskie, V. A., Rychko, V. A., Bachilo, T. M., and Lysyuk, A. G., "Basalt Melts for the Forming of Staple Fiber," *Steklo i Keramika*, No. 12, pp. 18–20 (Dec. 1968).

The most important properties of basalt melts for fiber production are viscosity and crystallization characteristics. Staple fibers can be produced from melts whose viscosity at 1400°C is at least 90 poise and at 1250°C below 350 poise. The upper limit of crystallization of the melts is 1250°C.

8.2.3 Darenskie, V. A., Yu, N. Dem'yanenko, Kozlovskii, P. P., Manhurnet, K. V., Kukarkin, A. I., Dzhugaryan, R. T., and Badalyan, D. S., "Producing Basalt Staple Fiber," *Steklo i Keramika*, No. 1, pp. 38–40 (Jan. 1968); *CA* **68,** 195933 (1968).

A description of a production plant in Armenia SSR gives some details of the equipment used. The temperature of the melt is 1490 ± 10°C. The fusion level in the melting tank is maintained at 195 mm. The level of the melt over the nozzle is 155–160 mm. The platinum–rhodium feeder is electrically heated and has 10 nozzles, each having a diameter of 3.0 ± 0.2 mm. The pressure of the compressed air fed to the blowing from the nozzle to the entry into the blowing head is 100 mm. The output is 2100 kg/day.

8.2.4 Subramanian, R. V., Austin, H. F., Raff, R. A. V., Sheldon, G., Dailey, R. T., and Wullenwaber, D., "Use of Basalt Rock for the Production of Mineral Fiber," Pacific Northwest Commission, Annual Report, June 1975, College of Engineering, Washington State University, Pullman, WA 99164.

A detailed description of an experimental unit for the production of continuous basalt fibers from basalt found in Washington is given. Equipment for testing physical properties of the fibers is also described. The latter half of the report gives a comprehensive economic analysis, market survery, and cost estimate for a plant to produce basalt fibers.

8.3 Properties of Basalt Fibers

8.3.1 Andreevskaya, G. D. and Plisko, T. A., "Some Physical Properties of Continuous Basalt Fibers," *Steklo i Keramika*, No. 8, pp. 15–18 (Aug. 1963); *CA* **60,** 1470 (1964).

Elastic modulus of 13–18 μm fibers was determined by a frequency method and was found to be 9230 kg/mm^2. Strength of fibers in relation to diameter were measured and given: 8 μm—260 kg/mm^2, 10 μm—200, 13 μm—125,

14 μm—105, 15 μm—100, 16 μm—98, 17 μm—92.

8.3.2 Dubovkaya, T. S. and Kosmina, N. E., "Chemical Stability of Staple Basaltic Fibers," *Strukt. Sostav. Svoistva Formovanie Steklyanykj Volokon*, pp. 1, 140, (1968), *CA* **73,** 112426 (1970).

Basalt fibers were found to be more stable in the presence of alkali than in acids.

8.3.3 Subramanian, R. V., Austin, H. F., and Wang, T. J. Y., "Use of Basalt Rock for the Production of Mineral Fiber," Pacific Northwest Regional Commission Annual Report (Sept. 1976), College of Engineering, Washington State University, Pullman, WA 99164.

A study of process parameters in the production of basalt fibers, properties and performance of basalt fiber reinforcements in composites, and modification of basalt fiber properties for reinforcement of polymer matrices.

8.3.4 Subramanian, R. V., Wang, T. J. Y., and Austin, H. F., "Reinforcement of Polymers by Mineral Fibers from Basalt Rocks" *SAMPE Quarterly*, pp. 1-10 (July 1977).

8.3.5 Subramanian, R. V. and Austin, H. F., U.S. Patent 4149866 (1979).

A method of improving the tensile strength of drawn fibers produced from molten basalt rock is patented. Strength is increased by minimizing the ferric iron content of the fibers, which is achieved by drawing them in an inert or reducing atmosphere.

8.3.6 Shu, K. H. "Interfacial Bonding in Basalt Fiber–Polymer Composite Systems," M.S. Thesis, Washington State University (1978).

Pull-out stresses were measured for basalt fibers embedded in epoxy or polyester matrix resins. The values are higher than those reported for E-glass fibers. The effects of iron and other metal oxides present in basalt fibers are manifested in the dependence of pull-out stress on pH of silane treatment. The contributions of radial compressive stresses caused by thermal mismatch and resin shrinkage during curing are also evident.

8.3.7 Subramanian, R. V. and Shu, Kuang-Hua H., in *Molecular Characterization of Composite Interfaces*, H. Ishida and G. Kumar (eds.), Plenum, New York (1985), p. 205.

The interfacial bond strength in basalt fiber–polymer systems has been investigated using a single fiber pull-out test method. A number of ionic and nonionic silane coupling agents were employed to study improvements in fiber–matrix adhesion. It is shown that silane treatment improves bonding in basalt fiber systems, and that basalt fiber has an excellent potential as a reinforcing fiber for polymer composites.

8.3.8 Subramanian, R. V. and Austin, H. F., *Int. J. Adhesion and Adhesives* 1: 50 (1980).

Basalt powder is used as a reinforcement for polyester. Retention of flexural strength after water boil is found to be excellent. The effects of various silane coupling agents on flexural strength are reported.

8.3.9 V. Velpari, Ph.D. Thesis, Washington State University (in preparation).

A study of (i) the strength properties of basalt fibers controlled by conditions of fiber drawing, and (ii) reinforcement of cement by basalt fibers.

8.3.10 Subramanian, R. V. and Velpari, V., "Strength and Durability of Basalt Fibers," American Ceramic Society 39th Pacific Coast Regional Meeting, Oct. 22, 1986, Seattle, WA.

The distribution of the strength of basalt fibers is investigated for fibers melt spun from basalt rock under different drawing conditions in oxidative and inert atmospheres. The effects of ferrous ion oxidation and nucleation of crystallization by ferric oxides are revealed in the statistics of fiber failure, and variation in mirror constants evaluated from the plots of fiber fracture stress versus inverse square root of the mirror zone radius, r, on the fracture surface. The in-situ strength of fibers in cement composites is followed over an extended period by measurement of r. Network dissolution, and interactions on the fiber surface in the composite are investigated by electron microprobe analysis.

Section V
High Modulus Filaments

16

CONTINUOUS METALLIC FILAMENTS

John J. Toon
Shielding/Textile Applications
National-Standard Co.
Corbin, Kentucky

Contents

1. INTRODUCTION

Metal fibers have been used by man for centuries, gold fibers having been used as decorative threads in garments of the Pharaohs. In 1981, the use of metal fibers as conductive fillers for plastics was introduced by Bekaert Steel Wire Corporation. Initial applications included the production of conductive plastics for Electromagnetic Interference (EMI) and Electrostatic Discharge (ESD) end-use applications.

A metal fiber is generally considered to be a metallic filament 50 μm or less in diameter. Metal fibers may be produced by a variety of means, the most common of which is the bundle-drawing process. Fibers as fine as 2 μm in diameter have been produced by this method, with 4 μm fibers available commercially.

Metal fibers are available in a variety of forms such as sliver, tow, chopped fibers, nonwoven fabrics, woven fabrics, and yarns, and in combination with textile fibers as yarn and fabric.

The advantage of using metal fibers as conductive fillers is that antistatic and shielding properties can be achieved with very low fiber loadings. In this way the physical properties of the plastics are largely maintained. Disadvantages in using metal fibers in plastics are in most cases cost considerations and the degree of expertise required to effectively utilize the fibers in minimal amounts.

Plastics are playing an ever increasing role in the semiconductor industry. Items such as business machine housings, tote-trays, storage bins, and work surfaces, as well as packaging and storage materials, are made from a variety of plastic materials. While plastic materials have been selected over other materials due to their advantages of weight, cost, and styling potential, problems exist with these materials. Unprotected plastic materials are susceptible to static electricity generation and are also transparent to electrostatic discharge and electromagnetic interference. It is in these two areas that metal fibers can provide practical and economical solutions.

Precompounded plastic pellets containing metal fibers can be used to injection-mold plastic parts that will exhibit electromagnetic interference shielding effectiveness in the range of 30 dB. The use of 5% to 10% by weight metal fibers has been found sufficient for this purpose. Electrostatic discharge problems may be solved with the same approach except that the metal fiber content can be reduced to 3% by weight. In order to achieve the maximum effectiveness, molding conditions must be such that enough metal fibers are present on the molded surface to form a conductive network.

Woven or knitted textile metal fiber fabrics can also be used to achieve EMI and ESD protection through lamination and thermoforming techniques. A further use would be in RIM type applications. Textile metal fiber veil-mats formed through nonwoven textile manufacturing techniques can also be used to provide EMI and ESD protection.

Advantages of using metal fibers for these applications include the following:

1. Owing to the low fiber loading requirements, there is minimal alteration of desirable resin system properties.
2. Low metal fiber loadings result in cost-effective shielding.
3. The shielding imparted to the resin system is permanent for the life of the molded part.
4. The use of metal fibers in injected-molded parts results in cosmetically acceptable surfaces without further processing or surface treatments.
5. Unlike other conductive fillers, plastic systems shielded with metal fibers can be pigmented to yield relatively light and bright colors.
6. Properly chosen metal fibers are not corrosive and will not react with normal resin system fillers and components.
7. The use of metal fibers will generally result in lower equipment contamination than occurs when other conductive fillers are used.
8. Metal fibers retain their shielding efficiency over a wide frequency range.
9. Excellent-quality parts with good shielding properties can be molded without equipment or mold modification.
10. Metal fibers can be combined with other conductive or nonconductive fillers or reinforcements to produce hybrid composites systems for special-purpose applications.

The price of metal fibers can vary greatly, depending upon manufacturing method, raw material costs, fiber diameter and length, fiber form, and end-use application. In general, bundle-drawn stainless steel fibers in forms suitable for conductive plastic applications will be priced in the \$40 to \$50 per pound range. Extruded nickel fiber pricing currently is not firm, but is thought to be in the \$15 per pound range. Drawn wire and steel wool–like structures are available at much lower prices, which are determined in much the same manner as those of the drawn and extruded fibers.

Properties of metal fibers are determined by raw materials, manufacturing process, or surface treatments. Cross section, surface properties, and surface reactivity can be influenced by the manufacturing process, while physical, chemical, and mechanical properties are more influenced by the raw material.

2. PATENT HISTORY OF METAL FIBERS

The technique of producing fine metal fibers by the bundle-drawing process was first patented in 1914 by Bergman-Elektrizitatswerke in Germany (Patent No. 286,717). Other patents relative to this process was granted to Everett (U.S. Patent Nos. 2,050,298 and 2,077,682) in 1936–1937. In 1940, Pipkin was awarded a patent for the production of fibers for use in photoflash bulbs (U.S. Patent No. 2,215,477).

Other patent development of interest in this field includes French patent No. 1,006,452 granted in 1952, the British patent No. 855,876 of 1960, and the Levi patent (U.S. patent No. 3,029,496) of 1962. A composite superconducting wire obtained by bundle-drawing is described in U.S. Patent No. 3,513,537 and several patents issued to The Brunswick Corporation.

One of the first applications for bundle-drawn metal fibers was antistatic textiles. In this connection a basic patent was issued to Burlington Industries (U.S. Patent No. 3,987,613). This patent was later assigned to NV Bekaert SA of Belgium.

Other U.S. patents specific to the manufacture of antistatic textile yarns were issued to Riegel Textile Corporation (U.S. Patent Nos. 3,838,543 and 3,703,073) and The Brunswick Corporation (U.S. Patent No. 3,678,675).

Patents related to the incorporation of metal fibers into security papers have been issued to the French company Ayomori-Prioux (U.S. Patent No. 4,265,703) and in the British patent No. 2,050,664, granted to NV Bekaert SA.

Patent applications relative to the use of metal fibers to produce conductive plastics and electromagnetic shielding have been made by NV Bekaert SA.

Additional patents that provide considerable detail relative to the bundle-drawing process are the Webber patents (U.S. Patent Nos. 3,277,564 and 3,379,000) originally assigned to the Roehr Products Company and subsequently to the Brunswick Corporation.

Later patents that describe alternative production methods are the Kobuku patent (U.S. Patent No. 3,643,304) and the Hamada patents (U.S. Patent Nos. 3,844,021 and 4,044,447). These patents have been assigned to Nippon Seisen Co., Ltd.

More recently, National-Standard Company has announced the availability of nickel fibers in the form of loose fibers as well as sintered or resin bonded web structures. These fibers are produced by a proprietary process that converts extruded nickel-oxide fibers into nickel fibers through an oxide-reduction process. This method of manufacture is described by U.S. Patent Nos. 3,671,228, 4,312,670, and 4,298,383.

3. METAL FIBER MANUFACTURING METHODS

3.1 Bundle Drawing

Bundle drawing is a modification of basic wire-drawing methods to allow a plurality of wires to be drawn simultaneously and continuously. Development of this manufacturing method was an outgrowth of the Apollo space program. Bundle drawing made the production of fine metal fibers practical and more economical than previously employed methods, producing fibers in the 2 to 50 μm diameter range.

Brunswick Corporation was a pioneer in the development of bundle drawing and still occupies a predominant position in the United States. NV Bekaert, SA in Belgium, and Nippon Seisen Co., Ltd. in Japan bundle draws metal fibers under license from Brunswick.

The bundle-drawing process begins with stringent raw material selection. In this case, wire rod of 6 to 8 mm diameter is selected, based on highest purity and absence of foreign particles that would cause a deterioration of properties in the finished fiber. This rod is reduced in diameter approximately ten times by a wire-drawing operation determined by the type of metal processed. The next step is to apply a coating or sheath to the surface of this drawn wire, the purpose of which is to prepare the wire for the composite formation that is the state in which the bundle drawing will actually take place.

The type of coating and its application must be carefully controlled and must meet some very stringent requirements:

- The coating or sheath must adhere perfectly to the wire and at the same time be easily removed by chemical, mechanical, electrical, or thermal means at the end of the process without detrimental effect to the fiber.
- The coating must be compatible with the reduction process, providing, uniform plastic flow necessary to produce uniform fiber.

The coated wire is then drawn to a fractional millimeter diameter.

The next stage represents the starting point of the bundle-drawing process, in that a plurality of the fine coated wires are assembled in a metal tube or sheath. The number of wires assembled in the tube will vary, depending upon the end-use application of the fiber, with approximately 100 wires being a practical minimum.

The composition of the metal tube does not necessarily have to be the same as that of the coating, but it must be compatible with the wire coating and meet the same performance criteria.

The composite formed by the fine coated wires and tube must be of a quality suitable for subsequent drawing to a fine diameter. These finely drawn composite wires may be combined with other composite wires in tubes or sheaths to form new composite wires that are then drawn to a fine diameter. This process is repeated until the final fiber size and fiber bundle are obtained. At intervals during the wire drawing and the composite drawing, heat-treating operations are necessary to soften the wire and composite so that further drawing operations can occur.

Once the final fiber/bundle size is achieved, it is then necessary to separate the fiber from the drawing matrix, which is composed of the metal tube or sheath and the wire coating. This can be accomplished by a combination heat and chemical treatment utilizing an acid such as nitric acid to dissolve the matrix. After the matrix is dissolved, the fiber is neutralized, rinsed, and dried. This operation is usually accomplished by immersion in a series of heated baths in a continuous manner. The metal fiber thus obtained is then ready for final processing into the required fiber form.

The intent of this process description has been to describe the manufacturing process in general and to convey an appreciation of the expertise, asset employment, time, and expense required to produce metal fibers. As an example, one ton of drawn composite wire may only yield 1000 pounds of metal fiber once the matrix material is removed. Another serious problem faced by the manufacturer is disposal of the effluent from the matrix removal baths.

Drawn metal fibers are somewhat expensive for obvious reasons. Please see Fig. 16-1 for a qualitative description of cost versus size.

The bundle-drawing process has been applied to a wide range of metals and alloys, a general listing of which follows:

- Ferrous base alloys
  - —stainless steels
- Nickel and cobalt base super alloys
- Titanium and titanium alloys
- Refractory metals
  - —tantalum, niobium
- Precious metals
  - —silver, gold, platinum

Fibers produced by the process are in a diameter range from submicron (less than 0.00004 in.) to 100 μm (approx. 0.004 in.), which is the upper limit corresponding to the economically competitive process of conventional wire drawing. The fibers are characterized by a degree of surface roughness and nonuniform shape. The product can be additionally characterized in terms of a coefficient of variation (standard deviation/mean) of area of cross section, which can be controlled to some degree by the producer. The coefficient of area variation depends on process parameters including:

- Volume fraction of fiber and matrix materials
- Mechanical property match of fiber and matrix materials
- Degree of working between anneals
- Fiber diameter required

In addition, certain materials such as steels must be of exceptionally high purity to prevent serious degradation in fiber properties resulting from nonductile inclusions. Such effects are dramatically illustrated in Fig. 16-2 for two steels of differing purity but processed identically. If similar data are presented in a different way, Fig. 16-3, it becomes obvious that the length of the fiber being tested is important, and

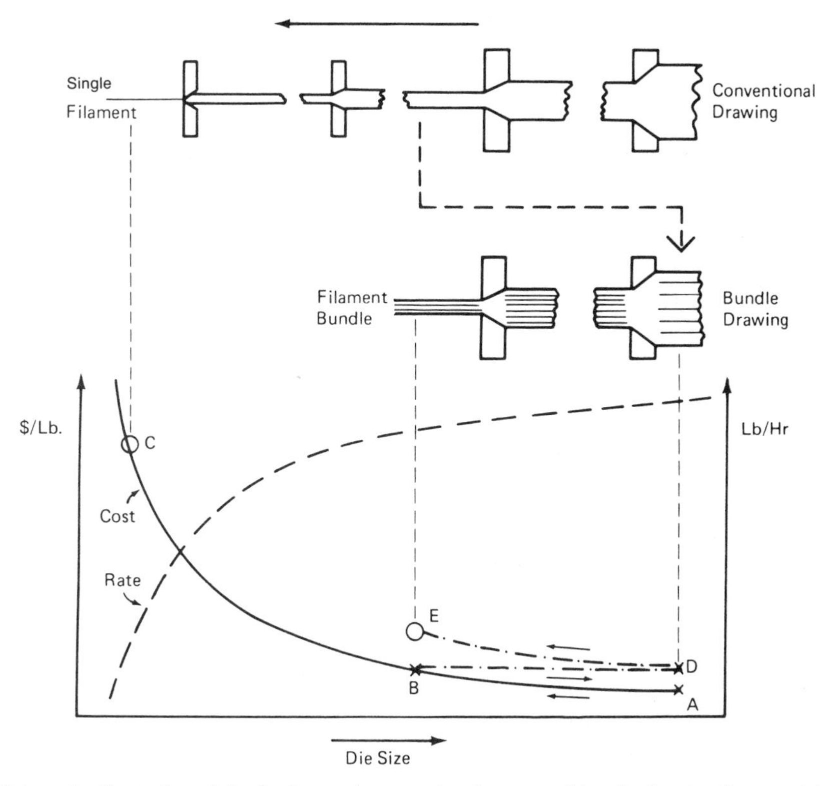

Fig. 16-1. Schematic illustration of the fundamental economic advantage of bundle-drawing fine metal filament. To arrive at the same filament diameter, conventional wire drawing follows the cost curve *ABC*, while bundle drawing operates only in the cost-insensitive region and follows the discontinuous curve *ABDE*.

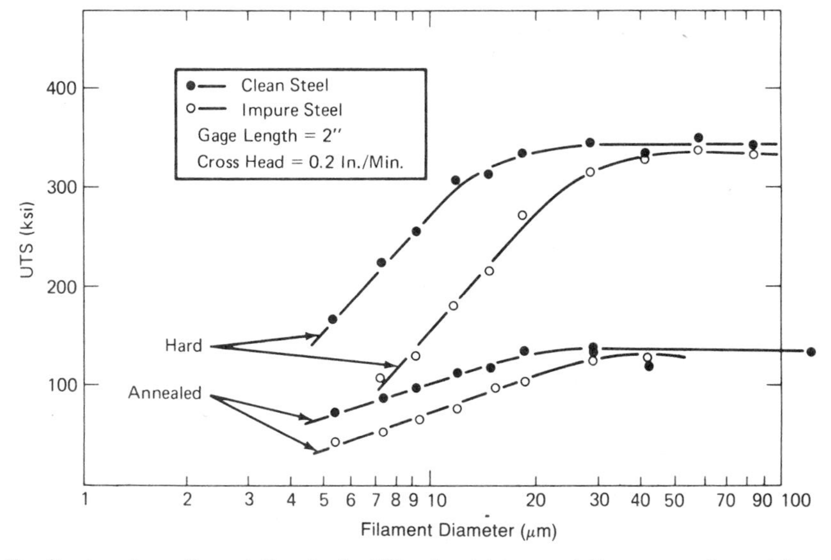

Fig. 16-2. Tensile strength vs. filament diameter for 300 series stainless steel filament manufactured from a clean heat of steel and a heat containing a relatively large number of nonmetallic inclusions.

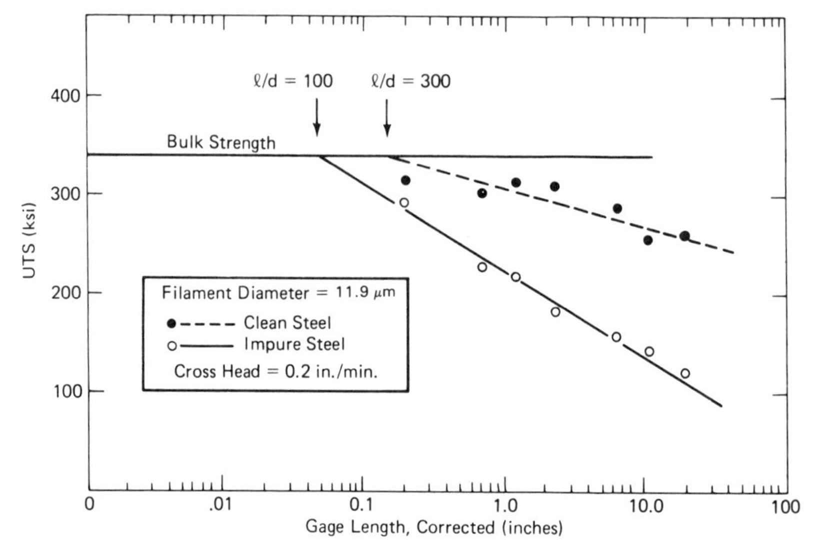

Fig. 16-3. Tensile strength vs. gage length for 12 μm stainless steel filament manufactured from the heats used to generate the data in Fig. 16-2.

a critical length can be defined at which the fiber effectively exhibits bulk properties.

3.2 Extrusion

Extrusion is a method of fiber forming that utilizes a metal powder or metal-oxide powder in shiny form along with a suitable carrier to form a fiber or filament by extrusion. This process is usually followed by a heat treatment or oxide-reduction step to sinter the metal particles together, thus improving the mechanical integrity of the fibers.

In late 1984, National-Standard Company of Niles, Michigan, commercialized a similar process for the production of nickel fibers when its fiber plant at Corbin, Kentucky, went on stream. This process utilizes nickel oxide to produce fiber, which is then converted through a proprietary oxide-reduction step to nickel fiber. While nickel fiber will comprise the initial production, in theory, any metal that can be oxidized and then reduced can be utilized by this method to produce fiber. Practical as well as economical constraints probably will limit production to a few metals.

Initial fiber forms will be fibers of 10, 20, and 40 μm diameter and sintered metallic webs using these fiber diameters.

Costs associated with this manufacturing method are substantially lower than those encountered in bundle drawing. As a consequence, pricing for extruded metal fibers is of a much lower order than the pricing normally associated with bundle-drawn metal fibers.

3.3 Miscellaneous Metal Fiber-Forming Processes

Attempts to obtain metals in fibrous form by an economical process or with unique properties have led investigators to explore many avenues in addition to those already described. Their efforts have met with mixed success. Examples of three such investigative or developmental areas will be taken as representative of this miscellaneous group of processes.

3.3.1 Electroplating. The electroplating approach to fiber forming is based on the reasoning that building up a fiber by deposition rather than attenuating a massive form would be a logical approach to economical filament manufacture. A number of novel designs for a suitable mandrel on which to plate the filament appear in the literature. Of particular interest is the concept described by Newton, which is a cylindrical drum constructed to define a helical, electrically conducting groove so that continuous deposition and wind-up can be achieved by rotating the drum. The method was used to produce nickel filament of roughly rectangular cross section, 0.001 in. × 0.0005 in. The mechanical properties reflect the extremely fine grain size, with filaments exhibiting a measured tensile strength of approximately 200,000 psi. Further development of the process is required, but the recent introduction of a commercial process for electroformed iron strip gives reason to direct new attention to the concept.

3.3.2 Vapor Plating. The application of vapor-plating techniques to fiber forming again uses the logic of the electroforming approach. There are two general process classifications, one where the composition of the deposited material is identical to the vapor phase (e.g., Al deposited from Al vapor), the second where the deposited material composition is the result of a vapor or gas phase reaction (e.g., Ni from Ni $(CO)_4$). In both cases, deposition takes place on a substrate, but in the latter process the substrate can be heated to promote decomposition, as might be the case in the reduction of $TiCl_4$ to Ti.

After deposition has occurred, the option exists of carrying the substrate material into the final product application or removing it by thermal chemical or mechanical means. The direction taken depends very much on the final application and the compatibility of the substrate material with the end-use environment. For example, if the substrate is an organic fiber and the plated product is intended for high-temperature filtration (as was the case with Ni fiber webs manufactured by Fram Corp.), it is necessary to pyrolyze the product prior to use in order to avoid possible gaseous contamination by the filter.

Vapor plating is used to prepare a variety of fibrous or fiberlike products, many of which are commerically available. These products include the metallized thread used in decorative textile application. Aluminum deposited on glass fiber or plastic film, which is subsequently slit to form a fiber or ribbon, is used for such diverse applications as radar reflection and tinsel for decorative purposes. In the case of the high-modulus boron filament (typically 0.004 in. diameter), the substrate material is generally 0.0005 in. diameter tungsten wire, but since the tungsten represents only a small fraction of the cross section ($<5\%$) and boron is nonmetallic in properties, this product will not be considered in the metal fiber category. Boron filaments are discussed in Chapter 19.

A highly specialized technology involving vapor deposition and decomposition processes has been developed for the production of metal and ceramic whiskers. These materials are fibers but composed essentially of one crystal. The conditions for producing whiskers require that growth on the substrate or condensing surface be highly directional and confined to limited regions. Most of the recent developments of the technology have related to the high-modulus ceramic materials. It does not appear that commercial processes for large-volume metal whiskers at economical manufacturing cost are available although the potential of such fibers, in view of their exceptional mechanical properties, justifies continued process development. Experimental samples of iron, copper, and tin whiskers have been produced. Ceramic whiskers are discussed in detail in Chapter 13.

3.3.3 In Situ Decomposition of Inorganic Compounds. Many attempts have been made to apply chemical decomposition, such as the reduction of oxides and salts, to form fiber directly. Union Carbide Corp. developed processes based on the concept of infiltrating an organic filament precursor with a solution of the desired metal salt and then chemically reducing the compound to form a metal filament. Claims were made for an exhaustive list of potential fiber products, with diameter, length, and form corresponding to the organic fiber precursor.

4. SUMMARY OF PROPERTIES

4.1 Fiber Properties

The chemical, physical, and mechanical behavior of metal fibers as a class of materials will generally approximate the bulk material properties. Exceptions may arise due to the influence of the fiberizing process or as a result of the diminished cross section (diameter) and increased length.

The outstanding characteristics of metals and alloys are summarized in Table 16-1. Detailed descriptive literature is available in the standard metal reference works, such as *Metals Handbook*, Vol. 1, Edition 8, published by the American Society for Metals.

Deviations from bulk material properties that occur as a result of the fiberizing process are briefly summarized in Table 16-2. The property alteration may or may not be advantageous, and in many cases the magnitude of the effect can be controlled by modifications of the processing technique employed. It is also conceivable that post-forming operations can be carried out that could radically change the property characteristic of interest. Heat treatment, for example, can be employed to cause recrystallization and hence eliminate the preferred crystalline orientation associated with wire drawing where such a fiberizing process is employed. In addition, heat treatment can promote diffusion and homogenization and thus change the reactivity of a fiber with minor surface chemistry alteration due to processing. Still further examples of the effect of heat treatment on fiber properties would be the transition of cold worked 302 or 304 type stainless steel from a ferromagnetic to a paramagnetic state, or the development of ductility in a cold worked fiber by inducing recrystallization.

Table 16-1. Outstanding characteristics of metals by general category.

Metal or Alloy	Outstanding Characteristics
Carbon and alloy steels	High strength, high modulus, low cost, ferromagnetic properties
Stainless steels	Good corrosion resistance
Nickel and cobalt base superalloys	Strength at elevated temperatures, oxidation resistance
Refractory metals and alloys, e.g., Ti, Ta, Nb	High melting points, good corrosion resistance
Copper and copper alloys	High electrical and thermal conductivity
Aluminum and aluminum alloys	Low density, high electrical and thermal conductivity, low cost; can be anodized to produce extensive color range
Precious metals, e.g., Pt, Ag, Au	High electrical and thermal conductivity; the noble metals are particularly corrosion-resistant

The effect of diminished cross section and increased length on a specific fiber property can have a far-reaching impact. The particular or intended application must be fully analyzed in order to choose the optimum fiber product. A brief listing of some of the properties on which physical size may have a major influence is presented in Table 16-3.

4.2 Fiber Geometry

In the foregoing discussion of fiber properties, the importance of the geometrical characteristics of the individual fibers is clear. The quantification of such characteristics is not simple, and no attempt will be made here to apply any detailed ranking. It is useful, however, when considering the application of fibers in composite products, to have a guide to cross-sectional shape and uniformity as well as surface topography. These characteristics can change dramatically from process to process and from material to material, and in some cases from one supplier to the next. An attempt is made in Fig. 16-4 to schematically illustrate fiber cross sections that would typify the various fiberizing processes listed. No size (diameter) comparisons between fibers of the different processes should be drawn from the figure. Where two or more shapes are shown for the same process, they are indicative of either the flexibility of the process to control the cross section or an inherent characteristic associated with the absolute physical dimenions. A series of scanning electron microscope photographs of representative samples of fibers produced by the various processes are presented in Figs. 16-5 through 16-16. Photomicrographs are shown for two levels of magnification in most cases in order to highlight the surface detail (high magnification) and provide a general field of view. The latter contains many fibers, allowing for a as-

Table 16-2. Property alteration associated with the fiberizing process.

Fiberizing Process	Possible Deviation from Bulk Material Properties
Conventional wire drawing	Strongly developed wire texture (preferred orientation of crystallographic directions with respect to wire axis)—particularly in unannealed or hard drawn filament. Such preferred orientation effects can give rise to directionality of the basic properties of the metal in the filament product
Bundle drawing	Preferred orientation effects can be expected with consequent directionality of properties. Surface chemistry alteration can occur and hence changes in initial reactivity. Small cyclical variations in cross section area can be present, resulting in an apparent property change, for example, in mechanical strength, which is based on the 'weakest link' (smallest cross section area) in the length being measured
Foil shearing or slitting	Fiber properties will reflect the parent foil properties which may or may not show preferred orientation. If preferred orientation is present, the fiber properties will depend on the direction of shear, that is, either parallel or perpendicular to the rolling direction in the foil
Shaving and related processes	The occurrence of partial fractures in fibers during processing results in weakened areas along the fiber length which constitute an apparent change in mechanical strength
Melt spin	Surface chemistry change associated with the product of the designed reaction to promote stabilization during fiber forming
Melt extraction	Extreme cooling rates produce fine grain size (particularly on the side of fiber in contact with extraction wheel) and can produce amorphous-like structures which will result in major property alteration. This latter phenomenon is exploited to the fullest in a product called 'Metglas' but which is not generally categorized as a fiber. The amorphous state is metastable and the unique properties disappear on heating

sessment of the range of size and shape and to some degree the "straightness" of the fibers. A brief description and the source of the sample is given in each case.

4.3 Product Properties Related to Form

The discussion to this point has dealt with the metal filament product as a raw fiber. This product form is not necessarily the most desirable for end-use application. Other fiber-based industries have evolved a vast array of first, second, and even third generation products by developing suitable fiber-manipulating technology. In many cases, the same technology can be applied to the metal fiber field. This is particularly true when the metal fiber geometry and properties approximate those of the natural,

Table 16-3. Size-related property alteration.

Property	Effect of Diminished Cross Section and Increased Length.
Corrosion	High surface to volume ratio provides for relatively large reaction surface and hence limited life in corrosive environment
Electrical conductivity	Small cross section provides efficient use of material for high frequency electrical conduction since the fiber radius can approach the skin depth
Mechanical properties	Sphere of influence of defects (e.g., inclusions in steel) associated with certain fibers approximates to the fiber diameter and results in reduced or degraded mechanical properties for the free fiber
	Whiskers (single crystal fibers) can be essentially dislocation-free and approach theoretical strength

Process	Fiber Cross Section (Schematic)
Conventional Wire Drawing	
Bundle Drawing	
Shaving	
Shearing	
Melt Spun	
Melt Extraction	

Fig. 16-4. Typical fiber cross sections.

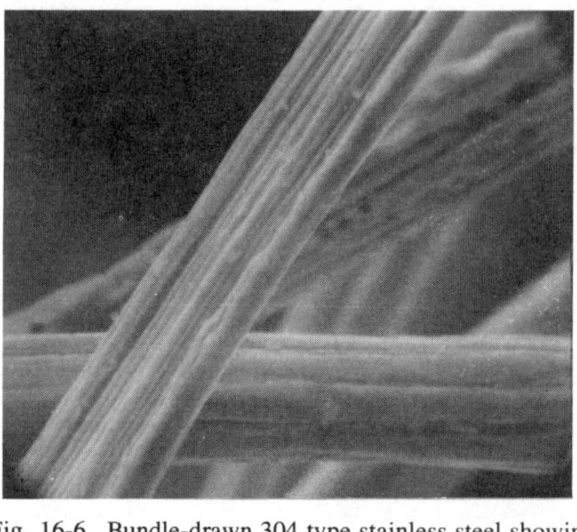

Fig. 16-6. Bundle-drawn 304 type stainless steel showing pronounced longitudinal markings and surface irregularities. Filament diameter 0.0003 in. (8 μm). Mag. 2200×.

Fig. 16-7. Grouping of bundle-drawn filaments illustrating good uniformity. Product manufactured by Brunswick Corp. and identified as BRUNSMET MF Al. Mag. 500×.

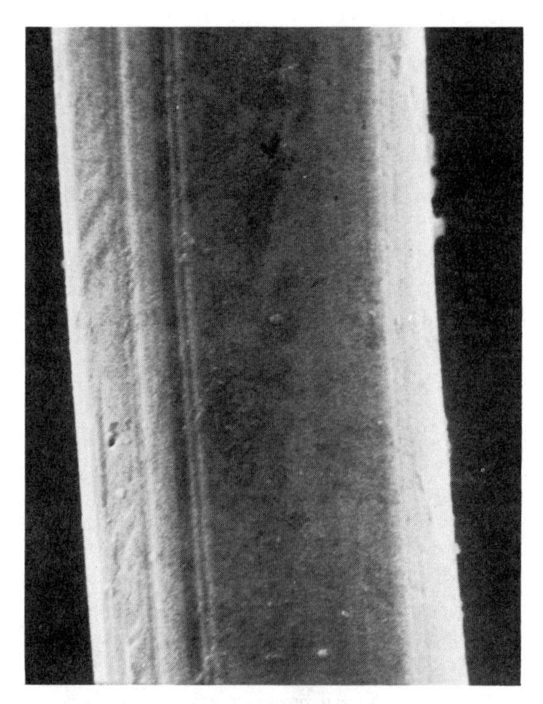

Fig. 16-5. Conventional wire-drawn 304 stainless steel, 0.001 in. diameter (nominal). Burnished surface with minor draw die defects (grooves). Mag. 1700×.

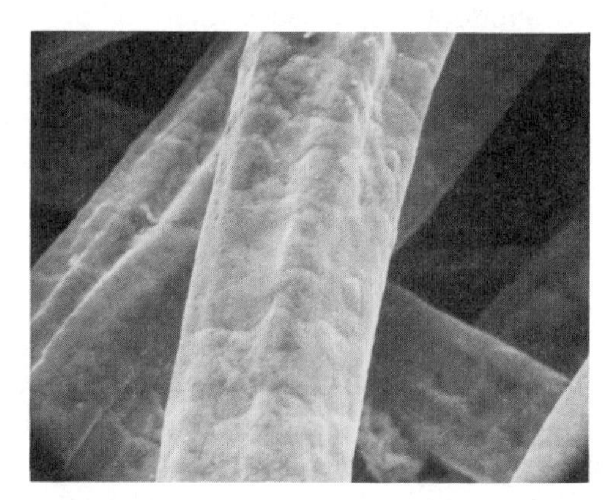

Fig. 16-8. Bundle-drawn 304 type stainless steel, diameter 0.00078 in. (19 μm). Annealed condition. Thermally etched grain boundaries are clearly visible. Mag. 1800×.

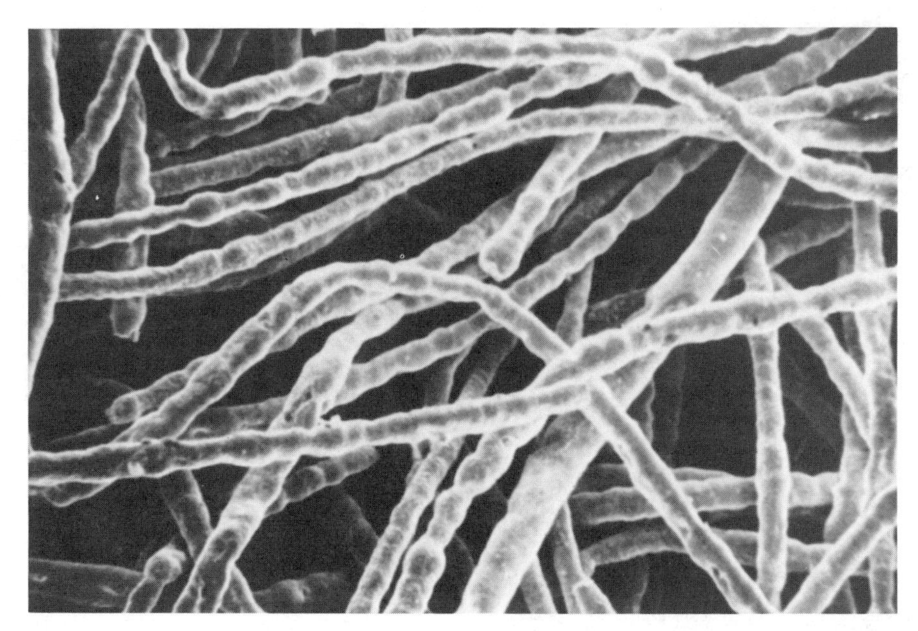

Fig. 16-9. Extruded nickel fibers. Product manufactured by National-Standard Company and identified as Fibrex fine nickel fibers.

synthetic, or glass fibers for which the manipulating techniques and machinery were originally developed. Fiber-handling methods can be broadly divided into two categories, primarily associated with fiber length. Short fibers (approximately $\frac{1}{2}$ in. or less) are generally handled in a slurry (paper-making techniques), while long fibers can be processed by a variety of machinery involving spinning, carding, and airlaying. Each machine produces a new product form or an intermediate form of a more complex final product. In some cases, minor modification of equipment may be necessary to accommodate the metallic properties of the fiber, such as hardness or abrasive character. Also, it is imperative that, in certain cases, completely new process steps such as annealing or heat treatment be incorporated to either facilitate processing or achieve the desired end-product properties.

Fig. 16-10. Grouping annealed bundle-drawn filament. Product manufactured by Brunswick Corp. and identified as BRUNSMET® MF Al. Mag. 450×.

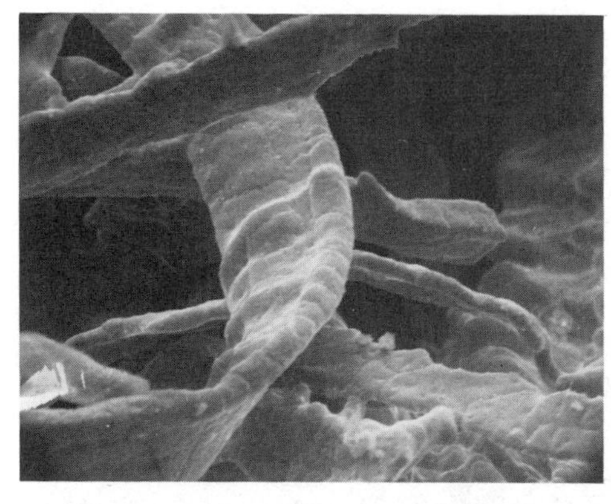

Fig. 16-11. Feltmetal fiber structure detailing surface characteristics of the Type A stainless steel fiber. Product manufactured by Brunswick Corp. and identified as Feltmetal FM 1100 series. Mag. 900×.

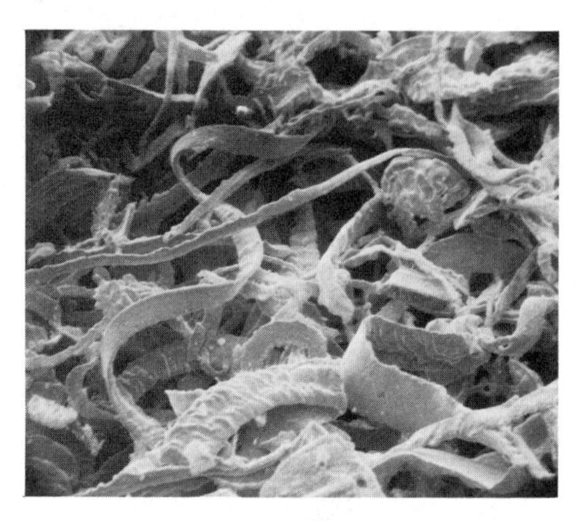

Fig. 16-12. Detail of the surface of Feltmetal FM 1100 series metal fiber structure showing wide range of fiber sections and dimensions. Mag. 180×.

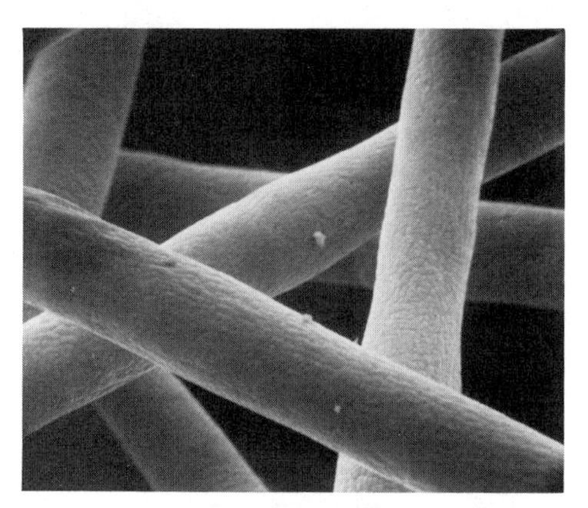

Fig. 16-13. Melt-spun aluminum filament showing characteristic surface detail associated with solidification process. Nominal diameter 0.002 in. Produced by Brunswick Corp. Mag. 450×.

Fig. 16-14. Array of melt-spun aluminum fibers showing regions of diminished diameter associated with liquid jet instability during the filament-forming process. Mag. 190×.

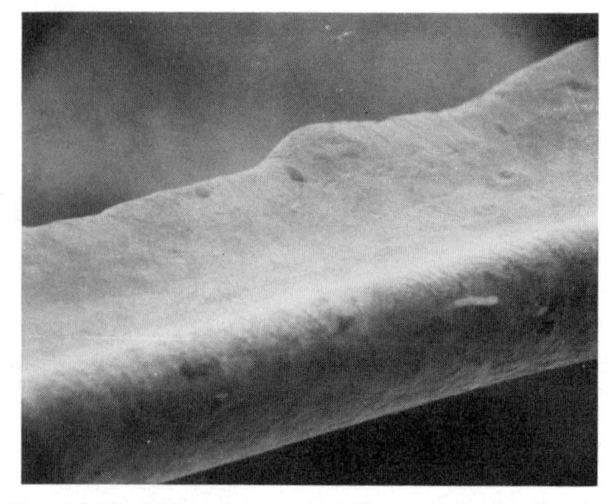

Fig. 16-15. 304 stainless steel fiberized by the Pendant Drop Melt Extraction Process. The crescent shape of the fiber caused by solidification of the filament on the disc mold is apparent. The dimensions of the fiber cross section are approximately 0.002 in. × 0.004 in. The sample was supplied by Battelle Memorial Institute. Mag. 600×.

Simplified flow diagrams to illustrate the process steps for the more significant product forms are shown in Figs. 16-17 and 16-18. Metal fibers made by a variety of filament-forming techniques have been successfully processed in the forms indicated, but, of course, the raw fiber must be consistent with the processing techniques to achieve a satisfactory result.

Properties associated with, or specific to, the major product forms are outlined below.

4.3.1 Yarn. Properties of continuous filament, staple, blends, and plies are a function of twist, staple length, where applicable, and

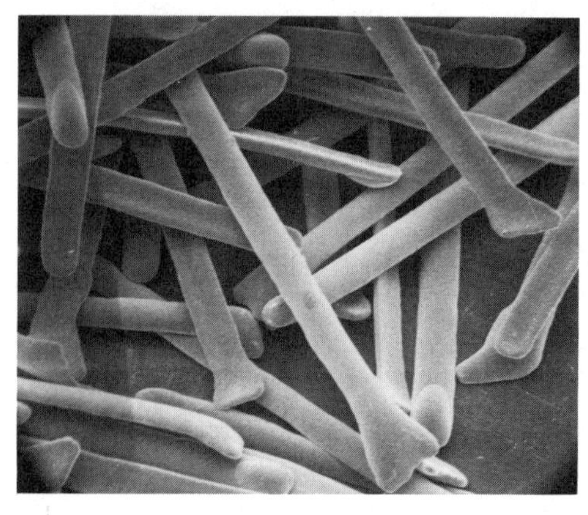

Fig. 16-16. Array of 0.125 in. long PDME process fibers showing dog bone leading end and rounded trailing end. (Nominal width 0.006 in.) Mag. 40×.

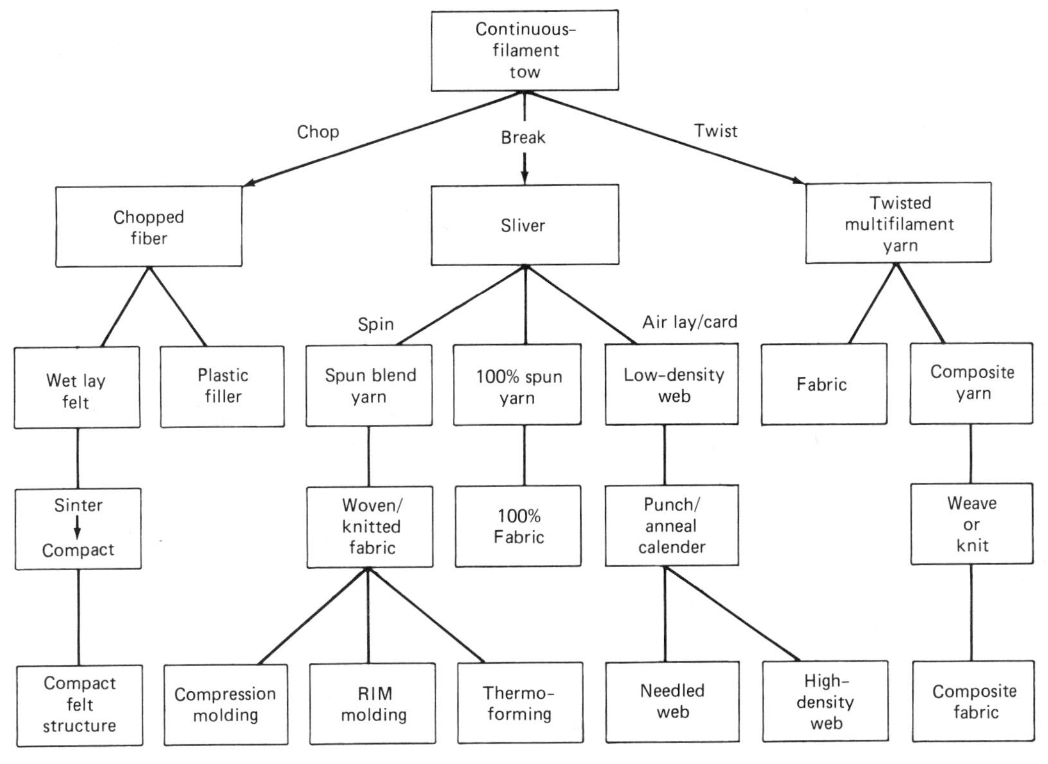

Fig. 16-17. Product form relationships.

interfiber friction. In a properly designed yarn, the breaking load can approach the cumulative breaking load of the individual filaments.

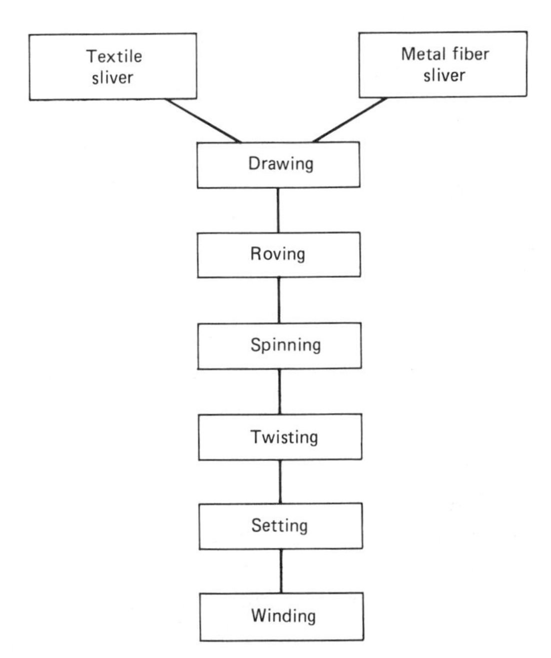

Fig. 16-18. Typical blended yarn process.

4.3.2 Woven and Knitted Structures. Properties can be expected to be highly anisotropic. Optimizing the principal property in the desired direction or directions for a particular end-use application can be achieved by design of the structure.

4.3.3 "Felted" or Web Structures (Nonwovens). Structures in the low-density range (>98% porosity) are available with long fiber lengths only and represent the ultimate in low-density porous metal structures. Mechanical properties are a function of density, but handleable products in the 99+% porosity range are practical. Typical web compressibility curves are shown in Fig. 16-19. It can be seen that annealing and sintering of the manufactured web improve the initial or low-load characteristics of the web, probably because of fiber bonding (welding). At high load levels, this tendency is reversed, and interfiber friction and the inherently high individual fiber strength, of the as-manufactured web, predominates.

In the intermediate and higher density range represented by the needled and compacted

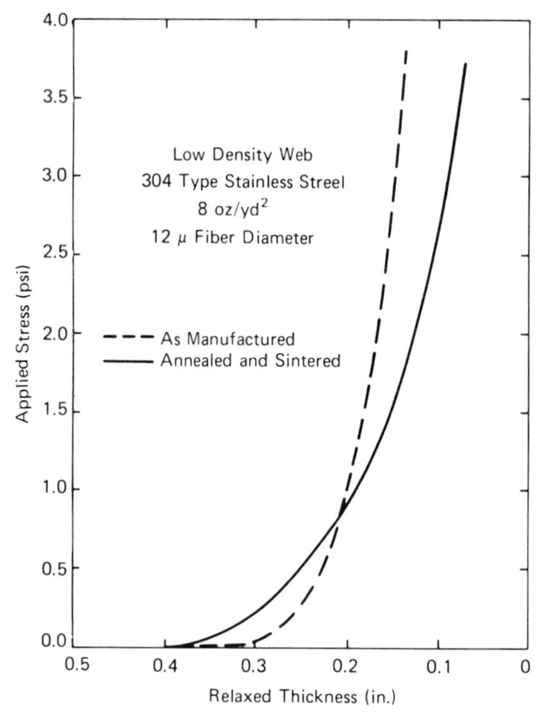

Fig. 16-19. Compressibility curves for low-density web (98+% porosity). Load applied normal to face of web.

structures, properties can vary widely in accord with the extremes of density and the type and condition of the parent fiber. By compacting (calendering or pressing) lower-density web with suitable annealing and sintering steps, controlled pore size, pore size distribution, and final density can be achieved reproducibly.

Certain properties of such structures as they relate to applications in composite materials are worthy of particular note. However, it should be remembered that once the structure is incorporated into a composite, the specific component property can be radically altered. For example, data on the thermal conductivity of a web in air (which can be important in composite fabrication processes) will require extensive correction once subsitution of the gas by a solid is made. Likewise, the apparent electrical conductivity of a compacted fibrous structure may be markedly altered, particularly if the second component of the composite has a significant electrical conductivity.

Electrical and thermal conductivity of fibrous web structures have been measured on Feltmetal® and compacted web of stainless steel and a nickel/chrome alloy fiber produced by the bundle-drawing process. The thermal conductivity data, measured through the thickness of the materials, is consistent with the relationship derived by Koh and Fontini:

$$\frac{\lambda}{\lambda_0} = \frac{1 - \xi}{1 + 11\xi^2}$$

where:

λ = porous structure thermal conductivity
λ_0 = bulk material thermal conductivity
ξ = porosity

A similar relationship holds true for the electrical resistivity of fibrous metal structures:

$$\frac{\rho_0}{\rho} = \frac{1 - \xi}{1 + 11\xi^2}$$

where:

ρ_0 = bulk material resistivity
ρ = porous structure resistivity
ξ = porosity

Electrical measurements made perpendicular and parallel to the porous structure surface of certain samples indicate a significant anisotropy. Therefore, it is important to be consistent in measurement techniques when transposing from electrical to thermal conductivity using the above equations.

A specific example of thermal conductivity data obtained for two samples of Feltmetal® composed of Type A nickel and stainless steel fiber with mean diameter of approximately 15 μm is presented in Table 16-4.

The tensile strength of fibrous porous structures will be dependent on a large number of parameters, including:

- Fiber diameter
- Fiber length
- Fiber mechanical properties
- Degree of sinter
- Degree of preferred orientation
- Density of structure
- Compacting process

Table 16-4. Thermal conductivity.

Material	Density (% bulk)	Specimen Thickness (in.)	Thermal[a] Conductivity Btu/hr/fr/°F	Thermal Conductivity (% bulk)
Ni	17	0.50	0.36	1
Stainless steel	46	1.06	0.28	2.3

[a]Measurement taken through thickness.

A series of measurements made on 304 stainless steel porous structures of varying density is shown in Fig. 16-20. The data points lie within a broad envelope, the upper bound representing the optimum conditions for high strength in the structure, e.g., high aspect ratio and hard sinter. The data serve to indicate the necessity of identifying and controlling material and process parameters in order to achieve reproducible properties.

5. APPLICATIONS

For many years metal fibers have been utilized in filter media, insulation, sound-deadening materials, and conductive elements in antistatic yarns and fabrics, as well as protective garments. A more recent development, the use of metal fibers to produce conductive plastics, shows far greater potential than any current applications.

The initial developmental thrust of this application was to compound stainless steel fibers directly into the plastic in order to form a conductive matrix as the part was molded. This process has presented numerous technical difficulties, depending upon the plastic used, the compounding and molding process, and the size and complexity of the part to be molded. More recent investigations have indicated that conductive webs and woven conductive fabrics can

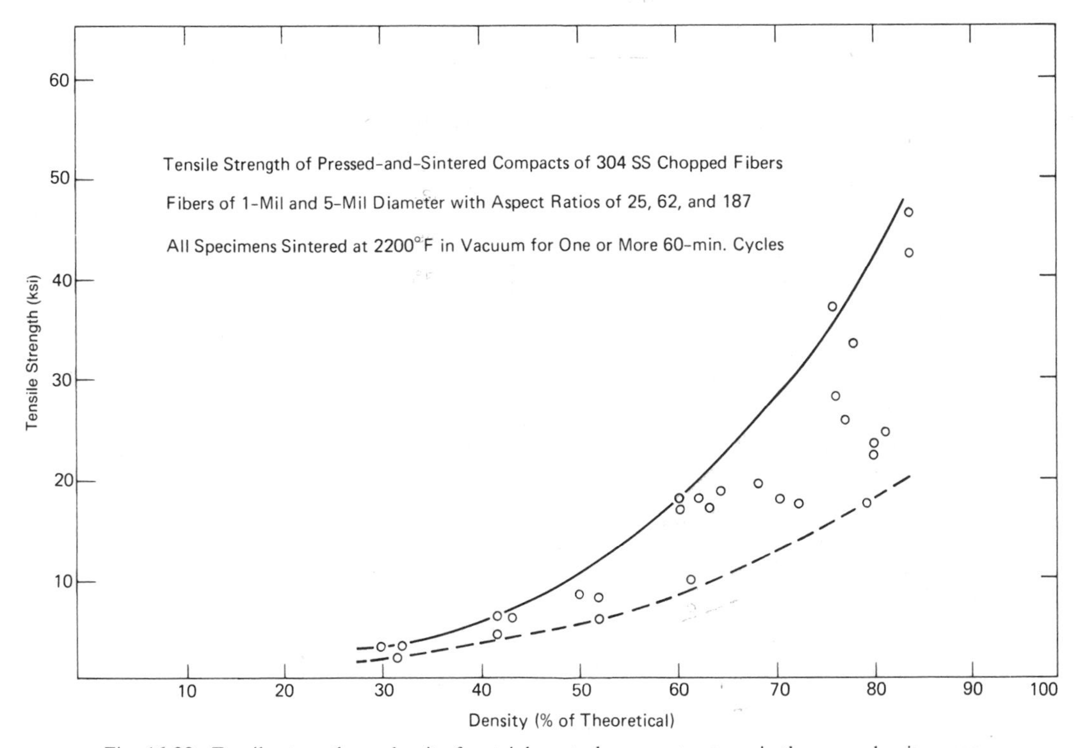

Fig. 16-20. Tensile strength vs. density for stainless steel porous structures in the upper density range.

be utilized to provide the required degree of conductivity in an economical manner.

The development of conductive plastics has been spurred by government regulations limiting the amount of electromagnetic radiation that may be emitted from an electronic device. Another reason for the development of conductive plastics has been the realization by the semiconductor industry of the economic consequences of electrostatic discharge damage to sensitive semiconductor parts.

The major market areas for conductive plastics are business machine housings, medical instrument housings and antistatic plastics for tote trays, work-surfaces, instrument housings, and other plastic applications that require the antistatic protection used in the semiconductor industry.

The primary reason for utilizing metal fibers to produce conductive plastics is that the required degree of conductivity may be obtained with relatively low weight loadings, which in turn have less effect on desirable physical properties of the plastic. The primary physical property affected by conductive fillers is impact resistance; metal fibers have a lower effect on this desirable property than other conductive fillers. Another physical property least affected by metal fibers is cosmetic appearance of the finished part. Figures 16-21 and 16-22 show the use of fine nickel fibers to improve the conductivity of plastics.

The initial plastic molding method to take advantage of the unique properties of metal fibers was injection molding. In this method, the metal fibers were first compounded into plastic pellets containing the requisite amount of metal fibers, in most cases 5% to 10% by weight. These compounded pellets were then fed through an injection-molding machine and molded into the final part. Engineering resins are the most widely used plastics in this system.

An attenuation level of 30 to 40 dB is normally required to meet mandated shielding levels. This can be achieved with the addition of 5% to 10% by weight metal fibers. The required amount of metal fibers is governed by many factors, among them the effective aspect ratio of fiber in the finished part, fiber distribution, type of plastic, equipment and compounding sequence, and, most important, techniques applied during molding and compounding.

Metal fibers for this application were initially supplied as chopped fiber bundles held

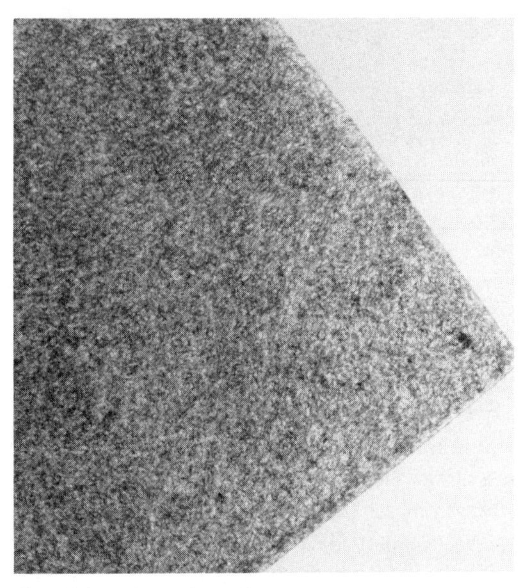

Fig. 16-21. Shows the uniform distribution of fine nickel fibers that improves conductivity of an engineering plastic.

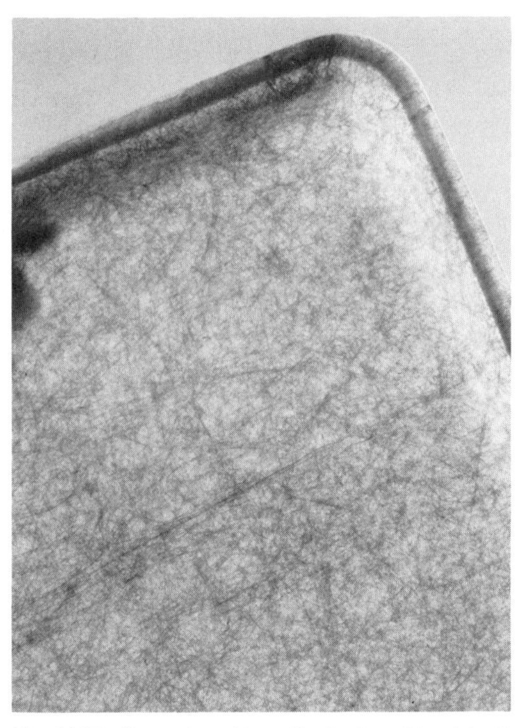

Fig. 16-22. Shows the uniform distribution of 5 weight % fine nickel fiber that improves the electrical conductivity of an engineering plastic.

together with a suitable binder. The type and amount of binder used is very important in that the binder performs a number of critical functions. It holds the fiber bundle together to prevent tangling, protects the fiber during chopping and compounding, governs the rate of dispersion, and acts as a fiber lubricant.

While stainless steel fibers of 6.5 to 8 μm diameters are the most widely used fibers for this application, recent evaluations have indicated that loose nickel fibers of 10 μm diameter may also be useful in the production of conductive plastics.

Antistatic plastics require lower amounts of metal fibers to achieve the required degree of conductivity. In general, 5% by weight should be sufficient to yield surface conductivities of 10, 5, 10^5 ohms or less. While resin-rich surfaces are desirable in the case of shielding conductive plastics for cosmetic purposes, fiber-rich surfaces are necessary for antistatic applications.

Other methods for achieving satisfactory shielding levels in plastics involve the use of woven or nonwoven conductive fabrics and webs. In a properly constructed conductive fabric or web, the conductive network is preformed prior to incorporation into the plastic part. In this way, the shielding effectiveness of the molded part is predictable before molding.

Preliminary evaluations have indicated that this shielding method will be the most effective in reaction injection molding, compression molding, and thermoforming applications. The high degree of effectiveness is due to the fabric and web adaptability as well as economic factors. Table 16-5 gives some results.

6. MATERIAL SUPPLIERS

The following list of suppliers is not meant to be exhaustive, but is representative of the vendors available.

Metal Fiber Suppliers

Company	*Address*
Bekaert Steel Wire Corp.	675 Third Avenue New York, NY 10017
Burnswick Corp.	2000 Burnswick Lane DeLand, FL 32720
National-Standard Co.	Industrial Blvd., P.O. Box 1620 Corbin, KY 40701-1620
Nippon Seisen Co., Ltd.	Kogin Bldg., Annex, 45, 5-chrome Korailashi, Higashi-ku Osaka, 541 Japan

Fine Wire Suppliers

Bekaert Steel Wire Corp.	675 Third Avenue New York, NY 10017
California Fine Wire Co.	338 S. Fourth St. Grover City, CA 93433
Everflex Products Inc.	35 Warwick Ave. Springfield, MA 01104
Ken-Tron Manufacturing Co.	P.O. Box 705 Owensboro, KY 42302

Table 16-5. Typical shielding results, ASTM draft method.

Shielding material \ Frequency MHz	30	100	300	1000
10 μm nickel web 0.56 oz/ft^2	73 dB	74 dB	64 dB	43 dB
10 μm nickel web 0.76 oz/ft^2	87 dB	72 dB	62 dB	50 dB
20 μm nickel web 0.56 oz/ft^2	76 dB	74 dB	65 dB	43 dB
20 μm nickel web 0.71 oz/ft^2	68 dB	72 dB	68 dB	50 dB

Company	*Address*
National-Standard Co.	Industrial Blvd., P.O. Box 162 Corbin, KY 40701-1620
Nestor Alloy Corp.	666 Passiac Ave. West Caldwell, NJ 07006
Owl Wire And Cable Inc.	Route 5, P.O. Box 187 Canastora, NY 13032
Technalloy Company Inc.	Route 113 Rahns, PA 19426

7. METAL FIBER AVAILABILITY

Fine metal fibers are commercially available in the following metals and alloys:

- 304 stainless steel
- 316 stainless steel
- 316L stainless steel
- 347 stainless steel
- Carpenter 20CB3®
- Inconel
- Hasteloy X
- Nickel
- 80/20 nickel chromium
- Titanum
- Tantalum

Typical metal fiber compositions are given in Table 16-6.

7.1 Commercial Physical Fiber Forms Suitable as Conductive Fillers

7.1.1 Tow. Tow is an untwisted bundle of continuous metal filaments available in bundle sizes from 1159 filaments per bundle to 29,400 filaments per bundle, depending upon fiber diameter. Available fiber diameters, in μm, are 4, 6.5, 8, 12, 22, and 25. However, not all alloys are available in all fiber diameters.

7.1.2 Sliver. Sliver is an untwisted bundle of discontinuous filaments broken to average lengths of 50 to 250 mm. Sliver weights are available in a range of 1–200 grains (g) per yard. All micron diameters are available with the exception of the 25 μm diameter. As with tow, not all alloys are available in all sliver fiber diameters.

7.1.3 Chopped Fibers. These fibers are formed by chopping water-soluble or solvent-soluble adhesive-coated tow into various cut lengths. Fiber diameter, bundle size, and cut lengths may be varied, depending upon application. Available fiber diameters are 4, 6.5, 8, 12, and 22 μm; cut lengths range from 1 to 20 mm; and bundle sizes range from 1 to 60 g/yd. Not all alloys are available in all fiber diameters, bundle sizes, and cut lengths.

The most widely used alloys for conductive fillers are the various stainless steel alloys. The properties and availabilities of the other alloys are mentioned in case special properties are desired or necessary.

Table 16-6. Typical composition elements.

Type	C %	Cl %	Ni %	Mo %	Fe %	Co %	Al %	Other %
316L	.03	16–18	10–14	2–3	bal	—	—	—
Carpenter 20CB3®	.06	20	34	2.5	bal	3.5		
Inconel	.05	23	bal	—	14	.5	1.5	
Hasteloy X	.1	22	bal	9	18	—	—	Co 1.5 W6
Nickel	<.15	—	>99		.4	<.25	—	Mn 35 Si 35
Titanium	≤.08				≤.25			Ti bal

8. FUTURE PRICING AND AVAILABILITY

Metal fiber prices have declined over the past few years. This trend will probably continue as additional suppliers of metal fibers enter the market. Improved production methods as well as economies of scale will also have a depressing effect on fiber prices.

Current production capacities appear to be adequate for near-term requirements. However, if a major new application requiring large quantities of metal fibers should develop, some shortages could be experienced due to the lead time required to construct new production facilities.

9. FUTURE UTILIZATION

Since interest in metal fibers has been increasing, it follows that utilization will also increase. The rate of increased utilization will be strongly influenced by fiber pricing and the rate at which technologies making use of metal fibers can be developed and introduced.

17 Polyethylene Filaments

John V. Milewski
Consultant
Santa Fe, New Mexico

Harry S. Katz
Utility Research Co.
Montclair, New Jersey

CONTENTS

1. INTRODUCTION

Spectra is the trade name of a new high-performance, continuous organic filament made of highly oriented polyethylene. Its high strength and high modulus plus low density give it specific strength and modulus properties far superior to most other structural reinforcement filaments such as glass, Kevlar, graphite, and boron. As a result, Spectra fiber is competitive in virtually every respect with the other high-performance industrial fibers. However, it cannot be used at high temperatures.

The technical merits of Spectra compared to other filamentary materials are illustrated in Fig. 17-1, in which specific tensile strength is plotted against specific tensile modulus.

The key product attributes listed in Table 17-1 depict a profile of a unique fiber remarkable for a range of performance properties, enhanced by a low density, that can rival the majority of today's specialty fibers. The technical information provided here enlarges on the attributes of Spectra. It will enable an assessment to be made of the potential of Spectra for new applications or its consideration as a cost-effective candidate to replace other fibers already in use.

2. PRODUCT DEVELOPMENT

Spectra-900™ filament was developed to overcome the limitations of existing filaments used for marine applications such as ropes and cables. Spectra-900 offers properties of good seawater corrosion resistance and hydraulic stability as well as resistance to creep under sustained loads.

A description of the molecular structure that

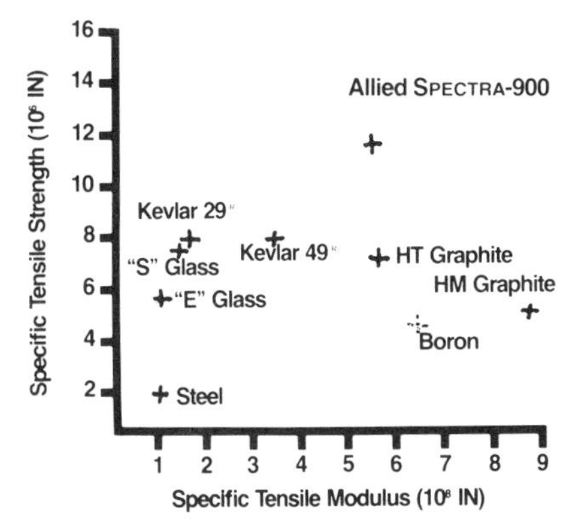

Fig. 17-1. Specific strength vs. specific modulus of reinforcing filaments.

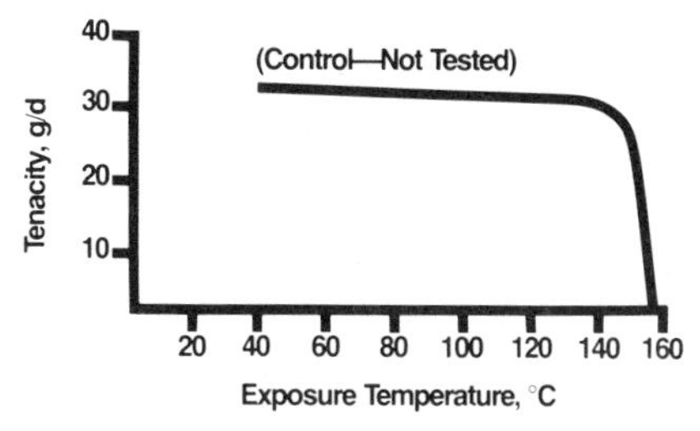

Fig. 17-2. Tenacity at room temperature after annealing.

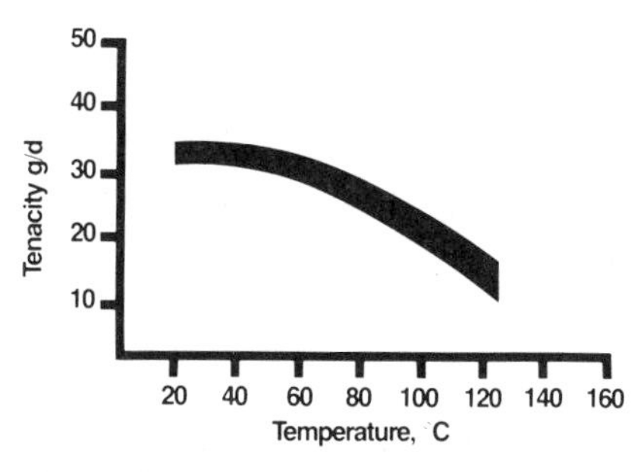

Fig. 17-3. Tenacity at various temperatures.

leads to these unusually high strength properties is given in Allied Corporation literature. The influence of molecular weight on the maximum strength of polyethylene fibers has been investigated using a stochastic Monte Carlo method based on the kinetic theory of fracture. A figure in reference 1 indicates that the tensile strength of a monodisperse polyethylene fiber with a molecular weight of 1 million will be 1.5 million psi. This indicates that there is room for much further improvement beyond the strength of Spectra 900.

3. PRODUCT AND PROPERTIES

3.1 Physical and Mechanical

The high strength of Spectra-900, better than 30 grams per denier tenacity at room temperature, is typically retained over a wide range of temperatures. For example, Fig. 17-2, which plots tenacity (strength) against exposure temperature, and Fig. 17-3, strength at various working temperatures, illustrate the broad applicability of Spectra-900. Additional temperature data are shown in Figs. 17-4 through 17-6. Figures 17-4 and 17-5 show typical stress–strain curves and the tensile properties at various temperatures, while Fig. 17-6 displays the effect of annealing temperatures on the tensile properties of the fiber and emphasizes its ability to withstand reasonable exposure and relatively high processing temperatures. Figures 17-7 and 17-8 show typical ultimate elongation and shrinkage measurements on Spectra-900 fi-

Table 17-1. Spectra product attributes.

Low density
High specific modulus
High specific strength
High energy to break
Low moisture sensitivity
High abrasion resistance
Excellent electrical properties
Good UV resistance
Excellent chemical resistance

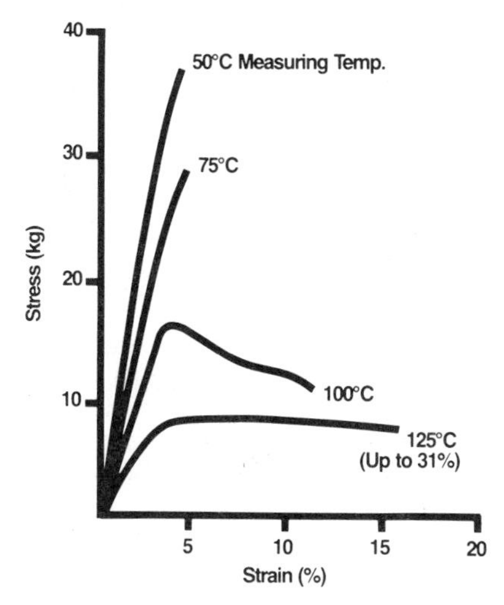

Fig. 17-4. Stress–strain curves at various temperatures.

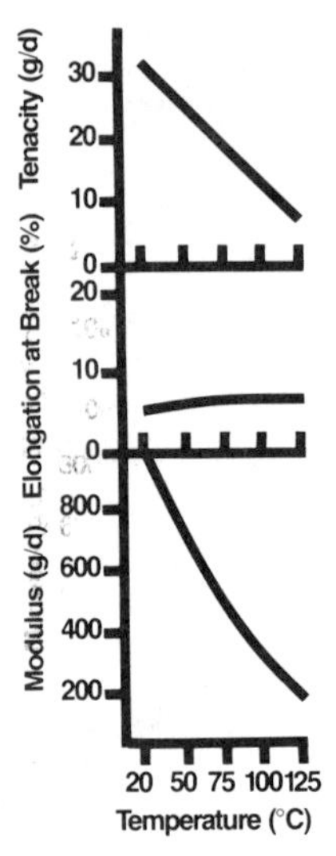

Fig. 17-5. Tensile properties at various temperatures.

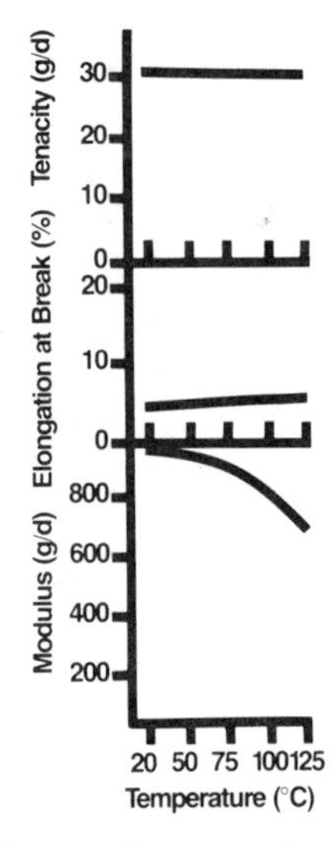

Fig. 17-6. Tensile properties at room temperature after annealing.

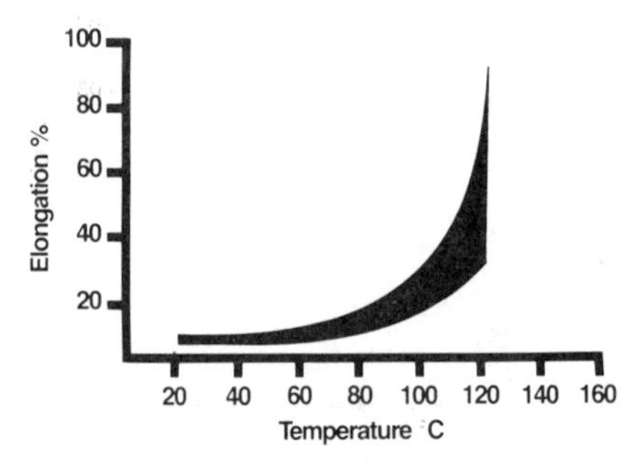

Fig. 17-7. Ultimate elongation at various temperatures.

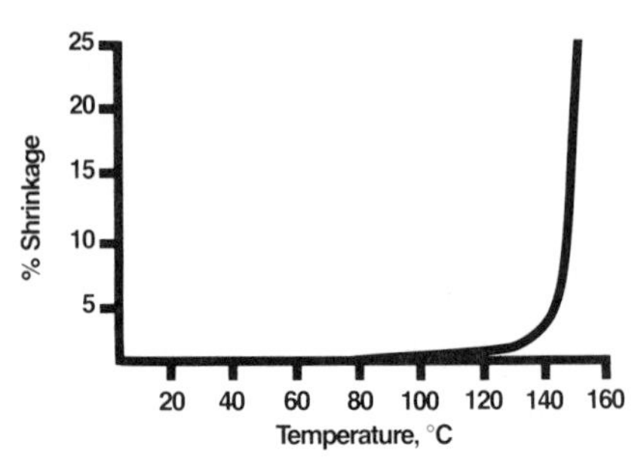

Fig. 17-8. Shrinkage at various temperatures.

ber measured over the same working temperature range, and attest to its superior dimensional stability.

The mechanical properties of Spectra-900 make it a strong contender as a reinforcing fiber for thermoset matrices. Table 17-2 summarizes the salient properties relevant to this application.

The tenacity of a new grade, Spectra-1000, at 160°F, is close to that of Spectra-900 at room temperature. In addition, there is less loss of properties during cure when Spectra-1000 is used in a thermoset composite, and it has better creep properties than Spectra-900.

3.2 Electrical

Spectra-900 fiber has excellent electrical insulating properties. The dielectric constant of the base polymer is 2.35, which is close to that of

Table 17-2. Spectra-900 as a reinforcing fiber.

Properties (In textile terminology)	
Density (lb/in.^{3})	0.035
Linear density (denier)	1200
Textile modulus (g/d)	1200
Tenacity (g/d)	30
Denier/filament	1200/118
General Properties	
Density (g/cm^3)	0.97
Filament diameter (μm)	38
Tensile Properties	
Strength (10^3 psi)	375
Modulus (10^6 psi)	17
Specific Tensile Properties	
Strength (10^6 in.)	10.7
Modulus (10^6 in.)	485

wax or paper, and the dielectric strength of about 700 kV/mm inhibits arcing or sparking.

3.3 Chemical

The chemical resistance of Spectra-900 is outstanding. It remained unaffected after three-month exposures to 10% detergent; hydraulic fluid; sea and distilled water; kerosene and gasoline; 1 AA hydrochloric acid; glacial acetate; ammonium hydroxide; perchloroethylene; hypophosphite solution; and toluene. These tests indicate that Spectra-900 may have chemical resistance greatly superior to that of other high-performance industrial fibers such as aramids. Its performance in harsh, corrosive environments is expected to be superior. Like other organic fibers, Spectra-900 absorbs liquids, which results in swelling. Its expected absorption behavior for a prolonged period of 30 days with a variety of organic solvents and with water is summarized in Table 17-3.

3.4 Textile

The textile properties of Spectra-900 meet requirements for most purposes. It can be successfully processed by regular textile methods on standard textile machinery (e.g., looms). In textile processing, a degree of twist is often inserted into a yarn to facilitate processing or confer certain properties on the final fabric. Figure 17-9 shows the effect of twist levels up to 10 turns per inch (TPI) on the tenacity and modulus of Spectra-900 fiber.

Table 17-3. Absorption of liquids by polyethylene; d = 0.96 g/cm^3 (after 30 days immersion at 20°C).

Solvent	% Wt. increase
Carbon tetrachloride	13.5
Benzene	5.0
Tetrahydrofuran	4.6
Lubricating oil	0.95
Acetone	0.79
Ethanol	0.4
Water	0.01

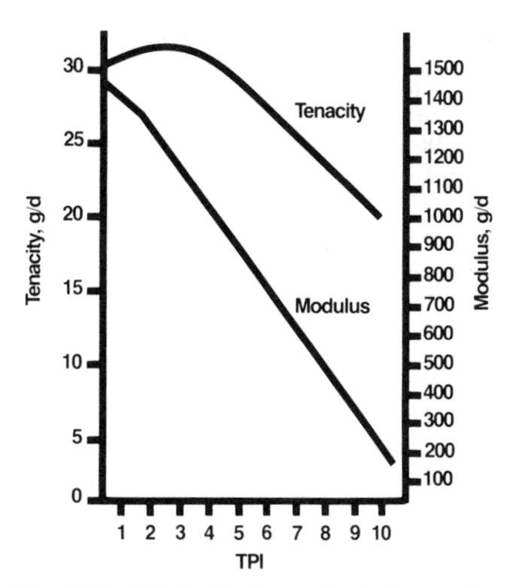

Fig. 17-9. Effect of twist on physical properties.

3.5 Stability

In many industrial fiber applications, and independent of the service environment, the long-term (several months) intrinsic stability of the fiber product or of the products of which it is a component is an important consideration.

A useful indicator of this stability is creep under specified loading. Figure 17-10 shows the percent length change undergone over a period of up to 300 days at ambient temperature. This behavior compares most favorably with that of other organic fibers.

Since Spectra-900 is a new-generation fiber, additional technical information is still being generated, which the manufacturer will share with customers as it becomes available. However, if some technical information is needed and is not supplied in this chapter, the reader may contact the manufacturer to request it.

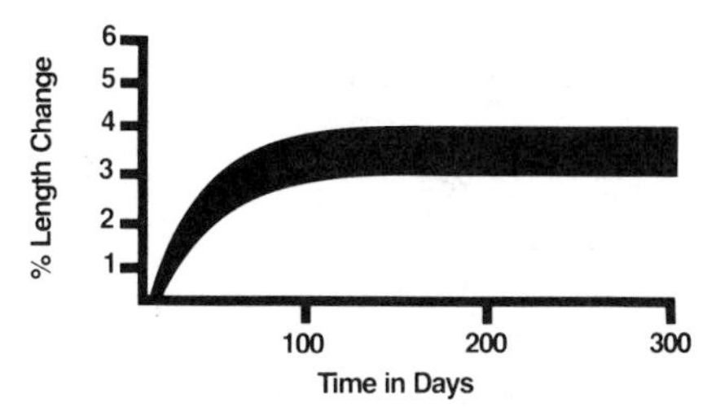

Fig. 17-10. Creep properties at 10% loading (22°C).

4. SUPPLIER, COST, AND AVAILABILITY

Spectra-900 and Spectra-1000 are products of the Allied Corporation.

4.1 Availability

Spectra-900 fiber is available in a 1200 denier version; in white only; and packaged on sleeves carrying 3 lb of yarn. It is shipped in standard industrial yarn boxes on a 3 3/16-in. I.D. × 13 in. long sleeve. Customers' special requirements will be acknowledged during the development stages of an application. Yarn quantity needs will be accommodated to the best of the manufacturer's ability.

Spectra-1000 fiber, with improved high-temperature properties, became available in late 1985.

4.2 Price

Since 1985, Spectra-900 was being sold on a commercial basis to the marine industry for ropes and cordage application. The prices in 1986 were $22 per pound for Spectra-900 and $28 per pound for Spectra-1000. More detailed information about cost and availability for composite applications can be obtained by contacting:

H. W. Chang
Manager, Composite Applications
Fibers Technical Center
P.O. Box 31
Petersburg, VA 23804
(804) 520-3265

5. PRESENT APPLICATIONS

Spectra-900 applications are discussed in the following paragraphs. While these applications reflect the broad versatility of Spectra-900, they should by no means be taken as limits to its possible end uses. Technical data and lists of applications are given to enable readers to evaluate where Spectra-900's unique set of properties qualify it for use in a particular industry and environment.

5.1 Marine

Spectra-900 fiber performs effectively and economically in many marine applications where its high strength and high modulus coupled with its resistance to harsh environments and its high abrasion resistance contribute to superior long-term performance.

In addition, Spectra-900 is the only high-performance filament that floats. Its low specific gravity of 0.97 is responsible for this unique property.

5.2 Ropes and Cordage

The low abrasion of Spectra-900 enables it to contrast sharply with other high-performance fibers in this application, and gives it a decided advantage over them, especially in braided ropes. The low moisture absorption and light weight of Spectra-900 result in highly desirable, durable ropes for a variety of industrial and recreational uses.

5.3 Composites

The compatibility of Spectra-900 fiber with most of the principal polymer resins used in composite systems and in such items as filament-wound pressure vessels qualify it for use in many different plastic structures. (Details related to interface bonding were not available when this chapter was written.) Its unique combination of high specific tensile strength and high specific tensile modulus, as shown in Fig. 17-1, along with its light weight and low creep, offer fabricators added advantages in manufacturing complex high-performance components such as radomes, where its electrical properties are favorable. Its energy absorbency and low density qualify Spectra-900 for ballistic protection applications.

5.4 Medical

Spectra-900 fiber meets all the requirements of a fiber product for hospital and health industry purposes. It is essentially inert, strong, stable, nonabsorptive and nonallergenic. It has potential for sutures, implants, prosthetic devices, and use in various kinds of hospital equipment.

5.5 Weaving

Spectra-900 fiber can be readily woven without sizing, flat or twisted, on both rapier and shuttle looms. As yet, there is only limited experience with Spectra-900 on the other loom types, but its developers believe that Spectra-900 will process satisfactorily on projectile looms. Weaving performance is improved by twisting both warp and filling yarns two to three turns per inch. Spectra-900 has been satisfactorily sized with polyvinyl alcohol, which may be especially appropriate for flat yarns. Special attention must be given to the sharpness of filling knives and scissors. Frequent sharpening will aid in trouble-free processing.

5.5.1 Twisting. Spectra-900 fiber can be twisted on conventional ring-twisters without machine modifications. Yarn thread-up should follow the most direct path from package to spindle. Yarn angles and yarn tensions should be kept to a minimum for best processing performance.

5.5.2 Repackaging. As in twisting, yarn angles should be kept to minimum values when Spectra-900 fiber is repackaged. Angles of 90 degrees should be avoided whenever possible. Yarn tension should be kept at a minimum.

References

1. Termonia, Y., Meakin, P., and Smith, P., "Theoretical Study of the Influence of the Molecular Weight on the Maximum Tensile Strength of Polymer Fibers," *Macromolecules* 18(11): 2246–2252, Nov. 1985.

18
ARAMIDS

W. S. Smith
E. I. DuPont
Wilmington, Delaware

CONTENTS

1. INTRODUCTION

The major commercially available "high-modulus" organic fibers are Kevlar® 29 and Kevlar® 49, which belong to a group of fibers called aramids, the term aramid being a generic description, granted by the Federal Trade Commission, for a group of aromatic polyamides that meet certain requirements. Aramid fibers differ from conventional organic fibers in having much higher strength and modulus in both tension and compression; they resemble inorganic reinforcing fibers (glass, graphite) in their tensile properties but have lower compressive strength, lower density, and considerably greater "toughness." They retain the processibility normally associated with conventional textile fibers while maintaining outstanding mechanical properties. This combination of strength, toughness, and textile processibility makes these fibers versatile reinforcements for a variety of applications—composites, ballistics, ropes, coated fabrics, papers, and so on.[1]

2. PRODUCT FORMS

2.1 Yarn and Roving

Commercially available yarns of Kevlar® 49 and Kevlar® 29 are usually designated by denier, a measure of a yarn's linear density that is numerically equal to the weight in grams of 9000 meters of the yarn. All but one of the yarns of Kevlar® 49 are based on the same filament denier (1.42 denier per filament or dpf), the yarn denier being equal to the dpf multiplied by the number of filaments in the yarn. The only yarn of Kevlar® 49 not based on the 1.42 dpf filament is the 2130 denier yarn, made up of 1000 filaments of 2.13 dpf. Kevlar® 29 yarns are based on a 1.5 dpf filament.

In contrast to yarns, which are designed primarily for weaving operations, rovings are usually designed for direct use in a fabrication process where it is desirable to have higher yields than are available in yarns; for example, filament winding, pultrusion, and rope manufacture. Rovings with yields from 980 to 197 yd/lb are available.

2.2 Fabrics and Woven Rovings

Fabric weights of Kevlar® aramid range from 1.8 to 6.8 oz/yd^2, while woven roving weights are as high as 15 oz/yd^2. The lighter fabrics, based on 195 and 380 denier yarns, offer lower per-ply thickness, but are more expensive than those fabrics based on 1140 and 1420 denier yarns. Plain, satin, and crowfoot weaves are available. Unbalanced fabrics ("unidirectional") with high percentages of fiber in the warp direction enable one to approach unidirectional properties in a laminate while retaining the processing and handling advantage of a fabric.

2.3 Pulp and Chopped Fibers

Pulp, a relatively new form of reinforcement, is a short, highly fibrillated reinforcement with a relatively high surface area of 8 to 10 m^2/g, or about 40 to 50 times that of chopped fibers. It is available in wet or dry form in lengths from about 2 mm to 5 mm. The wet form is suitable for wet-lay processes such as gasket sheeting or paper-making, while the dry form is suitable for dry or damp mixing processes such as drum brakes, molded clutch facings, or bulk molding compounds.

While Kevlar® 49 aramid fibers cannot be satisfactorily cut on conventional glass-cutting equipment, they can be chopped on equipment designed to apply a shearing force to the fiber; lengths from 1/8 in. to 2 in. or longer are available. In the shorter lengths, chopped fibers are suitable for friction products and plastics reinforcement; in longer lengths, they may be used in paper or mat products.

3. PROPERTIES

3.1 Physical

High modulus, continuous aramid yarns consist of multiple continuous single filaments of circular cross section brought together to form bundles of different sizes according to the desired denier. Dishevelment and catenary are carefully avoided in the process. Yarns and roving from the fiber manufacturer usually have no twist, but twist can be added to improve processibility. The diameter of the individual filaments is 0.00047 in. (12 μm). The density of the fiber is 1.44 g/cm^3. The fibers are highly crystalline and have a distinctive fibrillar morphology not observed in conventional fibers.[2]

The stress–strain curves of yarns of "Kevlar" 29 and 49 are compared to those of galvanized, improved plow steel wire and Dacron® polyester in Fig. 18-1. The high-modulus aramids have an order of magnitude higher tensile modulus than industrial nylon and up to two-thirds that of steel wire.[3] When one is reporting the strength of reinforcing fibers, it is important to designate the gage length, and whether single filaments, yarns, or impregnated strands were measured. This is particularly true when reporting the strength of brittle, inorganic fibers, whose strength can be strongly gage-sensitive. The bare yarn properties used for Fig. 18-1 reflect a gage length of 10 in. Yarn deniers of "Kevlar" 29 and 49 were 1500 and 1420 denier, respectively, tested at two turns per inch. The effect of twist on the tensile properties of high-modulus aramid fibers is shown in Figs. 18-2 and 18-3. If resin impreg-

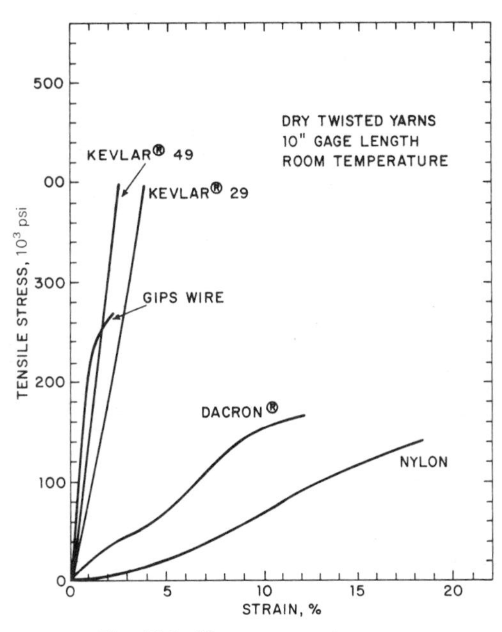

Fig. 18-1. Yarn stress–strain curves.

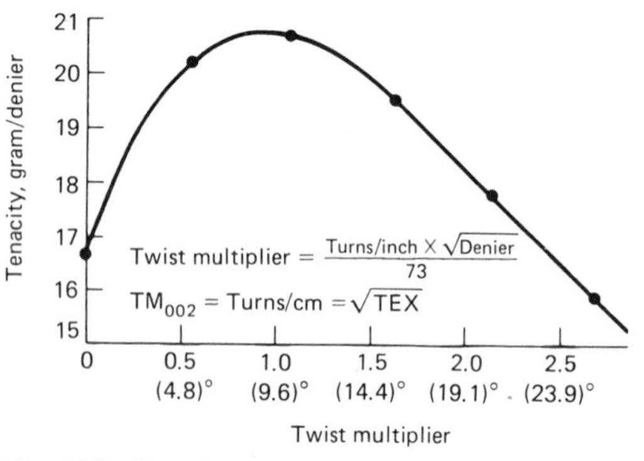

Fig. 18-2. The effect of twist on the strength of yarns of Kevlar® aramid.

nation as per the ASTM D2343 strand test is used, the tensile strength of "Kevlar" 29 and 49 will reach approximately 550,000 psi. This very high strength, combined with the low density of high-modulus aramid fibers, results in very high specific tensile strength. The specific tensile modulus of the organic fibers is intermediate between values of specific tensile modulus for glass fibers and those of the stiffer graphite and boron reinforcements. Both Young's modulus and fracture strain of Kevlar® 49 yarns have been reported to be the same at ballistic loading rates as at conventional rates.[4]

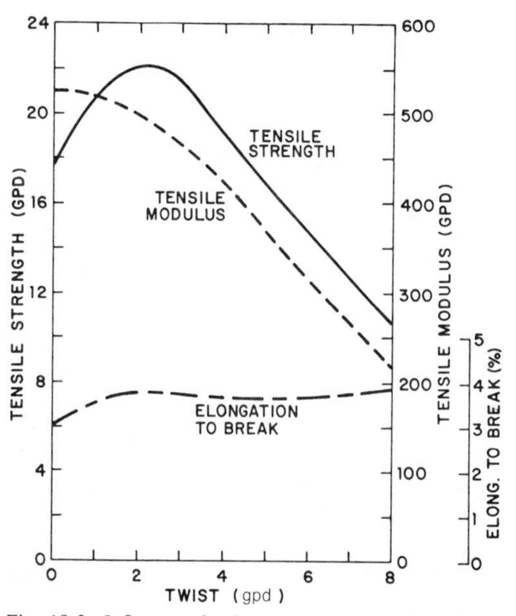

Fig. 18-3. Influence of twist on properties of Kevlar® 49 1420 denier yarn.

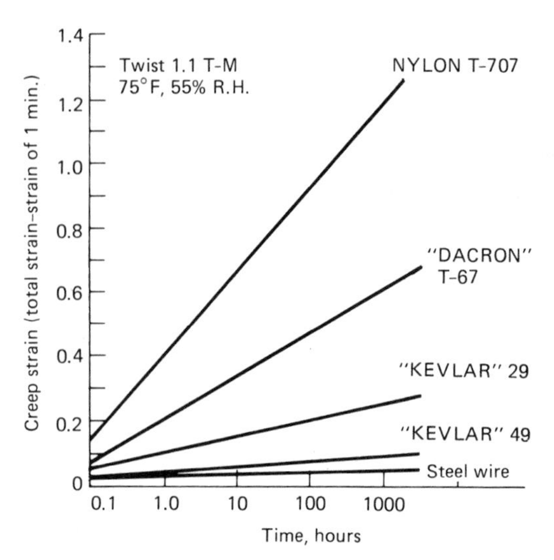

Fig. 18-4. Creep at 50% ultimate break strength.

Creep rates of the high-modulus aramid fibers are lower than for other organic fibers and approach the creep rate of steel. These rates are compared in Fig. 18-4 for fibers loaded to 50% of ultimate break strength.

The tension–tension fatigue performance of yarns and strands of high-modulus aramid fibers is excellent, as shown in Fig. 18-5 which compares fatigue of these fibers to steel and nylon fibers.

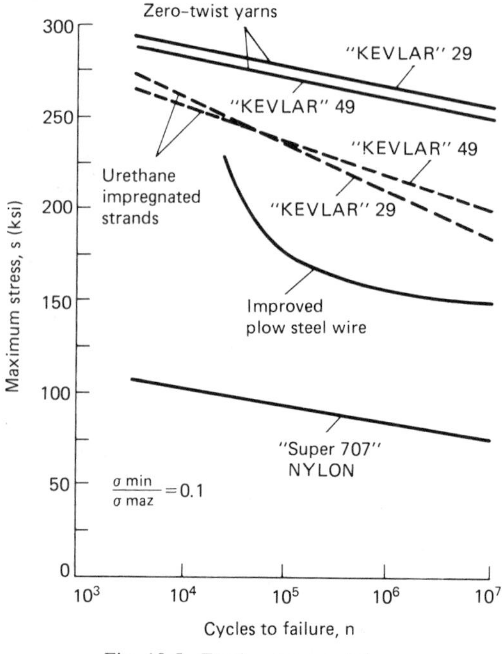

Fig. 18-5. Torsion–tension fatigue.

3.2 Thermal

Kevlar® aramid fibers have generally good thermal stability in both inert and oxidative environments; a comparison of break strength retention after air aging at 175°C is shown in Fig. 18-6 for aramid, polyester, and nylon fibers. The much better performance of the aramid fibers is due in part to their chemical structure, which contains only aromatic rather than aliphatic carbon–hydrogen bonds, a characteristic shared with other high-temperature polymers. Strength retentions after air aging from 150°C to 250°C are shown in Fig. 18-7. The activation energy calculated from air-aging exposures of bare yarn is about 19 to 20 kcal/mole.

The aramid fibers do not melt, but begin to decompose at higher temperatures; extrapolated zero-strength temperature for the yarn is reported to be about 625°C.[3] The fibers are capable of taking short exposures as high as 450°C in air. The fiber exhibits no T_g (glass transition temperature) up to at least 300°C, and no low-temperature transitions have been reported. Reference 5 reports several relaxations between 30°C and 270°C for noncommercial samples of poly (*p*-phenylene terephthalamide).

Mechanical properties of the fiber also vary with temperature. Plots of tensile strength versus test temperature are shown in Fig. 18-8 for the two high-modulus aramid fibers and several typical textile fibers. The curve is linear and is the same for both Kevlar® 29 and 49; tensile strength at 350°F (177°C) is typically about 75–80% of the value at room temperature. Figure 18-9 shows the corresponding plot for tensile modulus.

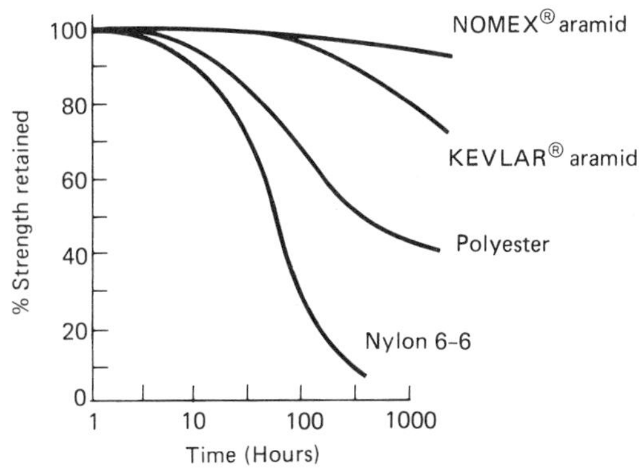

Fig. 18-6. Break strength retention (aged at 175°C).

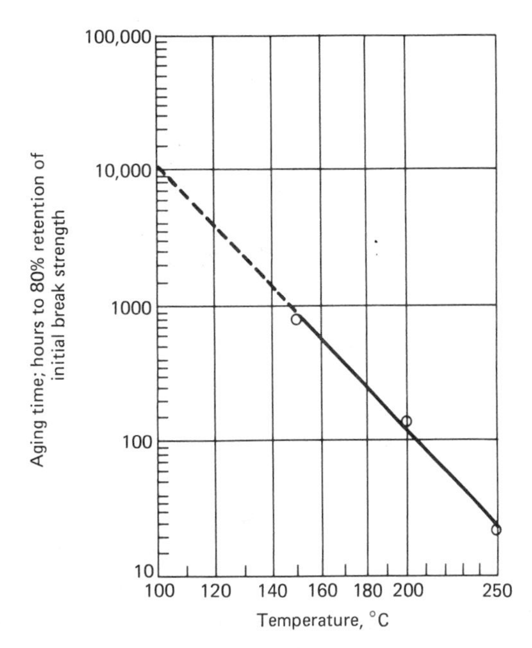

Fig. 18-7. Air aging of Kevlar® 49 yarn.

3.3 Chemical

Aramid fibers are generally more chemically resistant than most organic polymers, including many cross-linked polymers. They are essen-

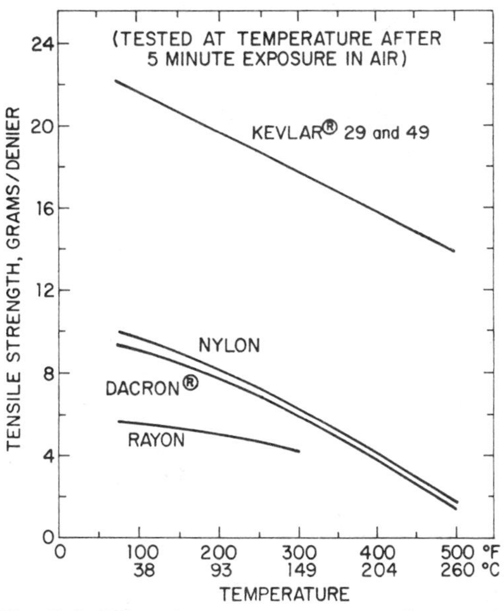

Fig. 18-8. Effect of temperature on yarn tensile strength. (Tested at temperature after 5-min. exposure in air.)

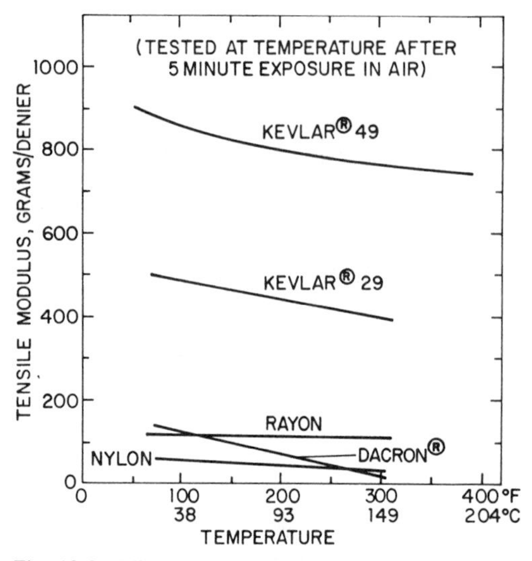

Fig. 18-9. Effect of temperature on yarn tensile modulus. (Tested at temperature after 5-min. exposure in air.)

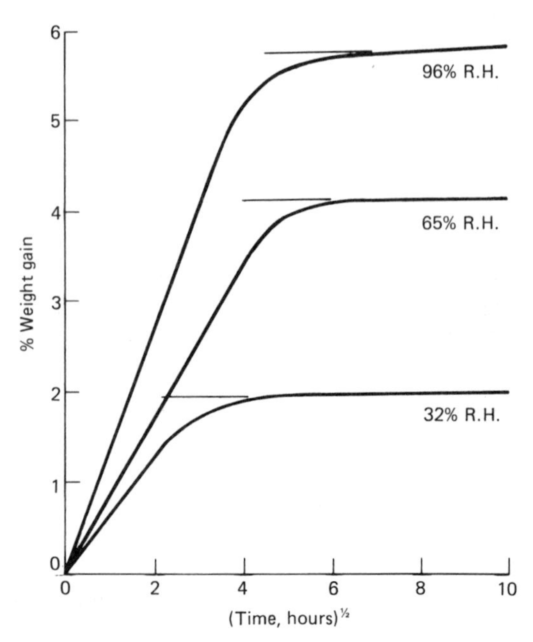

Fig. 18-10. Rate of moisture absorption of Kevlar® 49 yarns at various relative humidities.

tially inert to organic solvents, including chlorinated types, but are attacked by certain strong acids and bases at elevated temperatures. The resistance of aramid fibers to a variety of chemicals at varying temperatures and concentrations is shown in Table 18-1. Based on the limited number of environments tested, resistance of the bare fiber to a particular chemical is a good first approximation to performance of the fiber in a premium corrosion-resistant polyester resin laminate in that environment.

Moisture absorption characteristics of Kevlar® 49 are shown in Fig 18-10. The amount of moisture absorbed by the yarn at equilibrium varies with the relative humidity, as shown in the figure and can reach about 6% at high humidities. Figure 18-10 gives the rate at which the yarn absorbs moisture at room temperature for three humidity levels. In contrast to other fibers with similar moisture absorption, such as Nylon 66, Kevlar® 29 and 49 fibers do not lose stiffness after absorbing water. The stiffness and strength for the moisture-conditioned yarn are the same as for dry yarn.

Although the aramid yarns have excellent resistance to hydrolytic degradation at near-neutral pH, they are degraded at either high or low pH at elevated temperature.

3.4 Miscellaneous

The high-modulus aramid fibers absorb strongly in the ultraviolet (UV) region of the spectrum at about 340 nm; bare fibers suffer strength losses due to UV degradation. Because absorption is so strong ($\epsilon = \sim 25{,}000$), there is a substantial screening effect from the outer filaments in a bundle; this leads to a strong dependence of strength retention on bundle size. The effect is seen, for example, in unjacketed ropes or cables. In most applications, the fiber is covered or in a matrix resin that itself is UV-protected or painted, and fiber degradation is not a problem. (Long-term exposure results on composites are covered in section 4.)

The aramid fibers exhibit low flammability and smoke-generating characteristics in typical tests. Table 18-2 shows that the oxygen index value is 28, higher than that of most other organic fibers and of epoxy matrix resins; these fibers are rated self-extinguishing in the usual vertical flammability test. Bare fibers tested as fabric in the NBS smoke chamber at 2.5 watts/cm^2, flaming mode show smoke density values of about 1 to 7, compared to 100 to 200 for most matrix resins used in aircraft interiors and other parts; smoke performance of the aramid composite thus depends mainly on the matrix resin, as it does for glass or graphite reinforcement. Toxic gas emission tests on laminates with various epoxy or phenolic matrices show generally similar levels of hydrogen cy-

Table 18-1. Strength retention for Kevlar® yarns exposed to various chemical environments.

	Concentration, %	Temperature, °F	Time, hours	% Strength Retention	
				Kevlar 29	Kevlar 49
Hydrochloric acid	10	70	1000	17	35
	37	70	1000	12	18
	1	160	10	48	82
	1	160	1000	13	39
	10	160	10	15	34
	37	160	10	4	8
Hydrofluoric acid	10	70	100	90	94
Nitric acid	1	70	100	86	94
	10	70	100	21	26
	10	70	1000	10	18
	70	70	10	20	35
Sulfuric acid	1	70	1000	78	95
	10	70	100	91	88
	10	70	1000	40	69
	70	70	100	88	72
	70	70	1000	27	41
Sodium hydroxide	10	70	1000	26	46
	40	70	1000	34	57
	1	210	1000	25	38
	10	210	10	14	20
Ammonium hydroxide	28	70	1000	91	93
Chloroform	100	70	1000	100	100
Tetrachloroethane	100	70	1000	100	100
Perchloroethylene	100	210	10	100	100
Gasoline, leaded	100	70	1000	100	100
Transformer oil (Texaco #55)	100	140	500	95	100
Jet fuel (JP-4)	100	70	300	100	95
Seawater	100	—	1 year	98	98
Water	100	210	100	99	98

Table 18-2. Miscellaneous fiber properties.

Property	Value
Coefficient of linear thermal expansion	
Axial, PPM/°C	−4 to −5
Radial, PPM/°C	+58 to +61
Thermal conductivity	
Heat flow perpendicular to fibers, Btu · m./in. · ft²	0.285
Heat flow parallel to fibers, Btu · in./in. · ft²	0.334
Specific heat (at room temp), Btu/lb · °F	0.34
Heat of combustion, Btu/lb	15,000
Refractive index	
Parallel to fiber axis	2.0
Perpendicular to fiber axis	1.6
Oxygen index	28

anide whether the reinforcement is aramid fiber, glass, or graphite in the same matrix resin; the levels of nitrogen oxides and carbon dioxide are higher for the aramid-reinforced laminate, but much lower than called out in current aircraft specifications.

4. COMPOSITES

4.1 Fabrication

The high-modulus aramid fibers are fabricated into composites by methods similar, or identical, to those used for the inorganic reinforcing fibers. Although most of the common fabrication methods may be used, the most important are filament winding of roving, prepreg processing in autoclave or press, and hand lay-up with fabrics or woven roving. Pultrusion and matched-die molding of short fiber composites have also been used to some extent.

The high-modulus aramids are used in a variety of resin systems, depending on end-use requirements for the matrix. The most common matrix resins for these fibers are epoxies, which are used in most of the filament winding and prepreg applications. Polyester and vinyl esters are used in many of the hand lay-up applications such as boats, canoes, and ballistic armor, and in short-fiber molding compounds. Phenolic resins are used in fabric prepregs for aircraft interiors or military helmets; they have also been used in experimental short-fiber molding compounds. Polyimides, in general, have not been widely used with the aramids, but are candidates for use in printed circuit boards where high glass transition temperature and low thermal expansion are important; in addition, polyimides have occasionally been used with aramids in aircraft laminates that must survive short excursions to temperatures above the upper use-level of epoxy resins.

The aramid fibers are generally supplied as a finish-free roving for filament winding and as a weaving yarn with an appropriate weaving finish; woven fabrics are supplied finish-free or with a finish, applied by the weaver, to improve performance in a given matrix resin. Processing of aramid yarns for weaving is discussed in more detail in reference 6.

Because of moisture absorption by the fiber itself, it is always desirable to dry the fiber prior to impregnation and/or fabrication to obtain optimum laminate properties in the chosen matrix resin.[1] In practice, some resins are more susceptible than others to moisture in the fiber. The most common problem caused by fiber moisture is in anhydride-cured epoxies, where the use of undried fiber results in severe degradation of laminate properties. The same resins, used with well-dried fiber, give excellent laminates with high shear strengths.

4.2 Properties

The properties of unidirectional and fabric-reinforced composites made with Kevlar® 49

Table 18-3. Unidirectional composite lamina properties.

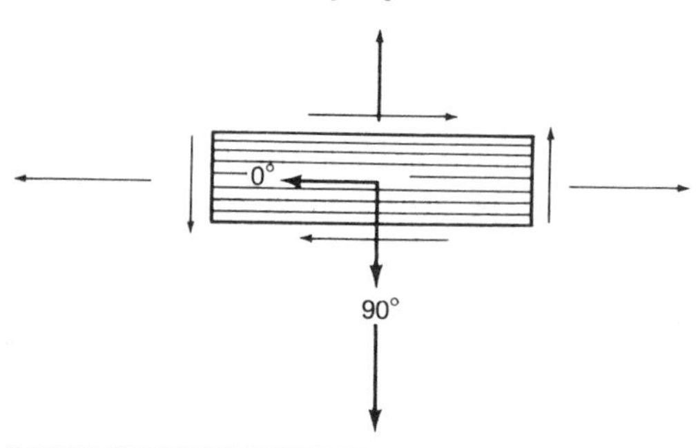

	Glass	"Kevlar" 49
Density, lb/in.^{3}	0.075	0.050
Tensile strength 0° (10^3 psi)	160	200
Compressive strength 0° (10^3 psi)	85	40
Tensile strength 90° (10^3 psi)	5.0	4.0
Compressive strength 90° (10^3 psi)	20	20
In-plane shear strength (10^3 psi)	9	6.4
Interlaminar shear strength (10^3 psi)	12	14
Poissons' ratio	0.30	0.34
Tens. and comp. modulus 0° (10^6 psi)	5.7	12
Tens. and comp. modulus 90° (10^6 psi)	1.3	0.8
In-plane shear modulus (10^6 psi)	0.5	0.3

are shown in Table 18-3. Although the tensile and compressive moduli in the fiber direction (0°) are the same for Kevlar® 49, the compressive strength is significantly lower than the tensile strength. The reason for this is that the fiber itself yields in compression at about 0.3 to 0.4% strain; this can be seen in Figs. 18-11 and 18-12, which show stress versus strain for unidirectional composites (60% fiber) in tension and compression. In tension the curve is linear to failure at about 1.90 to 2.0% strain, whereas in compression the curve bends at the strain corresponding to fiber yielding; the stress at which composite compressive yielding occurs is typically 30 to 40 ksi in unidirectional epoxy composites. It is important to note that after yielding the composite is still capable of carrying its yield load to much higher strains (3–

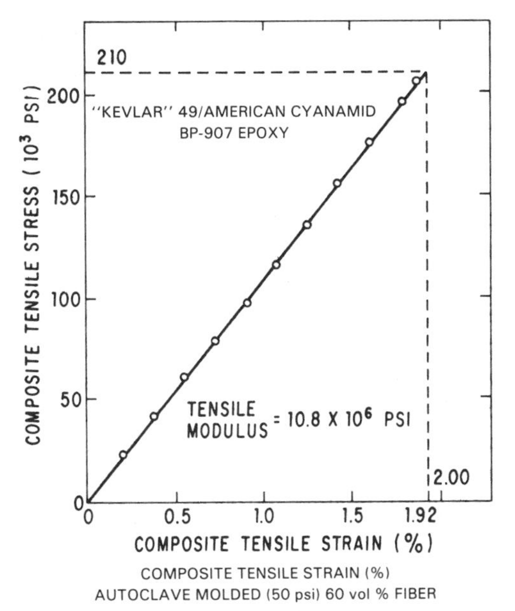

Fig. 18-11. Unidirectional composite tensile stress–strain curve.

4%) without actual fracture, and it is this failure mode which distinguishes the aramid fiber and composite from those of graphite or glass in compression.

The compressive yielding failure mode also affects flexural bending behavior of aramid composites, which typically fail by yielding on the compressive side at strains similar to those found for compression yielding of the fiber. A stress–strain curve for 4-point bending test is shown in Fig. 18-13.

Interlaminar shear strength of unidirectional epoxy composites of aramid fibers, as measured by the short beam shear test (ASTM D-2344), is typically in the 7 to 10 ksi range, and is dependent on matrix resin type and quality as well as the interface between resin and fiber. Short beam shear strengths in the 9 to 10 ksi range appear to be the highest attainable with aramid fibers, regardless of resin or interface improvements, because at this level adhesion to the fiber is sufficiently good to induce transverse failure in the filament, as evidenced by both fibrillation and filament splitting. In polyester and vinyl ester resins, typical shear strengths are 4 to 7 ksi for unidirectional composites; recent improvements in sizing technology show promise of increasing these values to the levels found for epoxies.

Shear strengths for fabric laminates, usually lower than for unidirectional laminates, are typically in the 4 to 8 ksi range for epoxies and 2 to 4 ksi for polyester, vinyl esters, polyimides, and phenolics.

The tension–tension fatigue performance of

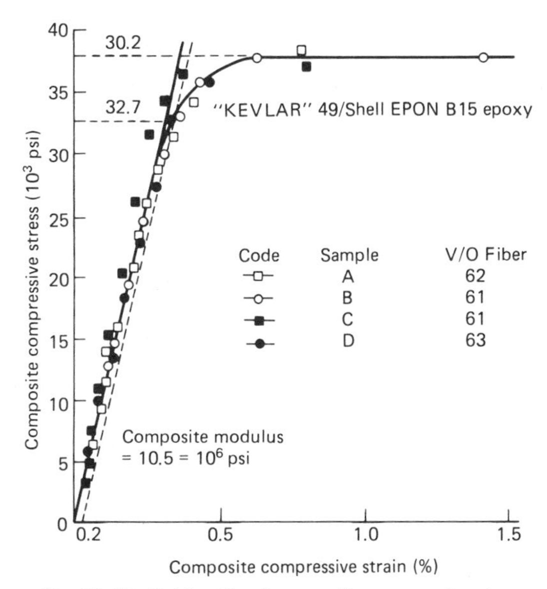

Fig. 18-12. Unidirectional composite compressive stress–strain curve.

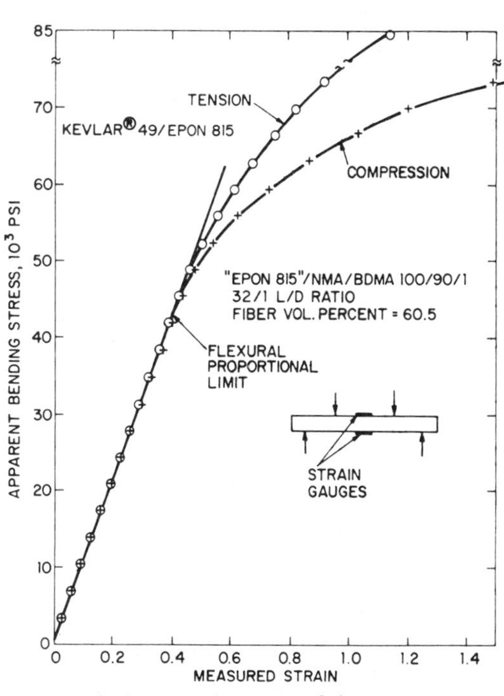

Fig. 18-13. Stress-strain curves of the outermost compressed and extended layers of a unidirectional composite in 4-point bending.

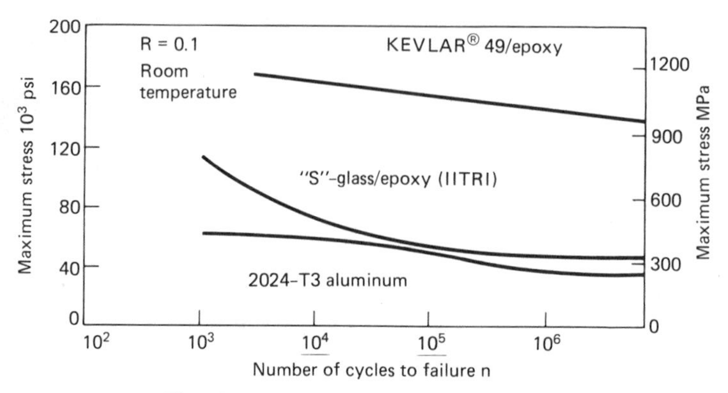

Fig. 18-14. Tension–tension fatigue behavior.

high-modulus aramid fiber composites, shown in Fig. 18-14, is generally better than that of glass-reinforced composites or aluminum, and equivalent to that of the best composites with other fibers.

Creep behavior of composites in the fiber direction is dominated by the fiber; the low fiber creep of the Kevlar® 49 aramid results in low composite creep. This is illustrated in Fig. 18-15, where the creep rates for Kevlar® 49/epoxy and S-glass/epoxy are seen to be very low and essentially equivalent. The stress is the same for both composites; so the higher modulus for the Kevlar® 49/epoxy results in lower initial strain.

The impact performance of solid panels reinforced with Kevlar® 29 and 49 fibers has been compared to that of the glass and graphite panels in an instrumented drop-weight test. Laminates of Kevlar® 29 and E-glass have the highest failure energy for a given panel weight, followed by those of Kevlar® 49 and graphite.[7] Other studies[8] have also shown composites reinforced with high-modulus aramid fibers to have significantly better impact properties than graphite composites; for example, fracture energies were higher (>3×), threshold energies for the onset of damage were higher (5×), and residual strength after impact was higher, as shown in Fig. 18-16.

The use of high-modulus aramid fibers in a hybrid construction with graphite fibers is a useful way to improve the impact and fracture toughness performance of the all-graphite composite; one example of this is shown in Fig. 18-17, where residual flexural strength after

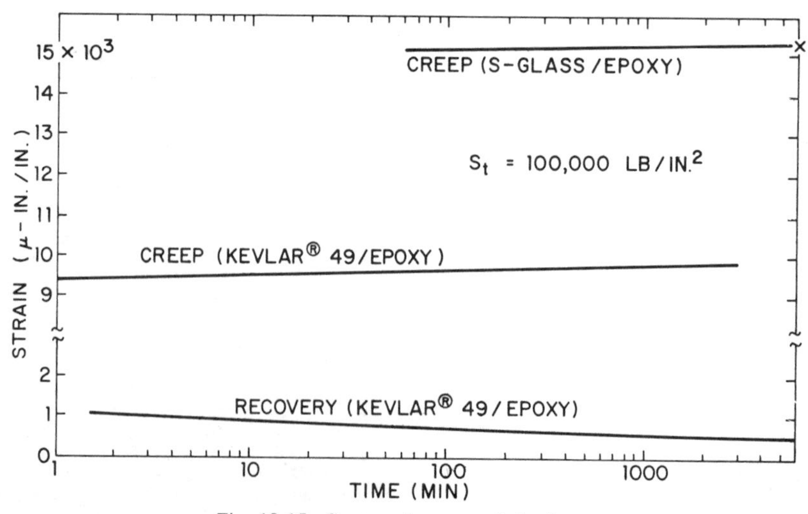

Fig. 18-15. Creep and recovery behavior.

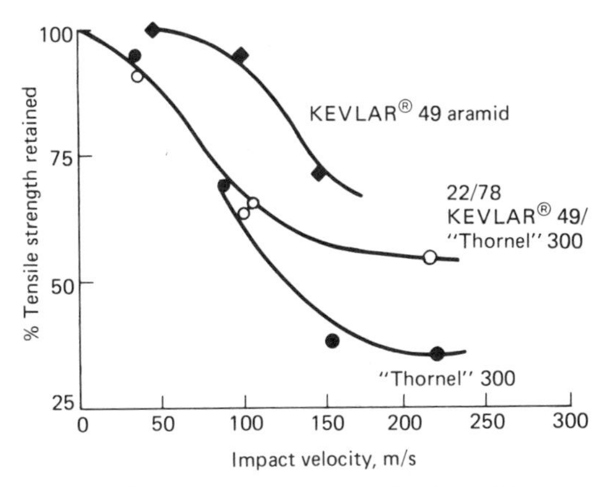

Fig. 18-16. Residual tensile strength of quasi-isotropic laminates.

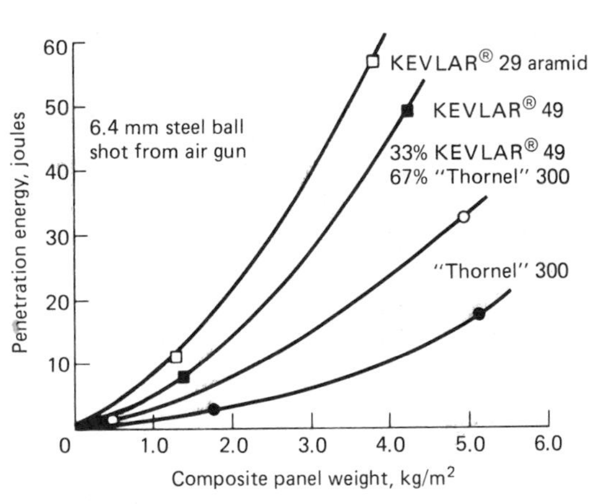

Fig. 18-17. Energy to penetrate epoxy quasi-isotropic laminates.

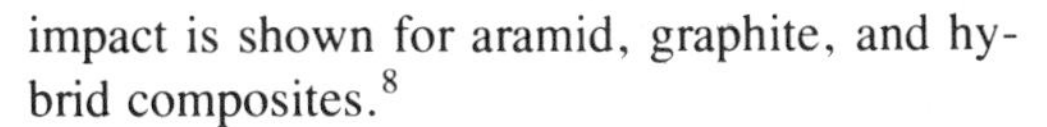

impact is shown for aramid, graphite, and hybrid composites.[8]

The coefficients of thermal expansion for unidirectional and fabric laminates reinforced with various fibers are given in Tables 18-4 and 18-5. Both high-modulus aramid fibers have slightly negative coefficients of thermal expansion in the axial direction, but strongly positive values in the radial direction. This enables one to control the laminate thermal expansion in the α_{11} direction (in-plane) by varying fiber volume; the corresponding α_{33} expansion (out-of-

Table 18-4. Thermal expansion coefficients (25°C to T_g), PPM °C^{-1}, unidirectional tape composites.

				Thermal Expansion Coefficient		
Reinforcement	Form	θ	V_F	α_{11}	α_{33}	Comments
None	—	—	0.00	49	51	3501-6 epoxy resin
Kevlar® 49	Tape	$[0]_{30}$	0.61	−3 to −4	59-66	$\alpha_{11}^F = -5, \alpha_{33}^F = +61$
Kevlar® 29	Tape	$[0]_{30}$	0.60	−2	53-60	$\alpha_{11}^F = -4, \alpha_{33}^F = +58$
Kevlar® 49	Tape	$[0/90]_{6S}$	0.56	1-2	75	
Kevlar® 29	Tape	$[0/90]_{8S}$	0.62	0-4	75	
Kevlar® 49	Tape	$[0, \pm 45, 90]_{3S}$	0.59	0-3	70	
Kevlar® 29	Tape	$[0, \pm 45, 90]_{3S}$	0.65	+3-6	75	

Table 18-5. Thermal expansion coefficients, 25°C to T_g, PPM, °C^{-1}, fabric-reinforced composites.

			Thermal Expansion Coefficient		
Material	Resin T_g, °C	V_F	α_{11}	α_{33}	Comments
Kevlar® 49 (120 fabric)	90-110	.44-.46	2-4	90-110	Hexcel F155
"	125-140	.60-.67	0 to −3	80-90	Epon 826/Ciba RD2
"	170-185	.54	2-3	65	Hercules 3501-6
"	205-215	.32-.42	2-6	48-63	Howe 7093
E-glass (? fabric)	205-215	?	11-15	44	Howe 7003

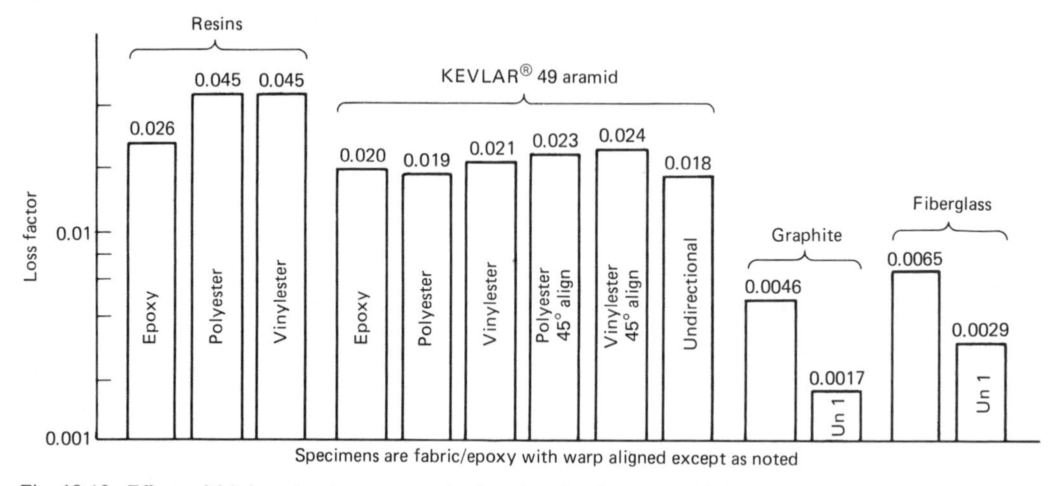

Fig. 18-18. Effects of fabric and resin on composite damping. Specimens are fabric/epoxy with warp aligned except as noted.

plane), however, is not restrained by the fibers because both fiber and resin have similar, positive expansion coefficients. The data in Table 18-5 show that α_{33} is highly dependent on the resin expansion coefficient, and that α_{33} values are larger for laminates reinforced with Kevlar® than for those reinforced with glass.

Vibration damping tests have shown that the damping of the Kevlar® fiber is much higher ($>10\times$) than that of glass or graphite fibers and approaches the damping of common matrix resins; refer to Figs. 18-18 and 18-19. The dominant source of damping in composites reinforced with Kevlar® is hysteresis in the fiber rather than interfacial slippage between resin and fiber.[9]

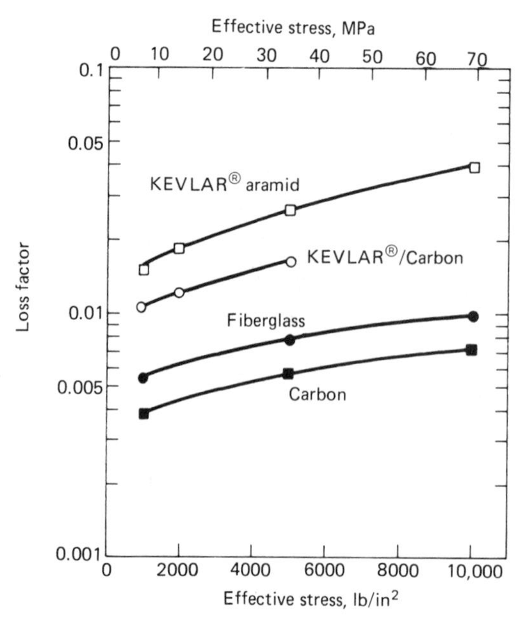

Fig. 18-19. Damping of epoxy resin composites.

Performance of composites under adverse environmental conditions is a concern in many applications. The matrix resin and its interface with the reinforcing fiber are usually the most important factors in determining performance in these cases, and this is also generally true with aramid fiber reinforcement. For example, in applications involving exposure to outdoor weathering conditions, it was found[10] that aircraft panel performance after eight years of flight service was essentially the same whether the panel was reinforced with Kevlar® 49 or E-glass. Concurrent worldwide coupon exposure tests showed that the aramid-reinforced panels absorbed about twice as much moisture as did their glass counterparts, but that the strength retention ratios were essentially the same for both. The conclusion from these and other coupons and part exposures from accelerated tests at higher temperature,[14] and from five-year tests of polyester resin laminates in water, is that the primary effect of moisture is plasticization of the resin matrix, leading to reductions in matrix-dominated properties. On a percent retention basis, performance of the aramid fiber composites is similar to that of glass or graphite composites.

Aramid fiber laminates also show good resistance to rain erosion when coated with typical, state-of-the-art rain erosion coatings (either urethane or fluorocarbon) and tested as solid panels or facings on honeycomb. These test results have been confirmed by the performance of aramid-reinforced radomes that have been in service for several years.

5. APPLICATIONS

The unique combination of properties found in the aramid fibers has led to their use in a wide variety of commercial applications. This section will highlight some of the applications in each major end-use area.

5.1 Aircraft/Aerospace

High-modulus aramid fibers are used extensively in aircraft/aerospace applications, primarily because of their light weight, good damage tolerance, and high strength and stiffness. In aircraft the modest compressive strength of the fiber has, so far, limited use mostly to nonstructural and semistructural parts, both interior and exterior, in large and small planes and helicopters. Some of the exterior parts on the de Havilland DASH-8 are illustrated in Fig. 18-20. Many other commercial aircraft utilize similar parts. Interior parts include partitions, storage bins, panels, ducts, cargo liners, and floor panels.[11,12]

Most of these applications use epoxy fabric prepregs in both 250°F and 350°F resin systems. Fabric weights are typically 1.8 to 5 oz/yd^2, and most constructions are sandwich panels with Nomex® aramid honeycomb cores and co-cured facings. For interior applications there is some use of phenolic resin systems to replace the typical 250°F curing fire-retardant epoxies to reduce smoke emission. In many exterior parts, hybird constructions of Kevlar®/graphite are used; these are usually ply-for-ply rather than intra-ply hybrids.

Other aircraft applications include radomes, passenger seats, seaplane floats, engine nacelles, and propellers. The Hartzell composite propeller is significantly lighter than the aluminum model it replaces, and is made from unidirectional Kevlar® 49/epoxy over a high-density urethane foam core. In addition to weight savings, the use of aramid fiber reinforcement gave better vibration damping, impact resistance, and fatigue life.

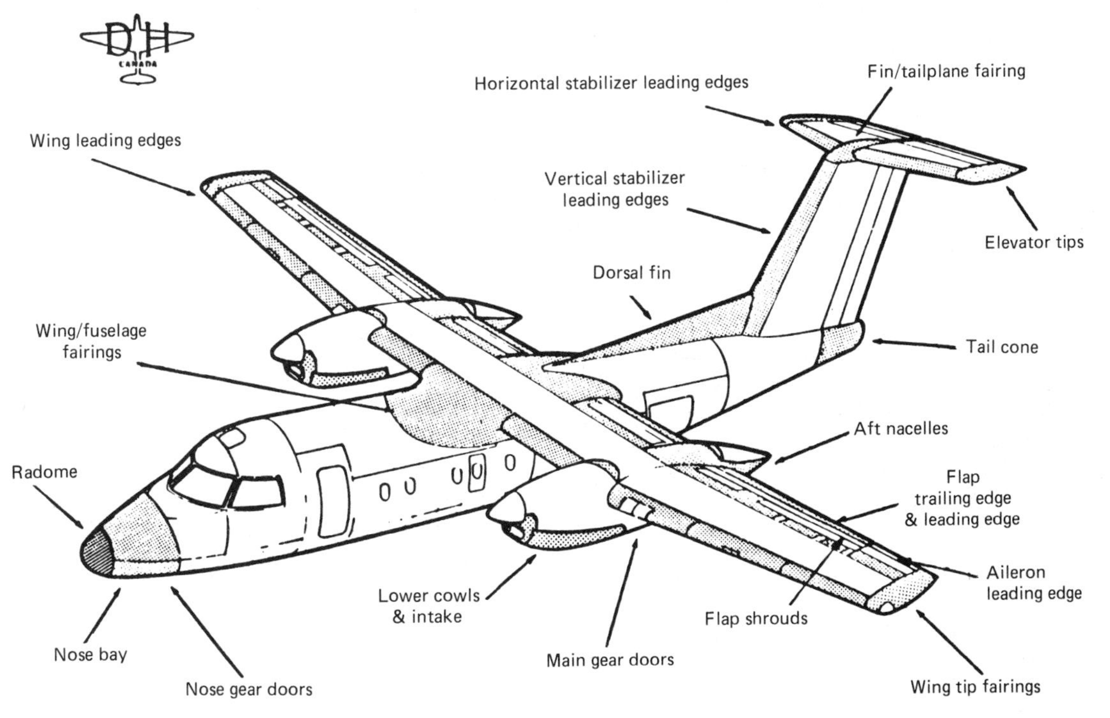

Fig. 18-20. Dash 8 composite structures.

The many helicopter applications include fairings, doors, canopies, cowlings, fuel pods, and floors.[13]

The major aerospace applications for aramid fibers have been in rocket motor cases and other pressure vessels. These are usually manufactured by filament winding, and the matrix resins are usually epoxies. In these applications the important fiber attributes are high strength and light weight, that is, high specific tensile strength; high specific modulus can be important in some designs. The first rocket motor case to be made of high-modulus organic fibers was the Trident, in which all three stages are filament wound from Kevlar® 49/epoxy, resulting in a weight savings of about 800 lb compared to a glass/epoxy construction. Motor cases for both the MX and Pershing II are also of Kevlar® 49/epoxy.

Pressure vessels for gas containment can be made lighter by the use of Kevlar® 49/epoxy overwrapped on a metal liner. A cylindrical model used by Boeing to inflate escape slides resulted in a weight reduction from 42 lb for the metal bottle to 20 lb for the composite replacement. A summary of properties for filament-wound pressure vessels of Kevlar® and S-glass is found in Table 18-6.

5.2 Marine

The use of high-modulus organic fiber reinforcement in marine applications offers the advantages of light weight, higher stiffness, good impact and repeated impact strength, good fatigue resistance, and vibration damping.

These advantages are typically utilized in canoes, kayaks, offshore racers, sport fishing boats, and patrol boats. Canoe and kayak constructions usually employ fabric reinforcement in polyester, vinyl ester, or epoxy matrix resins. The heavier constructions for larger boats are usually woven roving, either alone or alternating with glass chopped strand mat. Typical properties for woven roving laminates are shown in Table 18-7.

5.3 Printed Circuit Boards

The combination of its good dielectric properties, high thermal stability, high strength and stiffness, and low coefficient of thermal expansion has led to the use of Kevlar® as a candidate reinforcing fiber for printed circuit boards, especially those used for leadless ceramic chip carriers where the in-plane thermal expansion of the board must be matched to that of the ceramic (aluminum oxide).[15] Matching thermal expansions is accomplished by choosing the appropriate fiber volume for the specific matrix resin being used; an example is shown in Fig. 18-21 for two resins reinforced with a lightweight fabric of Kevlar® 49. The desired laminate expansion can be obtained with fiber volumes in the 35 to 45% range. Matrix resins for these applications are typically epoxy, poly-

Table 18-6. Filament-wound pressure vessel property summary.

Property	Reinforcement	
	Kevlar® 49 aramid	"S"-glass
Composite density, lb/in.^{3} (g/cm^3)	0.044 (1.22)	0.069 (1.91)
Fiber volume, %	65	65
Composite tensile stress, lb/in.^{2} (MPa)	223.300 (1540)	196,900 (1358)
Composite modulus, 10^6 lb/in.^{2} (MPa)	13.2 (91 014)	8.8 (60 676)
Specific strength, 10^6 in. (10^6 cm)	5.08 (12.90)	2.85 (7.24)
Relative specific strength	1.78	1.0
Specific modulus, 10^8 in. (10^8 cm)	3.00 (7.62)	1.28 (3.25)
Relative specific modulus	2.34	1.0
Relative cyclic performance 90% of ultimate	10	1
Performance factor, $\frac{PV*}{W}$, 10^6 in. (10^6 cm)	1.6 (4.1)	1.2 (3.0)

$* \frac{\text{Pressure \% Volume.}}{\text{Composite Weight}}$.

Table 18-7. Properties of marine laminates in flame-retardant resins.

	Kevlar® 49 Aramid		E-Glass
Construction	CSM $(KWR)_6$	CSM $(KWR)_6$	CSM $(GWR)_6$
Matrix resin	"Dion" 6692	"Derakane" 510-A40	"Dion" 6692
Resin content, % by weight	50.0	51.2	44.5
Areal density, lb/ft^2 (kg/m^2)	1.26 (6.16)	1.30 (6.36)	1.91 (9.34)
Specific gravity, g/cm^3	1.32	1.30 (1.40)	1.83
Tensile strength, 10^3 psi (MPa)	50.9 (351)	57.3 (395)	45.3 (312)
Tensile modulus, 10^6 psi (GPa)	3.81 (26.3)	3.41 (23.5)	2.88 (19.9)
Flexural strength, 10^3 psi (MPa) (mat up)	53.8 (371)	48.0 (331)	83.5 (576)
Flexural modulus, 10^6 psi (GPa) (mat up)	2.87 (19.7)	2.67 (18.4)	2.26 (15.6)
Interlaminar shear strength, 10^3 psi short beam method (MPa)	3.40 (23.4)	4.17 (28.7)	4.75 (32.8)

CSM is 1.0 oz/ft^2 (306 g/m^2) chopped strand glass mat.
KWR is 13.5 oz/yd^2 (458 g/m^2) woven roving of Kevlar® 49 aramid fiber.
GWR is 24 oz/yd^2 (815 g/m^2) woven roving of glass fiber.

imide, or modified versions of either. The effect of frequency and moisture content on the dielectric constant of a typical polyimide board reinforced with Kevlar® or E-glass is shown in Fig. 18-22.

5.4 Composite Armor

The excellent ballistic performance of Kevlar® aramid fibers has led to applications in both unimpregnated fabric and composite forms. In flexible, multilayered fabric vest, Kevlar® offers protection equivalent to that of nylon, but at half the weight; these vests are currently used by police and military personnel. Rigid armor, made from ballistic fabrics in an appropriate matrix resin, is used in vehicular, shipboard, aircraft/helicopter, and various military applications, replacing fiberglass, ceramic, or steel. A comparison of the ballistic performance of an aramid composite with glass and steel showed that at any given V_{50}, the aramid composite offers a 50 to 60% weight advantage. (V_{50}, a measure of stopping ability, is the projectile velocity at which 50% of the projectiles are stopped.) Another recent introduction is a new U.S. Army helmet that replaces the old steel helmet and provides substantially more ballistic protection at the same weight.

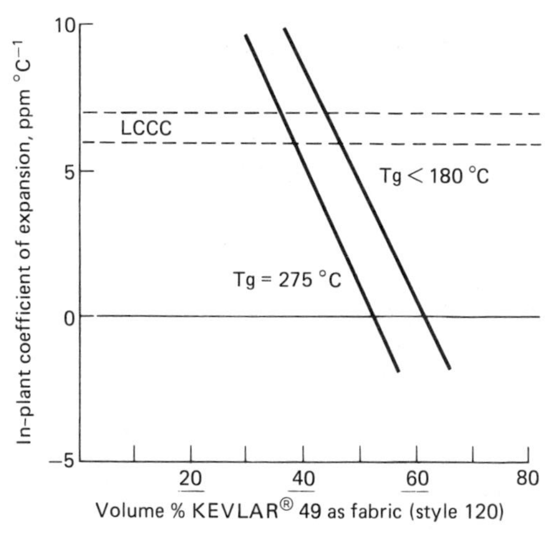

Fig. 18-21. In-plane thermal expansion coefficients of laminates reinforced by style 120 fabric of Kevlar® 49. (*DuPont registered trade mark.)

5.5 Pulp and Short Fibers

A combination of high strength, high-temperature capability, nonabrasiveness, and good wear resistance has led to the use of short fibers or pulp of Kevlar® aramid fibers in many asbestos-replacement applications. These include various types of brakes, clutch facings, gaskets, and papers. The good wear resistance of brake pads reinforced with Kevlar® is shown in Fig. 18-23. Because of the relatively small amounts of aramid fibers needed to replace asbestos, as well as good wear performance, lifetime costs of this material are sometimes lower than the cost of asbestos reinforcement.

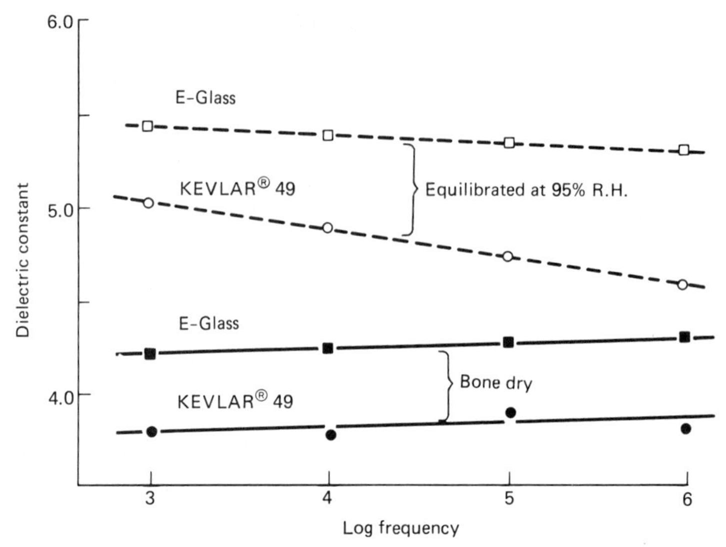

Fig. 18-22. The effect of frequency and moisture on the dielectric constants of laminates reinforced by Kevlar* 49 and E-glass. Resin: Kerimid[+] 601. (*DuPont registered trade mark; [+]Rhone-Poulenc Regd. T.M.).

5.6 Ropes and Cables

A wide variety of rope and cable structures have been demonstrated from Kevlar® 29 and 49, including three-strand, eight-strand, braided, parallel lay, and wire rope types of construction. An excellent compilation of comparative yarn and rope properties is contained in reference 16. As a general guideline, in properly made ropes of comparable construction, those made from Kevlar® will have about twice the strength of nylon or polyester ropes of equal weight. Marine rope and cable uses for Kevlar® 29 and Kevlar® 49 include not only mechanical ropes but also reinforcement for electromechanical cables and fiber optic cables.

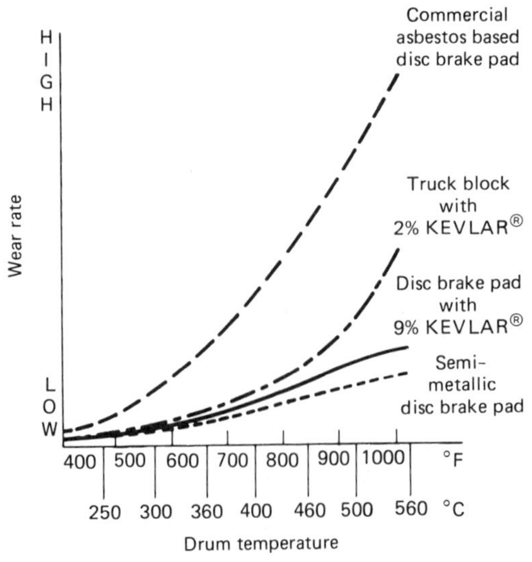

Fig. 18-23. Wear of brake pads at high temperatures.

5.7 Coated Fabrics

Resin-coated fabrics of Kevlar® aramid offer higher strength at low weight, low elongation, and good puncture, cut, and tear resistance, compared to other fabrics.[17] Applications include inflatable boats and rafts, oil containment booms, and a variety of other uses. Resins are usually urethane or neoprene, and the fabrics are typically based on Kevlar® 29 yarn.

REFERENCES

1. Technical Information Bulletin K-5, "Characteristics and Uses of Kevlar® 49 Armaid High Modulus Organic Fiber," E. I. Du Pont De Nemours & Company, Inc., Textile Fibers Department, Wilmington, DE 19898, Sept. 1981.
2. Panar, M. et al., *Journal Polymer Science, Polymer Physics Ed.* 21, 1955.
3. Wilfong, R. E. and Zimmerman, J., *Journal Polymer Science, Applicable Polymer Symposium* 31: 1–21, 1977.
4. Morrison, C. E. and Bowyer, W. H., "Advances in Composite Materials," *Proceedings* Third International Conference on Composite Materials, Aug. 1980, p. 233.

5. Kunugi, T. et al., *Journal Applicable Polymer Science* 24: 1039, 1979.
6. Technical Information Bulletin K-3, ''Processing Yarns of Kevlar® Aramid for Weaving,'' E. I. Du Pont De Nemours & Company, Inc., Textile Fibers Department, Wilmington, DE 19898, Dec. 1978.
7. Wardle, M. W. and Tokarsky, E. W., ''Composites Technology Review,'' pp. 4–10, Spring 1983.
8. Dorey, G. et al., *Composites*, pp. 25–31, Jan. 1978.
9. Pulgrano, L. J. and L. H. Miner, *SAMPE, National Symposium* 25: 557, 1980.
10. Stone, R. H., *SAMPE, National Technology Conference* 15, 1983.
11. Hammer, R. H., ''Advances in Composite Materials,'' *Proceedings*, Third International Conference on Composite Materials, Paris, Aug. 1980.
12. John, L. K., SAE Technical Paper 810640, Apr. 1981.
13. Crist, D., American Helicopter Society Preprint No. 79-31, 35th Annual National Forum, May 1979.
14. Baker, D. J., NASA TM 84637, Mar. 1983.
15. Packard, David C., *SAMPE J.*, p. 6, Jan./Feb. 1984.
16. Ferer, K. M., Swenson, R. C., NRL Report 8040, Washington, D.C., Nov. 1976.
17. Wardle, M. W., ''Coated Fabrics for Industrial Applications Seminar, Boston, June 1977.

19

Boron and Silicon Carbide Filaments

Raymond J. Suplinskas
AVCO Specialty Materials Textron
Lowell, Massachusetts

James V. Marzik
AVCO Specialty Materials Division
Lowell, Massachusetts

CONTENTS

1. INTRODUCTION

Boron fibers, first demonstrated at Texaco Experiments Inc. in 1959, were the first of a family of high-strength, high-modulus, low-density reinforcements developed for advanced aerospace applications. The initial development of boron was nurtured by the Air Force Materials Laboratory with the goal of producing higher-performance systems through the use of materials with higher specific properties (i.e., strength and stiffness divided by density). This Department of Defense interest continued and led to the achievement of routine commercial production of the fibers on a scale where they could be considered a viable structural material by the aerospace industry. As a result, boron/epoxy composites find substantial application in the F-14, F-15, and B-1 airplanes as well as numerous other aerospace uses and also sporting goods applications.

In the exploration of other possible applications it became evident that certain requirements demanded the consideration of alternate or improved fibers. These requirements relate to for example, fiber cost, processing cost, new matrix materials, temperature capability, and electrical properties. This need led to the development of a number of other ceramic fibers and their variations. However, none of these fibers has as yet achieved the same degree of commercial utilization as boron, although some have well-developed data bases and promise to become at least as significant as boron in the near future. Hence, this chapter will include a discussion of the known processes and properties, recognizing that further advances unquestionably will be made.

This chapter does not attempt to provide a complete history of the development of these fibers; such accounts already exist in the literature.[1,2] Rather, the aim is to provide an introduction to this technology through a description of the manufacturing process and a discussion of some current applications.

2. BORON MANUFACTURE AND PROPERTIES

The only commercial producer of boron fibers in the United States at present is Avco's Specialty Materials Division in Lowell, Massachusetts. Following sections describe the unique chemical vapor deposition process employed to manufacture the fibers, as well as fiber properties, preform manufacture, composite properties, and applications.

2.1 Chemical Vapor Deposition Process

A schematic diagram of the basic deposition unit is shown in Fig. 19-1. The reactor module consists of a glass deposition tube fitted with

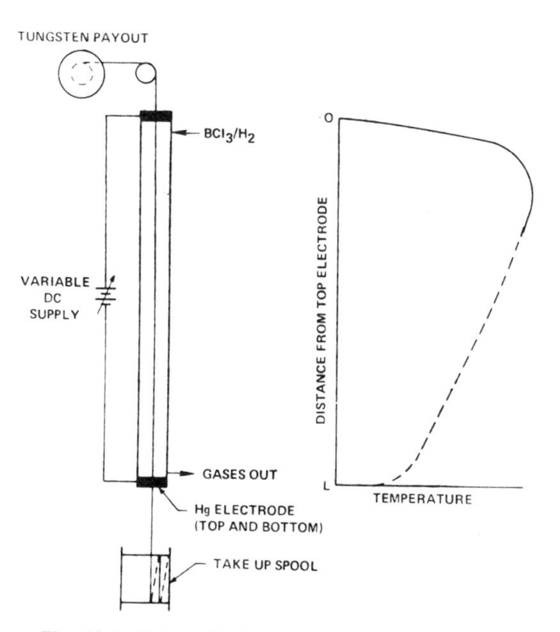

Fig. 19-1. Schematic diagram of boron filament reactor; temperature profile in single stage reactor.

gas inlet and outlet ports, two mercury-filled electrodes, a variable DC power supply connected to the two electrodes, a tungsten substrate payout system, and a boron filament take-up unit. The tungsten substrate, typically 0.0005 in. in diameter, is drawn through the reactor and heated to about 1300°C by the DC power supply. Prior to entering the deposition reactor, shown schematically in Fig. 19-1, the tungsten substrate is first passed through a short "cleaning" stage, not shown in the figure, in which the substrate is heated to candescence in a hydrogen atmosphere to remove surface contaminants and residual lubricants, which are used in the drawing of tungsten wire. A stoichiometric mixture of boron trichloride and hydrogen is introduced at the top of the reactor. At the 1300°C temperature level, a mantle of boron is deposited on the tungsten via the reaction:

$$BCl_3 + 3/2H_2 \rightarrow B + 3\ HCl$$

Exhaust gases, consisting of HCl, unreacted H_2, BCl_3, and intermediate species, are removed through the outlet port at the bottom of the reactor and processed to remove the unreacted BCl_3. Typical diameters of the filament exiting from the reactor are 0.004, 0.0056, or 0.008 in., depending upon the drawing rate.

A number of chemical reactions occur simultaneously during the formation of the filament. For example, boron is deposited via the halide–hydrogen reaction indicated above, and byproducts such as B_2H_6, $HBCl_2$, and B_2H_5Cl are formed in the gas phase, in addition to the HCl indicated by the stoichiometric relation. In the core of the filament, a solid-state reaction between tungsten and the boron mantle occurs to form a number of tungsten borides. Boride phases that have been identified in boron filament are δ-WB, W_2B_5, and WB_4.[3]

Figure 19-1 also shows the temperature profile of a filament that is heated with a DC power supply; the temperature rises rapidly and falls gradually as the filament passes through the reactor, and a combination of complex chemical and physical phenomena are responsible for the particular shape of the profile. The sharp rise in temperature as the substrate enters the reactor results from chemical reactions occurring in the core; that is, low-resistivity tungsten is con-

Fig. 19-2. Photograph of boron filament plant.

verted to higher-resistivity tungsten boride phases. The subsequent increase in diameter of the boron mantle causes the temperature to fall because of resistance considerations: the resistance decreases because the cross-sectional area increases and because the high-temperature resistivity of boron is about the same as that of tungsten boride. The highest temperature along the profile in Fig. 19-2 is called the deposition or ''hot spot'' temperature. The temperature at the reactor outlet has been measured to be about 300°C lower than the hot spot. A photograph of one hundred boron filament reactors, which produce about 9000 lb/yr in Avco's plant, is shown in Fig. 19-2.

Other fiber-forming processes were attempted, including the thermal decomposition of diborane and the drawing of fibers from molten boron. The CVD process described above, however, remains the only commercial process.

2.2 Conversion of Boron Filament into Preforms for Composites

The primary preform for organic matrix composites using boron is a prepreg tape. This is produced by collimating and coating large arrays of filaments with epoxy resin using either hot melt or film techniques. The tape typically is supported on a thin fiberglass backing to provide handleability during subsequent processing. The filament comprises about 65% by weight of the tape; for 4 mil fibers, the fiber density is a nominal 208/in. These parameters lead to composites with 50% fibers by volume. Figure 19-3 is a photograph of Avco's tape manufacturing equipment, and a photograph of 3-in.-wide boron/epoxy prepreg is shown in Fig. 19-4. This material is used to lay up composite structures using fabrication procedures developed in the 1950s for fiberglass structures and described elsewhere.[4]

Woven ''dry'' preforms can also be produced. This product consists of collimated arrays of fibers held in the transverse direction by a more flexible interweave material. There is a great deal of flexibility in choice of interweave; materials employed have included organic fibers, metallic wires, and ceramic yarns. The woven forms can be produced with the boron fibers in either the warp (continuous) or fill directions. These forms offer some advantages in shipping and later fabrication. For example, the

Fig. 19-3. Equipment for manufacturing boron/epoxy tape.

refrigerated storage that is necessary for boron/epoxy tape is eliminated. Also the woven preform can be placed in molds for casting or injection-molding compositing procedures. A photograph of Avco woven preform is shown in Fig. 19-5.

Fig. 19-4. Three-inch-wide boron/epoxy tape.

For metal matrix applications, the preform consists of either sheet or tape of boron/aluminum monolayer. The basic process consists of pressing an array of fibers between aluminum foils. At pressures of 5000 psi, the foils deform around the fibers and bond to the fibers and to each other. This preform can then be laid up to form structures in the same way that prepreg is used. Variations on the above process include plasma spraying of metal onto the fiber array and the step-pressing of continuous tape. In the latter, discrete segments of fiber-foil sandwiches are sequentially diffusion-bonded, producing preform that is continuous in the fiber direction.

2.3 Fiber Mechanical Properties

The strength of boron fibers is determined by the statistical distribution of flaws produced during the deposition process. The major flaw types are voids near the tungsten boride/boron interface, internal stresses "locked in" during deposition, and surface flaws, principally crystalline or nodular growth. A discussion of the flaws is given in reference 1. Because of these flaws, a histogram of fiber strengths does not

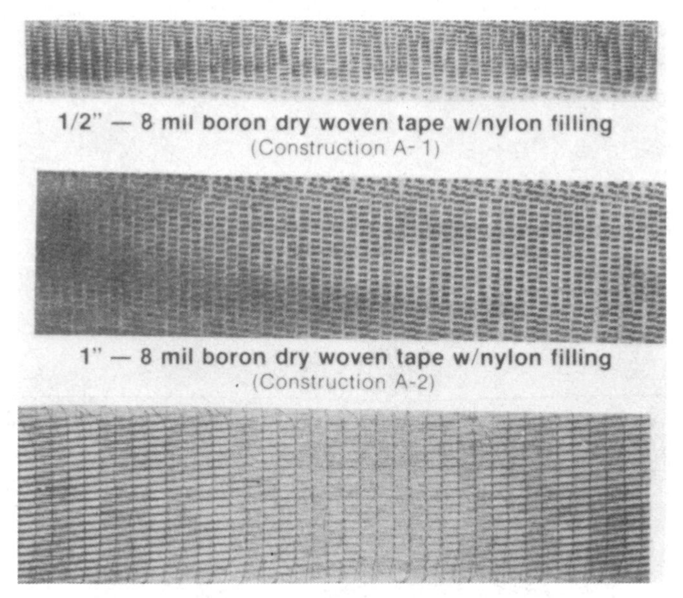

Fig. 19-5. "Dry-woven" boron tape.

follow a normal distribution, but rather is skewed by having a low-strength tail. The distribution is better described by Weibull statistics. Figure 19-6 shows a histogram of boron fiber strengths. It shows an average UTS of about 550 ksi, with a coefficient of variation of about 15%. These values have shown steady improvement over the nearly 20 years that these

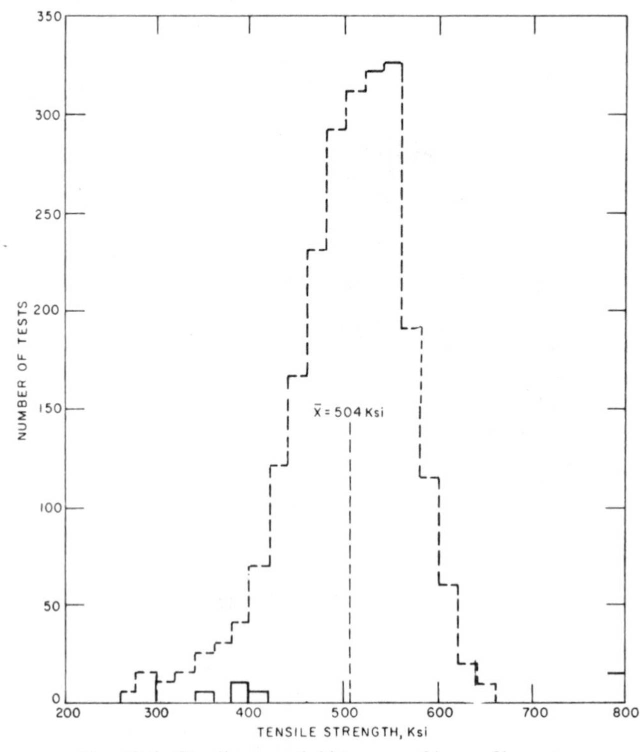

Fig. 19-6. Tensile strength histogram of boron filament.

fibers have been in commercial production. Early fibers showed average strengths of 450 ksi with 20% coefficients of variation.

It has been demonstrated by DiCarlo and Wagner[5] at NASA that post-production treatment of the fibers can reduce the severity of these flaws, leading to an average UTS of 800 ksi with a coefficient of variation of 5%.

Fiber modulus, on the other hand, is an intrinsic material property and shows very little variation. It is best regarded as a composite property depending on the (reacted) substrate and the volume fraction of pure boron in the fiber. For 4 mil boron on tungsten, the value is 55 Msi.

2.4 Composite Properties

The properties of boron composites depend in a critical way on the lay-up sequence. Properties parallel to the fiber direction are dominated by fiber properties, while those perpendicular to the fibers are determined by the matrix material. In a typical application, both ply orientation and number of plies vary across the final part. Tables 19-1 and 19-2 list the properties achieved in two particular lay-up sequences: unidirectional and ±45°.

Boron/aluminum composites exhibit similar properties, with the transverse properties dominated by those of the matrix.

2.5 Cost and Availability

The 1986 cost of boron fiber was approximately $350/lb (70,000 linear feet). The cost of the tungsten substrate contributes about one fourth of the price, and other raw materials add another fourth, with labor, depreciation, and overhead rounding out the figure. Much lower prices had been anticipated on the basis of large-volume production; however, demand has not exceeded 30,000 to 40,000 lb/yr.

Because of value-added processing, the cost of boron/epoxy prepreg is 75 to 100% of the cost of the fiber, depending on quantity and the matrix system. Economic improvements have been made over the years in prepreg production, but limited volume has kept the price relatively constant.

2.6 Applications

The primary use for boron composites is in the fabrication of weight and stiffness critical components for military aircraft, primarily the

Table 19-1. Properties of unidirectional boron/epoxy laminates (30% volume/fraction)

Property	Room Temperature Value	Value at 350°F
Longitudinal tensile strength, ksi	220	260
Transverse tensile strength, ksi	10.5	6.0
Longitudinal compressive strength, ksi	350	116
In-plane shear strength, ksi	15.3	5.5
Interlaminar shear strength, ksi	13	7
Transverse compressive strength, ksi	40	11
Longitudinal tensile modulus, Msi	30	29.9
Transverse tensile modulus, Msi	2.7	1.1
Longitudinal compressive modulus, Msi	30	29.9
Transverse compressive modulus, Msi	2.7	1.1
In-plane shear modulus, Msi	0.7	0.3
Longitudinal Poisson's ratio	0.21	0.3
Transverse Poisson's ratio	0.019	0.008
Physical Constants		
Density, lb/in^3	0.0725	0.0725
Longitudinal coefficient of thermal expansion, 10^{-6} in./in.·°F	1.3	3.0
Transverse coefficient of thermal expansion, 10^{-6} in./in.·°F	20.6	19.6

Table 19-2. Properties of a + 45°/boron/epoxy laminate (50% volume fraction).

Property	Room Temperature Value	Value at 350°F
In-plane shear strength, ksi	77	60
Ultimate longitudinal strain, %	2.6	5.0
Ultimate transverse strain, %	2.6	5.0
Longitudinal tensile modulus, Msi	2.6	1.16
Transverse tensile modulus, Msi	2.6	1.16
Longitudinal compressive modulus, Msi	2.6	1.16
Transverse compressive modulus, Msi	2.6	1.16
In-plane shear modulus, Msi	8.0	7.6
Longitudinal Poisson's ratio	0.848	0.927
Transverse Poisson's ratio	0.848	0.927
	Physical Constants	
Density, lb/in.^{3}	0.0725	0.0725
Longitudinal coefficient of thermal expansion, in./in.·°F	3.10	3.78
Transverse coefficient of thermal expansion, in./in.·°F	3.10	3.78

F-14, F-15, and B-1. A schematic diagram of the F-14 Tomcat is given in Fig. 19-7, where the shaded portions of the aircraft indicate the locations of boron/epoxy on the horizontal stabilizers of the empennage structure. Further, as the data in Fig. 19-7 show, if the stabilizers had been constructed of titanium, they would have weighed about 950 lb, but with the use of the stronger, stiffer, and lighter boron composite material, the equivalent performance stabilizer weights only 776 lb. The decrease of almost 200 lb on the tail structure provides

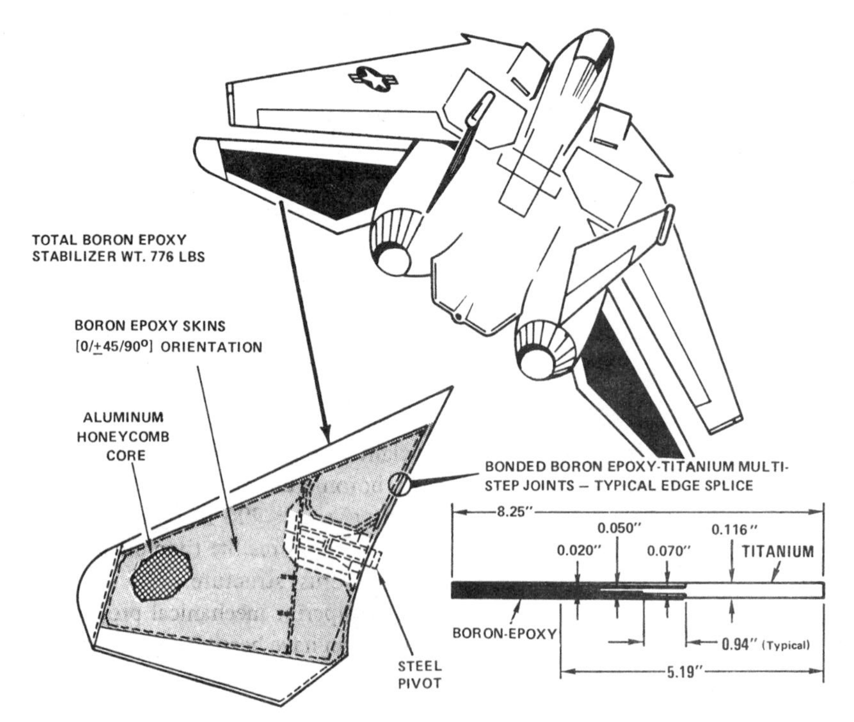

Fig. 19-7. Diagram of the F-14 and horizontal stabilizer.

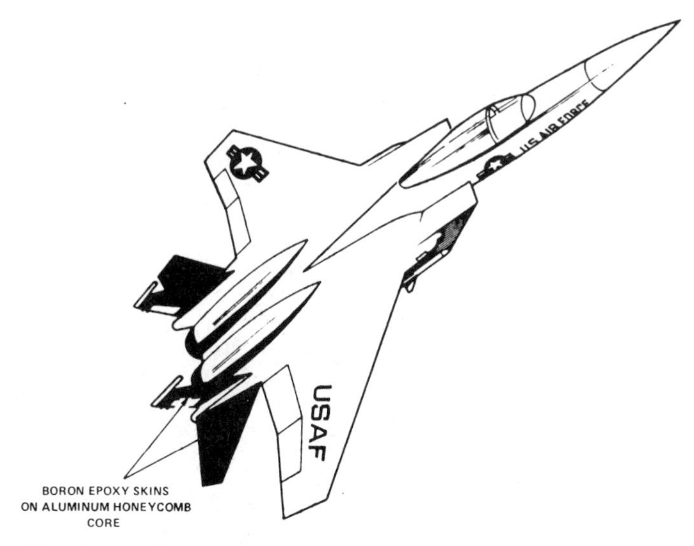

Fig. 19-8. Diagram of F-15, showing boron/epoxy horizontal and vertical stabilizers.

enhanced performance and maneuverability for the aircraft. Figure 19-8 shows a schematic of the F-15 Eagle, with the shaded portions pointing out the location of boron/epoxy on both the horizontal and the vertical stabilizers. Weight savings similar to those for the F-14 have been achieved on this aircraft. Figure 19-9 shows the locations of boron/epoxy components on the B-1 bomber. The principal use of the composite is in the main load-carrying longerons.

Another current application is in helicoptor rotor blades. The high specific mechanical properties of boron/epoxy find utilization in the Sikorsky Blackhawk series of vehicles.

In metal matrix applications, there is presently one production application of boron/aluminum. It is being used as the baseline material for the tubular truss members to reinforce the space shuttle structure. A photograph of a partially completed mid-fuselage structure is shown in Fig. 19-10. Over 200 tubes from 2 to 6 feet long are used in its construction, with a weight savings of 300 lb.

Sporting goods initially were a small fraction of the boron/epoxy market. Recently, however, numerous manufacturers have begun to utilize the high-performance characteristics of this material in their products. Domestic and foreign manufacturers of golf club shafts, tennis rackets, and fishing rods now account for roughly half the usage of boron/epoxy composites.

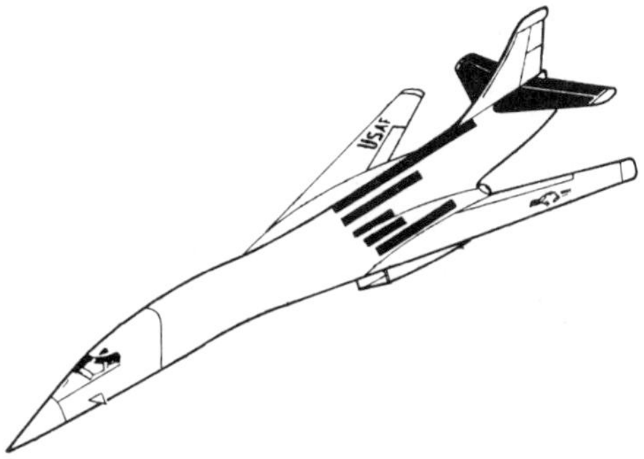

Fig. 19-9. Diagram of B-1, showing locations of boron/epoxy components.

3. SILICON CARBIDE MONOFILAMENT

The outstanding properties of boron/aluminum metal matrix composites spurred further interest in other applications for this material. Difficulties were encountered, however; at higher processing temperatures such as those used in casting, a severe reaction was noted between the filament and the aluminum matrix. Reaction was also a serious problem in matrices such as titanium. In response to this situation, development efforts were directed to using silicon carbide monofilament as an answer to these problems. The filament that these efforts pro-

Fig. 19-10. Photograph of partially completed space shuttle orbiter, showing boron/aluminum tubes in mid-fuselage structure.

duced also has shown superior properties in organic matrix systems.

3.1 Process Description

The CVD process used to manufacture Avco's SiC fiber is similar to that used for boron, which was described earlier. In the SiC process, a 1.3 mil (33 μm) carbon monofilament (CMF) substrate is fed through a glass reactor (Fig. 19-11) and resistively heated through mercury contacts. Initially a thin (1–1.2 μm) coating of pyrolytic graphite is deposited onto the carbon substrate. Hydrogen, propane, and silanes are subsequently fed into the reactor, partially consumed during the deposition process, and then exhausted to a gas separation and recycling facility. The deposition produced is primarily β-SiC, which exhibits a columnar structure with a growth pattern extending radially from the substrate (Fig. 19-12). The final region of the deposition reactor is used to coat the surface of the SiC fiber. The coating enhances the strength of the filament and can be tailored to enhance bonding to various matrix materials. After the application of the surface coating, the fiber exits the reactor and is continuously fed to a take-up spool. The diameter of SiC filament produced by Avco is 5.6 mil (140 μm).

From an economic standpoint, silicon carbide is potentially less costly than boron for three reasons: the carbon substrate used for silicon carbide is less expensive than the tungsten used for boron; raw materials for silicon carbide (chlorosilanes) are less expensive than boron trichloride, the raw material for boron; and deposition rates for silicon carbide are higher than for boron, so that more product can be made per unit time.

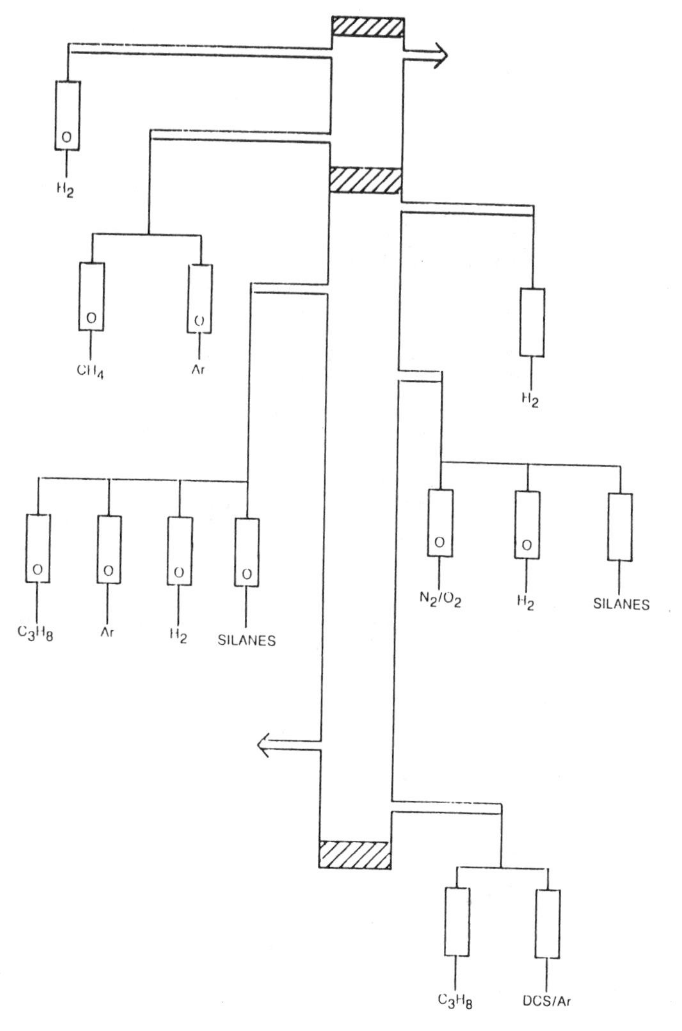

Fig. 19-11. Silicon carbide deposition reactor.

As in any vapor deposition or vapor transport process, temperature control is of utmost importance in producing CVD SiC fiber. The Avco process calls for a temperature of about 1300°C. Temperatures significantly greater than this cause rapid deposition and subsequent grain growth, resulting in a lower tensile strength. Temperatures significantly below the optimum cause high internal stresses in the fiber, which result in a degradation of composite properties upon machining.

Substrate quality is also an important consideration in SiC fiber quality. The carbon monofilament substrate, which is melt-spun from coal tar pitch, has a very smooth surface with occasional surface anomalies. If severe enough, these surface anomalies can result in a localized area of irregular deposition of pyrolytic graphite and silicon carbide (see Figs. 19-13 and 19-14), which is a stress-raising region that produces a strength-limiting flaw in the fiber. The carbon monofilament spinning process is controlled to minimize these local anomalies sufficiently to guarantee routine production of high-strength (>500 ksi) SiC fiber.

Another strength-limiting flaw that can result from an insufficiently controlled CVD process is the pyrolytic graphite (PG) flaw, which is due to irregularities in PG deposition. Two causes of PG flaws are: (1) disruption of the PG layer due to an anomaly in the carbon substrate surface; and (2) mechanical damage to

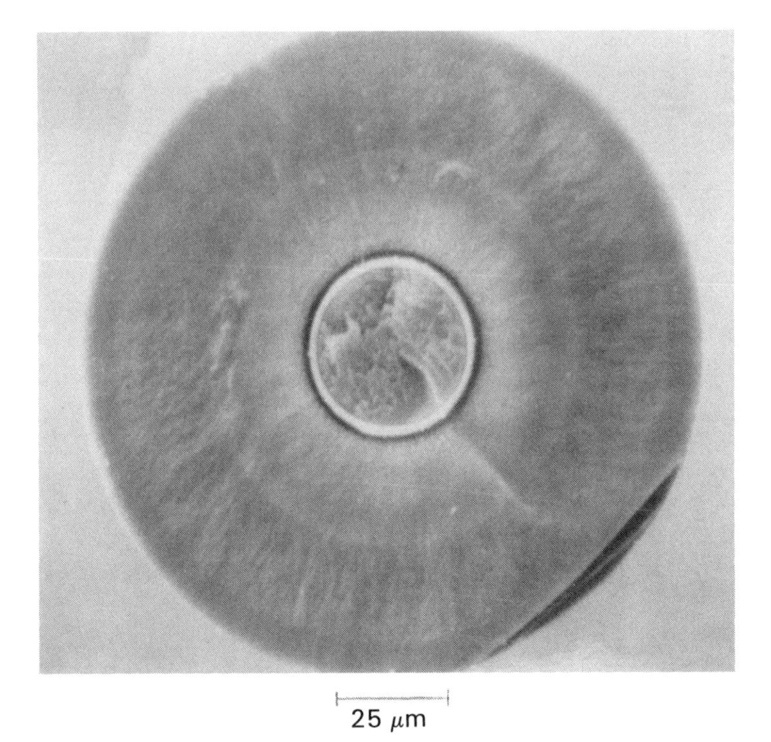

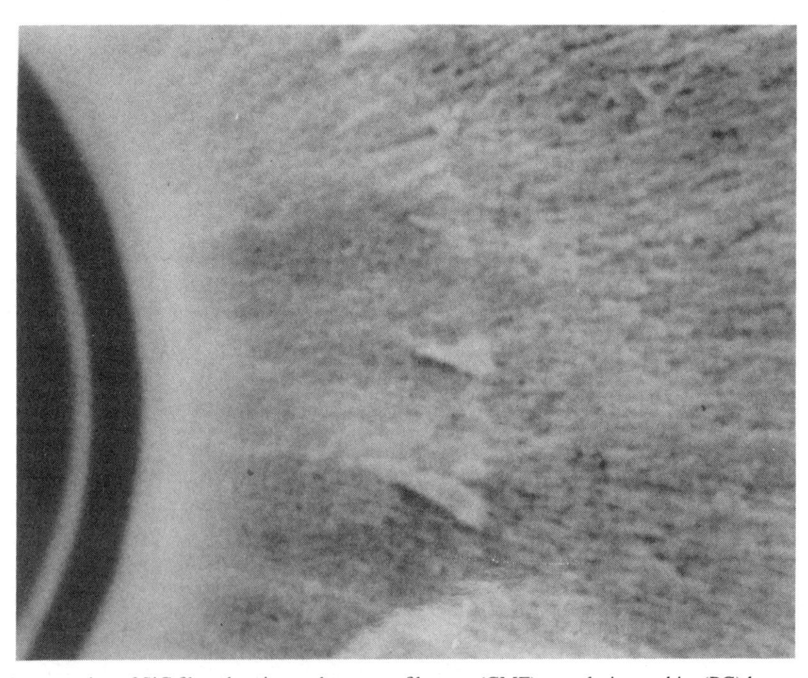

Fig. 19-12. Cross section of SiC fiber showing carbon monofilament (CMF), pyrolytic graphite (PG) layer, and columnar growth of SiC radially from the center.

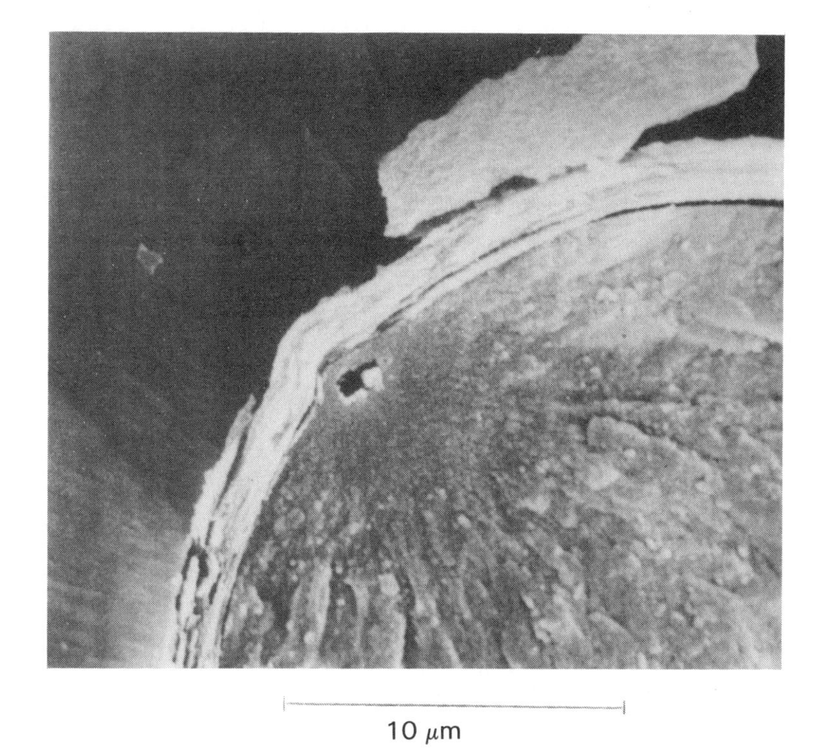

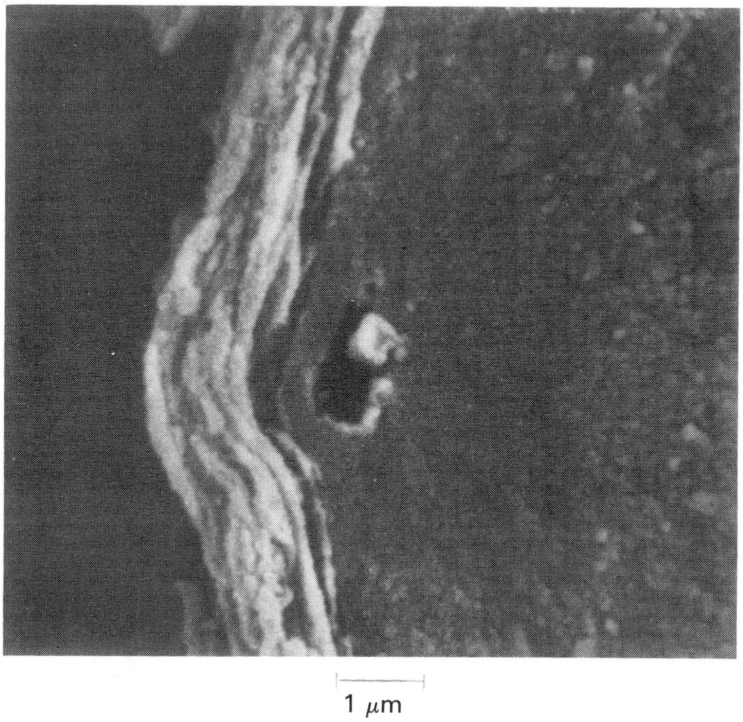

1 μm

Fig. 19-13. Strength-limiting flaw at the carbon monofilament/SiC interface, caused by void at the substrate surface. Layered morphology of the pyrolytic graphite (PG) is evident in the second photo.

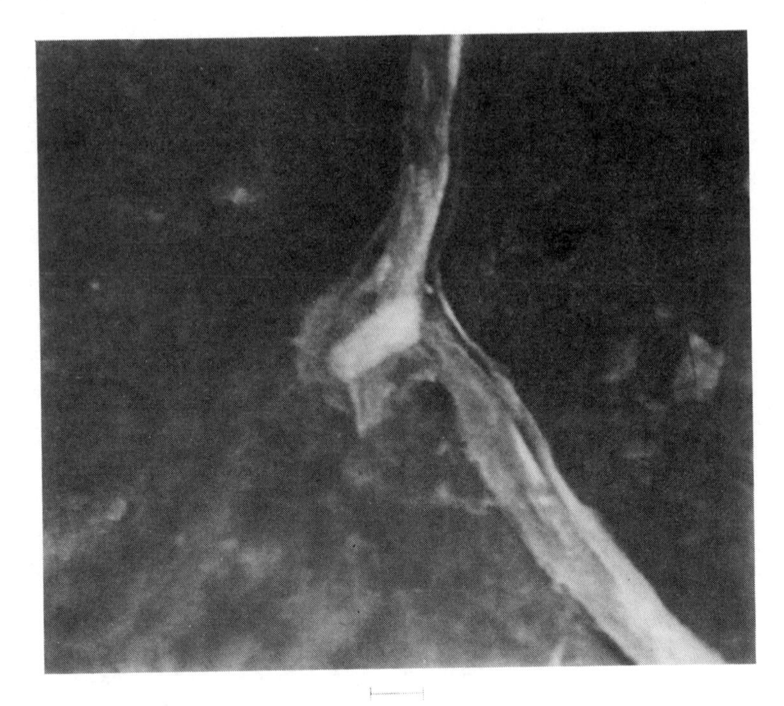

Fig. 19-14. Irregularity in the PG layer associated with anomaly in substrate surface.

the PG layer prior to SiC deposition. Examples of these types of flaws are shown in Figs. 19-14 and 19-15. PG flaws often cause a localized irregularity in SiC deposition, resulting in a bump on the surface (Fig. 19-15). Poor alignment of the reactor glass can produce mechanical damage to the PG layer through abrasion (Fig. 19-16a). A series of PG flaws results in what is called the "string of beads" phenomenon (Fig. 19-16b) at the surface of the fiber. The tensile strength of such fiber is severely reduced. These flaws are minimized by careful control of the PG deposition parameters, proper reactor alignment, and minimization of substrate surface anomalies.

The surface of Avco's SiC fibers is typically carbon-rich. This coating is important in protecting the fiber from surface damage and subsequent strength reduction. An improperly applied coating or mishandling of the fiber, for example, abrasion, can result in strength-limiting flaws at the surface. Surface flaws can be identified by optical examination of the fiber fracture face (Fig. 19-17). These flaws are minimized by proper handling of the fiber (minimizing surface abrasion).

Automated process control improves quality and consistency in any manufacturing process. A silicon carbide reactor in which many of the process parameters are computer-controlled has been demonstrated[6] by Avco Specialty Materials Division and is being implemented at its production facility. A schematic of the automated reactor is shown in Fig. 19-18. Computer-controlled parameters include reactor current, fiber diameter, fiber electrical resistance, drawing rate, and selected gas flows.

Typical mechanical properties for Avco CVD SiC fiber consist of average tensile strengths of 500 to 600 ksi and elastic moduli of 58 to 60 Msi. Figure 19-19 is a tensile strength histogram for a recent month of fiber production, which shows an average tensile strength of 580 ksi with a coefficient of variation of 15%.

3.2 SiC Organic Matrix Composites

There is not nearly so large a data base for SiC/epoxy as exists for boron. In order to define baseline properties, the Air Force funded a substantial program at the University of Dayton[7] to obtain test data from approximately

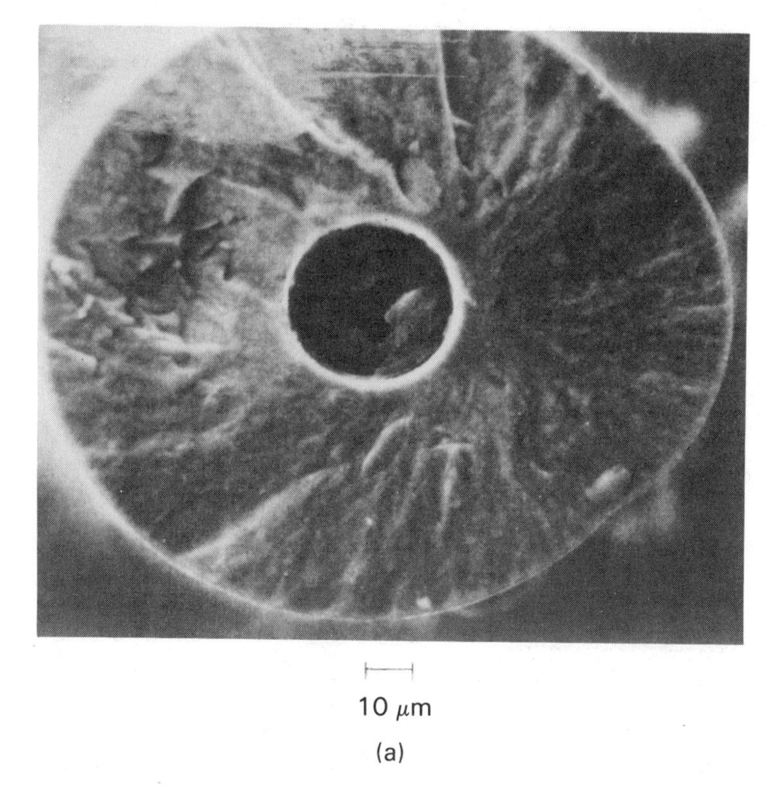

10 μm

(a)

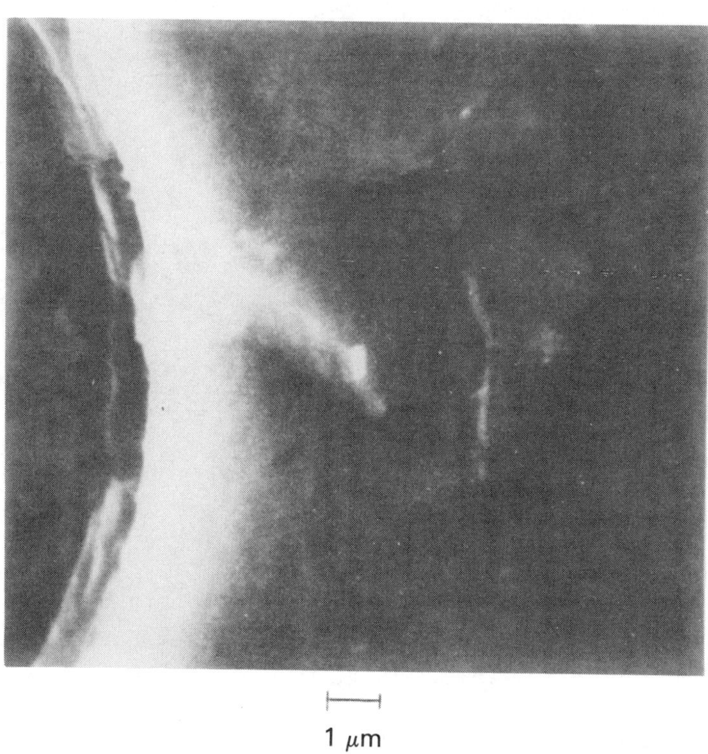

1 μm

(b)

Fig. 19-15. Surface bump in the SiC fiber (a), associated with damaged PG layer (b).

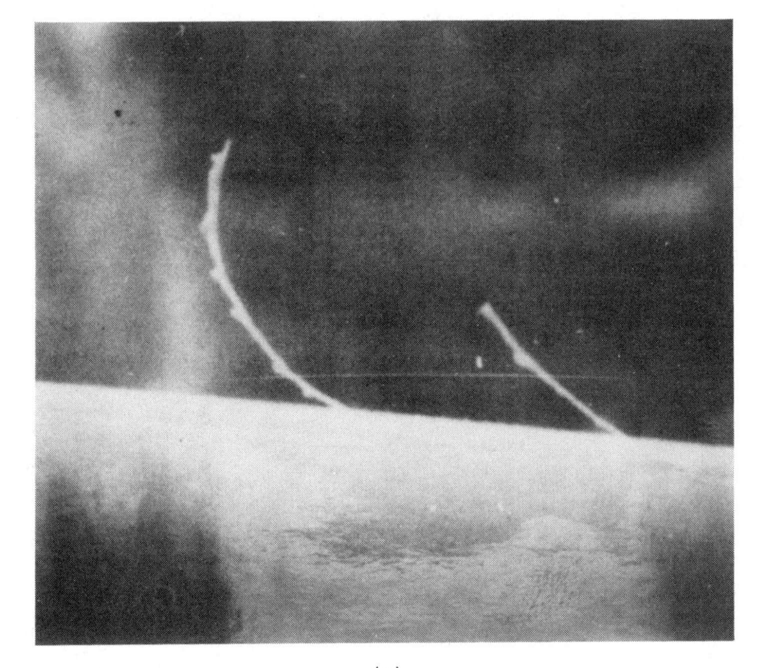

(a)

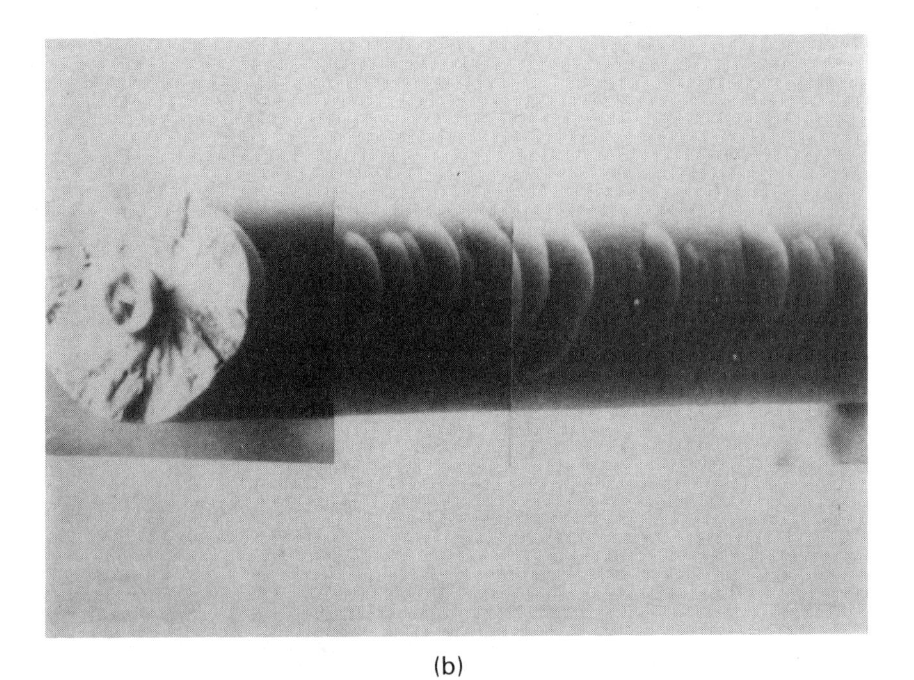

(b)

Fig. 19-16. Damaged pyrolytic graphite layer (a) and associated "string of beads" phenomenon (b).

Fig. 19-17. Fracture face of SiC fiber showing strength-limiting flaw initiated at the fiber surface (at 1 o'clock).

20 lb of composites. The SiC fiber used was a variation with a very carbon-rich surface; a very thin (<0.1 μm) layer of boron was added to enhance wetting by the matrix. A summary of the laminate properties is presented in Table 19-3. The results are at least as good and in some cases substantially better than those achievable with boron. This is especially encouraging, given that SiC is potentially much lower-cost than boron.

3.3 SiC Metal Matrix Composites

As noted earlier, the principal driving force for the development of SiC monofilament was the

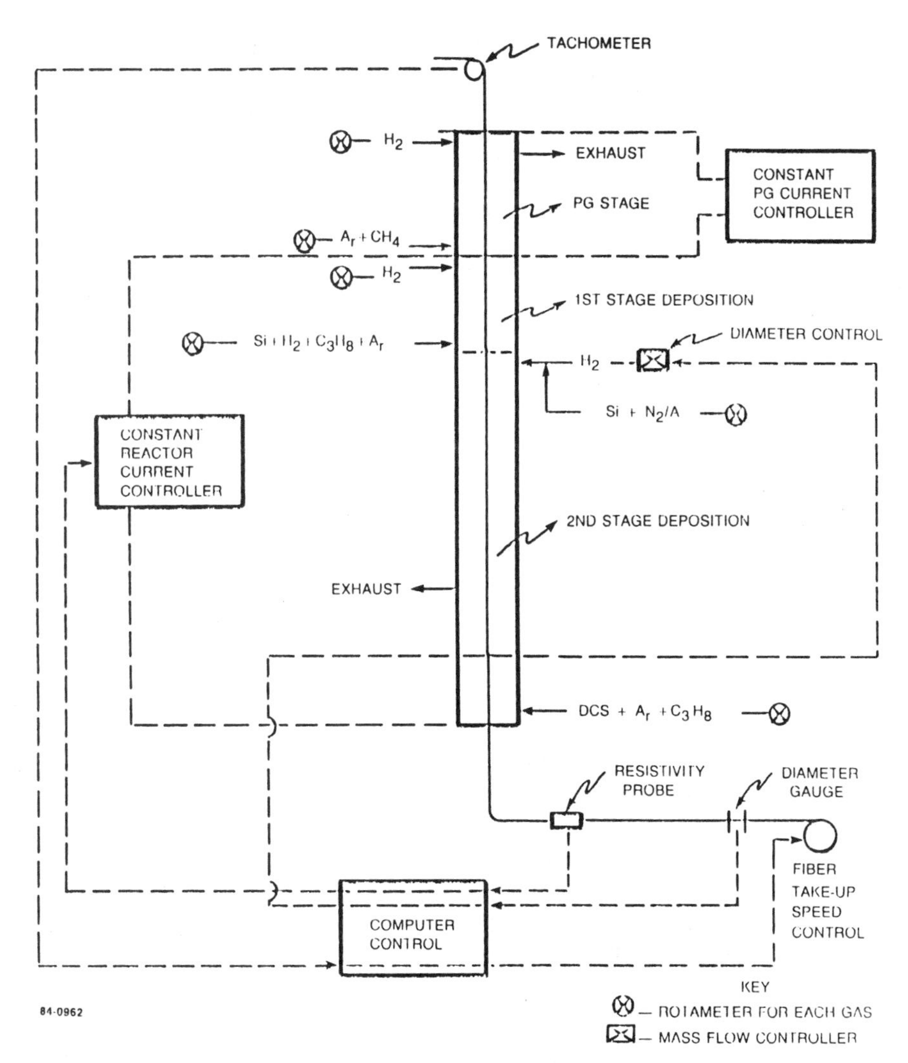

Fig. 19-18. Automated SCS-2 reactor.

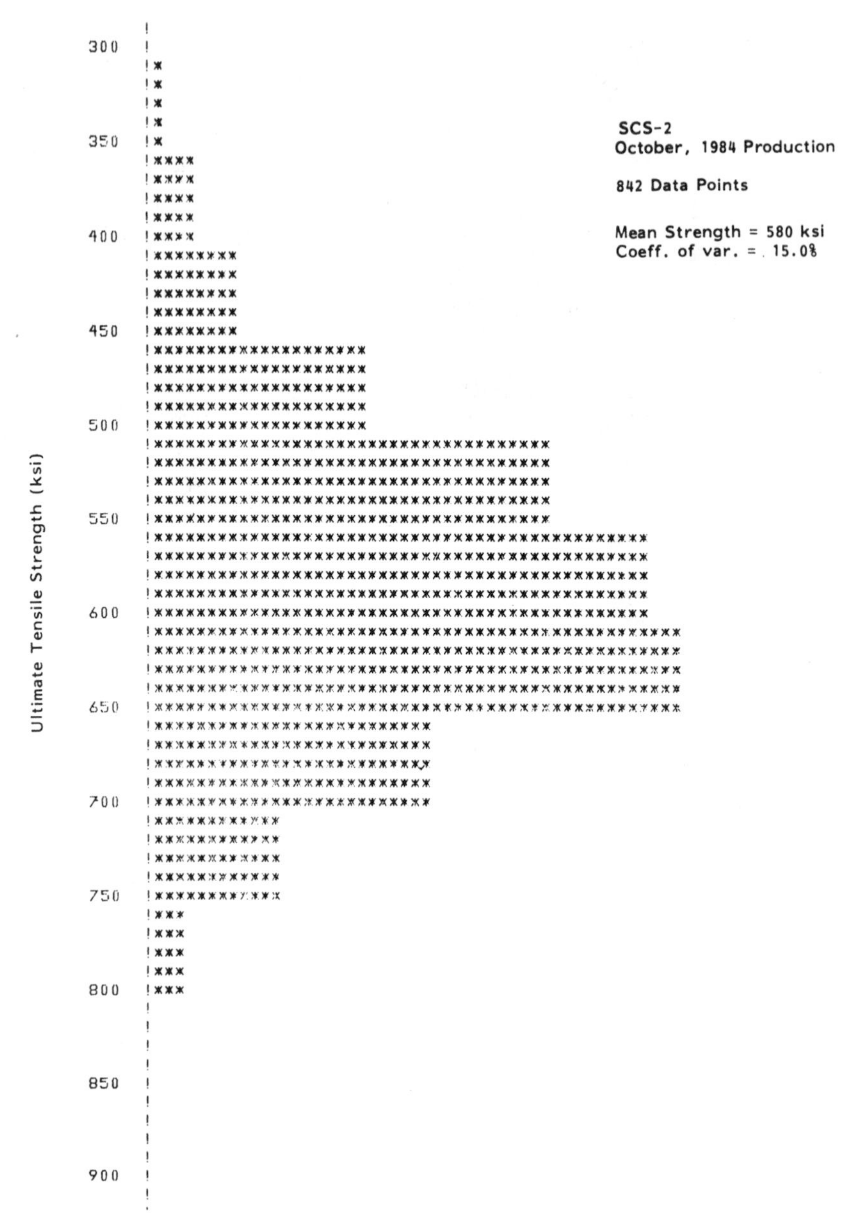

Fig. 19-19. Histogram of CVD SiC (SCS-2) fiber tensile strength.

Table 19-3. Typical properties of SiC/Epoxy laminates.*

Property	Room temperature	350°F
Longitudinal flexural strength, ksi	330	213
Longitudinal flexural modulus, Msi	30.6	25.5
Transverse flexural strength, ksi	18.3	11.3
Transverse modulus, Msi	3.1	1.2
Transverse tensile strength, ksi	11.0	—
Transverse tensile strain, %	0.48	—
Horizontal shear strength, ksi	17.2	9.1

*50% volume fraction.

requirement for higher-temperature fabrication and higher application temperature in contact with metal than could be provided by boron.

The surface coating of the SiC fiber must be tailored to the matrix. Figure 19-20 shows schematics of the surface composition of three fiber types. SCS-2 has a 1 μm carbon-rich coating that increases in silicon content as the outer surface is approached. This fiber has been used to a large extent to reinforce aluminum. SCS-6 has a thicker (3 μm) carbon-rich coating in which the silicon content exhibits maxima at the outer surface and about 1.5 μm from the outer surface. SCS-6 is primarily used to reinforce titanium.

SCS-8 has been developed as an improvement over SCS-2 to give better mechanical properties in aluminum composites transverse to the fiber direction. The SCS-8 coating consists of 6 μm of very fine-grained SiC, a carbon-rich region of about 0.5 μm, and a silicon-rich region of 0.5 μm that becomes pure silicon at the outer surface.

3.3.1 SiC/Aluminum. There is a large data base in SCS-2/aluminum composites, a large

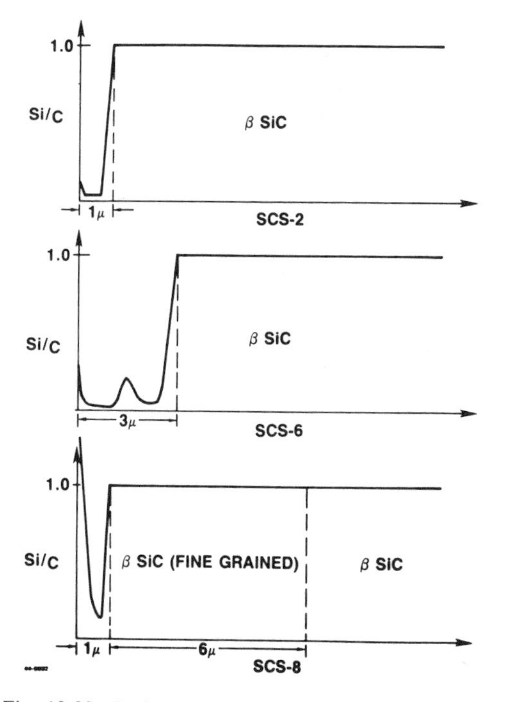

Fig. 19-20. Surface region compositions of Avco silicon carbide.

part of which has been generated under a bridging components program with the U.S. Army.[8] Axial strengths of over 200 ksi and axial moduli of over 30 Msi have consistently been achieved at 48% fiber volume fraction, as shown in Table 19-4. Typical transverse properties consist of 8 to 10 ksi tensile strength, moduli of 16 to 20 Msi, and strain-to-failure of 600 to 1000 μin./in. The reinforced aluminum panels were fabricated by hot-molding preforms of SCS-2 onto which 6061 aluminum had been plasma-sprayed. Temperatures were held between the solidus and liquidus temperatures of the alloy, enabling pressure to be applied via an autoclave (< 1000 psi) rather than a press.

The data base for SCS-8/6061 aluminum is not so large as that with SCS-2 fiber; however, significant improvements in transverse composite properties have been achieved. With axial properties remaining above 200 ksi in strength and 30 Msi in modulus, the transverse strength has improved to 13 to 16 ksi, and the transverse strain-to-failure has significantly improved to 3000 to 8000 μin./in.

CVD SiC with a silicon-rich surface has been used to reinforce cast aluminum 356 alloys. An Avco-developed woven fiber preform is used to reinforce the casting. Strengths of 160 to 170 ksi have been achieved at 35% volume fraction, and subscale cylinders have been cast for a torpedo shell program with NOSC.

3.3.2 SiC/Titanium. SCS-6/Ti 6-4 composites were developed under an NRL program[10] to withstand extended exposure at high temperatures (7 hr at 1660°F). As seen by the data in Table 19-5, composite strengths remained over 200 ksi after this extended heat treatment. There has been a successful program to reinforce the more inexpensive Ti alloy 15-3-3-3 with SiC fiber, and superior composite properties have been achieved, for example 230 to 280 ksi tensile strengths, as shown in Table 19-5. Fabrication of titanium parts has been accomplished by diffusion bonding and hot isostatic pressing (HIP). The HIP technique has been particularly successful in the forming of shaped reinforced parts, such as tubes, by the use of woven SiC fabric as a preform. The high-strength, high-modulus properties of SCS-6/Ti represent a major improvement over B_4C-B/Ti

Table 19-4. SCS2/Al 6061, bridging program data.

Panel No.	Fiber Direction	No. of Plies	Fiber V/O, %	Test Direction	Ultimate Strength (ksi)	Modulus (Msi)	Total Strain (%)
82A-675A	0°	6	48	Long.	235.47	32.97	0.936
					214.48	31.45	0.824
82A-675B	0°	6	48	Long.	229.08	33.50	0.924
					226.59	33.18	0.903
82A-692	0°	6	48	Long.	229.04	32.16	0.902
					215.58	29.04	0.884
					222.41	30.98	0.826
					222.35	31.24	0.824
				Trans.	10.21	19.02	0.057
					12.04	17.15	0.074
					12.27	15.38	0.090
82A-693	0°	6	47	Long.	227.10	30.14	0.894
					229.39	32.20	0.900
					222.77	27.73	0.892
					210.09	31.02	0.826
				Trans.	8.19	18.01	0.046
					8.72	16.41	0.053
					9.51	16.33	0.055
82A-702	0°	6	47	Long.	202.71	33.46	0.750
					204.17	33.74	0.750
				Trans.	11.10	20.02	0.064
					10.06	20.13	0.061
					11.08	20.13	0.063

composites, in which the modulus is increased relative to the matrix, but the tensile strength is not. Development components of SiC/Ti include engine discs and drive shafts.

3.3.3 SiC/Mg and SiC/Cu. SCS-2 has been successfully cast in magnesium. Properties are listed in Table 19-6. Under a recent NSWC program,[11] Naval Surface Weapons Command

Table 19-5. SCS-6/Ti data. Physical Properties of SiC/Ti-6-4 (35% v / o) Sample size: 62 panels

	As Fabricated (1680°F, 30 min, 6 ksi)		After Heat Treating (7 hr @ 1660°F)	
	Mean	Standard Deviation	Mean	Standard Deviation
Ultimate tensile strength, ksi	245	17.3	208	15.8
Modulus, Msi	27.0	1.1	27.6	1.2
Strain to failure, %	0.96	0.091	0.86	0.087

Physical Properties of SiC/Ti-15-3-3-3 (38–41% v/o)

	As Fabricated (1800°F, 30 min, 8 ksi) 11 samples		After Heat Treating (16 hr @ 900°F) 13 samples	
	Mean	Standard Deviation	Mean	Standard Deviation
Ultimate tensile strength, ksi	228	20	283	14
Modulus, Msi	28.7	0.9	30.9	0.7

Table 19-6. SCS/magnesium cast rods (ZE 41 at 1250°F).

Sample #	Exposure time, min.	Ultimate tensile strength, ksi	Strain to failure, %	Modulus, (Msi)	Fiber vol. fraction, %
VIR 67	5	145	.83	24.6	34%
VIR 69	10	221	.88	30.4	46%
VIR 72	10	193	.78	33.4	50%
VIR 77	10	200	.95	26.2	37%

Table 19-7. SCS/copper panels.

Panel	Fiber v/o %	Axial ultimate tensile strength (ksi)	Axial modulus (Msi)
84-014	23	100	25.0
84-153	33	140	29.3
84-377	33	130	27.2

development of SiC-reinforced copper has been initiated. At present, about 85% of rule-of-mixture strengths have been achieved at volume fractions of 20 to 33%. Typical data are presented in Table 19-7.

3.4 Other Matrices

Although relatively little work has been done in reinforcing glasses or ceramics with CVD SiC, there are some encouraging data. Table 19-8 is a comparison of the properties of reinforced borosilicate glass with CVD SiC and Nippon Carbon's SiC yarn, Nicalon. It was determined that CVD SiC had superior axial properties,[12] but the same investigators[13] found that its transverse properties and general integrity were inferior to those of composites reinforced with Nicalon. Work at Ohio State[14] indicates that CVD SiC has the ability to deflect and to arrest cracks in a glass–ceramic matrix. Both these properties are highly desirable, in that they tend to inhibit the catastrophic failure that is one of the principal drawbacks of using ceramics as an engineering material.

3.5 Summary

CVD fibers have shown great promise as reinforcements for a wide variety of matrix materials. SiC fibers, in particular, have shown great promise in reinforcing aluminum and titanium alloys, and have demonstrated both amenability to a wide range of processing techniques and resistance to high-temperature degradation. These characteristics, combined with potential low cost and superior mechanical properties, make CVD SiC fibers a promising class of materials for use as reinforcement in composites.

Table 19-8. Properties of silicon carbide reinforced 7740 glass (Prewo and Brennan, 1980).

	Nicalon	SCS	
Fibre content, vol. %	40	35	65
Density, lb/in.^{3}	0.087	0.094	0.105
Axial flexural strength, ksi			
70°F	42	94	120
662°F	52	—	135
1112°F	75	120	180
Axial elastic modulus (70°F), Msi	17	27	42
Axial fracture toughness, ksi $\sqrt{in.}$			
70°F	10.5	17.1	—
1112°F	6.4	13.0	—

4. POLYMER-BASED SILICON CARBIDE FIBERS

Silicon carbide ceramics from organometallic polymer precursors were first reported by Yajima et al.[15] in 1975. This development is of considerable interest because the use of polymers offers potential cost advantages over the CVD process while at the same time achieving similar properties. This goal has not yet been reached but substantial progress has been made. The following sections describe the commercial variations available and experimental efforts that are expected to lead to commercial products in the near future.

4.1 Nicalon

The process first reported by Yajima has been brought to commercial production level by the Nippon Carbon Co. The basic process consists of four discrete steps. First, organochlorosilanes are converted to polysilanes by alkali metal reduction:

$$n\,(CH_3)_2\,Si\,Cl_2 + 2n\mathrm{Na} \rightarrow$$

$$\left(-\underset{CH_3}{\overset{CH_3}{Si}} - \underset{CH_3}{\overset{CH_3}{Si}}- \right)_{n/2} + 2n\mathrm{NaCl}$$

Next, the intermediate polysilane is rearranged by thermolysis under pressure:

$$\left(-\underset{CH_3}{\overset{CH_3}{Si}} - \underset{CH_3}{\overset{CH_3}{Si}}- \right)_{n/2} \xrightarrow[\Delta,\, p]{} \left(-\underset{H}{\overset{CH_3}{Si}} - CH_2- \right)_m$$

The polycarbosilane intermediate, after purification, can be melt-spun into fibers. The fibers are next thermoset by oxidative cross-linking. Finally, pyrolysis yields silicon carbide fibers. The fibers contain both excess carbon and oxygen; their composition is best represented by Si_3C_4O. Variations on the above process have been reported in which some fraction of the methyl groups is replaced by phenyl or vinyl groups, or in which borate groups have been added to assist cross-linking. The final composition is generally unaffected by these changes, although intermediate processing can be affected. The nominal characteristics of the fiber are shown in Table 19-9. Progress is still being made in reducing diameter variability and in raising tensile strengths. The high-temperature properties are less than optimum because of two phenomena: the outgassing of SiO and CO because of the high oxygen content, and the recrystallization of the amorphous SiC.

The fibers have been used in organic, metal, and ceramic matrix composites. Table 19-10 lists typical properties.[16]

4.2 Tyranno Fiber

A variation of the above process has also achieved commercial status. Ube Industries of Japan employs a process in which a titanium alkoxide is added to the polymer to assist cross-linking.[17] The polymer structure is a mixture of the following functional groups:

$$-\underset{H}{\overset{CH_3}{Si}}-CH_2- \qquad -\underset{CH_3}{\overset{CH_3}{Si}}-CH_2-$$

$$-\underset{\substack{|\\ O\\ |\\ -O-Ti-O-\\ |\\ O\\ |}}{\overset{CH_3}{Si}}-CH_2-$$

A typical composition for both the polymer and the pyrolyzed fiber is shown in Table 19-11. The addition of titanium is postulated to retard the recrystallization of the silicon carbide at high temperatures, a process that leads to strength degradation. It is claimed that an

Table 19-9. Typical properties of Nicalon.

Property	Value
Filament diameter	10–15 μm
Density	2.5 g/cm³
Tensile strength	360–470 ksi
Tensile modulus	26–29 Msi
Strain to failure	1.5%

Table 19-10. Properties of Nicalon and SiC monofilament organic matrix composites.

	Nicalon			SiC Monofilament		
Property	V_f, %	Ultimate tensile strength, MPa	Modulus GPa	V_f, %	Ultimate tensile strength, MPa	Modulus, GPa
0° Tension	48	875	106	57	1410	222
0° Compression	50	1200	117	59	1990	253
90° Tension	48	45.5	17.4	57	61.3	22.2
90° Compression	48	107	17.6	57	161	20.6

Table 19-11. Composition of Tyranno polymer and fiber.

	Polymer		Fiber	
Element	Weight %	Rel. Atomic %	Wt. %	Rel. Atomic %
Si	30.8	1	47.4	1
Ti	4.23	.08	7.35	0.09
C	39.4	3.00	28.3	1.40
O	4.0	0.23	16.0	0.59
H	8.88	8.10	0.15	0.09
B	0.06	0.005	0.07	0.004

amorphous microstructure is maintained to 1300°C.

The fiber is marketed under the trade name Tyranno by Ube Industries, Ltd. Advertised properties are shown in Table 19-12. This material is in early stages of production; improvements in its properties can be expected with process improvements and production experience.

4.3 Experimental Fibers

Numerous variations on the polymer route to ceramic fibers are possible, based on choice of starting materials, cross-linking strategy, and pyrolysis treatments. A recent paper[18] reviewed the state of the art through 1984. At the time of this writing, two of these processes appear on the verge of commercial exploitation. The first of these is a process being explored at Dow Corning under government funding to produce preceramic polymer from disilane by-products of silane manufacturing. The second is a cooperative effort between Bayer Chemical Co. and Avco Corporation to produce polymer by the thermolysis of silazanes. The approaches and preliminary properties are discussed below.

Table 19-12. Mechanical properties of Tyranno fiber.

Tensile strength	250 kg/mm^2
Modulus	12,000 kg/mm^2
Strain to failure	2,0–2.3%
Average diameter	10–13 μm
Diameter variation	$\pm$ 1.5 μm
Density	2.3–2.5 g/cm^3

4.3.1 Dow Corning Process. The manufacturing process for the production of silane monomers produces a mixture of chloromethyldisilanes as by-product. A typical structure is:

$$\begin{array}{ccccc} & CH_3 & & Cl & \\ & | & & | & \\ CH_3- & Si & - & Si & -Cl \\ & | & & | & \\ & Cl & & CH_3 & \end{array}$$

As described in reference 19, the polymerization process consists of three basic steps. First is the initial amination of Si–Cl bonds on methylchlorodisilane by hexamethyldisilazane:

$$\equiv SiCl + (Me_3Si)_2NH \rightarrow \equiv SiNHSiMe_3 + Me_3SiCl \uparrow$$

The second step is a condensation polymerization to increase molecular weight:

$$2 \equiv SiNHSiMe_3 \rightarrow \equiv SiNHSi\equiv + (Me_3SiN)_2NH$$

The final step is further amination to reduce chlorine content:

$$\equiv SiCl + (Me_3Si)_2NH \rightarrow \equiv SiNHSiMe_3 + Me_3SiCl \uparrow$$

The properties of the intermediate polymer can be affected by the addition of compounds such as vinyl- or phenyl vinyl-silanes to control cross-linking. After being spun into fibers, the polymer can be stabilized (rendered infusible) by oxidative, hydrolytic, or radiation treatment. High-temperature pyrolysis then yields the ceramic fiber. As with the Nicalon process, the principal concern appears to be in limiting the oxygen content of the material at all stages of the process. Initial data show that tensile strengths of 300 ksi, modulus of 25 to 30 Msi, and density of 2.5 g/cm^3 can be achieved.

4.3.2 Bayer–Avco Process. During the 1970s, Mansmann and Verbeek at Bayer researched processes to produce preceramic polymers from silazanes.[20] The silazanes were produced by the reaction between ammonia or amines and pure monosilanes. A typical reaction is illustrated below:

$$H_2N-CH_3 + CH_3SiCl_3 \rightarrow CH_3Si(NHCH_3)_3 + CH_3NH_3Cl$$

The tris-silazane is then polymerized by thermolysis to yield a polymer of cross-linked six-membered rings:

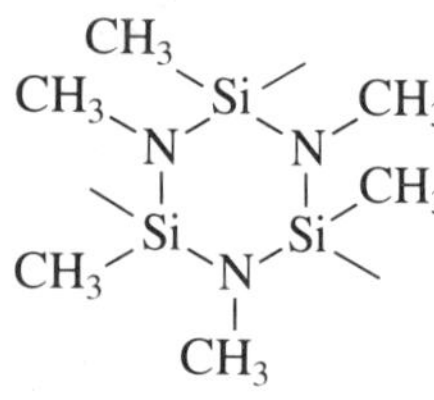

CH3 CH3 Si CH3 N N CH3 Si Si CH3 N CH3

where the indicated free bonds on Si can connect to either methyl groups or other chain components. Upon pyrolysis, Si–C bonds are thermodynamically favored over Si–N bonds; hence the product is almost pure (98%) SiC. Avco Specialty Materials Division has recently reinstituted development efforts in this approach in cooperation with Bayer. Preliminary results indicated that SiC fibers can be made with greater than 500 ksi tensile strength and 40 to 50 Msi modulus. Fiber diameters are 6 to 15 μm.

REFERENCES

1. DeBolt, Harold, "Boron and Other High Strength, High Modulus, Low Density Filamentary Reinforcing Agents," in *Handbook of Composites*, George Lubin (ed.), Van Nostrand Reinhold Co., New York, 1982, p. 171.
2. Bracke, P., Schurmans, H., and Verhoest, J., *Inorganic Fibers and Composite Materials, A Survey of Recent Developments*, Pergamon Press Ltd., Oxford, 1984.
3. Wawner, F. E. Jr., "Boron Filaments," in *Modern Composite Materials*, L. Broutman and R. Krock (eds.), Addison-Wesley, Reading, Mass., 1967, pp. 244–269.
4. Sampson, R. N., "Laminates, Reinforced Plastics and Composites," in *Handbook of Plastics and Elastomers*, C. A. Harper (ed.), McGraw-Hill, New York, 1975, pp. 5-1–5-111.
5. DiCarlo, J. and Wagner, T., "Oxidation-Induced Contraction and Strengthening of Boron Fibers," NASA Technical Memorandum 82599, Lewis Research Center, Cleveland, Ohio, 1981.
6. Suplinskas, R. J., "Manufacturing Technology for Silicon Carbide Fiber," AFWAL-TR-84-4005, 1983.
7. Askins, D. Robert, "Development of Engineering Data on Advanced Composite Materials," Air Force Report No. AFWAL-TR-82-4172, University of Dayton Research Institute, 1981.
8. Grant, W. F. and Henshaw, J., "Fabrication of Low-Cost SiC/Al Metal Matrix Bridging Components," DAAG46-82-C-0033, 1984.

9. Kumnick, A. J. and Burke, J. T., "Reinforced Cast Aluminum Structures," N66001-83-C-0183, 1983.
10. Kumnick, A., Suplinskas, R., Grant, W. F., and Cornie, J. A., "Filament Modification to Provide Extended High Temperature Consolidation and Fabrication Capability and to Explore Alternative Consolidation Techniques," N00019-82-C-0282, 1983.
11. Marzik, J. V. and Kumnick, A. J., "The Development of SCS/Copper Composite Material," N60921-83-C-0183, 1984.
12. Prewo, K. M. and Brennan, J. J., *J. Mater. Sci.* 25: 463, 1980.
13. Prewo, K. M. and Brennan, J. J., *J. Mater. Sci.* 17: 1201, 1982.
14. Jarmon, D. C., "Study of the Fiber/Matrix Interface in Fiber-Reinforced Glass and Glass–Ceramics by Indentation Techniques," M.C.E. thesis, Ohio State University, Columbus, 1984.
15. Yajima, S. et al., *Chem. Lett.* 9: 931, 1975.
16. Strife, J. R. and Prewo, K. M., *J. Mater. Sci.* 17: 65, 1982.
17. Yamamura, T. et al., U.S. Patent Nos. 4342712, 4347347, 4359559, and 4399232, 1982.
18. Wynne, K. J. and Rice, R. W., "Ceramics via Polymer Pyrolysis," *Ann. Rev. Mater. Sci.* 14: 297–334, 1984.
19. "Advanced Ceramics Based on Polymer Processing," Dow Corning Corporation, Midland, Mich.; Contract F33615-83-C-5006, U.S. Air Force, Wright Aeronautical Laboratories, Wright-Patterson AFB, Ohio 45433.
20. Winter, G., Verbeek, W., and Mansmann, M., U.S. Patent 3892583, 1975.

20
CARBON-GRAPHITE FILAMENTS

Harry S. Katz
Utility Research Co
Montclair, New Jersey

CONTENTS

1. INTRODUCTION

A brief note in the October 1985 issue of *Modern Plastics*[1] reported some details of a market survey on advanced carbon fibers by SRI International. The study reported a 24% annual growth in U.S. sales of advanced carbon reinforcements from 1980 to 1984, and predicted 16% growth through 1987, commenting that "The leading market sector for the product is aerospace, where growth is forecast at 19%."

This high growth rate justifies the enthusiasm this material generated nearly a decade ago.[25] More evidence of the excellent potential of this class of materials is found in the many recent publications on this subject.[2–7] Advanced carbon-graphite filaments have the potential of dominating the composite field. Their superior modulus-to-weight and strength-to-weight ratios, combined with the possibility of much lower costs, will lead to their increased use in preference to metals and fiberglass composites. The successful flight of the all-composite Voyager experimental aircraft, which circled the globe without a refueling stop, was achieved by the use of graphite fiber reinforcements.

There are a large number of different types of products within this class of materials, ranging from fiber mats, with relatively low strengths and moduli, to continuous filaments, with extremely high tensile strengths and moduli. During late 1985, a Japanese firm announced that it would be marketing a graphite whisker.

Recent progress in the improvement of commercially available continuous filaments has been remarkable. In order to improve the toughness of these filaments, improve their performance, and avoid damage during processing and use, it was necessary to increase the elongation at break. There was also a need for filaments with even higher strengths and moduli than were available during the early 1980s. In 1985, filaments became available with elongations at break of 2%, rather than the former value of below 1.5%; tensile moduli of 120 million psi, rather than the former 70 to 80 million psi; and tensile strengths of 800,000 psi, rather than the previous values of about 500,000 psi. Of course, these improvments in properties are not all available in any one filament, but this combination of properties would not be required for a specific end use. Some continuous graphite filaments have been produced with a modulus of about 130 million psi, which is amazingly close to the modulus of graphite whiskers.

Prices of high-performance filaments dropped rapidly from the initial level of $350/lb in the early 1960s to a current level of about $20/lb to 30/lb for a general-purpose grade of continuous filament. It was predicted in the late 1970s that a quality filament derived from a pitch precursor would be available in the near future at less than $10/lb, but this price does not appear to be realistic now. The lower-priced mat materials and chopped or milled fibers have been selling at about $9/lb and have been used because of their unique properties, such as high electrical conductivity and low thermal expansion. Among the outstanding properties of carbon-graphite filaments are high modulus and strength, low density, high electrical conductivity, low thermal coefficient of expansion, low coefficient of friction, and excellent resistance to most environmental exposure conditions and chemicals. The combination of desirable properties of these materials will lead to their widespread use in most industries.

Although high-performance carbon-graphite filaments have proved cost-effective in many applications, their relatively high current cost must be considered for any potential application. Another possible disadvantage is the fact that this is a new material, and many materials engineers or designers may not have the experience to take full advantage of its potential. Due consideration must be given to the low interlaminar shear strength of many types of filaments. Also, the differential thermal coefficients of expansion of the resin and the filament may cause crazing problems in some end products, especially those molded at very high temperatures or exposed to both high temperature and high humidity conditions. The impact strength of the composite may be low. These and other problems can be considered birth pains to a revolutionary new material, and will undoubtedly be resolved by technical progress in carbon-graphite filament composites.

In many applications, carbon-graphite filaments may be combined with another filament, such as Kevlar 49 or glass, in order to form a structure that takes advantage of the best features of both types of filaments. The Kevlar or glass can provide higher impact resistance and lower cost without significant reduction of the performance of the carbon-graphite filament composite. These hybrid structures have been utilized in aerospace components and sporting goods.

2. PRODUCTION METHODS

Carbon filaments have been produced for more than 100 years. In 1879, Thomas Alva Edison produced the first practical, electric, incandescent lamp, which had a filament of carbonized sewing thread in an evacuated glass globe. The original lamp element consisted of a relatively fragile carbon filament, 1 mm in diameter. These lamps were marketed in 1880. U.S. Patents 223,898 and 230,309 describe the production of carbon filaments in lamp manufacturing. At about the same time, an English scientist, Sir Joseph Wilson Swan, produced lamp filaments by carbonizing cotton threads that had been parchmentized in sulfuric acid. In 1883, Swan was granted a British patent for a process to produce fine cellulose filaments, which he had developed as a means toward smaller-diameter carbon filaments. After about 1910, the lamp industry turned to the use of tungsten filaments, and the production of carbon filaments for this application was terminated.

The renewed interest in carbon-graphite filaments started in the late 1950s. The advent of jet engines, proposed faster and larger aircraft, and other requirements of advanced technology led to an intensive search for improved structural materials. Laboratory tests of graphite whiskers showed values of 3 million psi tensile strength and 140 million psi tensile modulus.[8] This is a strength-to-weight ratio, specific strength, over 30 times greater and specific modulus over 17 times greater than the structural metals, as shown in Table 20-1.

Research aimed toward the production of continuous graphite filaments was initiated by the Air Force Materials Laboratory. In a parallel effort during 1957, Dr. A. Shindo of the Government Industrial Research Institute, Osaka, Japan, developed a process for manu-

Table 20-1. Properties of graphite whiskers vs. structural metals.

	Density lb/in.^{3}	Modulus 10^6 psi	M/D 10^6 in.	Tensile Strength 10^3 psi	TS/D 10^6 in.
Graphite whiskers	0.081	142	1750	3000	37
Titanium 6A1-4V	0.16	15	94	155	0.97
Aluminum 7075	0.10	10	103	82	0.81
Maraging steel	0.26	27	93	319	1.1

facturing carbon-graphite filaments from polyacrylonitrile (PAN) filaments. Two patents, Showa 37-4405 (1962) and Showa 38-12375 (1963), evolved from this research and were licensed to Tokai Electrode Co., Nippon Carbon Co., and Nitto Spinning Co. Also, during the 1960s, patents were granted to other groups for methods to produce carbon-graphite filaments. British Patent 1,034,542, granted to Union Carbide in 1966, described a process for continuously graphitizing a textile thread, and discussed the method of strengthening the yarn by stretching at a temperature of at least 2200°C. Rayon was the preferred precursor, but PAN was also mentioned. British Patent 1,110,791, granted to the National Research Development Corporation (NRDC) in 1968, stressed the use of PAN precursor and cited values of 260,000 psi tensile strength and 60 million psi Young's modulus. Claim 1 of this invention is: "Carbon fiber made by the conversion of organic polymer fiber, and which has been given a high ultimate tensile strength and a high value of Young's modulus by having been submitted at some stage of the conversion to the combined effect of heat and longitudinal tension." There have been a large number of licensing and cross-licensing arrangements. A list of pertinent patents is given in Table 20-2.

The basic elements for producing carbon-graphite filaments from rayon are shown in Fig. 20-1 (U.S. Patents 3,305,315 and 3,235,323).

The rayon may be subjected to a chemical treatment before the first-stage exposure in the oxidizing furnace at about 500°F. The chemical bath can be an aqueous ammonium chloride solution or a dilute solution of phosphoric acid in denatured ethanol, and serves to reduce the time for the low-temperature step from hours to about 5 minutes. Note that the rayon-precursor process requires that the filament be stretched after the intermediate carbonizing step.

In the PAN-precursor process, which was the method stressed by the British manufacturers, it was found preferable to restrain or stretch the filaments during the low-temperature oxidation step in order to improve the modulus.

Table 20-3 from British Patent 1,110,791, shows the effect of the initial stretch and the graphitizing treatment on the final modulus of the PAN-precursor filament.

The oxidation of the PAN filaments in air is a diffusion-controlled process. During this pro-

Table 20-2. Selected carbon-graphite filament patents.

United States	Other
223,898	Japan
230,309	Showa 37-4405
2,957,956	38-12375
3,011,951	38-20609
3,011,981	39-16681
3,053,775	41-15728
3,107,152	43-893
3,116,975	43-4550
3,235,323	44-21175
3,285,696	United Kingdom
3,305,315	894,458
3,313,596	1,016,351
3,313,597	1,034,542
3,378,345	1,110,791
3,392,216	Belgium
3,399,252	678,679
3,404,061	France
3,412,062	1,430,803
3,443,899	
3,449,077	
3,454,362	
3,461,082	
3,462,289	
3,462,340	
3,503,708	
4,005,183	

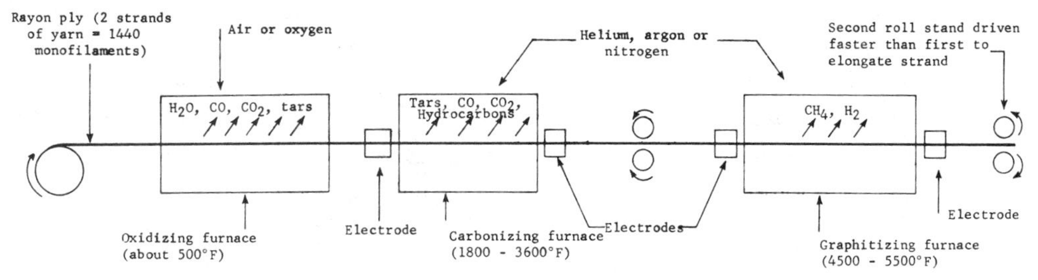

Fig. 20-1. Production line for carbon-graphite filaments from rayon.

cess, the filaments add about 8% by weight of oxygen, which may cross-link the parallel oriented chains of PAN and prevent their relaxation or contraction during the later, high-temperature exposures.

The effect of flaws on the strength of carbon filaments produced from acrylic precursors has been reported by D. J. Thorne, Rolls-Royce Limited. The tensile fracture of a carbon filament occurs almost exclusively at a detectable flaw, which may be at the surface or internal. Surface flaws, which produce the greatest reduction in observed filament strength, may be removed with controlled surface etching by ion bombardment, wet chemical, or gaseous oxidation etching.

A study that supports the theory that the practical tensile strength is determined by flaws involved the treatment of commercial carbon fibers, which had been made from a PAN precursor, with liquid bromine or bromine solutions in nitromethane.[9] Upon comparison of the original treated fibers, definite changes were evident in the macrostructure of the treated fibers. Some of the treated fibers showed tensile strength improvements up to 50%.

Internal flaws often originate in the organic precursor, which can contain voids and contaminating inorganic and organic particles. Because of the high degree of stretching of the polymer during carbon filament production, the internal defects are all elongated. The presence of these flaws results in a great decrease in fiber breaking strength with increasing length. The closed micropore volume of the carbon filament can be as high as 15%, while that of the original PAN is less than 1%. If internally flaw-free carbon filaments could be prepared with a flaw-free surface, strengths of 600,000 to 800,000 psi and breaking strains of 2 to 3% at usual lengths could be anticipated. An ultimate "ideal tensile strength of 26 × 10^6 psi for carbon fiber" as an optimistic value has been discussed in theoretical terms.[9] This upper limit appears unattainable in light of the more usual assumption that a level of strength may be attained that is about one-tenth that of the modulus. Since graphite whiskers have a modulus

Table 20-3. Factors affecting PAN-precursor graphite filament modulus.

Load Applied to a Yarn of 100 Filaments of 2½ Denier Polyacrylonitrile Fibers, for 24 hr at 220°C for Initial Treatment by Complete Oxygen Permeation	Length Change During 220°C Treatment	PROPERTIES OF THE FIBER AFTER SUBSEQUENT CARBONIZING TO 1000°C IN AN INERT ATMOSPHERE		PROPERTIES OF THE CARBON FIBER AFTER HEAT TREATING TO 2500°C IN AN INERT ATMOSPHERE	
		Tensile Strength 10^3 psi	Youngs Modulus Axially of Fiber 10^6 psi	Tensile Strength 10^3 psi	Youngs Modulus Axially of Fiber 10^6 psi
Nil	-40%	100	13	80	30
10 g	-12%	100	16	100	38
20	+2%	120	20	120	47
30	+15%	200	21	200	53
40	+36%	200	21	200	60

of 140 million psi, it would seem justifiable to assume that the optimistic upper limit for the strength of graphite filaments is 14 million psi.

A large number of precursors have been considered for the production of carbon-graphite filaments. The list includes cotton, PAN, rayon, polyvinylidene chloride (saran), aromatic polyamide, polybenzamidazole, polyoxadiazole, polyphenylene, lignin, cross-linked polyethylene, and various pitches from PVC, petroleum, or coal tar. At the present time, commercial high-modulus graphite filament is produced primarily from PAN precursor, and rayon-derived material has a minor share of the market. Low-modulus carbon fibers from a low-cost pitch precursor have been produced in Japan. Union Carbide Corporation has been marketing a high-modulus mat and continuous filaments that have been made from a pitch precursor.[10] This line of Thornel products was acquired by Amoco. Figure 20-2 is a photograph of these filaments.

Fig. 20-2. Pitch precursor carbon filaments-240×. (Courtesy Amoco.)

The cross-sectional shape of graphite filaments from rayon precursor is irregular or crenulated, as shown in Fig. 20-3, in contrast to the circular cross-section of the pitch or PAN precursor materials, as shown in Fig. 20-4. The latter filaments are about 250 to 1000 Å in diameter, and the turbostratic graphite crystallites are about 50 Å in size.

A serious problem that was noted with early high-modulus graphite filament composites was extremely poor interlaminar shear strength (ILS), in the order of 2000 to 4000 psi. Surface treatments for the filaments were developed in order to improve this important property to values in the order of 10,000 psi, and usually involved surface oxidation or halogenation.

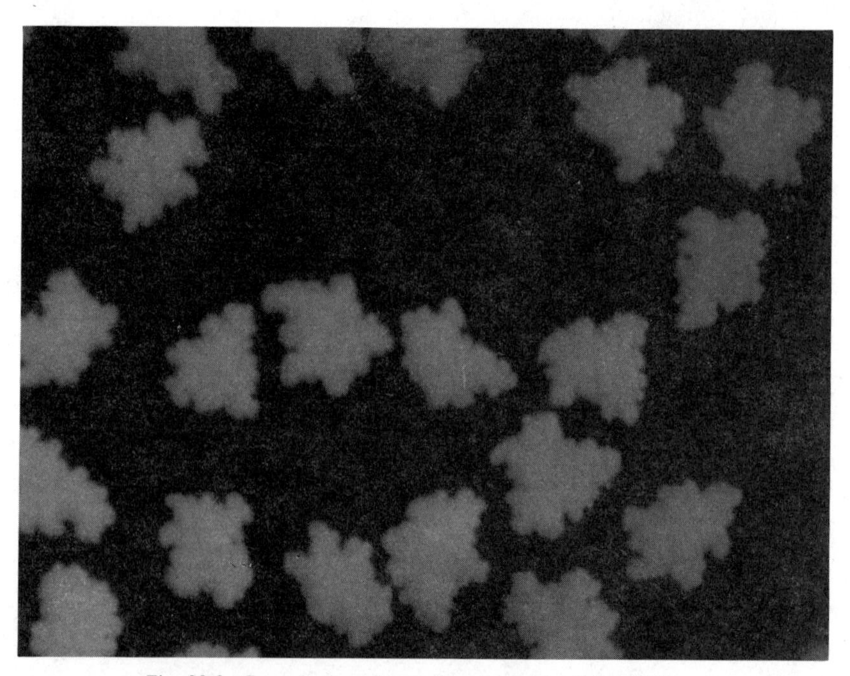

Fig. 20-3. Cross section of rayon-based graphite fiber—2000×.

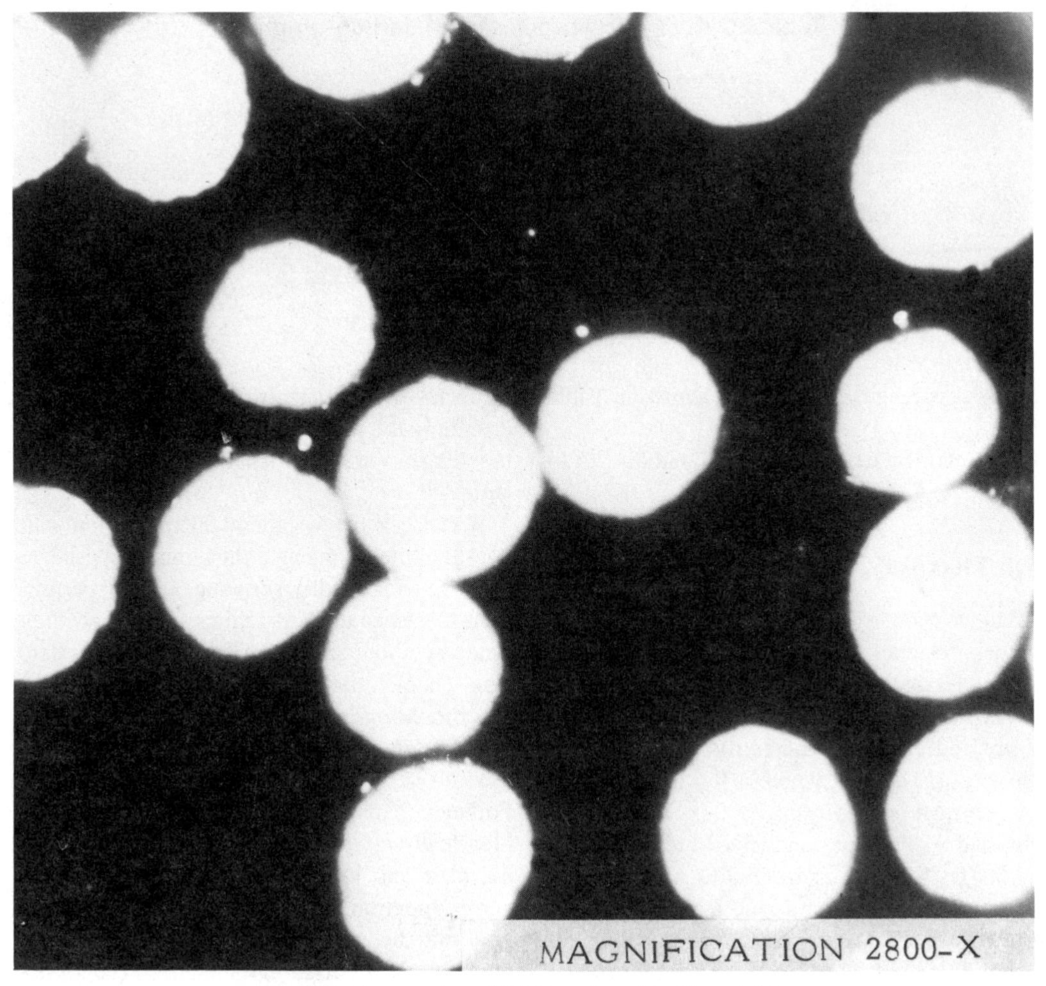

Fig. 20-4. Cross section of PAN-based graphite fiber—1120×.

3. PROPERTIES OF CARBON-GRAPHITE FILAMENTS

3.1 Chemical

Carbon-graphite filaments are relatively inert to most chemicals and resistant to corrosion under ambient conditions. The surface is subject to attack by strong oxidizing agents and halogens, especially at elevated temperatures. This has been a means for surface treatment of the high-performance filaments in order to improve the interlaminar shear strength, which was a severe problem with early graphite-reinforced composites. In general, carbon and graphite are resistant to common alkaline solutions at all concentrations and temperatures, and to the aqueous solutions of most inorganic salts up to their boiling point. Immersion of carbon filaments in the following reagents showed no significant loss in tensile strength or modulus at 50°C: 50% v/v hydrochloric acid, sulfuric acid, and nitric acid; 50% w/v sodium hydroxide solution: 50% w/w hypophosphorous acid; and orthophosphoric acid. At 20°C, the fibers are resistant to glacial acetic acid, 90% w/w formic acid, 32% w/v benzene sulfonic acid, and 0.880 N ammonium hydroxide. The fibers react with most molten metals.

3.2 Physical

Many types of carbon-graphite filaments are commercially available. The main types of high-performance materials are usually divided into three classes, designated as high-modulus, high-strength, and Type A or Type III filaments. Typical properties of these materials and the recently developed ultrahigh-modulus and

Table 20-4. Typical properties of carbon-graphite filaments.

Fiber type	Tensile strength, 10^3 psi	Tensile modulus, 10^6 psi
Ultrahigh modulus	325–350	100–125
High modulus	250–325	50–80
Ultrahigh strength	700–800	41–37
High tensile	350–450	35–42
A or III	275–325	28–35

ultrahigh-strength materials are shown in Table 20-4.

The filament diameter is usually about 7.5 to 8 μm.

3.3 Electrical

A unique property of carbon-graphite filaments is their relatively high electrical conductivity. Union Carbide's pitch precursor "Thornel" Mat Grade VMA filaments have a diameter of 9 μm and electrical resistivity of 12×10^{-4} ohm · cm. The mat, which is 0.4 in. thick, has an electrical resistivity of 7000×10^{-4} ohm · cm, and its thermal conductivity is 0.24 BTU in./hr/ft^2°F. Because of its low electrical resistivity, the mat can be used as a surface veil in a glass fiber sheet molding compound where it provides a conductive surface that can be electrostatically painted. The filamentary form of the product also makes it applicable for resistance heating, since the conductivity is almost the equivalent of Nichrome heating wire.

3.4 Thermal

The thermal expansion coefficient of graphite filament is approximately -0.55×10^{-6} in./in./°F in the direction of the filament axis and about 9.3×10^{-6} in./in./°F in the transverse direction. The low, slightly negative longitudinal expansion can be an advantage in applications such as a composite for a space antenna, which must be held to a close dimensional tolerance. However, in some polymer matrix composites, especially in the cross-ply configuration, the difference between the two thermal expansion coefficients can cause an undesirable stress pattern that can result in microcracking of the laminate. Thermal expansion characteristics must also be considered when designing the method for attachment of a high-modulus graphite filament composite to a metal structure.

Because of the low thermal expansion of carbon-graphite filaments, they may be used to produce zero mold shrinkage. This permits tight-tolerance thick and thin sections to be fabricated without sinks or warpage. The standard sink characteristic of an automotive grade, sheet molding compound is about 0.5 mil, but the addition of Thornel P carbon fibers has been used to reduce this to less than 0.1 mil.[11] Polymer matrix graphite fiber composites exhibit relatively high thermal conductivity along the fiber and low conductivities in the transverse direction. Conductivities equivalent to steel may be reached along the fiber if high fiber volumes are used in the matrix. The ability to dissipate heat is thought to be an important characteristic in increased fatigue life of components such as compressor blades.

Table 20-5 shows the thermal conductivity of graphite composites as compared with fiberglass, aluminum, and steel.

Table 20-5. Thermal conductivity.

Type Material	CONDUCTIVITY (Btu-ft/hr-ft^2-°F) 0°	0°, ±45, 90°
Graphite AS	6–10	2–3
Graphite HTS	12–20	3–6
Graphite HMS	28–35	6–12
Fiberglass	2	0.2
Aluminum	80–125	
Steel	9–27	

Courtesy: *Hercules, Inc.*

Table 20-6. Creep rupture resistance of composites.

Material	Orientation (Degrees)	Tensile Strength 10^3 psi	Creep[a] Load (%)	Creep Rupture (hr)
S-Glass	0	260	85	0.01
			60	60
			51	10^4
Kevlar-49	0	200	80	2
			70	150
			60	10^4
AS graphite	0	200	80	>1000[b]
HTS graphite @ 250°F	0/90	83	90	455
7075-T6 aluminum		70	96	100
		70	93	350

[a]Percent of static ultimate.
[b]No failure in 1000 hr.

Table 20-7. Damage tolerance levels, notched charpy impact.

Material	Impact Energy (ft-lb), Onset of Plastic Deformation	Total Impact Energy (ft-lb)
2024-T6 aluminum	0.05	65
7075-T6 aluminum	0.60	3.0
17-7 PH stainless steel	0.75	67.0
4130 steel (R_C20)	1.60	102.0
6AL-4V titanium	2.10	10.5
4340 steel (R_C55)	3.40	6.0
3501/AS Hercules composite	3.40	6.0

Courtesy: *Hercules, Inc.*

3.5 Miscellaneous

Carbon-graphite filaments exhibit excellent resistance to creep. The creep rupture resistance of several materials is compared in Table 20-6. The HTS graphite filament epoxy 0°, 90° laminates have longer creep rupture times at 250°F than those of 7075-T6 aluminum at room temperature and at the same stress level.

The impact resistance of carbon-graphite filament composites is generally considered low. This factor should be given consideration in the design of any component that requires high impact strength. A comparison of the impact characteristics of graphite composites with those of metals is shown in Table 20-7.

Carbon-graphite filament composites have vibration damping superior to that of aluminum and steel. They have low coefficients of friction, and excellent dry lubrication characteristics. This property, in combination with the high thermal conductivity that permits dissipation of frictional heat, leads to excellent, nonlubricated bearings with low wear rates.

Table 20-8 shows the coefficient of friction of various materials. Note that the value for steel on graphite–epoxy composite is even lower than that for lubricated steel on steel.

Table 20-8. Coefficient of friction of various materials.

Material (No Lubrication)	μ
Steel on HTS epoxy composite	0.15–0.18
Steel on UHM[a] epoxy composite	0.13–0.16
Steel on steel	0.50
Steel on nylon 6/6	0.25
Steel on nylon 6/6 (20% glass)	0.18
Steel on nylon 6/6 (20% graphite)	0.10
Steel on steel (lubricated)	0.20

[a]75×10^6 psi modulus graphite fiber.
Courtesy: *Hercules, Inc.*

Table 20-9. Wear resistance of various materials.

Material (No Lubrication)	K[a] (10^{-7})
Copper on copper	300,000
Zinc on zinc	200,000
Copper on low carbon steel	10,000
Low carbon steel on copper	5,000
Steel on steel	5,000
Bakelite on bakelite	200
Graphite composite on steel	10
Steel on graphite composite	1
Metal on metal (lubricated)	1-10

[a]Wear coefficient $K = \frac{\Delta V(\sigma)}{L(P)}$.
Courtesy: *Hercules, Inc.*

A study[12] of wear coefficients has shown that nonlubricated graphite composites can be as resistant to wear as lubricated metals, as shown in Table 20-9. Therefore, carbon-graphite filament composites are outstanding for use as self-lubricating bearings and seals, and liners for journal bearings. The use of a high-temperature polymer matrix, such as a polyimide, makes the composite suitable for high-temperature bearings.[13]

4. MATERIALS SUPPLIERS

The *World-wide Carbon Fibre Directory*[7] provides an excellent summary of suppliers throughout the world. The following is a list of the key U.S. suppliers.

- *Filaments*

Amoco Performance Products, Inc.
P.O. Box 849
Greenville, SC 29602
800/222-2448
(former Union Carbide product line)

Ashland Petroleum Company
Carbon Fibers Division
P.O. Box 391
Ashland, KY 41114
606/329-3333

AVCO Corporation
Specialty Materials Division
2 Industrial Avenue
Lowell, MA 01805
(617) 452-8961

BASF Structural Materials, Inc.
Celion Carbon Fibers
11501 Steele Creek Rd.
P.O. Box 7687
Charlotte, NC 28217
704/529-8280

Great Lakes Carbon Co.
Fortifil Division
320 Old Briarcliff Road
Briarcliff Manor NY 10510
(914) 941-7800

Hercules Incorporated
910 Market Street
Wilmington, DE 19899
(302) 575-6500

Hitco Materials Division
Armco Steel Corporation
1600 West 135th Street
Gardena, CA 90249
(213) 321-8080

Stackpole Fibers, Co., Inc.
Foundry Industrial Park
Lowell, MA 01852
(617) 454-0409

- *Fabrics*

Barber-Coleman Company
Textile Machinery Division
1300 Rock Street
P.O. Box 1240
Rockford, IL 61105-1240
(815) 968-6833

Composite Reinforcements Business
P.O. Box 3199
Tuscaloosa, AL 35404
(205) 553-6200

Hitco Materials Division
Armco Steel Corporation
1600 West Gardena, CA 90249
(213) 321-8080

Polycarbon Inc.
7418 Fulton Avenue
North Hollywood, CA 91605
(213) 875-0226

- *Prepregs, thermosetting*

AVCO Corporation
Specialty Materials Division
2 Industrial Avenue
Lowell, MA 01850
(617) 452-8961

Ferro Corporation
34 Smith Street
Norwalk, CT 06852
(203) 853-2123

Fiberite Corporation
Composites Division
501 W. Third Street
Winona, MN 55987
(507) 454-3611

Fiber-Resin Corporation
P.O. Box 4187
Burbank, CA 91503
(213) 849-4608

Fothergill Composites, Inc.
317 Northside Drive
P.O. Box 618
Bennington, VT 05201
(802) 442-9964

Great Lakes Carbon Corporation
Fortafil Fibers
P.O. Box 727
360 Rainbow Blvd. South
Niagra Falls, NY 14302
(716) 278-7844

Hercules Incorporated
910 Market Street
Wilmington, DE 19899
(302) 575-6500

Hexcel Corporation
11711 Dublin Blvd.
Dublin, CA 94566
(415) 828-4200

McCann Manufacturing
Box 429
Route 14A
Oneco, CT 06373
(203) 564-4046

3M Company
Industrial Specialities Division
3M Center
St. Paul, MN 55101
(612) 733-1110

Narmco Materials Inc.
Division of BASF
1440 N. Kraemer Rd.
Anaheim, CA 92806
(714) 630-9400

- *Prepregs, thermoplastic*

ICI Americas Inc.
Concord Pike & New Murphy Road
Wilmington, DE 19897
(302) 575-3000

Phillips Petroleum Company
Thermoplastic Composites
Building 71-C, PRC
Bartlesville, OK 74004
(918) 661-1984

Polymer Composites, Inc.
5217 Wayzata Blvd.
Suite 220
Minneapolis, MN 55416
(612) 544-0768

RTP Company
P.O. Box 439
580 East Front Street
Winona, MN 55987
(507) 454-6900

- *Thermoplastic molding compounds*

Fiberite Corporation
Composites Division
501 W. Third Street
Winona, MN 55987
(507) 454-3611

LNP Corporation
412 King Street
Malvern, PA 19355
(215) 644-5200

Thermofil Inc.
P.O. Box 489
Brighton, MI 48116-0489
(313) 227-3500

Wilson Fiberfil International
P.O. Box 3333
Evansville, IN 47732
(812) 424-3831

- *Specialty items: Metal-coated-fibers*

American Cyanamid Company
Polymer Products Division
One Cyanamid Plaza
Wayne, NJ 07470
(201) 8931-3148

5. COST AND AVAILABILITY

The great expenditures required to develop the manufacturing processes and the need for good quality control contributed to the high prices of the initial high-modulus carbon-graphite filaments. In 1970, the average price for continuous filaments was about $325/lb, and the price for meter-length materials about $175/lb. In 1975, there was a wide spread of prices because of the many different types of products; while some types of filaments were still above the $300/lb level, the average selling price of high-modulus filaments was about $50/lb.

Between 1975 and 1987, a large number of new types of carbon-graphite filaments became available. It would not be useful to provide detailed cost and availability data on each of these products. Instead, the reader should ask the supplier (see section 4) who handles the type required for a specific application for such information.

Costs vary widely, depending on the type of material and quantity. Some low-strength and low-modulus mat and milled or chopped fiber materials cost less than $10/lb. There are grades of good-quality continuous filaments available at about $20 to $40/lb. However, some grades still sell at very high prices, such as the over $1000/lb. for some ultrahigh-modulus grades.

6. APPLICATIONS

Carbon-graphite filaments have been utilized in every usual type of organic matrix.

6.1 Thermoplastic Composites

Although most of the early industrial activity on carbon-graphite filament composites involved thermoset resins, such as the phenolic matrices for ablative components, and epoxy matrices for ultrahigh-modulus and high-strength aerospace components, the anticipated predominance of thermoplastic composites has begun to emerge. The high volume and low cost of thermoplastic thermal forming, injection molding and extrusion processes have been an incentive for development of engineering thermoplastic resins.

It can be anticipated that thermoplastic systems will grow at an even faster rate than thermosetting systems for composite applications. This is a result of the lower processing cost, faster processing, higher impact resistance, and possibly better resistance to moisture penetration and damage, of thermoplastics compared to thermosetting plastics. Advances in this field are aided by the development of new high-temperature polymer matrix materials, including ICI's PEEK resin, Phillips Petroleum's Ryton resin, and various polyimide resins. These thermoplastics are available in both continuous-fiber prepregs and injection-molding compounds.

Table 20-10 shows the wear properties of reinforced thermoplastics.

Compared with glass-fiber reinforced compounds, the carbon fiber composites generally have greatly improved tensile and flexural strengths and moduli, but the impact strengths are lower. Thermal expansion of the carbon formulations is lower, and mold shrinkage ranges from about one-half that of glass-reinforced compounds to one-fifth that of the nonreinforced resin. Thermal conductivity of the carbon composites is about twice that of the equivalent glass-reinforced formulation. The much higher electrical conductivity of the carbon composites can be advantageous for many applications, such as when electrostatic painting is required or for electromechanical components when it is desirable to dissipate static charges. The electrical conductivity makes these composites unsuited for applications that require a good insulator.

Carbon fiber reinforcement reduces the coefficient of friction of thermoplastic resins to an extent that is determined by the type of resin. However, the higher the degree of graphitization of the fiber, the lower will be the coefficient of friction, with the type of matrix material being less of a factor. The coefficient of friction of carbon fibers against steel is about 0.25 as compared with 0.8 for glass fibers. Also, polymers reinforced with glass fibers cause at least ten times more wear on a relatively soft counterface, such as mild steel, than the equivalent carbon fiber composite.

Table 20-10. Wear, frictional, and PV properties of reinforced thermoplastics.

Base Resin	Fiber Content (%) and Type	Wear Factor[a] At Equilibrium	Coef of Friction Static	Coef of Friction Dynamic	Limiting PV 10 fpm	Limiting PV 100 fpm	Limiting PV 1,000 fpm
Nylon 6/6	0	200	0.20	0.28	3,000	2,500	2,500
	30 Glass	75	0.21	0.20	12,500	10,000	7,500
	30 Carbon	20	0.16	0.20	21,000	27,000	8,000
Polysulfone	0	>2,500	–	–	5,000	5,000	3,000
	30 Glass	90	0.22	0.18	–	–	–
	30 Carbon	75	0.17	0.14	8,500	8,500	6,000
Polyester	30 Carbon	24	0.12	0.15	11,000	12,500	3,000
Polyphenylene sulfide	0	540	0.30	0.24	2,500	3,000	4,000
	30 Carbon	160	0.23	0.20	8,000	20,000	8,000
ETFE	0	6,000	0.32	0.40	–	–	–
	30 Glass	20	0.38	0.36	7,500	10,000	13,000
	30 Carbon	17	0.24	0.22	14,000	28,000	14,000
Vinylidene fluoride-PTFE	30 Carbon	13	0.10	0.13	–	–	–

[a]A wear factor of $K = 1$ indicates a test condition that produces a wear volume of 1 in.^{3} of material in 1 hr at a load of 1 psi and a velocity of 1 fpm. Wear data presented here (to be multiplied by 10^{-10}) were determined on a 0.900-in. ID $\times$ 0.125-in. OD plastic thrust washer operating at 50 fpm under a 40-psi load, against a 1040 steel washer with a 12-rms finish.
Courtesy: *LNP Corp.*

6.2 Thermoset Composites

A list of companies that supply carbon-graphite filament prepregs was given in section 4 of this chapter. These prepregs have usually involved the use of an epoxy resin, because epoxies have excellent properties, such as superior adhesion, low shrinkage during cure, and versatile formulation capabilities that result in a wide choice of processing characteristics and final properties, and they are readily "B"-staged for long-term storability of the prepreg. The epoxy matrix can be cured as low as 250°F. As the high-performance-composite technology advances, and prices become lower and more competitive, there will probably be increased usage of the lower-priced thermoset polyester matrices. Also, there may be increased usage of high-temperature-capability thermoset matrices, such as the polyimide and bismaleimide resins.

The epoxy composites have generally been limited to a maximum use temperature of 350°F. Polyimide resins have been proposed for long-term durability at about 500°F and short-term about 1000 hr or less, up to 600°F.

The recent activity with bismaleimide resins indicates that this matrix will provide superior performance at 600°F, and higher temperatures. There are a number of other candidate high-temperature matrix resins.

One of the main early uses of carbon-graphite filaments was in ablative components for the rocket industry. In these applications, a phenolic matrix was usually employed because of the hard char formed and low erosion rate during the extremely high temperatures encountered in rocket nozzles and heat shields.

The ultimate in high-temperature capability is afforded by carbon–carbon composites, where carbon is formed as a matrix for the carbon-graphite filament reinforcement. The aerospace industry has made use of this type of composite in projects such as the space shuttle.

Fabrication and properties of carbon–carbon composites are indicated in the following summary.[14] PAN precursor carbon fibers were used to mold a composite with a char-yielding resin, preferably a phenolic resin, which was then carbonized by heating to 1000°C in an inert atmosphere. High-modulus fiber composites, densified by resin reimpregnation, produced composites with a modulus of 200 GN/m^2, 29 million psi, and strength of 0.8 GN/m^2, 116 ksi, after heat treatment at 2600°C. High-strength fibers gave similar strengths after heat treatment at 1600–1800°F. Densification by chemical vapor deposition of carbon led to much higher strengths, greater than 217 ksi. The mechanical properties of the carbon–carbon composites were virtually unaffected by temperatures in the order of 1600°C, and the composites exhibited very little creep at stresses of 70% of the ultimate strength.

In the preparation of carbon–carbon composites, various char-forming resins may be used, such as petroleum pitch or furfuryl resin[15] and any form of carbon-graphite filament, fabric, or felt.[16]

7. CURRENT AND PROJECTED UTILIZATION OF CARBON-GRAPHITE COMPOSITES

Carbon-graphite filament composites were introduced and initially used primarily in the aerospace industry, where the high price could be justified because each pound of weight reduction resulted in great take-off and in-service cost savings. For example, a pound saved in the fan blade of an aircraft engine eventually saved 5 to 8 lb as lighter supporting structures became feasible and led, in turn, to lighter pylons, wings, and fuselage.[18] Benefits that can result are increased range, increased payload, lower power requirements, and lower fuel consumption. Table 20-11 is an estimate of the value/lb saved in aerospace structures in 1978.

The most frequently used metals in the aerospace industry are aluminum and titanium. The initial cost per pound of these materials is much lower than the current cost of high-performance graphite filament composites. However, when an aircraft part is fabricated from aluminum or titanium, high-strength forgings are generally required, and a large amount of the metal, often from four to twelve times the final part weight, is machined away as scrap. In contrast, the scrap rate for the prepreg tape fabrication process may be in the order of 10 to 30%. Therefore, the actual cost per fabricated pound of aerospace structural metals and the graphite filament composites will be much closer than a comparison of raw materials costs would indicate.

The chief factor in the rapid growth of carbon-graphite filament composites in the aerospace industry is the great improvement that these composites offer in specific strength and specific modulus as compared with the structural metals. This comparison is shown in Fig. 20-5.

Table 20-11. Value of weight saved in aerospace structures (1978).

Value/lb Saved ($)	Application
10,000–15,000	Space Shuttle
10,000	Synchronous orbit satellite
1,000	Near orbit satellite
200–500	SST
150–200	Fighter plane
150–200	Boeing 747
100–200	Aircraft engines
100	Commercial planes
50–75	Transport type aircraft

The many advantages of carbon-graphite filaments have led to their use in a variety of applications, including military and commercial aircraft components, satellites, space antennas, ultralight aircraft, automotive components, tooling materials systems, fishing poles, golf clubs, tennis rackets, skis, canoes, and racing yachts.

Figure 20-6 shows a 12-foot-diameter filament-wound case segment for the space shuttle. Figure 20-7 shows the case as it is lowered into a test stand. This graphite filament composite case weighs about 70,000 lb, which is

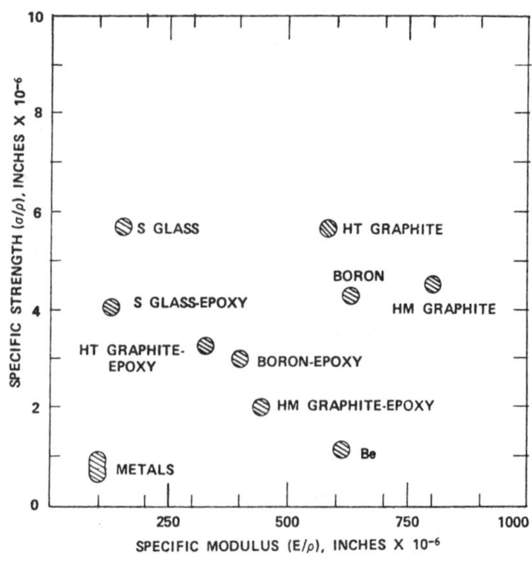

Fig. 20-5. Comparison of specific strength and specific modulus of aerospace materials. (Courtesy Hercules, Inc.)

Fig. 20-6. Twelve-foot-diameter filament-wound case (FWC) segment for space shuttle solid rocket motors in manufacture on one of two computerized winding machines at Hercules Composite Structures Center in northern Utah. (Courtesy of Hercules, Inc.)

Fig. 20-7. Filament-wound case readied for tests. A solid rocket booster test article is lowered into a test stand at the Marshall Space Flight Center in Huntsville, Ala. The article contains two filament-wound solid rocket motor case segments (black in photo). These cases increased the Shuttle's payload carrying capacity by about 4,600 pounds. The motor case makes up most of the nearly 150-foot length of a shuttle solid rocket booster. (Courtesy of Hercules, Inc.)

approximately 30,000 lb less than the original steel motor case.

Figure 20-8 shows the Navy's F-18 airplane, which has more than 1000 lb of graphite filament composite in structural components. The F-18A fighters use 1326 lb of graphite–epoxy composite for the horizontal and vertical stabilizers, wing and tail control surfaces, speed brakes, and doors.

Figure 20-9 is a photo of the Lear Fan, which is a multi-passenger, pressurized-fuselage aircraft constructed entirely of graphite–epoxy composite. This high-performance lightweight business plane has excellent fuel economy.

Figure 20-10 is a photo of a graphite composite racing car monocoque, which was used in a car that won two consecutive world championships. Ford Motors has built a prototype graphite filament composite automobile.[19] Polimotors Corp., N.J., has built graphite filament composite auto engines that have been used successfully in racing cars.

Cessna Aircraft Co. produces structural parts for the Army/Hughes AH-64 Apache helicopter, with a production method that involves the use of graphite–epoxy tooling. The tools demonstrate needed flexural and tensile strengths during autoclave cycling at temperatures to 350°F, and eliminate heat sink considerations

Fig. 20-8. Navy's F-18, which has more than 1000 lb of Hercules graphite materials in structural areas. (Courtesy of Hercules, Inc.)

Fig. 20-9. Lear Fan, constructed entirely of graphite-epoxy composite.

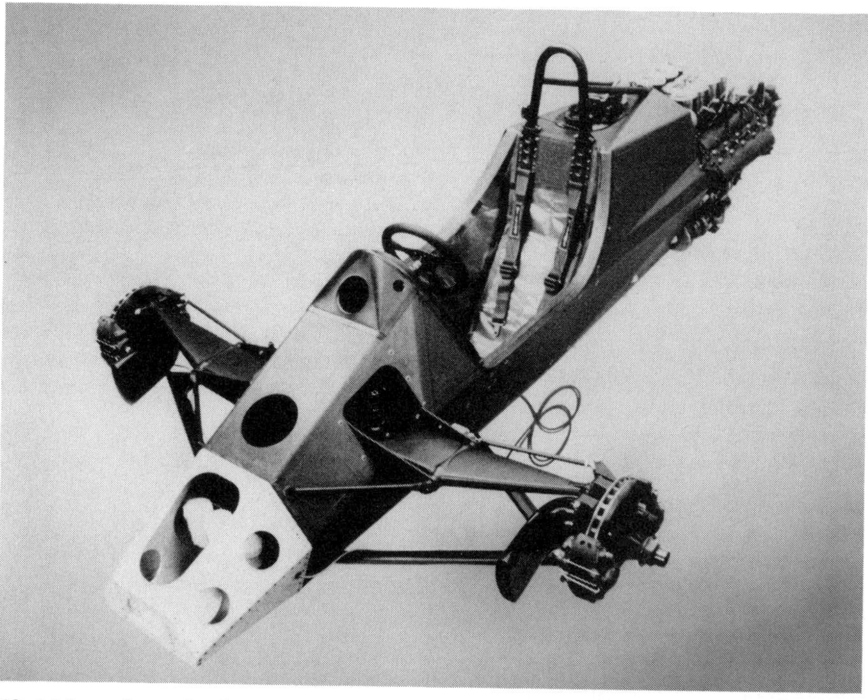

Fig. 20-10. McLaren International Formula One racing car monocoque, built by Hercules at its Bacchus Works. McLaren has combined hi-tech graphite filament composites with a new turbo engine to win two consecutive world championships. (Courtesy of Hercules, Inc.)

during tool curing and subsequent molding of parts, thus cutting autoclave time and reducing manufacturing costs.

The excellent properties of the carbon-graphite filament composites have attracted the attention of the sporting goods industry, and have resulted in their rapid acceptance and growth rate in this field. In 1973, about 140,000 lb of graphite–epoxy prepreg was used in the manufacture of golf club shafts.[20] A golf club with a composite shaft retails for about $75 as compared to $40 for one with a steel shaft. However, the reduced weight of the composite shaft, 2.5 oz versus 4 oz, permits a lower overall weight and a little more weight in the head, which leads to a possible increased drive of 5 to 10 yards.

Fishermen will pay a premium for fly rods. As costly as carbon fibers are, only about $2 worth is used in a fishing rod for which anglers will pay $150.[21] About 35,000 graphite fishing rods were made and sold during 1975.

A 1976 issue of a tennis magazine had several ads advertising the superior performance of graphite filament, reinforced rackets. During 1975, about 50,000 tennis rackets were produced, representing about 12,000 lb of graphite fiber prepreg.

Other applications where carbon-graphite filament composites are being utilized are racing bicycles, skis, canoes, kayaks, racing yachts, and bows and arrows.

The widespread use of carbon-graphite filaments in the sporting goods field has been a great help in increasing sales volumes and decreasing price levels.

In many applications, there is a choice of the type of starting material to be used, regarding whether the unidirectional tape or a woven fabric should be chosen. For attaining the highest strength or modulus, the unidirectional tape had been recommended because of concern that the bends and fiber-to-fiber contact in a fabric would lead to greatly reduced properties. However, the fabric approach presents some economies in materials and fabrication costs. Recent studies have shown that it may be advantageous to use fabric prepreg in many applications.[22]

Much recent work has been spent on the development of combinations of carbon-graphite filaments with another filament, such as Kevlar 49, glass, or boron. The dual structure can be synergistic in exhibiting the better properties of each of the filaments. This type of hybrid structure has been used in many aerospace and industrial parts. A particularly beneficial combination consists of graphite filaments and Kevlar 49, which has thermal expansion properties compatible with those of graphite. The Kevlar 49 contributes lower cost, lower density, and superior toughness and damage tolerance. This hybrid has light weight, stiffness, and low cost; is particularly promising for aircraft applications;[23,24] and has also been used for the manufacture of golf club shafts.

In the future, graphite filament composites will find applications in all industries, including textile equipment, automotive components,[26] packaging containers and equipment, and household furniture. This will be especially useful for any equipment involving oscillating, reciprocating, and rotating parts. Therefore, it will be essential for anyone involved in the fabrication or use of plastics components to be completely familiar with the latest progress in carbon-graphite filament reinforced composites.

REFERENCES

1. "Advanced Carbon Fibers: 'Finally a Market Success,'" *Modern Plastics*, p. 120, Oct. 1985.
2. Delmonte, John, *Technology of Carbon and Graphite Fibers*, Van Nostrand Reinhold Co., New York, 1981.
3. Donnet, J. and Bansal, R. C., *Carbon Fibers*, Marcel Dekker, New York, 1984.
4. Sittig, M., *Carbon and Graphite Fibers*, Noyes Data Corp., Park Ridge, N.J., 1980.
5. NASA Langley Research Center, *Tough Composite Materials: Recent Developments*, Noyes Publications, Park Ridge, N.J., 1985.
6. Hilado, Carlos J. (ed.), *Carbon Reinforced Epoxy Systems, Parts IV and V, Materials Technology Series*, Vols. 12 and 13, Technomic Publishing Co., Lancaster, Pa., 1984.
7. Pamington, David (ed.), *World-wide Carbon Fibre Directory*, 2nd ed., Pammac Directories Limited, Slough, England, 1982.
8. Bacon R., *Applied Physics* **31**:383, 1960.
9. Deitz, R. D. and Vaughan, W. H., "The Fracture and Breaking Strengths of Single Carbon Fibers Before and After Bromine Treatment," SPI Reinforced Plastic/Composites Institute 21-A, 1974.
10. Forsyth, R. B., "Low Cost Continuous Fibers From a Pitch Precursor," 20th National SAMPE Symposium, 1975.
11. "Zero Mold Shrinkage Made Possible with Carbon Fibers," *Plastics Design & Processing* 6, May 1976.
12. Berg, C. A., Barta, S., and Tirosh, J., "Friction and Wear of Graphite Fiber Composites," *Research of the N.B.S.—C. Engineering and Instrumentation*, Vol. 76C, Jan.-June, 1972.
13. Sliney, H. E. and Jacobson, T. P., "Graphite Fiber-Polyimide Composite Rod End Bearings for High-Temperature High-Load Applications," NASA Tech Brief B75-10151, Oct. 1975.
14. Hill, J., Thomas, C. R., and Walker, E. J., "Advanced Carbon-carbon Composites for Structural Applications," Plastics Polymer Conference *Supplement* **6**:122–130, 1974.
15. Adams, D. F., "Transverse Tensile and Longitudinal Shear Behavior of Unidirectional Carbon–Carbon Composites," *Material Science English*, **17**(1):139–152, 1975.
16. Pierson, H. O. and Northrop, D. A., "Carbon-Felt, Carbon-matrix Composites," *Composite Materials*, **9**:118–137, 1975.
17. Sattar, S. A., Stargardter, H., and Randall, D. G., "Development of JT8D Turbofan Engine Composite Fan Blades," *J. Aircraft* **8**:648, Aug. 1971.
18. Brandmauer, H. E., Katz, H. S., and McInnis, W. F., "Molding Composite Compressor Blades from Preimpregnated Composite Materials," *SAMPE Quarterly*, **1**(3):1–14, Apr. 1970.
19. Katz, Harry S., "Fillers and Reinforcements for Improved Plastics," 13th National SAMPE Conference, Oct. 13–15, 1981, pp. 32–37.
20. "Now Fiber Composites Also Reduce Costs," *Iron Age*, June 16, 1975.
21. Aerospace Fibers Score High in Sports," *Business Week*, 28H, 28J. Aug. 9, 1976.
22. Shibata, N., Nishimura, A., and Norita, T., "Graphite Fiber's Fabric Design and Composite Properties," *SAMPE Quarterly*, pp. 25–33, July 1976.
23. Zweben, C. and Norman, J. C., "Kevlar 49/Thornel 300 Hybrid Fabric Composites for Aerospace Applications," *SAMPE Quarterly*, pp. 1–10, July 1976.
24. Hamersveld, J. V. and Fogg, L. D., "Producibility Aspects of Advanced Composites for an L-1011 Aileron," SAMPE Journal, pp. 6–13, May/June 1976.
25. Katz, H. S. and Milewski, J. V. (eds.), *Handbook of Fillers and Reinforcements for Plastics*, Van Nostrand Reinhold Co., New York, 1978, pp. 562–563.
26. Katz, H. S., "Carbon/graphite: reinforcements are coming of age," *Plastics Compounding*, March/April 1979, pp. 18–25.

21
CERAMIC FIBERS AND FILAMENTS

John V. Milewski
Consultant
Santa Fe, N.M.

CONTENTS

1. INTRODUCTION

Ceramic fibers are continuous fibers of metal oxides that feature combinations of properties not previously available. Their major advantages are resistance to extremely high temperatures (2500–3000°F), coupled with higher modulus and excellent compression strength. Their ceramic composition gives them exceptionally good chemical resistance, and their small fiber diameter lends flexibility and workability to the fiber. Because of these features, they can be combined into a strand and woven or braided into fabrics. Many of these fibers were developed specifically for the insulation market, but because of unusual properties, they are being considered for unique reinforcing applications.

They are not noted for high tensile strength, which ranges from 50 ksi for the weakest types to about 250 ksi for the stronger grades.

Some typical applications are high-temperature-resistant continuous conveyor belts, jackets, heat shields, high-temperature filtration systems, and reinforcement of metals and ceramics, as well as reinforced plastics in such products as radomes, sporting goods, chemical ware, aircraft components, and brake lining or friction materials.

2. SUMMARY OF PROPERTIES

2.1 DuPont's Fiber FP

DuPont's fiber is a round, continuous, multifilament, polycrystalline alumina yarn having essentially 100% alpha alumina composition. The 20 μm diameter, permits bending of the extremely stiff, yet flexible fiber over small radii, while its yarn form can be handled easily. The properties are given in Table 21-1.

Figures 21-1 and 21-2 show the variations of fiber strength and modulus with temperature.

Table 21-1. Properties of fiber FP.

- Tensile strength 200–230 ksi (1/4 in. filament gage length)
- Tensile modulus 55×10^6 psi
- Strain to failure ~0.35%
- Melting point 2045°C
- Density 3.90 g/cm^3 (0.141 lb/in.^{3})
- Fiber diameter 15–25 μm
- Form continuous multifilament
- Strength retention after 1000°C, air atmosphere, 92%

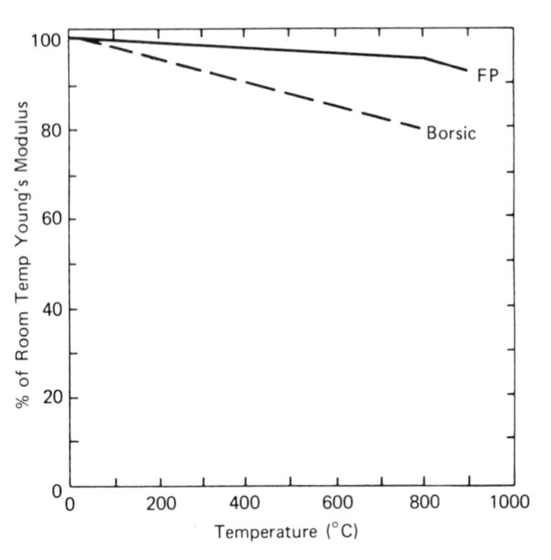

Fig. 21-2. Young's modulus at high temperature. (Data from reference 1.)

2.2 3M's Ceramic Fiber Products

3M ceramic fibers are continuous fibers of metal oxides. Major advantages, in addition to the continuous nature of 3M ceramic fibers, include strength and resistance to extremely high temperatures. They are flexible, dense, transparent, and essentially chemically resistant. They have excellent compressive strength and can be internally colored as opposed to being dyed or color-coated.

3M ceramic fibers can be tailor-made, within limits, to meet specific requirements. Because the product is continuous and strong, it can be easily combined into a strand and woven or braided into fabrics and sleeving on conventional fiberglass production equipment.

The properties of three ceramic fibers currently under investigation are given in Tables 21-2, 21-3, and 21-4.

3M ceramic fibers are also available in noncontinuous blanket form. Their properties are given in Table 21-5.

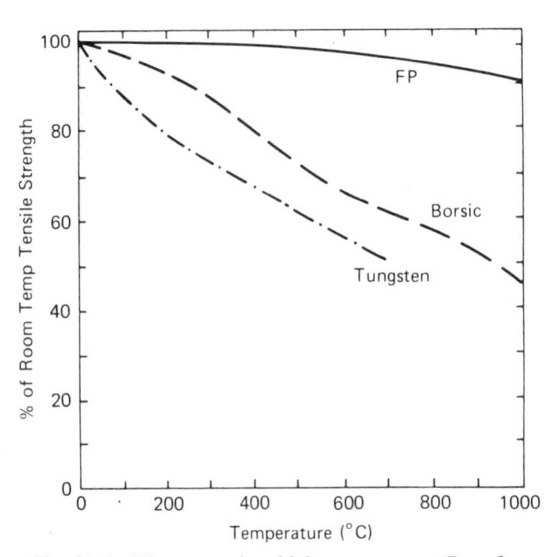

Fig. 21-1. Fiber strength at high temperature. (Data from reference 1.)

2.3 ICI's Saffil Fibers

"Saffil" fibers are unique. No other fiber combines uniform fine diameters, virtual absence of nonfibrous impurities, excellent flexibility, and resilience with such high degrees of refractoriness and chemical inertness. They are microcrystalline and hence do not undergo thermal devitrification. Individual fibers are microporous, and this property enhances the adsorptive and related surface effects characteristic of alumina and zirconia. The bulk material has a nonirritant, silky handle. Initial screening tests on animals have shown no demonstrable toxic effects.

The "Saffil" fiber manufacturing process permits close control over fiber diameter, and a mean diameter has been selected to optimize key properties. "Saffil" fibers have mean diameters around 3 μm and a relatively narrow range compared with other inorganic fibers, as illustrated by the photographs (Fig. 21-3).

Fine fibers have important technological advantages such as exceptionally good thermal insulation properties (more interstices per unit weight), a large surface area for bonding if the fibers are to be used in a composite refractory material, and, in general, better handling and resilience.

Table 21-2. 3M brand ceramic fiber data, NEXTEL 312 fiber.

Composition	Alumina-boria-silica
Forms	Continuous yarn, roving, chopped fibers, fabric, braided products
Appearance	Smooth, dense, transparent, continuous
Average denier (390 filament strand)	900
Color	White
Density	0.0995 lb/in.^{3} (2.70 g/cm^3)
Average diameter	.44 mil (11 μm)
Surface area	<4.8 × 10^3 ft^2/lb (<1.0 m^2/g)
Tensile strength	256 × 10^3 psi (1720 mN/m^2)
Tensile modulus of elasticity	22 × 10^6 psi (152 × 10^3 mN/m^2)
Elongation at break	1.2%
Specific tensile strength	2.7 × 10^6 in. (6.85 × 10^6 cm)
Specific tensile modulus of elasticity	225 × 10^6 in. (615 × 10^6 cm)
Continuous-use temperature	2200°F (1204°C)
Short-term use temperature	2600°F (1427°C)
First liquid phase	3090°F (1700°C)
Liquidus temperature	>3270°F (>1800°C)
Thermal conductivity	
@1000°F, 3 lb/ft^3 density	2.5 Btu in./hr ft^2°F
@1600°F, 3 lb/ft^3 density	6.2 Btu in./hr ft^2°F
Suggested uses	Thermal insulation, reinforced plastics, heat shielding, continuous high-temperature conveyor belts

Available Grades and Forms. Four basic grades of fiber are available to meet the wide range of possible needs:

"Saffil" zirconia fiber:
  Standard grade (S)
  High temperature grade (HT)
"Saffil" alumina fiber:
  Standard grade (S)
  High temperature grade (HT)

The standard and high temperature grades differ in the nature and amount of crystal phase stabilizer.

See Table 21-6 for typical "Saffil" properties.

Table 21-3. 3M brand ceramic fiber data, NEXTEL 440 fiber.

Composition	Alumina-silica-boria
Forms	Continuous yarn, rovings, chopped, nonwoven
Fiber appearance	Smooth, dense transparent
Average denier (390 filament strand)	900
Color	white
Density	0.111 lb/in.^{3} (3.10 g/cm^3)
Average diameter	0.44 mil (11 μm)
Tensile strength	300 × 10^3 psi (2064 mN/m^2)
Tensile modulus of elasticity	28 × 10^6 psi (193 × 10^3 mN/m^2)
Elongation at break	1.2%
Extended use temperature	2500°F (1370°C)
Short-term use temperature	2800°F (1540°C)
Suggested uses	Thermal insulation, reinforcements for plastics, ceramics, and metals

Table 21-4. 3M brand ceramic fiber data, NEXTEL ZS-11 fiber.

Composition	Zirconia-silica
Forms	Continuous yarn, roving, chopped fibers, fabric
Appearance	Smooth, round, dense, transparent, continuous
Chemical resistance	Inorganic alkali and acid, organic solvents
Average denier (390 filament strand)	1200
Color	White
Density	0.134 lb/in.^{3} (3.70 g/cm^3)
Average diameter	0.44 mil (11 μm)
Surface area	<4.8 × 10^3 ft^2/lb (<1.0 m^2/gm)
Tensile strength	190 × 10^3 psi (1505 mN/m^2)
Tensile modulus of elasticity	14.0 × 10^6 psi (96.5 × 10^3 mN/m^2)
Elongation at break	1.1%
Extended-use temperature	1830°F (1000°C)
Short-term temperature use	1927°C (3500°F)
First liquid phase	3090°F (1700°C)
Liquidus temperature	>3630°F (>2000°C)
Suggested uses	Thermal insulation, reinforced ceramics and concrete, high temperature filtration, ablative

Table 21-5. 3M brand ceramic fiber data, NEXTEL 312 and 440.

Composition	Alumina, boria, silica		
Forms	Nonwoven blanket, chopped, needle punched, stitchbonded		
Appearance	Smooth, round, dense, transparent continuous		
Average diameter	0.14 mils (3.5 μm)		
Thermal conductivity			
312 @ 1800°F, 4 lb/ft^3	1.8 Btu in./hr ft^2°F		
312 @ 1800°F, 8 lb/ft^3	1.3 Btu in./hr ft^2°F		
Linear shrinkage (24 hours)	312	1600°F	0%
		2200°F	3.6%
		2600°F	15.1%
	440	2800°F	1.6%
		5000°F	3.9%

2.4 AVCO/Sumitomo Alumina Fiber

Sumitomo Chemical has developed a new ceramic reinforcing filament. AVCO Speciality Material Division is the exclusive U.S. and Canadian distributor. This new ceramic fiber can be used as a reinforcing agent in polymer, metal, and ceramic matrices.

The Sumitomo alumina fiber has a unique combination of properties:

- Is available in continuous filaments.
- Has high strength and high modulus.
- Is an electrical insulator.
- Is stable in molten metals.
- Has excellent high-temperature properties in air; at 1250°C, its strength is 90% of its room-temperature value.
- Has good handleability; can be woven.
- Is radar-transparent.
- Undergoes no galvanic corrosion.

The product specifications are the following:

- *Chemical composition*
  Al_2O_3: 85 ± 1 wt %
  SiO_2: 15 ± 1 wt %
- *Filament characteristics*
  Diameter: (17 ± 2) μm
  Filament length: 300 m
  Filaments/yarn: 1,000
- *Mechanical properties*
  Tensile strength: 1.5 G Pa
  Tensile modulus: 200 G Pa

Table 21-6. Typical "Saffil" properties.

Summary of Typical Properties.

"Saffil" zirconia fiber		
Fiber density	g/cm^3	5.6
Melting point	°C	>2500 (4500°F)
Maximum use temperature		
HT grade	°C	1600 (2910°F)
S grade	°C	1400 (2550°F)
Specific heat	cal/g°C	0.14
Tensile strength	MN/m^2	0.7 × 10^3 (100 × 10^3 psi)
Specific tensile strength	m^2/s^2	13 × 10^4
Young's modulus	MN/m^2	1 × 10^5 (15 × 10^6 psi)
Specific modulus	m^2/s^2	2 × 10^7
Mean diameter	μm	3
Surface area	m^2/g	5
Hardness (Mohs)		6
Shot content		Negligible
"Saffil" alumina fiber		
Fiber density	g/cm^3	2.8
Melting point	°C	2000 (3600°F)
Maximum use temperature		
HT grade	°C	1400 (2550°F)
S grade	°C	1000 (1830°F)
Specific heat	cal/g°C	0.25
Tensile strength	MN/m^2	1 × 10^3 (150 × 10^3 psi)
Specific tensile strength	m^2/s^2	40 × 10^4
Young's modulus	MN/m^2	1 × 10^5 (15 × 10^6 psi)
Specific modulus	m^2/s^2	4 × 10^7
Mean diameter	μm	3
Surface area	m^2/g	100
Hardness (Mohs)		6
Shot content		Negligible

"Saffil" is a trade mark of Imperial Chemical Industries Limited for inorganic fibers.

- *Sizing*
  - None
  - Polyvinyl alcohol
  - Poly methylmethacrylate
  - Epoxy resin
- *Spooling*
  I.D.: ca. 77 mm
  O.D.: ca. 88 mm
  Length: ca. 300 mm

2.5 High Silica and Quartz Fibers

(Abstracted from reference 4 by Hugh Shulock)

High Silica. The term high silica can be used to describe any high-purity glass. However, for use in reinforced plastics, high silica may be defined as a high-purity glass of 95%-plus purity SiO_2, produced by a leaching process. High-silica fibers and fabrics are flexible materials similar in appearance to, and produced

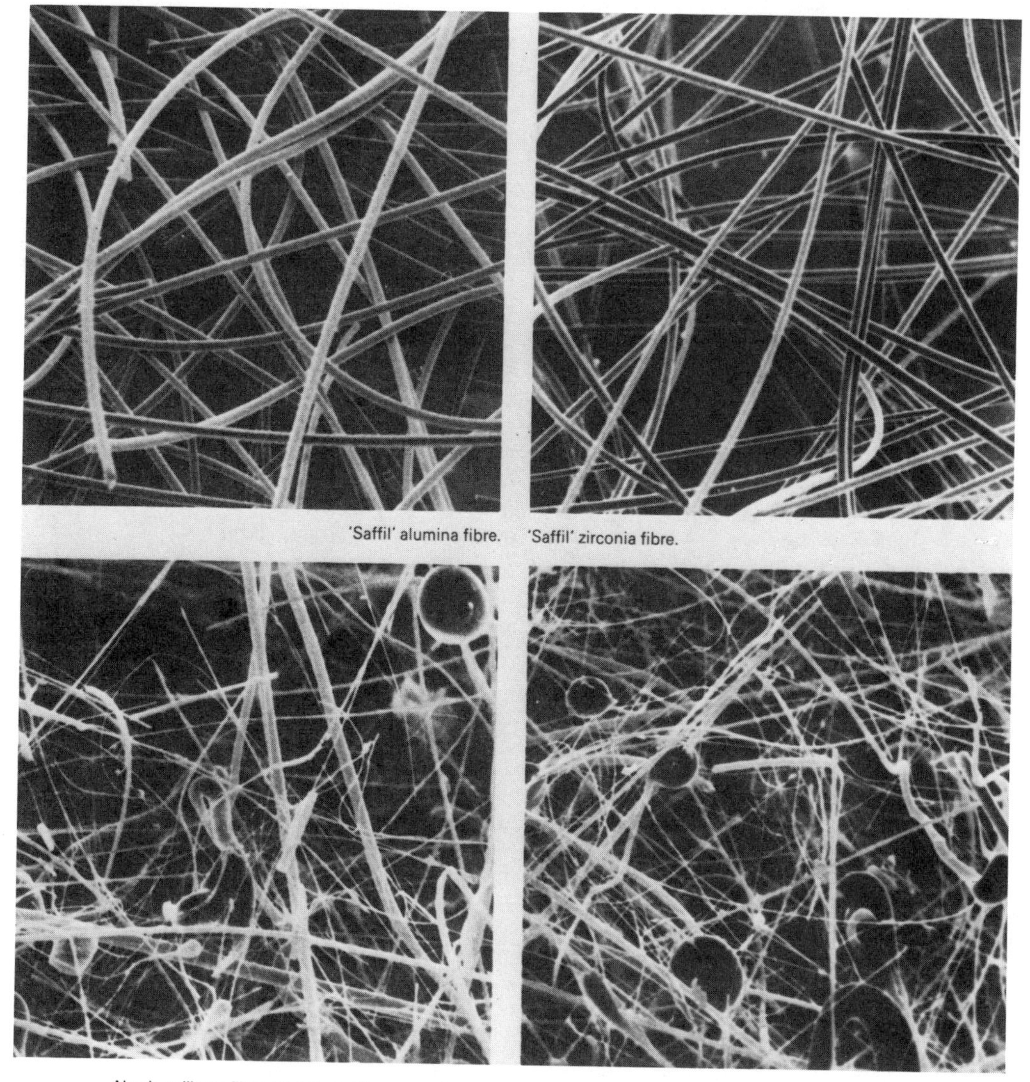

Fig. 21-3. Photomicrographs (magnification ×600) of "Saffil" alumina and "Saffil" zirconia fibers, for comparison with some alumina-silicate fibers. These demonstrate the uniform diameter of "Saffil" fibers and their freedom from shot. (Courtesy ICI—USA.)

from, fiberglass. Fiberglass, which has a silica content of 65%, is subjected to a hot acid treatment that removes virtually all of the impurities and leaves the silica intact. This is commonly called the leaching process.

Most textile forms (chopped fiber, mat, yarn, and fabric) are available in high silica. Fabrics are available in various weights and thicknesses; with purities from 95% to 99.4% SiO_2.

Quartz. The word quartz can also mean high-purity glass, but we will define quartz fibers as those fibers produced from high-purity (99.95% SiO_2) natural quartz crystals. The crystals are formed into rods from which filaments one-fifth the diameter of human hair are drawn. Up to 240 filaments are combined to form a flexible high-strength fiber, which is made into yarn and woven into fabric. All textile forms (including chopped fiber, mat rovings, cordage, sleevings, tapes, and fabrics) are available in quartz.

Quartz fiber with a purity of 99.95% SiO_2, exclusive of binder, retains virtually all of the characteristics and properties of solid quartz.

Physical and Mechanical Properties. High

silica and quartz materials have higher strength to weight ratios than most other high temperature materials, with quartz having approximately five times the tensile strength of silica. High silica and quartz yarns are perfectly elastic, and their elongation at break is approximately 1%.

The physical properties of the fibers, yarns, and fabrics made from high silica and quartz are given in Table 21-7, and a typical chemical analysis is given in Table 21-8.

Thermal Properties. Since high silica and quartz are so similar in nature, they also have similar thermal characteristics. The major variable is a higher melt viscosity for quartz, because of its higher silica content (see Fig. 21-4). High silica and quartz do not melt or vaporize until the temperature exceeds 3000°F.

Table 21-7a. Physical properties—fibers.

Property	High Silica	Quartz
Filament diameter (in.)		
Yarn and fabric	0.0004	0.0004
Mat	0.00005-0.0004	0.00005-0.0006
Filament tensile strength (psi)	0	0
@ Room temperature		130.000
@ 400°F		99.000
Specific gravity	1.74	2.2
Hardness—Mohs scale		5-6
Young's modulus (psi)		10×10^6

Table 21-7b. Physical properties—yarns.

Property	High Silica		Quartz	
Type number	#100	300–2/2	#300–2/4	#300–4/4
Nominal diameter (in.)	.020	.005	.010	.014
Approximate yd/lb	3050	7,500	3750	1875
Minimum breaking strength (lb)	2.5	3.0	5.0	11.0
Linear shrinking	12%	1%	1%	1%

Table 21-7c. Physical properties—fabrics.

Property	High Silica		Quartz	
Type Number	#82	#84	#581	#570
Wt–oz/sq yd	10.3	18.5	8.4	19.5
Thickness (in.)	.013	.026	.011	.027
Thread count–warp		50	57	38
fill		40	54	24
Breaking strength (lb)–warp	30	70	185	480
fill	25	45	170	400
Weave	8H satin	8H satin	8H satin	8H satin
Areal shrinkage % (max)	5	5	1	1
Moisture content %	5	5	1	1
pH	4.0–7.0	4.0–7.0	6.0–8.0	6.0–8.0
Silica content %	98–99.2	98–99.2	99.9+	99.9+
Surface area sq m/g	20–150	20–150	1–5	1–5

Table 21-8. Typical chemical analysis, % by weight.

Element	High Silica (high purity)	Quartz
SiO_2% (exclusive of binders)	99.23	99.95
Phosphorous % (ppm)	0	3
Sodium (ppm)	4	9
Potassium (ppm)	3	5
Lithium (ppm)	1	1
Boron (ppm)	205	10
Calcium (ppm)	5	23
Magnesium (ppm)	6	2
Strontium (ppm)	0	0
Iron (ppm)	16	3
Titanium (ppm)	2,800	12
Aluminum (ppm)	900	100
Manganese (ppm)	1	2
Copper (ppm)	2	1
Cadmium (ppm)		.5
Antimony (ppm)		.5
Chromium (ppm)	150.	0

From J. P. Stevens and Co., Inc., Central Research Laboratory, Garfield, N.J.

At continuous temperatures in excess of 1800°F, both forms of silica will begin to devitrify into a crystallized form known as christobalite. This conversion tends to stiffen the material, but causes no change in its physical form or insulating properties.

Resistance to thermal shock is excellent. Silica products can be heated to 2000°F and rapidly quenched in water without any apparent change.

Chemical Properties. High silica and quartz have similar chemical properties. They are not affected by halogens or common acids in the liquid or gaseous state, with the exception of hydrofluoric and hot phosphoric acids. Silica products are not recommended for use with alkalies, either hot or cold. Weak alkali solutions can be used for certain applications. Silica is not soluble in water or organic solvents.

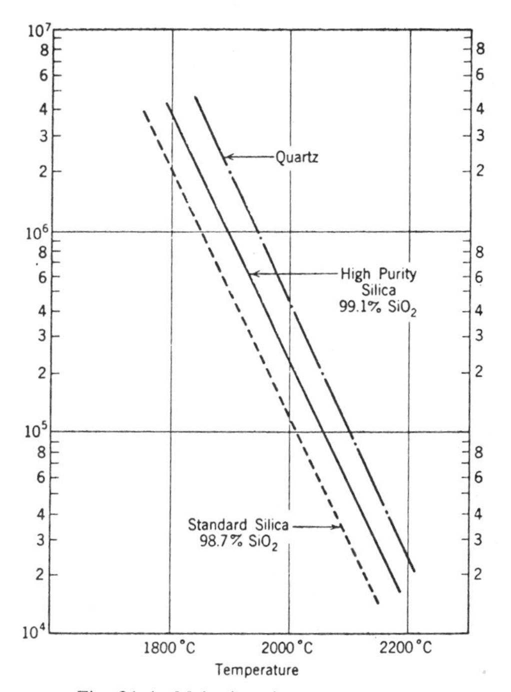

Fig. 21-4. Melt viscosity vs. temperature.

3. SUPPLIERS AND COST

3.1 Fiber FP

At present, fiber FP is at an experimental stage and hence availability is limited. The initial price is expected to be about $200/lb. The ultimate price will depend on such factors as market volume and process improvements. Current availability and cost information can be obtained by contacting E. I. DuPont de Nemours and Co., Pioneering Research Laboratory, Textile Fiber Department, Experimental Station, Wilmington, DE 19898, (302) 722-1201.

3.2 Ceramic Fibers

The 3M NEXTEL 312 and 440 continuous ceramic fibers are commercially available. The NEXTEL 312 price is in the $45/lb range, and the NEXTEL 440 is in the $120/lb range. Information on availability and current prices can be obtained by contacting the 3M Company—Ceramic Fiber Products, 3M Center, St. Paul, MN. 55101, (612) 733-1986.

The 3M NEXTEL 312 and 440 noncontinuous, blanket materials are at a development stage, and hence availability is somewhat limited. Information on availability and current prices can be obtained from the 3M Company—New Products Department, 3M Center St. Paul, MN 55144-1000, (612) 736-3014.

3.3 Saffil Fibers

ICI's Saffil Fibers are being produced in semi-commercial quantities in low-density mat form. The 1985 price range was $20 to $25 / lb, FOB Wilmington, DE. Detailed information on current prices can be obtained from: ICI America Inc., New Ventures Development, Wilmington, DE 19899, (302) 575-3000.

3.4 AVCO/Sumitomo Alumina Fiber

Development quantities of AVCO/Sumitomo alumina fiber are being produced in yarn form. It is available from:

AVCO Specialty Material Division
2 Industrial Avenue
Lowell, MA 01851

at $335 / lb. For more information, contact Thomas Foltz at (619) 452-5650.

3.5 High Silica and Quartz Suppliers, Materials and Cost

The suppliers of these silica products and the specific forms that are available are given in Table 21-9.

Addresses of these companies are listed below:

Alpha Associates, Inc.
2 Amboy Avenue
Woodbridge, N.J. 07075
(609) 634-5700

General Electric Co.
24400 Highland Road
Richmond Heights, OH 44143
(216) 266-2424

Haveg Division
900 Greenbank Road
Marshallton, DE 19808
(302) 995-0400

Hitco (Subsidiary Owens-Corning Inc.)
840 Newport Center Drive
Newport Beach, CA 92715
(714) 720-9300

Johns Manville Corp.
P.O. Box 5108
Denver, CO 80217
(303) 978-2000

J. P. Stevens and Co. Inc.
1185 Avenue of the Americas
New York, NY 10036
(212) 930-2000

Table 21-9. Silica products and suppliers.

Form	High Silica	Quartz
Woven fabrics	Haveg Corp. Hitco Corp. J. P. Stevens and Co., Inc.	J. P. Stevens and Co., Inc.
Woven tapes	Hitco Corp.	Alpha Associates, Inc.
Yarn	Hitco Corp.	Alpha Associates, Inc. General Electric Co. J. P. Stevens and Co., Inc.
Chopped fibers	Hitco Corp.	J. P. Stevens and Co., Inc.
Roving	–	General Electric Co. J. P. Stevens and Co., Inc.
Cordage	Hitco Corp.	Alpha Associates, Inc.
Sewing thread	Hitco Corp.	Alpha Associates, Inc. Dodge Industries, Inc.
Mat	Haveg Corp. Hitco Corp. Johns Manville Co.	General Electric Co. J. P. Stevens and Co., Inc. Johns Manville Co.
Wool		General Electric Co. J. P. Stevens and Co., Inc.
Braided sleeving	Hitco Corp.	Alpha Associates, Inc. Thermo Electric Co.
Knit fabrics	–	Alpha Associates, Inc.
Knit tapes	–	Alpha Associates, Inc.

Costs and Pricing. The principal raw material used to produce fiberglass fibers from which high silica is made is sand, available abundantly at approximately 15¢/lb. Quartz fibers are produced from natural pure fused quartz crystals that cost approximately $5/lb. The great disparity in raw material costs is not always reflected in the final product. There are great differences in manufacturing costs. High silica is sold in a price range of $4 to $50/lb, depending on the form and the quantity purchased. Quartz products are priced from $25 to $95/lb, again depending on quantity and form. It is sometimes less expensive in the final product to use the more expensive but stronger, more flexible, and more easily processible quartz product than the less expensive high silica. Because of the great difference in strength, it is sometimes possible to use considerably less reinforcement with quartz than silica. If thermal protection is all that is required, then high silica is the obvious choice. Comparative prices on the most widely used products in the largest volume categories are shown in Table 21-10.

3.6 Refractory Fibers

Although the prime objective of this book is reinforcements, the author believes that the subject of ceramic fibers would not be complete without mention of the kinds and sources of the principal refractory fibers available in the United States. These fibers are primarily used for insulation, but could be used as a moderate reinforcement or secondary filler to complete a packing requirement as the art of packing becomes more of a science in the future.

Suppliers of other ceramic fibers used as insulation are listed below.

Product name: Kao Wool
Description: Aluminum silicate fiber wool and other insulation products
Supplier: Babcock and Wilcox Co.
Refractories Div.
Old Savannah Road
Box 923
Augusta, GA 30903
(404) 798-8000

Product name: Fiberfrax
Description: Aluminum silicate fiber wool
Supplier: Carborundum Co.
Research and Development Div.
Box 337, Niagara Falls, NY 14302
(716) 278-2000

Product name: Cerafiber
Description: Aluminum silicate fiber wool
Supplier: Johns Manville Corp.
Greenwood Plaza
Denver, CO 80217

Table 21-10. Cost comparison.

HIGH SILICA		QUARTZ	
Product Description	Price	Product Description	Price
Silica fabric–heavy Wt–18.5 oz/sq yd Thickness–0.26 in. Strength–50 lb/in.	$6.15/sq yd	Quartz fabric–heavy Wt–19.5 oz/sq yd Thickness–0.27 in. Strength–480 lb/in.	$48.25/sq yd
Silica fabric–light Wt–10.3 oz/sq yd Thickness–.013 in. Strength–30 lb/in.	$4.00/sq yd	Quartz fabric–light Wt–8.4 oz/sq yd Thickness–0.11 in. Strength–185 lb/in.	$21.00/sq yd
Silica yarn Teflon binder Yield 3000 yd/lb Breaking strength average 2.5 lb	$18.00/lb or $.006/yd	Quartz yarn Teflon binder Yield 7500 yd/lb Breaking strength average 3.0 lb	$29.00/lb $.004/yd

Product name: Zircar Product Inc.
Description: Zircona and alumina insulation
Supplier: Zircar Product
110 North Main Street
Florida, NY 10921
(914) 651-4481

4. PROCESSING DATA, CURRENT AND PROJECTED APPLICATIONS

4.1 Fiber FP Application Data

The following properties are important features of DuPont's FP fibers:

- Excellent retention of modulus and strength at elevated temperatures.
- High compression strength.
- Extremely high temperature stability.
- Chemically resistant, hard refractory ceramic.
- Fabricability—in continuous yarn form.

The continuous forms of FP fiber make it amenable to low-cost processing techniques for making forms such as rods, tubes, billets, beams, and engine blades containing high fiber volume loading in the order of 60%.

4.2 3M Brand Ceramic Fibers Application Data

The unique combination of properties of 3M Ceramic Fibers suggest many applications not satisfied by existing fibers. The NEXTEL 312 fiber, for example, can withstand continuous temperatures of up to 2200°F and short-term temperatures of up to 2600°F. It has a modulus elasticity of 22–25 $\times$ 10^6 psi and a tensile strength of about 250 $\times$ 10^3 psi.

3M Ceramic Fibers are sold in the industrial, electrical, and aerospace markets. They are used in thermocouple wire insulation, high-temperature-resistant continuous conveyor belts, gaskets, heat shields and high-temperature filtration, reinforcement of metals and ceramics, and reinforcement of plastics in such products as radomes, sporting goods, and aircraft structural components.

Composite property data for NEXTEL-312 fiber is given in Tables 21-11 through 21-13. Excellent translation of fiber properties in the composite is evident in all cases.

4.3 ICI's Saffil Fiber Application Data

A wide variety of applications are expected. Areas of particular importance are high-temperature thermal and acoustic insulation and high-efficiency filtration. Other applications include use as a catalyst support and in the reinforcement of composite products such as friction materials.

The fibers are produced in staple form with a length of several centimeters. The primary physical form of "Saffil" fibers is a loose wool

Table 21-11. Composite data—NEXTEL 312 Fibers in PR 286 epoxy resin.[a]

Property	Value
Volume fiber loading	50%
Composite density	0.073 lb/in.^{3} (2.02 g/cm^3)
Young's modulus in the fiber direction	11.0 $\times$ 10^6 psi (75.8 $\times$ 10^3 mN/m^2
Specific modulus	150 $\times$ 10^6 in. (380 $\times$ 10^6 cm)
Young's modulus transverse to the fiber direction	1.7 $\times$ 10^6 psi (11.7 $\times$ 10^3 mN/m^2)
Tensile strength in the fiber direction	135 $\times$ 10^3 psi (862 mN/m^2)
Specific tensile strength	1.7 $\times$ 10^6 in. (4.30 $\times$ 10^6 cm)
Tensile strength transverse to the fiber direction	3 $\times$ 10^3 psi (20.7 mN/m^2)
Interlaminar shear strength	14 $\times$ 10^3 psi (96.5 mN/m^2)
Compressive strength	>135 $\times$ 10^3 psi (>862 mN/m^2)
Flexural strength in the fiber direction	180 $\times$ 10^3 psi (1240 mN/m^2)
Coefficient of thermal expansion in the fiber direction	3.5 $\times$ 10^{-6} in./in./°F (6.3 $\times$ 10^{-6} cm/cm/°C)
Coefficient of thermal expansion transverse to the fiber direction	17.0 $\times$ 10^{-6} in./in./°F (30.6 $\times$ 10^{-6} cm/cm/°C)

[a]3M PR 286 epoxy resin system.
A data source: 3M Ref. 2.

Table 21-12. Composite data—NEXTEL 312 fabric in PR 286 epoxy resin.[a]

Weave	Unidirectional[b]	Bidirectional[c]
Flexural modulus warp direction	8.8×10^6 psi (60.6×10^3 mN/m^2)	7.0×10^6 psi (48.2×10^3 mN/m^2)
Flexural strength warp direction	110×10^3 psi (7580 mN/m^2)	80×10^3 psi (5510 mN/m^2)
Volume percent fibers in composite	48.5	54.0
Volume percent fibers in warp direction	40.0	27.5

[a]3M PR 286 epoxy resin system.
[b]Fabric contains 83% of the fibers in the warp direction.
[c]Fabric contains 51% of the fibers in the warp direction.
Data Source: 3M Ref. 2.

that is capable of further conversion to other end forms, including:

- Chopped fiber—a range of short staple fibers of aspect ratio 10–1000.
- Blankets—in a range of thicknesses and bulk densities.
- Needled blankets—blankets needled between organic scrims.
- Strip—narrow sections of blanket.
- Paper—conventionally manufactured papers of various weights containing organic and inorganic binders alone or in combination.
- Boards—rigid or semirigid, containing various proportions of ''Saffil'' fiber with inorganic and/or organic binders and possibly fillers.
- Vacuum formed shapes—rigid or semirigid shapes such as cones and tubes with inorganic and/or organic binders and possible fillers (see Fig. 21-5).
- Temporarily compressed forms—novel forms of densified boards and shapes containing organic binders that burn out at a high temperature allowing the fiber to recover much of its original volume.
- Textile forms—yarns, rope, cloth, etc., under development.

Not all of these forms are commercially available. ''Saffil'' fibers offer users a material that can be formed by novel processes, and can provide solutions to unusual product requirements using previously available materials.

The author sees increasing use of Saffil filler in plastic composites as the price becomes more

Table 21-13. Composite Data—NEXTEL 312 Fibers in PR 286 epoxy resin[a]—unidirectional fatigue properties.[b]

$\frac{\sigma \text{ Max}}{\sigma \text{ Ult}}$	N Cycles
0.71	660,000
0.51	9,000,000[c]
0.40	6,000,000[d]

[a]3M PR 286 epoxy resin system.
[b]$R = \frac{\sigma \text{ Max}}{\sigma \text{ Min}} = 0.1$, 20 hz, 40% fiber volume fraction.
[c]95% Static ultimate residual strength.
[d]100% Static ultimate residual strength.
Data Source: 3M Ref. 2.

Fig. 21-5. ''Saffil'' fibers formed shapes. (Courtesy ICI—USA.)

competitive with other reinforcements. Its combination of fine diameter, high modulus, and long lengths is unavailable in any current reinforcing fiber. This combination can be used in applications not possible with short microfibers or other large-diameter, high-modulus fibers.

One suggested application is the utilization of an extremely low-cost forming technique such as fiber vacuum forming. This forming method permits the rapid formation of large complex shapes on a very low-cost screen mold. (See Fig. 21-5.) The formed parts are subsequently impregnated with a low-viscosity resin solution, and the excess solution is allowed to drain from the highly porous felt material. Curing the resin then forms a strong, low-density product, which is a resin-bonded, trusslike structure.

As Saffil prices become competitive with other reinforcement filaments, they will provide the potential for creative ideas in new processing and new product development.

4.4 AVCO/Sumitomo Alumina Fiber Applications

The AVCO/Sumitomo fiber is recommended as a reinforcement in metallic, polymer, ceramic, and glass matrices. Some data have been developed in this fiber in both epoxy resin and aluminum. These data are given in Tables 21-14 through 21-17.

Table 21-14. Properties of unidirectional alumina/epoxy composites.

Property	Unit	Value
Density	(Mg/m^3)	2.4
Tensile strength (0°)*[1]	(GPa)	1.35 (1.7)*[2]
(45°)	(GPa)	0.21
(90°)	(GPa)	0.068
Tensile modulus (0°)	(GPa)	137
(45°)	(GPa)	18
(90°)	(GPa)	18
Flexural strength (0°)	(GPa)	1.56 (1.8)*[2]
Flexural modulus (0°)	(GPa)	122
Flexural fatigue strength at 10^7 cycles (0°)	(GPa)	0.63
Compressive strength (0°)	(GPa)	1.7
Compressive modulus (0°)	(GPa)	120
Poisson's ratio	—	0.25
Modulus of rigidity*[3]	(GPa)	7.25
Interlaminar shear strength	(GPa)	0.13
Izod impact (0°)	(J/cm^2)	20
Rockwell hardness	—	E85
Thermal expansion (0°)	(1/K)	0.4×10^{-5}
(90°)	(1/K)	3.7×10^{-5}
Thermal conductivity (0°)	(W/m·K)	2.32
(90°)	(W/m·K)	0.85
Volume electrical resistivity (90°)	(Ω·m)	10^{13}
Arc resistance time	(s)	130–180
Dielectric strength	(kV/mm)	16–20
Dielectric constant at 10^6 Hz	—	4.9
at 10^{10} Hz	—	5.1
Tangent loss at 10^6 Hz	—	0.015
at 10^{10} Hz	—	0.016

Fiber loading—60 vol %.
Resin: DDM type epoxy (Sumitomo's SUMIEPOXY® ELM-434).
Hardener: DDS.
*[1] 0°, 45°, and 90° in the parentheses indicate the angles measured from the fiber direction.
*[2] For FRP made from laboratory-scale fibers.
*[3] $-e_y/e_x$, where e_x and e_y are longitudinal and transverse strains, respectively, when the specimen is loaded along the fiber direction.

Table 21-15. Properties of alumina/epoxy fabric.

		293K	453K
Tensile strength (0°)*[1]	(MPa)	405	—
(45°)		206	—
Tensile modulus (0°)	(GPa)	55	—
(45°)		28	—
Flexural strength (0°)	(MPa)	673	474
(45°)		242	133
Flexural modulus (0°)	(GPa)	55	47
(45°)		24	15
Compressive strength (0°)	(MPa)	568	—
(45°)		342	—
Compressive modulus (0°)	(GPa)	63	—
(45°)		25	—
Interlaminar shear strength (0°)	(MPa)	71	55

Fiber loading = 47 vol %.
Fabric: Plain weave, areal weight = 390 g/m^2, warps/fills = 6.8 × 6.8 yarns/cm, 330 filaments/yarn.
Resin: DDM type epoxy.
Hardener: DDS.
*[1] 0° and 45° in the parentheses indicate the angle between warp direction and load direction.

Table 21-16. Properties of unidirectional alumina/aluminum composites.

			Alumina/Al	Alumina/Al + 5% Cu
Density		(Mg/m^3)	2.9	2.9
Tensile strength (0°)*[1]	at 293K	(MPa)	860	590
(45°)				440
(90°)			98	230
(0°)	at 723K		860	670
(90°)			31	86
Tensile modulus (0°)	at 293K	(GPa)	150	150
(90°)			110	110
(0°)	at 723K		140	140
Tensile fatigue strength at 10^7 cycles (0°)	at 298K	(MPa)	100*[2]	250
(90°)			—	80
Flexural strength (0°)	at 293K	(MPa)	1,100	1,000
(45°)			400	800
(90°)			180	480
(0°)	at 723K		800	950
Flexural modulus (0°)	at 293K	(GPa)	135	135
(0°)	at 723K		—	100
Flexural fatigue strength at 10^7 cycles (0°)	at 298K	(MPa)	300*[2]	
Compressive strength (0°)	at 293K	(MPa)	1,430	2,200
(45°)			—	580
(90°)			—	540
(0°)	at 723K		—	580
Compressive modulus (0°)	at 293K	(GPa)	140	140
Poisson's ratio*[3]	at 293K	—	—	0.33
Modulus of rigidity	at 293 K	(GPa)	—	35

Table 21-16. (*Continued*)

			Alumina/Al	Alumina/Al + 5% Cu
Charpy impact (0°)	at 293K	(J/cm^2)	9.4	2.2
(45°)			—	3.0
(90°)			—	2.0
(0°)	at 723K		—	3.0
Stress for minimum creep rate of 0.0001%/hr (0°)	at 623–723K	(MPa)	—	260
Specific heat		(kJ/kg·K)	0.88	—
Thermal expansion (0°)		(1/K)	7.6×10^{-6}*2	—
(90°)			14.0×10^{-6}*2	—
Thermal conductivity (0°)	at 300–673 K	(W/m·K)	105	—
(90°)			75.3	41.9
Fiber loading—50 vol%				

*1 0°, 45°, and 90° in the parentheses indicate the angles measured from the fiber direction.
*2 According to T. Furuta of National Aerospace Laboratory, Japan.
*3 $-e_y/e_x$, where e_x and e_y are longitudinal and transverse strains, respectively, when the specimen is loaded along the fiber direction.

4.5 High Silica and Quartz Fiber Application and Compositing Data

Properties of Composites. High silica and quartz can be used with most resin systems, and are easily impregnated on most conventional coating equipment, either vertical or horizontal. Preimpregnated fabrics are easily cut, either straight or on a bias, for use in tape winding or in lay-up techniques. The lighter-weight fabrics are usually more flexible, and are normally used on highly contoured parts. Fabricated parts can also be made from chopped squares or with molding compounds made from shredded fabrics or chopped fibers. High silica materials are usually used without a finish or a binder. However, many types of resin-compatible finishes and modifying additives can be applied to silica fabrics.

Quartz fabrics are generally supplied with chemical finishes compatible with the resin system to be used. Resin-compatible finishes substantially improve laminate physical properties and reduce moisture absorption. Most finishes, binders, or additives that are available on fiberglass fabrics can be applied to quartz. Standard commercial binders are available for phenolics, epoxies, polyimides, silicones, polyesters, and fluorocarbon resin systems.

All property data presented in Tables 21-18 through 21-21 were produced under laboratory conditions using flat panels prepared in accordance with MIL R-9300 with unfinished silica fabrics. These values are for comparative reference purposes only.

Uses. High silica and quartz are both used in a wide variety of similar products. The choice

Table 21-17. Properties of alumina fabric aluminum composites.

		293K	573K	723K
Tensile strength (0°)*1	(MPa)	304	—	—
(45°)	(MPa)	343	—	—
Flexural strength (0°)	(MPa)	627	500	323
(45°)	(MPa)	637	480	167

Fiber loading = 39 vol %.
Fabric: Plain weave, areal weight = 390 g/m^2.
warps/fills = 6.8 × 6.8 yarns/cm, 330 filaments/yarn.
Matrix: Al + 5% Cu.
*1 0° and 45° in the parentheses indicate the angles measured from the warp direction.

Table 21-18. Composite data—unidirectional impact characteristics.

Materials[a]	Apparent Flex Strength	Total Energy per Unit Area	Ductility Index
	(ksi)	$\frac{\text{ft lb}}{\text{in.}^2}$	
E-glass[c] epoxy	73	114	0.4
Kevlar[c] epoxy	142	124	1.6[b]
HMS graphite[c] epoxy	125	3.8	0
3M fiber[d] epoxy	126	39	0.47

[a]Specimen thickness ranged from 0.11 to 0.15 in.
[b]Based on maximum stress, not point of initial nonlinearity.
[c]Beaumont, P. W. R., et al., "Methods for Improving the Impact Resistance of Composite Materials", to be published in ASTM STP 568, 1975.
[d]AB-312 fibers in 3M PR 286 epoxy resin system.
Data Source: 3M Ref. 2.

of type of raw material to use is generally dictated by a combination of performance requirements, manufacturing needs, and cost. (See section 3.5 on costs and pricing.)

Table 21-22 lists the most widely known end uses and product recommendations, but should not be treated as a complete listing. New products and end uses are being developed continually, and the product manufacturers should be contacted for the most current information.

Table 21-19. Typical laminate properties—high silica and quartz filaments in phenolic resin.

	HEAVY WEIGHT FABRIC		LIGHT WEIGHT FABRIC	
Property	High Silica	Quartz	High Silica	Quartz
Fabric description				
Oz/sq yd	18.5	19.5	10.3	8.4
Finish	None	A–1100	None	A–1100
Weave	8H satin	5H satin	8H satin	8H satin
Laminate preparation				
Resin–phenolic	V-204[a]	91-LD[b]	SC-1008[c]	V-204[a]
Cure and molding pressure	60 min at 325°F		60 min at 325°F	
Post cure	180 min at 325°F pressure 250 psi		180 min at 325°F pressure, 250 psi	
Laminate properties				
Flexural strength, psi $\times 10^{-3}$				
Average at 75 ± 5°F	33.3	71.7	36.6	95.4
Average at 500 ± 10°F ½ hr	24.8	35.5	21.1	60.0
Tensile strength psi $\times 10^{-3}$				
Average at 75 ± 5°F	23.1	56.8	22.7	72.1
Average at 500 ± 10°F ½ hr	19.1	50.0	15.3	47.0
Specific gravity	1.72	1.73	1.62	1.80
% resin	30.3	34.0	37.0	30.7
Thickness–in.	.124	.127	.152	.118
Number of plies	6	6	12	12

From J. P. Stevens and Co., Inc., Central Research Laboratory, Garfield, N.J.
[a]Barrett Division Allied Chemical Co.
[b]Cincinnati Testing Labs, Division Studebaker-Packard
[c]Monsanto Chemical Co.

Table 21-20. Laminate comparison of mechanical and electrical properties for various resin binders.

QUARTZ FABRIC

Property Fabric description	Phenolic[a] V-204	Silicone[b]	Epoxy[c] Epon 828/CL	Teflon[d]	PBI AF-R-100
Oz/sq yd	8.4	8.4	8.4	8.4	8.4
Weave	8H satin	8H satin	8H satin	8H satin	8H satin
Finish	A1100	None	A1100	Teflon	A1100
Mechanical					
Flexural, psi × 10^{-3}					
dry at 75° ± 5°F	95.4	34.0	98.6	–	–
Tensile, psi × 10^{-3}					
dry at 75 ± 5°F	72.1	34.0	79.2	–	–
Electrical					
Dielectric constant					
X-band, dry-room temp. (9,375 megacycles)	3.86	2.93	3.470	2.47	3.360
Loss tangent					
X-band, dry-room temp.	0.040	0.00098	0.0092	0.0007	0.0034

[a]U.S. Polymeric Inc. data.
[b]Coast Manufacturing and Supply Co. data.
[c]Ferro Corp. and Grumman Aircraft Engineering Corp. data.
[d]Custom Materials, Inc. data.

Table 21-21. Comparison of mechanical and electrical properties, silicone resin laminates.

QUARTZ VS D GLASS VS E GLASS

Property	E Glass[a]	D Glass[b]	Quartz[c]
Style number	181	181	581
Oz/sq yd	9.0	9.0	8.4
Finish	None	None	None
Mechanical			
Tensile strength psi, dry–room temp.	39,500	25,000	34,000
Compressive strength psi, dry–room temp.	27,300	18,300	24,000
Flexural strength psi, dry–room temp.	–	41,800	36,000
Flexural modulus psi × 10^{-6}, dry-room temp.	–	2.90	2.80
Electrical			
Dielectric constant X-band, dry–room temp.	3.946	3.60	2.93
Loss tangent X-band, dry–room temp.	0.0082	0.002	0.00098

[a]Summary Card, "Mechanical Properties of Aerospace Fiberglas Fabric Laminates," Owens Corning Fiberglas Corp. Office of Aerospace and Defense, April 1966.
[b]Trevarno, F., "130 Silicone Data Bulletin #54b," Coast Manufacturing and Supply Co., January 1965.
[c]Trevarno, F., "Structural Materials Data Bulletin #109," Coast Manufacturing and Supply Co.

Table 21-22. Product recommendations.

End Use	High Silica	Quartz
Ablative		
Re-entry heat shields		
Tape wrapped	Fabric–18.5 oz/sq yd	Fabric–19.5 oz/sq yd
Filament wound		Roving–20 end
Woven-three-dimensional		Yarn–to order
Nose cones		
Compression molded	Fabric–18.5 oz/sq yd –10.3 oz/sq yd	Fabric–19.5 oz/sq yd –8.4 oz/sq yd
Filament wound		Roving–20 end
Woven–three-dimensional		Yarn–to order
Nozzles and exit cones		
Tape wrapped	Fabric–18.5 oz/sq yd	Fabric–19.5 oz/sq yd
Filament wound	Cordage–1/8 in. diameter	Roving–20 end Cordage–1/8 in diameter
Compression molded	Fabric–18.5 oz/sq yd –10.3 oz/sq yd Chopped squares–1/2 in. squares Chopped fiber–1/4 in. length	Fabric–18.5 oz/sq yd –8.4 oz/sq yd Chopped squares–1/2 in. squares Chopped fiber–1/4 in. length
Structural		
Fins and struts		
Lay-up (vacuum or autoclave molded)		Fabric–8.4 oz/sq yd
Filament wound		Roving–20 end
Compression molded		Fabric–8.4 oz/sq yd
Electrical		
High pressure laminates		Fabric–5.6 oz/sq yd
Antennas		
Lay-up (vacuum or autoclave molded)		Fabric–8.4 oz/sq yd
Filament wound		Roving–20 end
Radomes		
Lay-up		Fabric–8.4 oz/sq yd
Filament wound		Roving–20 end
Thermal		
Blast shields		
Compression molded	Fabric–18.5 oz/sq yd Mat Chopped fiber–1/4 in. length Chopped squares–1/2 in. squares	Fabric–19.5 oz/sq yd Mat Chopped fiber–1/4 in. length Chopped squares–1/2 in. squares
Separators		
Compression molded	Fabric–18.5 oz/sq yd	

REFERENCES

1. Dhingra, A. K., Champion, A. R., and Krueger, W. H., "Fiber Reinforced Aluminum and Magnesium Composites," E. I. DuPont presented, 1st Institute for Defense Analysis Workshop on Metal Matrix Composites, Washington, D.C., July 1975.
2. 3M Company's Product Bulletin "New 3M Ceramic Fiber Products Continue Where Others Leave Off."
3. ICI's Product Bulletin #NV/84/IED/103/274 "Saffil Fibers As High Temperature Materials."
4. Shulock, H., "High Silica and Quartz," *Handbook of Fiberglass and Advanced Plastics Composites*, Chapter 9, G. Lubin (ed.), Van Nostrand Reinhold, New York, 1969.

Section VI
Procedures—Equipment and Utilization Technology

22

Procedures, Equipment, and Utilization Technology

Benjamin M. Walker

Walker Engineering Associates
Madison, Connecticut

CONTENTS

1. INTRODUCTION

In order to take advantage of the benefits derived from the use of fillers and reinforcements, the molder must be aware of the special considerations that are involved, compared with the processing of unfilled plastics. Molders who have not been involved with filled plastics may be concerned with their higher melt viscosity and abrasiveness, which may lead to molding problems, lower production rates, and equipment erosion. The uniform dispersion of the filler can also cause difficulty. During recent years, there has been great progress in the effective utilization of fillers and reinforcements, and fast and economical production rates have been achieved in both the compounding and molding of composite materials. In some cases, this has been accomplished by minor modifications of existing equipment, such as the use of hard surface coatings on wear surfaces. In other cases, special equipment has been developed for more effective utilization of these materials. Examples are the new multicomponent spray guns that can spray polyester resin, catalyst, filler, and chopped fibers simultaneously, and the high-production-rate pultrusion lines.

Compounding and processing objectives are generally the same whether fillers, reinforcements, or combinations of them are added to a resin. Both types of fillers must be uniformly dispersed in the polymer, and good wetting of fillers by the polymers is also desired. However, the processing problems with fillers and reinforcements are different in several important respects, and the compounding and processing equipment and procedures may be different. These two classes of fillers are therefore treated separately as to equipment and utilization procedures.

Types of compounds included are plastisols, filled thermoplastics, filled thermosets, and epoxy, polyester, and other liquid resins. Topics discussed included criteria for equipment selection, types and sources of equipment, and special notes regarding equipment design and use.

2. CRITERIA FOR SELECTION OF EQUIPMENT

2.1 Mixing Equipment

2.1.1 General Discussion. In compounding with fillers or reinforcements, the objectives are generally the same. Both types of additives must be uniformly dispersed in the polymer, and good wetting of additives by the polymers is also desired.

The type of equipment best suited for mixing an additive with a resin will depend upon the nature and proportions of the ingredients used and the physical characteristics of these materials and of the final mix, including particle size, viscosity of resin, and viscosity of the mixture. Temperature is also important, as room-temperature blending of a solid powder resin with a filler may require only a dry tumbling operation. The same resin at a higher temperature may be a viscous liquid, and require a more intensive kneading or milling type of mixing.

The forming process used with the compound will also affect the thoroughness of mixing and type of mixer required. A preliminary dry blending of granules and filler may be sufficient if the compound is later to be screw-extruded or injection-molded in a reciprocating screw machine, because these processes will provide a great deal of additional mixing. Compression or transfer molding, or plunger injection molding will require a more intimate mixture of materials to ensure a uniform product.

2.1.2 Fillers. Resins may be dry-mixed at room temperature, and this procedure may vary from a dry tumbling of powder and pellets to fine grinding or milling of the pellets for more thorough mixing. The same resin pellets or powder may be heated to an elevated temperature where they become viscous fluids, and a more intimate dispersion of filler and resin may then be obtained. The type of mixer required for this higher temperature mixing will depend upon the viscosity of the melt and the amount of power required to mix it.

With thermosetting resins, such as phenolics, alkyds, epoxies, and polyesters, it is important to control the temperature during mixing. Room-temperature mixing is used, and it is important to avoid heat build-up in grinding and mixing, which would cause precure of the resins.

2.1.3 Reinforcement Fibers. When fiber reinforcements are mixed with resins, different procedures may be required from those used for powdered fillers. Fibers are most effective in providing reinforcement in compounds when they maintain a high length to diameter ratio. Therefore, it is desirable to minimize the breaking up of fiber length during mixing. Such intensive, high-shear mixers as Banbury, roll mills, or screw extruders must be used judiciously with such compounds. Some breakdown in fiber length by such mixers may be acceptable, but excessive breakdown may reduce the effectiveness of the fiber reinforcement. (See Fig. 22-1, photo of fibers with mixing, and Fig. 22-2, curve of properties vs. mixing). An optimum balance is required between good mixing and wetting of fibers with the resin and retention of fiber length.

For retention of maximum fiber length, special processes have been devised. For example, continuous strands of glass fibers may be fed through an extruder die, and thermoplastic melt extruded over the fiber. Other processes, especially adapted to liquid resins, involve combining fibers in thread, cord, mat, or fabric form with liquid resins applied with spray guns, rolls, or doctor blades.

Fig. 22-1. Effect of mixing equipment on glass fiber length. Upper—using twin screw extruder. Lower—using intensive mixer. (Courtesy Baker Perkins, Inc.)

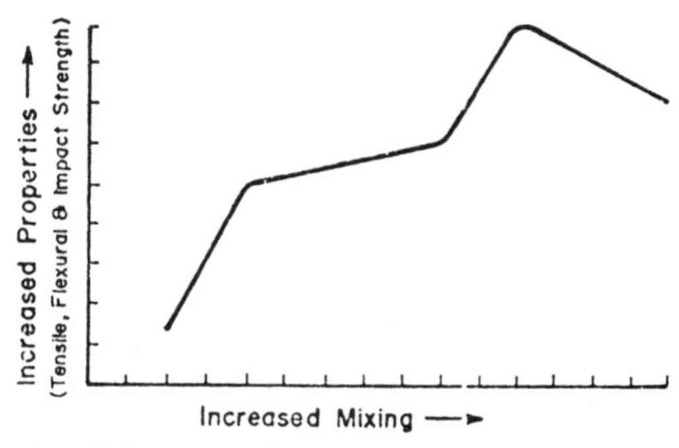

Fig. 22.2. Effect of degree of mixing on properties. (Courtesy Liquid Nitrogen Processing Co.)

Note: Patent litigation is presently in process on this subject of glass-filled, thermoplastic resins and their processing, and its status should be checked.[23]

2.2 Processing Equipment

2.2.1 General Notes. Plastics compounds containing fillers and/or reinforcements may be processed with the complete range of processing equipment used for unfilled plastics. This includes screw extruders and compression, transfer, and injection-molding machines, and other forming methods. Processes are essentially the same, but the details of conditions may be somewhat different. These differences are due to such factors as increased viscosity of the compound melt at processing temperatures, with corresponding decrease in flow rate and the need to use higher pressures.[1,2,19,20,22,24,25]

2.2.2 Wear. Wear of equipment or molds is essentially the same in processing filled or unfilled compounds, provided the compound contains a well-dispersed filler and the molds are properly designed. If some abrasive filler, such as glass fibers, is not well dispersed, or if mold gates are too small, excessive wear may occur.

2.2.3 Feeding. Feeding of fillers, reinforcing and nonreinforcing, to the plasticating unit may present problems due to bridging of material in the feed hopper. More positive feed equipment and procedures may be necessary. Because of the abrasive nature of some of the fillers, as well as the need for retention of the aspect ratio of reinforcing fillers, special techniques of downstream feeding into the melts are frequently used.

2.2.4 Drying and Venting. Drying and venting to ensure a void-free product is even more important with filled and/or reinforced materials than with unfilled compounds. The fillers and reinforcements used are of small particle size, and the resulting large surface area of the particle favors adsorption of moisture. This, combined with the natural tendency of many such materials to absorb moisture, makes it especially important to use well-dried material and/or to provide for venting of the processing equipment.

3. MIXING EQUIPMENT

3.1 Description of Types of Mixers

3.1.1 Intensive Internal Mixers. *Banbury* (name registered by Farrel Co.). This high-shear mixer consists of two contrarotating, spiral-shaped rotors encased in segments of cylindrical housings, intersecting so as to have a ridge between the rotors. Blades may be cored for circulating heating or cooling media. It is used for a wide range of plastics and rubber compounds on a batch basis.

Plasticator (Plastimat by Day Mixing Co.). This high-shear machine is designed to accomplish mixing similar to the Banbury.

Turbulent Mixer. This is a high-speed cylindrical blender for very rapid mixing of solids/solids and solids/liquids, useful for combining liquid polyesters and glass fibers.

Roll Mills. Different-speed rolls are placed close to each other to provide high shearing action to the plastics materials and compounding ingredients being mixed.

DISKPACK™ Compounder. This variable-intensity shear mixer consists of a series of specially shaped disk profiles on a common shaft. Channels formed by the disk profiles are of four basic types: melting, mixing, venting/devolatilizing, and pumping. A combination of channel blocks and transfer passages causes the polymer to flow from one channel to another. The intensity of shear can be controlled independently of the production rate by varying the channel block clearance and placing mix pins in the melt path. Fillers and reinforcements may be added with the polymer or added downstream, after melting has occurred. (See Figs. 22-3 and 22-4.)

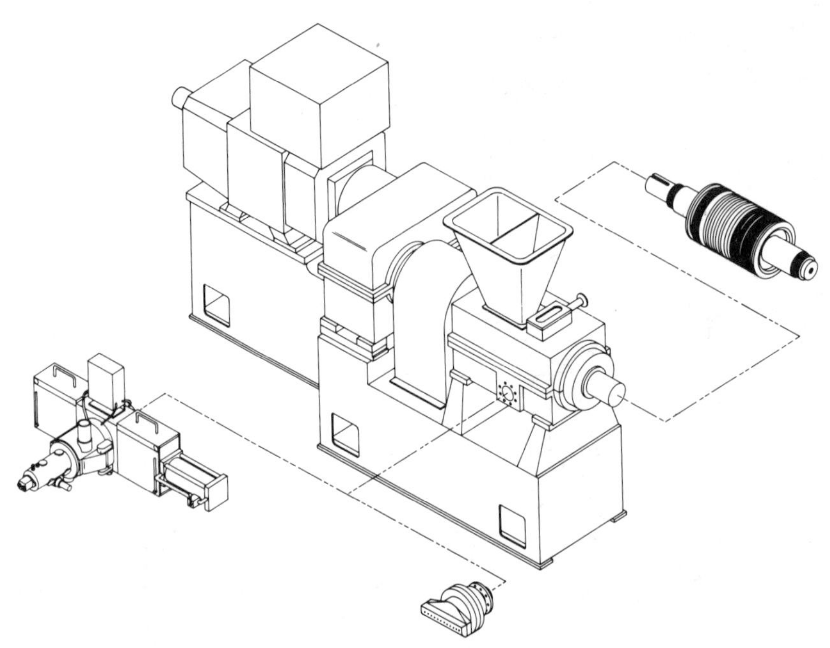

Fig. 22-3. DISKPACK compounder showing rotor (disk) configuration and alternate discharge arrangements: strand die or underwater pelletizer. (Courtesy Farrel Company, Emhart Machinery Group.)

Fig. 22-4. 350 mm, DISKPACK compounder with H150 underwater pelletizer mounted on discharge.

3.1.2 Dry Mixers and Blenders. *Ribbon Blender.* Consisting of helical, ribbon-shaped blades rotating close to the edge of a U-shaped vessel, these blenders are used for relatively high-viscosity fluids and dry blends such as PVC calendering and extrusion compounds.

Cone Blender. A cone-shaped vertical tank has a screw flight agitator that rotates on its own axis while orbiting the periphery of the tank. It is used for mixing plastisols and organisols, and as a mixer/dryer for powdered and pelletized plastics.

Kneader. This device mixes with a pair of intermeshing blades, often in the shape of the letter Z, and is used for working plastics masses of semi-dry or rubbery consistency.

Double Arm Mixer. Two agitators rotate in a tank. Tangential or overlapping types of action may be used, depending upon the viscosity of the materials being mixed.

Vortical Intensive Mixers. These mixers consist of a propeller-like impeller rotating at high speed in the bottom of a stationary container, continuously recirculating the materials between closely spaced stationary and rotating pins. They are used for dry-blending resins such as PVC with plasticizers and other additives.

Ball Mill. A cylindrical or conical shell rotates about a horizontal axis, partially filled with a grinding medium such as flint pebbles, ceramic pellets, or metal balls. Material to be ground is added, the shell is rotated, and the cascading pellets reduce the particle size by impact on the material.

Double Planetary Mixer (Ross). Two rectangularly shaped blades revolve around the tank on a central axis. Simultaneously, each blade revolves on its own axis at twice the speed of central rotation. This is effective for liquid/solid mixes.

3.1.3 Rotary Tumbers. *Drum Tumbler.* A cylindrical drum rotates end over end or about an inclined axis to thoroughly blend materials. They are commonly used to mix plastic pellets with color concentrates and/or regrind.

Conical Dry Blender. This blender consists of two hollow cones joined at their bases by a short cylindrical section, mounted on a horizontal shaft passing through the sides of the cylindrical section. Material is mixed by cascading, rolling, and tumbling actions as the cones rotate.

3.1.4 Liquid Mixers. *Vertical Cone Blender.* This blender has a vertical conical tank with screw flight agitator that rotates on its own axis while orbiting the periphery of the tank. It is especially effective for liquid/solid mixes.

Paddle Mixer. This mixer has single or double propellers or paddles on a shaft rotating inside a tank at low or high speeds. It is used for suspending solids in a liquid medium.

Intensive Stirrer. A high-speed impeller blade in a baffled mixing tank produces high-shear mixing in dispersions of solids in liquids.

3.1.5 Extruders. *Single Screw Extruder.* With a screw designed for compound mixing, this extruder is commonly used for compounding fillers and resins.

Twin-Screw Extruder (Fig. 22-5). This is an extruder with two, co-rotating screws having sectionized barrel and screw components. The process section may be modified so that materials can be fed or devolatilized at any desired location. The screw configuration may be tailored for optimum shear and mixing intensity. High-performance engineering plastic compounds are generally produced using twin-screw extruders.

Mixer Extruder. This extruder combines mixing, such as kneader blades, shear cone, or special mixing screw, with an extruder for continuous delivery of the mixed compound. (See Fig. 22-6.)

3.1.6 Other Types. *Motionless Mixer.* This is a low-shear device with no moving parts in which melted polymer is pushed by a ram or screw feed through a baffled section that divides and recombines the melt flow. Its effectiveness is a function of material melt viscosity and design of the mixing section, which often results in improved quality of mixing. (See Figs. 22-7 and 22-8.)

Medium-Intensity Mixer (Farrell). This mixer combines the action of blades and screw to furnish variable-intensity mixing, depending upon requirements. It may be used for dry and liquid materials. (See Fig. 22-9).

3.1.7 Special RP Equipment. *Pultrusion Equipment.* This equipment is used in a process for combining long continuous strands of liquid, resin-impregnated, reinforcing material, such as polyester/glass fibers, which are pulled through a steel die and heating chamber to yield

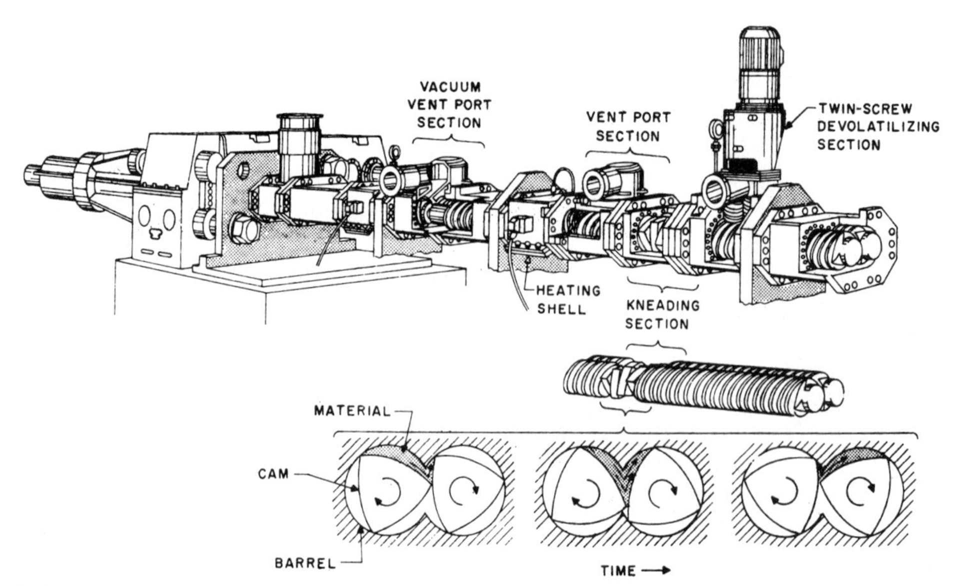

Fig. 22-5. Twin-screw compounding extruder with co-rotating intermeshing screws. (Courtesy Farrell Company, Emhart Machinery Group.)

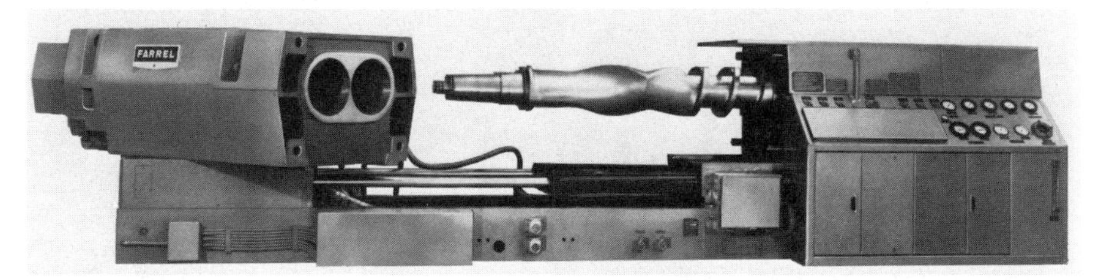

Fig. 22-6. Continuous mixer extruder (patented). (Courtesy Farrel Company, Emhart Machinery Group.)

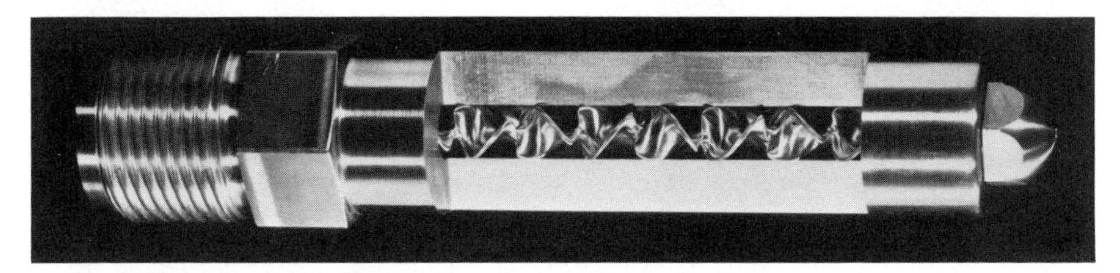

Fig. 22-7. Motionless mixer—injection-molding "super nozzle." (Courtesy Kenics Corp.)

continuous lengths of material with high unidirectional strength. (See Fig. 22-10.)

Tape Lay-up Equipment. This type of equipment is used primarily in the aerospace industry for automatic and semi-automatic lay-up of unidirectional prepreg tape components such as wing and control surface skins. (See Fig. 22-11.)

Filament Winding Equipment. This equipment is used for combining continuous, resin-impregnated reinforcing fibers by coating and winding them onto the final product shape in a continuous process. The method yields high reinforcement/resin binder ratios and very high structural strength. (See Figs. 22-12 through 22-14.)

Fig. 22-8. Motionless mixer—injection-molding "mixing head." (Courtesy Koch Engineering Co.)

On-Site Tank Erection System. This equipment produces a double tongue/groove pultrusion profile, spiral-wrapped and bonded together to produce a storage tank. (See Fig. 22-15.)

Pulformer™. A curved pulformer is a type of

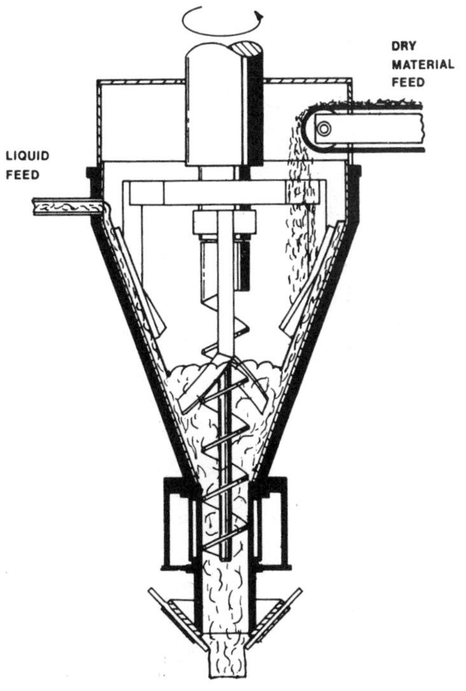

Fig. 22-9. Medium-intensity mixer (patented). (Courtesy Farrel Company, Emhart Machinery Group.)

Fig. 22-10. LSP: Large size pultrusion machine with 18 in. × 36 in. cross-section envelope and 20,000 lb pulling capacity.

Fig. 22-11. Tape-laying head with digital shear allowing cutoff while in motion; used for composite wing and control surface skins. (Courtesy Goldsworthy Engineering.)

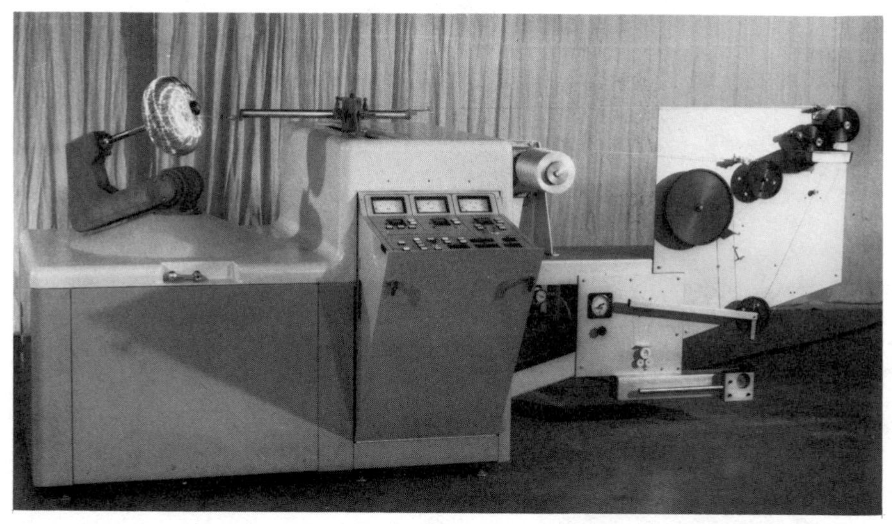

Fig. 22-12. Laboratory filament winder. (Courtesy Goldsworthy Engineering, Inc.)

equipment derived from pultrusion equipment that can continuously form and cure varying-cross-section parts, such as automobile leaf springs. (See Fig. 22-16.)

Spray-up Equipment. This refers to equipment used in spray-up, a general term for several processes using a spray gun, including simultaneous mixing and spraying of resin and chopped reinforcing fibers and/or fillers onto a mold or mandrel. (See Fig. 22-17.)

Matched Metal Die Molding Equipment. This equipment is used in a molding process for forming reinforced plastics articles with high fiber/resin ratios, similar to metal stamping. One variation uses precut mats of reinforcing material placed in the mold, with liquid resin then poured over the mat. Another variation uses preimpregnated fibrous mat in the mold. This may be referred to as "SMC" or "prepreg." Still another variation uses "premix," which has a high content of chopped fibers compounded with polyester resin.

Fig. 22-13. Filament winder–polar orbital winder with gantry traverse. (Courtesy Goldsworthy Engineering, Inc.)

Fig. 22-14. Computer controlled, 4-axis helical filament winder. (Courtesy Goldsworthy Engineering, Inc.)

Fig. 22-15. On-site tank erection system. (Courtesy Goldsworthy Engineering, Inc.)

Fig. 22-16. Curved Pulformer™, an automated system for malsing fiberglass automatic leaf springs. (Courtesy Goldsworthy Engineering, Inc.)

Fig. 22-17. Dual spray gun and chopper. (Courtesy Binks Manufacturing Co.)

3.2 Summary of Mixer Types and Uses

Table 22-1 gives a summary of the various types of mixers and their uses.

3.3 Suppliers of Mixing Equipment

These listings are based upon information supplied by equipment manufacturers. More extensive listings of suppliers are given in references 5, 16, 17, and 18.

3.3.1 Intensive Internal Mixers

J. H. Day Co.—Plastimat, high-speed turbulent mixers, roll mills

Farrel Co.—Banbury, FCM continuous mixers, roll mills

Charles Ross & Son Co.—Roll mills

Teledyne Readco—High intensity mixers

3.3.2 Dry Mixers and Blenders

Paul O. Abbe Inc.—Ribbon blenders, RotaCone blenders, double shaft mixers, ball mills

J. H. Day Co.—Ribbon blenders, Nautamixer

Farrel Co.—Medium intensity mixers

Feeco International, Inc.—Ribbon mixer, rotary mixer

General Machine Co. of N.J.—Cone blenders, V-shape blenders

S. Howes Co., Inc.—Ribbon mixers, vertical and horizontal batch mixers

Patterson Industries, Inc.—Double cone Thoroblender, ribbon, and paddle mixers

Pfaudler Sybron Corp.—Conical dryer-blender

Prodex, HPM Div., Koehring Co.—High intensity mixer

Charles Ross & Son Co.—Kneader extruder, double planetary mixer

Teledyne Readco—Double arm mixers

3.3.3 Tumblers

Feeco International, Inc.—Rotary mixers

General Machine Co. of N.J.—Cone blenders, V-shape blenders

Patterson Industries, Inc.—Double cone Thoroblender

Table 22-1. Summary of mixer types and uses.

	1	2	3	4	5	6	7
Intensive Internal Mixers							
Banbury		X	X	X	X	X	
Farrel continuous mixer		X	X	X	X		X
Plasticator		X	X	X	X	X	
Vertical intensive mixer		X				X	X
Turbulent mixer	X	X	X			X	
Roll mill			X		X	X	
Dry Mixers and Blenders							
Ribbon blender	X	X	X			X	
Cone blender		X				X	
V-shape blender		X				X	
Kneader, double arm mixer	X	X	X			X	
Vertical high intensity	X	X				X	
Ball mill				X		X	
Double planetary	X					X	
Tumblers							
Drum tumbler		X				X	
Conical dry blender		X				X	
Liquid Mixers							
Vertical cone blender	X					X	
Paddle mixer	X					X	
Intensive stirrer	X					X	
Extruders							
Single screw			X		X		X
Twin screw			X		X		X
Mixer extruder			X		X		X
Other Types							
Motionless mixer			X				X

KEY: 1. Liquid resins/solids (plastisols, organosols, thermosets)
2. Solids, cold (dry blends)
3. Solids, hot
4. High shear mixing, grinders, cold
5. High shear mixing, hot
6. Batch mixing
7. Continuous mixing

NOTES:

1. *Dry blending* of fillers and/or pigments may give sufficient mixing if some additional mixing takes place in subsequent processing, as in extrusion or injection molding with a reciprocating screw type molding machine.

2. *Dry blending* of reinforcements may be desirable to minimize shear in mixing and reduce breakdown of fiber length.

3. Heat sensitive or thermosetting materials usually require room temperature mixing conditions. The equipment used must be capable of being maintained at ambient temperature during mixing.

4. The many variations of these types of equipment by many manufacturers require that the specific end use required be thoroughly discussed and explained with the manufacturers before purchase.

3.3.4 Liquid Mixers

Conn and Co.—Intensive stirrers, vertical mixers

J.H. Day Co.—Nautamixer, Daymax

LFE Corp.—Portable paddle mixers
Farrel Co.—Medium intensity mixers
Feeco International—Paddle mixers
S. Howes and Co., Inc.—Vertical and horizontal batch mixers
Kinetic Dispersion Corp.—Kady mill—high-speed rotor dispersion
Oakes Machine Corp.—Continuous mixers for plastisols and latices
Patterson Industries, Inc.—Ribbon and paddle mixers
Pfaudler Sybron Corp.—Paddle mixers
Charles Ross & Son Co.—Double planetary mixers

3.3.5 Extruders and Mixer Extruders

Baker-Perkins Inc.—Twin screw continuous mixer
Charles Ross & Son Co.—Kneader extruder
Teledyne Readco—Continuous processor
Werner & Pfleiderer Corp.—Twin screw compounding extruder

3.3.6 Motionless Mixers

Armorflite-Northeast & Southeast Welding & Machine Co.—Intromix for extruders
Kenics Corp.—Thermogenizer
Koch Engineering
Charles Ross & Son Co.—Melt homogenizer for extrusion, IsoNozzle for injection molding

3.4 Names and Addresses of Mixer Manufacturers

Abbe, Paul O., Inc., 509 Center Ave., Little Falls, NJ 07424
Baker-Perkins, Inc., 1000 Hess St., Saginaw, MI 48601
Conn and Co., James St., Warren, PA 16365
Day Mixing Div. of LeBlond Inc., 4932 Beech St., Cincinnati, OH 45212
Farrel Co., Div. Emhart Machinery Group, Ansonia, CT 06401
Feeco International, Inc., 3913 Algoma Road, Green Bay, WI 54301
General Machine Co. of N.J., Inc., 55 Evergreen Ave., Newark, NJ 07114
Howes, S. Co., Inc., 138 Howard St., Silver Creek, NY 14135
Chemineer Kenics Corp., 45 S. Main St., Dayton, OH 45401
Kinetic Dispersion Corp., 88 Beacon St., Buffalo, NY 14220
Koch Engineering Co., Inc., 4111 E. 37th St. N., Wichita, KS 67220
Northeast Welding & Machine Co., Inc., 662 Cross St., Malden, MA 02148
Oakes Machinery Corp., 235 Grant Ave., Islip, NY 11751
Ross, Charles & Son, Inc., 710 Old Willets Path, P.O. Box 2200, Hauppage, NY 11787
Werner & Pfleiderer Corp., 663 E. Crescent Ave., Ramsey, NJ 07446

4. PROCESSING EQUIPMENT

4.1 General Notes on Utilization

4.1.1 Materials of Construction. The same equipment, such as screw extruders or injection molding machines, that is used for unfilled materials may also be used for processing filled or reinforced compounds. However, these filled materials have higher viscosities and often higher processing temperatures and pressures than the unfilled materials. Also, processing under nonoptimum conditions of mixing uniformity and tool design may lead to increased rates of wear of equipment. It is therefore desirable to use tougher and more resistant materials of equipment construction than are otherwise sometimes used. Some examples are:

Barrels:

- Nitrided medium alloy–medium carbon steels, heat treated to hardness of Rc50-63.
- Nitriding steels, such as nitraloy 135 or 135M, heat treated to bore hardness of over R15N-90.
- Bimetallic barrel designs, combining a high alloy, hardened steel sleeve and replaceable liner of more resistant compositions. Liners of Xalloy, Wisconsin Centrifugal, and other such metals are used, and may be shrunk or cast in the barrel. This construction is superior in corrosion and abrasion resistance, but its cost is also greater. The choice will be made in a particular application after balancing initial cost and overall processing economics.

Screws:

- Nitrided steels, the same as for barrels.
- Medium alloy–medium carbon steels with Union Carbide "Stellite" applied over the diameter of the screw flights where they are most subject to wear.
- Stainless steels, hardened, sometimes used for improved corrosion resistance, but more expensive than nitrided steel screws.

Screw Tips, Barrel Heads, Extrusion Dies:

- Medium carbon steels, polished or plated.
- Nitrided steels.

4.1.2 Injection Molding Machine Requirements. *Machine Types*. Both plunger and screw type injection molding machines may be used for filled and/or reinforced plastics compounds. However, a screw type machine, either two-stage screw or a reciprocating screw, is recommended for best uniformity and control of material properties and product tolerances.

Injection Pressure. An injection pressure of at least 20,000 psi should be provided for the machine to accommodate the higher melt viscosities encountered with filled materials. Injection speed control is also desirable.

Screw. A general-purpose screw is acceptable for all resins except PVC. For PVC, a standard low-shear PVC screw with 2:1 compression ratio should be used.

Clamp. A clamping force of 5 to 10 tons/in.^{2} of projected area is used to accommodate the higher injection pressures used.

Machine Capacity. With the higher pressures and temperatures used, and the lower rates of flow encountered, machine capacity will be only 50 to 75% of the capacity as rated for unfilled materials such as polystyrene.

Drying Materials. Facilities should be provided for drying hygroscopic filled materials before processing. A *dehumidifying oven* with dry hot air passing through it may be used prior to processing. Material is spread in a thin layer over trays and baked for 2 to 4 hours at a temperature of at least 250°F. A *hopper drier* may also be used.

Mold Temperature. Mold temperatures for filled and/or reinforced compounds are usually higher than for unfilled resins. Higher mold temperatures result in better flow of the melt and a better surface finish. Standard mold temperature control equipment is satisfactory.

Nozzle. A large diameter, heated nozzle with reverse taper of 2° to 3° is recommended. A positive cutoff valve should be used with low-viscosity resins such as nylons, or when molding in a vertical machine. Nozzle diameter may be $\frac{7}{16}$ to $\frac{1}{2}$ in. for most fiber-reinforced thermoplastics. A hot nozzle extension will aid molding efficiency by maintaining melt temperature as close as possible to that of the part being molded.

4.2 Special Notes on Injection Mold Design

4.2.1 Sprues. These should be large, as short as possible, and highly polished, with a good draft angle of at least 3°. Radius of the sprue bushing should be $\frac{1}{8}$ to $\frac{1}{4}$ in. greater than the radius of the nozzle to provide a contact line rather than a broad surface.

4.2.2 Runners. Cold slug wells should be used where possible to trap the leading cold resin from the nozzle, and permit hot resin to enter the runner and cavity. They should be made with a back taper to permit pulling the part. Runners should be full round, with larger diameters the longer they travel. As reinforced resins tend to solidify at the walls of the runner, this design will maintain a maximum free flow path through the center of the runner. A diameter of $\frac{1}{4}$ to $\frac{3}{8}$ is recommended for most parts. Streamlined flow should be provided as nearly as possible, but when right angles are used, a cold slug well should be provided. The runner system should be as short as possible and highly polished. In multicavity molds, a balanced runner system is important to obtain part to part uniformity. Where possible, cavities should be bypassed with the main runner, and branch runners should be as short as possible.

4.2.3 Gates. Gates should be located to achieve a balance of flow in all directions, if possible. When flow is primarily in one direction, reinforcing fibers will tend to become oriented in that direction. The result will be differences in properties and shrinkage in different directions in the part, and a resulting tendency toward poor dimensional tolerance control and warping. Where uniform flow is not possible, the gate should be located so that flow direction is along the axis of the most critical dimension.

Gates should be as large as possible, but may be as small as 0.040 in. in diameter. Short land lengths from 0.020 to 0.040 in. are recommended. Gates should be located at thick sections of the part to minimize sinks or voids. The number of gates used should be kept to a minimum to avoid unnecessary weld lines. Gates are usually round or rectangular, but diaphragm or flash gates may be used. The size of round or rectangular gates should be from two-thirds to full width of the cavity wall. Small pinhole gates should be avoided.

4.2.4 Weld Lines. Weak or prominent weld lines in filled materials may be minimized by *proper mold design*. One technique is the use of overflow wells in the mold, allowing *venting* and cold surface material to be removed from the cavity, and promoting the intermixing of fibers and resins: Another procedure is the use of a cold well bypass extension of the main runner, allowing a branch runner to deliver hot resin into the cavity. Hot runner molds will also promote good bonds.

4.2.5 Venting. Good venting of the cavity is desirable for a rapid and uniform fill. Vents should be located wherever air may be trapped in the cavity as the resin flows in.

4.3 Special Processing Equipment

4.3.1 Pultrusion. This process, developed by Goldsworthy Engineering, Inc., combines long continuous lengths of reinforcing fibers, such as glass, with liquid resins such as polyester or epoxy resins. It provides a high reinforcement/resin ratio, and the result is a very strong, resilient continuous shape, which has found many unique applications, such as fishing poles. Graphite fibers and fibers of DuPont's Kelvar are also used with, or in place of, glass fibers.

4.3.2 Fiberglass Spray-Up Systems. Several systems are marketed for combining catalyzed polyester resins with glass fibers in a continuous spray operation. As a spray gun mixes and dispenses the mixed liquid resin, this resin is combined with chopped glass fibers in the desired proportions. The mixture is applied to a surface or into a mold to cure in the desired final shape. This process is well adapted and widely used for making structural parts, such as boat hulls, shower stalls, and other large, integral structures.

4.3.3 Matched Metal Die Molding. This is done with reinforcing fiber/liquid resin mixtures in which liquid resin, catalysts, glass fiber, and any other additives are combined into a one-compound molding material. This may be done using sheet molding compounds (SMC) or a dough-like *premix* material. Applications are widespread, and are similar to those for spray-up systems. This process may have advantages of closer control of structural dimensions, tolerances, and finish, but may be more expensive, depending upon specific applications.

4.3.4 Filament Winding. This consists of winding a continuous, resin-coated strand of reinforced fiber or roving on a mandrel having the structural shape desired. By winding the fibers in a predetermined pattern, a very strong, thin, light structure may be obtained. Here again, the liquid resins may be polyester, vinyl ester, or epoxy. The very high structural strength and stiffness values resulting are due to the high glass fiber/resin ratio, and the precise orientation of these fibers within the structure to obtain maximum utilization of their strength.

4.3.5 RIM and RRIM Processing. The Reaction Injection Molding (RIM) process has been steadily growing in use in recent years, but still accounts for only a relatively small percentage of the total use of plastics. This growth is based on an expanding list of applications for which it is especially well suited.

RIM is a process of injection molding in which the monomeric reactive fluid components of a polymer are mixed as they are injected into a mold cavity, where they react to form the finished polymeric product form desired. The process has been largely used with polyurethanes, but has also been applied to other resin types, including epoxy, nylon, and polyester.

RIM compounds, like other plastics compositions, may benefit from the use of fillers and reinforcements to modify their properties (RRIM), and many of the same additives have been used as for other plastics, including such materials as glass fibers, glass flake, wollastonite, barytes, and celestite at levels of 10% to 30% or more. The additives may be premixed with one of the reactive components, or, more usually, by metering with the liquid resin components just before injection into the mold. Much of this work with additives has been done by material suppliers to develop proprietary systems, or by individual process users. (See reference 41.)

4.4 Suppliers of Special Processing Equipment

4.4.1 Pultrusion Equipment

Gatto Machinery Dev. Corp., 45 Rabro Drive, Hauppage, NY 11787

Goldsworthy Engineering, Inc., 23930 Madison St., Torrance, CA 90505

4.4.2 Spray-Up Equipment

Binks Mfg. Co., 9201 W. Belmont Ave., Franklin Park, IL 60131

Cook Paint & Varnish Co., P.O. Box 389, Kansas City, MO 64141

Lincoln St. Louis, Div. McNeil Corp., 4010 Goodfellow, St. Louis, MO 63120

Miles, A. L. Co., 4060 Wyne St., Houston, TX 77017

Ransburg Electrostatic Equipment, Box 88220, Indianapolis, IN 46208

Shyodu Instrument Co., 197 King St., Brooklyn, NY 11231

Venus Products, Inc., 1862 Ives Ave., Kent, VA 98031

4.4.3 Matched Metal Die Equipment (SMC and Premix)

Blue, E. B. Co., 651 Connecticut Ave., S. Norwalk, CT 06854

Brenner, I. G. Co., 100 Manning St., Newark, OH 43055

Dake Div., JSJ Corp., 641 Robbins Rd., Grand Haven, MI 49417

Engineering Technology, Inc., 145 W. 2950 S., Salt Lake City, UT 84115

Finn & Fram Inc., 13231 Louvre St., Arleta, CA 91331

Littleford Bros., Inc., 7451 Empire Dr., Florence, KY 41042

4.4.4 Filament Winders

American Barmag Corp., 1101 Westinghouse Ave., P.O. Box 7046, Charlotte, NC 28217

Goldsworthy Engineering, Inc., 23930 Madison St., Torrance, CA 90505

Vermont Instrument Co., Inc., 62 Overlake Park, Burlington, VT 05401

5. PROCESSING FILLED COMPOUNDS

5.1 Injection Molding Conditions

The general procedures and conditions for molding are similar for filled and unfilled plastics compounds. However, filled materials have higher viscosities and poorer melt flow than unfilled materials, requiring higher processing temperatures. Also, fiber-reinforced compounds may show orientation of fibers as the melt flows into the mold cavity, resulting in anisotropy, or properties that are different in different directions in the part. This orientation may result in problems such as poor dimensional stability, warpage, and undesirable effects upon properties of the part. Special attention to molding conditions may be needed.

The effect of *melt temperature* depends upon the polymer, properties sometimes increasing with temperature, in other polymers decreasing with temperature, and, in still other polymers, independent of temperature. High temperatures will also affect different polymers in regard to polymer degradation, melt viscosity, shear sensitivity, and crystallinity. The final properties obtained are a composite result of all of these factors. Generalities are difficult, and the supplier of the particular polymer should be consulted for his processing recommendations.

Dimensional instability due to excessive orientation may be influenced by molding conditions, and minimum orientation and maximum dimensional stability will be favored by the following conditions:

Material temperature: hot
Mold temperature: hot
Part cooling time: slow
Injection pressure: low
Ram forward time: short
Part thickness: thick
Gate size: small
Fill rate: fast

REFERENCES

1. Theberge, John E., "How to Process Glass Fiber Fortified Thermoplastics," *Plastics Design and Processing* 13(1): 14–17, Jan. 1973; and (2): 19–20, Feb. 1973.
2. Jakopin, Stan, "Compounding of Fillers," *Advances in Chemistry Series*, Number 134, Am. Chem. Soc., 1974.
3. Mack, Wolfgang, "Continuous Compounding—Where Twin Screws Fit," *Plastics Technology*, Feb. 1975.
4. Quillen, S. C., "Mixing," *Chem. Eng.* 61: 204, 1954.
5. Updegrove, L. B., "Selecting Mixers and Compounders," *Plastics World*, pp. 53–57, Nov. 18, 1974.
6. Seamon, R. G. and Merrill, A. M. "Machinery and Equipment for Rubber and Plastics," 1952.
7. Bolen, W. R. and Colwell, R. E., "Intensive Mixing," SPE 14th ANTEC, 1958, p. 1004.
8. Bergen, J. T., "Mixing and Dispersing Processes," *Processing of Thermoplastic Materials*. Bernhardt (ed.), Van Nostrand Reinhold, New York, 1959, pp. 409 ff.
9. Schlick, W. R., Hagen, R. S., Thomas, D. P., and Musselman, K. A., "Critical Parameters for Direct Injection Molding of Glass Fiber Thermoplastic Powder Blends," SPE Tech. Paper, 1967, p. 929.
10. Goldsworthy, B., "Continuous Pultrusion," *Modern Plastics*, p. 28, June 1975.
11. Skoblar, Sandra M., "What's All the Fuss About Motionless Mixing," *Plastics Technology* 11: 37–43, 1974.
12. Dietz, G. R., "The Art of Formulating Plastisols," *Plastics Eng.*, pp. 50–51, Apr. 1974.
13. DuBois, J. Harry, "Equipment for Compounding Plastics Materials," *Plastics Machinery and Equipment*, Part I, 4(7): 17–20, July 1975; Part II, 4(8): 19–20, Aug. 1975.
14. Hall, Alan, "In-house Compounding," *Plastics World*, pp. 34–36, Apr. 21, 1975.
15. DeLuca, Robert M, "Mixing and Compounding Equipment," *Modern Plastics Encyclopedia*, pp. 500–501, 1974–75.
16. "Compounding Equipment," *Modern Plastics Encyclopedia*, pp. 845–846, 1974–75.
17. "Mixing and Blending Equipment," *Plastics Technology, Manufacturing Handbook and Buyers Guide*, pp. 237–238, 1974–75.
18. Listing of Mixing Equipment Mfrs., *Chemical Engineering Equipment Buyers Guide*, pp. 388–389, 1974–76.
19. "Fortified Polymers, Processing," Liquid Nitrogen Processing Co., Bulletin 203-769, p. 3.
20. "Molding and Engineering Data for Reinforced Thermoplastics Processors, Equipment Recommendations," Fiberfil Corp., Form 34-7154, p. 11.
21. Pickens, S. H., "Pultrusion—The Accent is on the Long Pull," *Plastics Engineering*, pp. 16–21, July 1975.
22. Olmsted, B. A., "How Glass-Fiber Fillers Affect Injection Machines," *SPE Journal* 26(2): 42–43, Feb. 1970.
23. Brandt, R., (to Fiberfil), "Glass-Reinforced Thermoplastic Injection Molding Compound and Injection Molding Process Employing It," U.S. Patent 2,877,501, Mar. 1959.
24. "Molding Filled Material Is a Lot Easier Now," *Modern Plastics*, No. 7, pp. 44–45, July 1975.
25. Avery, Donald, "Wear—The Unseen Causes," *Plastics Technology*, pp. 31–36, Oct. 1975.
26. Saltzman, Gilbert A., "Wear Data Air Selection of Barrel-Screw Combination," *Plastics Design and Processing*, pp. 16–18, Oct. 1975.
27. *Plastics Compounding*, Redbook Industry Media Inc., 1983/84.
28. Maris, Judy, "Guide to Mixers," *Chemical Processing*. pp. 57–64, July 1979.
29. "Update on Continuous Compounding Equipment, Part II: Twin-Screw Extruders," *Plastics Compounding*, pp. 38–49, July/Aug. 1979.
30. "Update on Continuous Compounding Equipment, Part III: Two-Stage Mixer/Extruders," *Plastics Compounding*, pp. 33–47, Sept./Oct. 1979.
31. Byrnes, Daniel, "Guide to Mixing," *Chemical Processing*, pp. 59–66, July 1980.
32. Kraus, T. J. and Copus, J. E., "A Primer, Compounding PVC, Part II: Methods," *Plastics Compounding*, pp. 23–35, Nov./Dec. 1980.
33. Bacchetti, Jerome A. and Bonady, Frances M., "Guide to Mixing," *Chemical Processing*, pp. 45–54, July 1983.
34. Cheng, C. Y., "Extruder-Screw Design for Compounding," *Plastics Compounding*, pp. 29–40, Mar./Apr. 1981.
35. "Update on Batch-Compounding Equipment, Part I: Fluxing and Nonfluxing High-Intensity Mixers," *Plastics Compounding*, pp. 20–32, Jan./Feb. 1981.
36. White, James Lindsay, "Rheological Behavior of Highly Filled/Reinforced Polymer Melts," *Plastics Compounding*, pp. 48–62, Jan./Feb. 1982.
37. "NPE '82 Report: Part II" (including Compounding Equipment, Reinforced Plastics and RIM), *Plastics Technology*, pp. 49–55. Sept. 1982.
38. Naitove, Matthew H., "At RP Meeting: An Upbeat

Mood, Modest Advances in Technology,'' *Plastics Technology*, pp. 48-53, Mar. 1983.

39. ''News Update: Compounding Machinery,'' *Plastics Technology*, pp. 24-32, July 1983.
40. Colangelo, Michael and Naitove, Matthew H., ''Pultrusion Process Technology: Beyond Infancy, Not Yet Mature,'' *Plastics Technology*, pp. 49-53, Aug. 1983.
41. Frados, Joel, ''Reaction Injection Molding,'' *Plastics Engineering Handbook*, pp. 519, 522-523, Van Nostrand Reinhold, New York, 1976.
42. ''FRP Use in Cars Gains Speed,'' *Plastics World*, pp. 51-53, Sept. 1982.
43. ''RIM and Urethane Processing,'' *Plastics Technology*, pp. 52-54, Sept. 1982.
44. Ferrarini, James and Cohen, Stuart, ''Reinforcing Fillers in RIM PUR,'' *Modern Plastics*, pp. 66-72, Oct. 1982.
45. von Hassell, Agostino, ''RIM Materials Take Quantum Leap Forward,'' *Plastics Technology*, pp. 37-44, Mar. 1983.
46. ''Nylon RIM Unit Eliminates Premixing,'' *Plastics World*, p. 58, July 1983.
47. ''Fillers/Reinforcements in RRIM,'' *Plastics Technology*, p. 105, July 1983.
48. Galli, Ed, ''RIM Extends Its Versatility,'' *Plastics Design Forum*, pp. 17-24, Sept./Oct. 1983.
49. Eise, Kurt, Curry, John, and Nangeroni, James, (Werner & Pfleiderer Corp.), ''New Compounding Developments Improve Polyblend Production,'' *Polymer Engineering and Science* 23(11), Mid-Aug. 1983.

Appendix
METRIC CONVERSIONS

Harry S. Katz
Utility Research Co.
Montclair, New Jersey

Harold E. Brandmaier
Consultant
Harrington Park, New Jersey

The conversion of the United States from U.S. Customary units to metric units is inevitable. The United States is the only major industrial country in the world that has not fully adopted the metric system. The metric system is more logical than the present U.S. system of measurement and is, for this reason, employed in many laboratory procedures.

In 1954, the General Conference on Weights and Measures adopted the MKSA system of units based on the meter, kilogram, second, ampere, degree Kelvin, and candela. In 1960, the system was named the International System of Units or SI. The major advantage of SI is that there is only one unit for each physical quantity. As an example, such diverse volume units as barrel, fluid ounce, gallon, liter and cubic foot are replaced by meter^3. Note that while meter and liter are the preferred spellings in the United States, metre and litre are also used.

Table A1 lists the nine fundamental SI quantities and the corresponding unit and SI symbol for each. From these, the mechanical, thermal, electrical and other quantities shown in Table A2 were derived. Tables A3, A4, and A5 contain representative quantities considered useful to those using this handbook. More extensive information can be found, for example, in ASTM standard E-380-84, "Using the SI system of Units," F. S. Conant, *Rubber Chemistry and Technology*, March/April 1975; and "Brief History of Measurement Systems with a Chart of the Modernized Metric System," National Bureau of Standards Special Publication 304A, October 1972.

That the units in Table A2 are derivable from the fundamental units in Table A1 can be readily demonstrated. Thus, the SI unit of force is the newton (N), equal to kg $\cdot$ m/s^2. From this, the unit of energy, the joule (J), is defined as N $\cdot$ m. Continuing, the SI unit of power, the watt (W), is given as J/s, and finally the unit of electromotive force, the volt (V), is

Table A1. SI units.

Quantity	Unit	SI Symbol
Basic units		
Length	meter	m
Mass	kilogram	kg
Time	second	s
Electric current	ampere	A
Thermodynamic temperature	Kelvin	K
Amount of substance	mole	mol
Luminous intensity	candela	cd
Supplementary units		
Plane angle	radian	rad
Solid angle	steradian	sr

Table A3. SI prefixes.

Multiplication Factor	Prefix	SI Symbol
10^{-18}	atto	a
10^{-15}	femto	f
10^{-12}	pico	p
10^{-9}	nano	n
10^{-6}	micro	μ
10^{-3}	milli	m
10^{3}	kilo	k
10^{6}	mega	M
10^{9}	giga	G
10^{12}	tera	T
10^{15}	peta	P
10^{18}	exa	E

Table A2. Examples of derived units.

Quantity	Unit	SI Symbol	Formula
Force	newton	N	$kg \cdot m/s^2$
Pressure, stress, elastic modulus	pascal	Pa	N/m^2
Tear strength, adhesion	newton per meter	N/m	—
Frequency	hertz	Hz	1 /s
Celsius temperature	degree Celsius	°C	K − 273.15
Energy, quantity of heat, work	joule	J	N · m
Power, radiant flux	watt	W	J/s
Specific heat	joule per kilogram kelvin	J / (kg · K)	—
Thermal conductivity	watt per meter kelvin	W / (m · K)	—
Dynamic viscosity	pascal second	Pa · s	—
Kinematic viscosity	square meter per second	m^2/s	—
Luminous flux	lumen	lm	cd · sr
Illuminance	lux	lx	lm/m^2
Electromotive force, electric potential	volt	V	W/A
Dielectric strength	volt per meter	V/m	—
Quantity of electricity, electric charge	coulomb	C	A · s
Electric capacitance	farad	F	C/V
Electric inductance	henry	H	Wb/A
Electric conductance	siemens	S	A/V
Electric resistance	ohm	Ω	V/A
Magnetic field strength	ampere per meter	A/m	—
Magnetic flux	weber	W/b	V · s
Magnetic flux density	tesla	T	Wb/m^2
Activity of a radionuclide	becquerel	Bq	1/s
Absorbed dose	gray	Gy	J/kg
Dose equivalent	sievert	Sv	J/kg

Table A4. Frequently used U.S. to SI conversion factors.

Quantity	U.S. System	Multiplied by	= SI
Length	ft	$3.048\ 000 \times 10^{-1}$	m
	mil	$2.540\ 000 \times 10^{-5}$	m
Velocity	in./min	$4.233\ 333 \times 10^{-4}$	m/s
Acceleration	ft/s^2	$3.048\ 000 \times 10^{-1}$	m/s^2
Area	ft^2	$9.290\ 304 \times 10^{-2}$	m^2
Volume	ft^3	$2.831\ 685 \times 10^{-2}$	m^3
	ounce (U.S. fluid)	$2.975\ 353 \times 10^{-5}$	m^3
	gallon (U.S. liquid)	$3.785\ 412 \times 10^{-3}$	m^3
Section moment	in.^{4}	$4.162\ 314 \times 10^{-7}$	m^4
Mass	lb (avoirdupois)	$4.535\ 924 \times 10^{-1}$	kg
	short ton	$9.071\ 847 \times 10^{2}$	kg
	slug	$1.459\ 390 \times 10^{1}$	kg
Density	lb/ft^3	$1.601\ 846 \times 10^{1}$	kg/m^3
	slug/ft^3	$5.153\ 788 \times 10^{2}$	kg/m^2
	lb/gallon (U.S. liquid)	$1.198\ 264 \times 10^{2}$	kg/m^3
	oz (avoirdupois)/yd^2	$3.390\ 575 \times 10^{-2}$	kg/m^2
Force	lbf	4.448 222	N
Pressure, stress, elastic modulus	lbf/in.^{2}	$6.894\ 757 \times 10^{3}$	Pa
Pressure	inch of mercury (60°F)	$3.376\ 85 \times 10^{3}$	Pa
	atmosphere (standard)	$1.013\ 250 \times 10^{5}$	Pa
Tear strength, adhesion	lbf/in.	$1.751\ 268 \times 10^{2}$	N/m
Impact strength	ft · lbf/in.	$5.337\ 866 \times 10^{1}$	J/m
Bending moment	lbf · in.	$1.129\ 848 \times 10^{-1}$	N · m
Energy	Btu*	$1.055\ 056 \times 10^{3}$	J
	kW · h	$3.600\ 000 \times 10^{6}$	J
Power	Horsepower (550 ft · lbf/s)	$7.456\ 999 \times 10^{2}$	W
Specific heat	Btu*/(lb · °F)	$4.186\ 800 \times 10^{3}$	J/(kg · K)
Thermal conductivity	Btu* · in./(s · ft^2 · °F)	$5.192\ 204 \times 10^{2}$	W/(m · K)
Linear coefficient of thermal expansion	in./(in. · °F)	1.800 000	m/(m · °K)
Luminance	footlambert	3.426 259	cd/m^2
Illuminance	footcandle	$1.076\ 391 \times 10^{1}$	lx
Dynamic viscosity	lbf · s/ft^2	$4.788\ 026 \times 10^{1}$	Pa · s
Kinematic viscosity	ft^2/s	$9.290\ 304 \times 10^{-2}$	m^2/s
Plane angle	degree	$1.745\ 329 \times 10^{-2}$	rad
Dielectric strength	V/mil	$3.937\ 008 \times 10^{4}$	V/m
Electric resistivity	ohm · circular mil ft	$1.662\ 426 \times 10^{-9}$	Ω · m

*International Table.

Table A5. Frequently used metric to SI conversion factors.

Quantity	Metric System	Multiplied by	= SI
Length	micron	$1.000\ 000 \times 10^{-6}$	m
	angstrom	$1.000\ 000 \times 10^{-10}$	m
Area	barn	$1.000\ 000 \times 10^{-28}$	m^2
Volume	liter	$1.000\ 000 \times 10^{-3}$	m^3
Mass	metric ton	$1.000\ 000 \times 10^{3}$	kg
Density	grams per cubic centimeter	$1.000\ 000 \times 10^{3}$	kg/m^3
	denier	$1.111\ 111 \times 10^{-7}$	kg/m
Force	kilogram-force	9.806 650	N
	dyne	$1.000\ 000 \times 10^{-5}$	N
Pressure	torr (mm hg, 0°C)	$1.333\ 22 \times 10^{2}$	Pa
	bar	$1.000\ 000 \times 10^{5}$	Pa
	dyne per square centimeter	$1.000\ 000 \times 10^{-1}$	Pa
Energy	calorie (International Table)	4.186 800	J
	erg	$1.000\ 000 \times 10^{-7}$	J
	electron volt	$1.602\ 19 \times 10^{-19}$	J
Dynamic viscosity	centipoise	$1.000\ 000 \times 10^{-3}$	Pa · s
Kinematic viscosity	centistoke	$1.000\ 000 \times 10^{-6}$	m^2/s
Electrical resistivity	ohm centimeter	$1.000\ 000 \times 10^{-2}$	Ω · m
Magnetic flux	maxwell	$1.000\ 000 \times 10^{-8}$	Wb
Magnetic flux density	gauss	$1.000\ 000 \times 10^{-4}$	T
Magnetic field strength	oersted	$7.957\ 747 \times 10^{1}$	A/m
Illuminance	Phot	$1.000\ 000 \times 10^{4}$	lx
Brightness	lambert	$3.183\ 099 \times 10^{3}$	cd/m^2
Radiation absorbed	rad	$1.000\ 000 \times 10^{-2}$	Gy
Radiation equivalent in man	rem	$1.000\ 000 \times 10^{-2}$	Sv
Radioactivity	curie	$3.700\ 000 \times 10^{10}$	Bq
Ionizing radiation dosage	roentgen	2.58×10^{-4}	C/kg

watt/ampere (W/A). The remaining mechanical, thermal, electrical, and magnetic quantities in Table A2 can be derived from these four quantities. Generally, very large or very small quantities are written in the technical literature as a number greater than 1 and less than 10 multiplied by a suitable power of 10, e.g., 3×10^6 V for 3,000,000 V. SI facilitates expression of these quantities in written and oral communications by the use of the prefixes given in Table A3. Thus 3×10^6 V becomes 3 megavolts or simply 3 MV. Adjacent prefixes are related by a factor of 1000. The basic number can then be between 1 and 1000. Other prefixes such as centi for 10^{-2} should be used only where customary in specific applications.

Table A4 gives conversion factors for a number of useful quantities. The first column is the physical quantity, and the second column gives the customary units in the U.S. system. Multiplying by the factor in the third column yields the physical quantity expressed in SI units. As an example, consider a resin with an elastic modulus of 5×10^5 psi. Multiplying by the pressure or stress conversion factor of $6.894\ 757 \times 10^3$ gives 3.4473785 GPa. This example illustrates two problems.

The first problem is the difficulty of thinking

in SI; unless one is experienced using SI, it would be difficult even to categorize a material with an elastic modulus of 3 GPa; i.e., is it similar to steel or similar to polyethylene? The transition from the U.S. system to SI is less difficult if a physical quantity is expressed in both systems during the early years of this transition.

The second problem is the number of significant figures after the conversion—8 compared to 1—which implies an accuracy greater than that of the given quantity. To be correct, the SI quantity should be rounded to the same number of significant digits as the U.S. quantity, e.g., 3 GPa. The conversion factors in Table A4 are precise to the number of significant figures shown.

Table A4 does not contain the temperature conversion from degree Fahrenheit (°F) in the U.S. system to Kelvin (K) in SI. This conversion is not a simple multiplication but uses this equation:

$$K = (°F + 459.67)/1.8$$

Finally, Table A5 contains conversions from commonly used metric units to SI. Although not evident in Table A5, SI considerably simplifies the existing metric system. This is particularly true for electrical units where, in addition to the MKSA (meter, kilogram, second, ampere) system of units, there coexisted CGS (centimeter, gram, second) electrostatic units, and CGS electromagnetic units.

Index

Index